Materials for Land, Air, and Space Transportation

As transportation systems for land, air, and space vehicles continue to grow increasingly sophisticated, more advanced materials are needed to support their development and commercialization. *Materials for Land, Air, and Space Transportation* details new materials development for these transportation applications, emphasizing physical properties, research theories, and cutting-edge processing technologies, as well as advanced high-precision inspection methodologies.

- Covers materials, design, and manufacturing for lightweight vehicles, high-speed trains, fuel cell vehicles, and aerospace and aeronautical use.
- Focuses on the newest material function and processing technologies, covering nanotechnology, modern additive manufacturing techniques, and physical and mechanical studies of structure-sensitive properties of materials.
- Describes theoretical deduction, numerical simulation, and experimental studies of various materials.

This reference will be of interest to engineers and researchers in the disciplines of materials, mechanical, and transportation engineering, and related areas.

Advances in Materials Science and Engineering

Series Editor

Sam Zhang

Advances in Magnetic Materials: Processing, Properties, and Performance
Sam Zhang and Dongliang Zhao

Micro- and Macromechanical Properties of Materials
Yichun Zhou, Li Yang, and Yongli Huang

Nanobiomaterials: Development and Applications
Dong Kee Yi and Georgia C. Papaefthymiou

Biological and Biomedical Coatings Handbook: Applications
Sam Zhang

Hierarchical Micro/Nanostructured Materials: Fabrication, Properties, and Applications
Weiping Cai, Guotao Duan, and Yue Li

Biological and Biomedical Coatings Handbook, Two-Volume Set
Sam Zhang

Nanostructured and Advanced Materials for Fuel Cells
San Ping Jiang and Pei Kang Shen

Hydroxyapatite Coatings for Biomedical Applications
Sam Zhang

Carbon Nanomaterials: Modeling, Design, and Applications
Kun Zhou

Materials for Energy
Sam Zhang

Protective Thin Coatings and Functional Thin Films Technology, Two-Volume Set
Sam Zhang, Jyh-Ming Ting, Wan-Yu Wu

Fundamentals of Crystallography, Powder X-ray Diffraction, and Transmission Electron Microscopy for Materials Scientists
Dong ZhiLi

Materials for Devices
Sam Zhang

Materials in Advanced Manufacturing
Yinquan Yu and Sam Zhang

For more information about this series, please visit: www.routledge.com/Advances-in-Materials-Science-and-Engineering/book-series/CRCADVMATSCIENG

Materials for Land, Air, and Space Transportation

Edited by
Yinquan Yu and Sam Zhang

CRC Press
Taylor & Francis Group
Boca Raton London New York

CRC Press is an imprint of the
Taylor & Francis Group, an **informa** business

Designed cover image: Yinquan Yu and Sam Zhang

First edition published 2025
by CRC Press
2385 NW Executive Center Drive, Suite 320, Boca Raton FL 33431

and by CRC Press
4 Park Square, Milton Park, Abingdon, Oxon, OX14 4RN

CRC Press is an imprint of Taylor & Francis Group, LLC

Library of Congress Cataloging-in-Publication Data
Names: Yu, Yinquan, editor. | Zhang, Sam, editor.
Title: Materials for land, air, and space transportation / edited by Yinquan Yu, Sam Zhang.
Description: First edition. | Boca Raton, FL : CRC Press, 2025. |
Series: Advances in materials science and engineering |
Includes bibliographical references and index. |
Identifiers: LCCN 2024027447 (print) | LCCN 2024027448 (ebook) |
ISBN 9781032478548 (hbk) | ISBN 9781032478555 (pbk) | ISBN 9781003386223 (ebk)
Subjects: LCSH: Motor vehicles–Materials | Motor vehicles–Design and construction. |
Airplanes–Materials | Airplanes–Design and construction. | Space vehicles–Materials |
Space vehicles–Design and construction.
Classification: LCC TL240 .M3628 2025 (print) | LCC TL240 (ebook) |
DDC 629.04/6–dc23/eng/20241023
LC record available at https://lccn.loc.gov/2024027447
LC ebook record available at https://lccn.loc.gov/2024027448

ISBN: 978-1-032-47854-8 (hbk)
ISBN: 978-1-032-47855-5 (pbk)
ISBN: 978-1-003-38622-3 (ebk)

DOI: 10.1201/978-1-003-38622-3

Typeset in Times
by Newgen Publishing UK

Contents

Contributors

Ze Chai
Shanghai Jiao Tong University
Shanghai, China

Huabin Chen
Shanghai Jiao Tong University
Shanghai, China

Xiaoqi Chen
Shanghai Jiao Tong University
Shanghai, China
South China University of Technology
Guangzhou, Guangdong, China

Laifei Cheng
Northwestern Polytechnical University
Xian, Shanxi, China

Xilun Ding
Beihang University
Beijing, China

Bosheng Dong
University of Wollongong
Sydney, New South Wales, Australia

Gang Gao
Harbin Institute of Technology
Harbin, Heilongjiang, China

Tingting Gao
Hebei University of Technology
Tianjin, Hebei, China

Yingchun Guan
Beihang University
Beijing, China

Wenxing Guo
Shanghai Jiao Tong University
Shanghai, China

Haichao Hu
Hebei University of Technology
Tianjin, Hebei, China

Xiaokang Huang
Shanghai Jiao Tong University
Shanghai, China

Yilong Jing
Hunan University
Changsha, Hunan, China

Hongxia Li
Yangzhou University
Yangzhou, Jiangsu, China

Huijun Li
University of Wollongong
Sydney, New South Wales, Australia

Jinfeng Li
Yangzhou University
Yangzhou, Jiangsu, China

Xinxin Li
Beihang University
Beijing, China

Yuxing Li
University of Wollongong
Sydney, New South Wales, Australia

Yue Liu
Hebei University of Technology
Tianjin, Hebei, China

Yan Ma
University of Wollongong
Sydney, New South Wales, Australia

Zengxi Pan
University of Wollongong
Sydney, New South Wales, Australia

Zhijun Qiu
University of Wollongong
Sydney, New South Wales, Australia

Mingxue Shen
East China Jiaotong University
Nanchang, Jiangxi, China

Chaokun Song
Northwestern Polytechnical University
Xian, Shanxi, China

Wei Tao
Australian Nuclear Science and Technology Organisation
Sydney, New South Wales, Australia

Bo Tian
Yangzhou University
Yangzhou, Jiangsu, China

Langping Wang
Harbin Institute of Technology
Harbin, Heilongjiang, China

Quanjie Wang
Hefei Innovation Research Institute, Beihang University
Hefei, Anhui, China

Qiang Wei
Hebei University of Technology
Tianjin, Hebei, China

Qunhong Weng
Hunan University
Changsha, Hunan, China

Chaoqun Xia
Hebei University of Technology
Tianjin, Hebei, China

Yelong Xiao
East China Jiaotong University
Nanchang, Jiangxi, China

Kuai Yang
Tsinghua University
Beijing, China

Fang Ye
Northwestern Polytechnical University
Xian, Shanxi, China

Yinquan Yu
Singapore Manufacturing Intelligent Technologies
Singapore

Yonghao Yu
Harbin Institute of Technology
Harbin, Heilongjiang, China

Sam Zhang
Harbin Institute of Technology
Harbin, China
HIT Zhengzhou Research Institute
Zhengzhou, Henan, China

Zhen Zhang
Tsinghua University
Beijing, China

Tianhao Zhou
University of Wollongong
Sydney, New South Wales, Australia

Jiaqi Zhu
Harbin Institute of Technology
Harbin, Heilongjiang, China

Series Preface

Materials form the foundation of technologies that govern our everyday life, from housing and household appliances to handheld phones, drug-delivery systems, airplanes, and satellites. Development of new and increasingly tailored materials is key to further advancing important applications with the potential to dramatically enhance and enrich our experiences. The *Advances in Materials Science and Engineering* series by CRC Press/Taylor & Francis is designed to help meet new and exciting challenges in Materials Science and Engineering disciplines. The books and monographs in the series are based on cutting-edge research and development, and thus are up to date with new discoveries, new understanding, and new insights in all aspects of materials development, including processing and characterization and applications in metallurgy, bulk or surface engineering, interfaces, thin films, coatings, and composites, just to name a few. The series aims at delivering an authoritative information source to readers in academia, research institutes, and industry. The Publisher and its Series Editor are fully aware of the importance of Materials Science and Engineering as the foundation for many other disciplines of knowledge. As such, the team is committed to making this series the most comprehensive and accurate literary source to serve the whole materials world and the associated fields. As Series Editor, I would like to thank all authors and editors of the books in this series for their noble contributions to the advancement of Materials Science and Engineering and to the advancement of humankind.

Sam Zhang
Harbin Institute of Technology
Harbin, China
HIT Zhengzhou Research Institute
Zhengzhou, Henan, China

Preface

Embarking on exploration of the future of transportation through revolutionary advances of materials science and engineering, *Materials for Land, Air, and Space Transportation* dives into eleven key areas where these advancements are propelling the world towards a cleaner, safer, and more efficient transportation landscape unlike anything the world has ever seen.

This comprehensive volume explores the science behind technology enabling clean-burning hydrogen fuel cell vehicles (Chapter 1) to ensuring the safety of high-speed trains through innovative brake friction materials (Chapter 2). Chapter 3 looks into the critical science of wheel–rail contact through tribology, and Chapter 4 explores the intricate relationship between grinding processes and the performance of superalloy turbine blades of the workhorses of modern jet engines. Extending the service life of critical components is an on-going pursuit of materials research in the transportation industry, which is explored in depth. Chapter 5 details strategies for extending the service life of components in aero engines, while Chapter 6 tackles the scientific challenges associated with infrared anti-reflection protective films, a technology with far-reaching applications. Looking towards the future of manufacturing itself, Chapter 7 introduces a novel laser micro-processing that promises to revolutionize precision and efficiency in the transportation industry. Meanwhile, Chapter 8 explores Plasma Immersion Ion Implantation and Deposition as a powerful technique for enhancing surface properties of components used in everything from cars to airplanes. The journey doesn't stop on Earth. Chapter 9 ventures beyond our atmosphere, exploring the fascinating interplay between the space environment and materials. It examines how the harsh realities of space can both damage and create new materials, pushing the boundaries of what's possible. Finally, Chapters 10 and 11 shed light on the impact of radiation on polymer materials used in aerospace applications (Chapter 10) and introduce Wire Arc Additive Manufacturing, a promising technique for fabricating complex bimetallic components (Chapter 11).

Whether you're a seasoned engineer, a curious student, or simply someone fascinated by the future of transportation and the universe beyond, this book offers a comprehensive and insightful look at the materials science and engineering advancements that are propelling our current world forward. Prepare to be driven by the potentials these advancements hold for a future filled with cleaner, safer, and more efficient transportation on land, in the air, and even in the vast expanse of the space.

This book is written for engineers with senior bachelor's background, preferably in materials science or mechanical engineering. No prior knowledge of manufacturing is required. This book is also suitable as a textbook at the senior undergraduate or graduate levels. The contents of this book can be taught as a one-semester advanced course in the transportation area.

The editors would like to take this opportunity to thank all chapter authors for their zest in participation in this book project and for their hard work to make this book a reality, and to thank my daughter, Shihui Yu, a young artist from Singapore for designing the book cover. Moreover, thanks to Allison Shatkin and Gabrielle

Vernachio, the editors at CRC Press, for their guidance and assistance in the formulation and publication of the book.

Finally, Prof. Yu would like to acknowledge the funding of the National Natural Science Foundation of China (52067006), Foreign Expert Project of the Ministry of Science and Technology (G2021022002L), Jiangxi Science and Technology Innovation High-end Talent Project (jxsq2019101027). Prof. Zhang would like to thank Harbin Institute of Technology for all the conveniences provided in working on this project.

Yinquan Yu
Singapore Manufacturing Intelligent Technologies
Singapore
School of Electrical Engineering and Artificial Intelligence
Xiamen University Malaysia
Kuala Lumpur, Malaysia
School of Mechatronics and Vehicle Engineering
East China Jiao Tong University
Nanchang, China

Sam Zhang
School of Aeronautics
Harbin Institute of Technology
Harbin, China
HIT Zhengzhou Research Institute
Zhengzhou, China

Editors

Professor Yinquan Yu (余银犬) earned a Bachelor of Engineering in Mechanical Engineering at Chongqing University, China and a Master of Engineering in Mechatronics, and a PhD in Electrical and Computer Engineering at the National University of Singapore. He was a Scientist at the Agency for Science, Technology and Research (A*star), Singapore. In January 2019, he joined the School of Mechatronics and Vehicle Engineering, East China Jiaotong University, Nanchang, China, and assumed duties of the Director of the Institute of Intelligent Manufacturing and Precision Research, the Director of the Intelligent Transportation Tools and Power Control Research Center and Deputy Director of the Key Laboratory of transport tools and equipment of the Ministry of Education. Professor Yu was elected as a Fellow of SAS (2020), a member of ASME, Singapore executive committee member ASME (2017–2018), a Senior member of IES (2018) and IEEE (2019), and a mentor of the Lean Launch Pad Singapore (2018). He was also elected into the Double Thousands Program of Jiangxi Province (2019) and the Top Overseas Experts Program by the Ministry of Science and Technology of the PRC (2021). Since 2024, he has been as a director of Singapore Manufacturing Intelligent Technologies and an adjunct Professor in East China Jiaotong University and that of Xiamen University Malaysia.

Professor Sam Zhang Shanyong (張善勇), FRSC, FTFS, FIoMMM, academically known as Sam Zhang, was born and brought up in the famous "City of Mountains" Chongqing, China. He received his Bachelor of Engineering in Materials in 1982 from Northeastern University (Shenyang, China), Master of Engineering in Materials in 1984 from Iron & Steel Research Institute (Beijing, China) and a PhD in Ceramics in 1991 from The University of Wisconsin-Madison, USA. He was a tenured full professor (since 2006) at the School of Mechanical and Aerospace Engineering, Nanyang Technological University Singapore. Between 2018 and 2022, he was Professor at the School of Materials and Energy, Southwest University, China and served as the founding director of the Centre for Advanced Thin Films and Devices. Since 2023, he has been a Professor at the School of Aeronautics, Harbin Institute of Technology, and he is also affiliated with HIT Zhengzhou Research Institute.

1 Porous Boron Nitride Materials of Hydrogen Storage for Fuel Cell Vehicles

Yilong Jing and Qunhong Weng

1.1 INTRODUCTION

The level of human civilization relies heavily on the development and utilization of energy. Since the Industrial Revolution, traditional fossil fuels such as coal, oil, and natural gas have laid the foundation for the development of modern industry (Figure 1.1). However, with population growth and social development, human demand for energy is increasing. The Paris Energy Agency predicts that, according to current conditions and policies, by 2035 global energy demand will increase from about 12 billion tons of oil equivalent in 2009 to 18 billion tons of gasoline equivalent, or 17 billion tons of gasoline equivalent. The use of so much fossil fuel is bound to aggravate environmental pollution. According to current policies and projections, global carbon dioxide emissions are expected to increase from 2.9 billion tons per year to 4.3 billion tons per year or 3.6 billion tons per year [1]. In 2022, the global average atmospheric concentration of carbon dioxide has reached an unprecedented level of 421 ppm. As the population and economy continue to rapidly grow, the increasing demands for energy will lead to more and more environmental and climate issues. Therefore, seeking pollution-free and sustainable green energy is not only related to the ecological environment of the Earth but also to the development of all mankind. As an important part of energy consumption, the energy and power technology innovation in the transportation industry puts forward specific requirements for promoting the development of renewable energy, which is of great significance.

In recent years, renewable energy has received increasing attention from experts and the public. By using renewable energy such as hydrogen, solar energy, and wind energy to replace traditional fossil fuels, it is hoped to solve problems such as fossil fuel depletion and global warming [3]. Currently, the fuels we use include two types: fossil fuels and renewable energy. Although we are all working to promote the replacement of fossil fuels with renewable energy, as of 2021, fossil fuels still account for 77.1 percent of global energy consumption, while renewable energy accounts for only 18.9 percent, reflecting human society's dependence on fossil fuels.

DOI: 10.1201/9781003386223-1

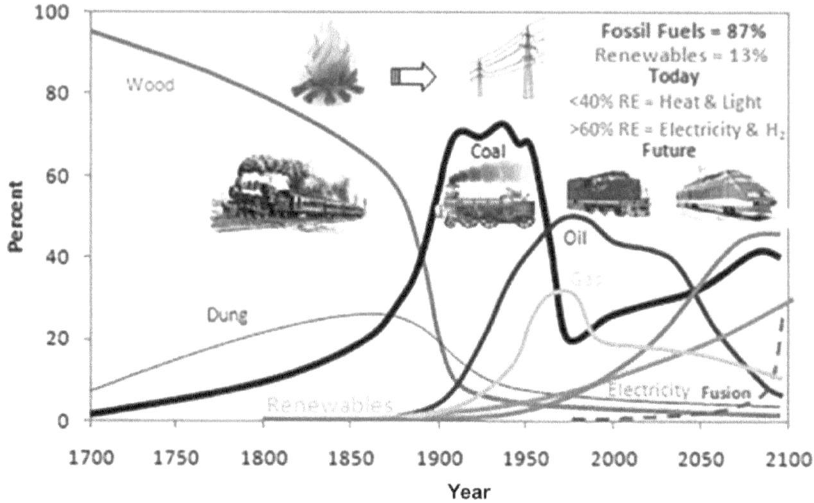

FIGURE 1.1 Illustrative image of energy consumption for human society, the horizontal coordinate of the Figure is the year. It is reproduced from Ref [2].

The development of human society cannot be separated from energy, and we need to increase the development and utilization of renewable energy to achieve sustainable development of human society.

Renewable energy is a broad concept – anything that can be artificially created and is non-polluting to the environment can be considered as renewable energy, including wind, hydro, geothermal, and hydrogen energy. Among these renewable energies, hydrogen is considered as a clean and sustainable energy source that can achieve carbon neutrality, known as the "energy of the future" [4–6]. This is because, compared to conventional fuels, hydrogen has many irreplaceable natural advantages.

(1) Hydrogen is a non-toxic and pollution-free clean energy source. Its mass energy density is three times that of gasoline (the mass energy densities of hydrogen and gasoline are 143 MJ/kg and 44 MJ/kg, respectively).
(2) Hydrogen can be produced using various methods such as electrolysis of water, steam reforming of natural gas, and gasification of coal or other heavy hydrocarbons. It can also be produced from various sources of energy, including renewable energy [7].
(3) When hydrogen is burned, it only produces water as its byproduct, and is extremely environmentally friendly. It can also alleviate the greenhouse effect caused by excessive CO_2 emissions.
(4) Hydrogen can be used in various aspects of the society and economy, such as a fuel for cars (hydrogen-powered fuel cell vehicles FCV) [8, 9] and for power generation through fuel cells [10, 11].

It is because of these advantages, hydrogen energy has attracted so much attention. Hydrogen combustion produces only water without emissions of carbon dioxide, and hydrogen is widely used in fuel cell vehicles to alleviate the environmental problems caused by greenhouse effects faced by the planet today [12]. Vehicles are a great invention of industrial society, not only providing us with great convenience in daily life, but also playing an irreplaceable role in various industries. At the same time, automobiles are one of the largest users of fossil fuels, and the use of fossil fuels will greatly burden the environment [4]. The large-scale use of traditional cars has already caused a certain amount of damage to the ecological environment, but modern society cannot be developed without automobiles. Therefore, we need to limit the use of fossil fuels in vehicles and use more environmentally friendly energy sources. In recent years, electric vehicles have been widely put into production. These vehicles use electricity as power and are more environmentally friendly. However, charging electric vehicles often takes several hours and their range is also insufficient compared to fossil fuels, which limits their large-scale development. Hydrogen fuel cell vehicles are similar to electric vehicles in terms of pollution and are very environmentally friendly. They also have advantages that electric vehicles do not have. The energy density per mass of hydrogen is three times that of gasoline, and the charging time is only a few minutes, which demonstrates the potential of hydrogen fuel cell vehicles. Once sustainable hydrogen production methods are developed, hydrogen fuel cell vehicles will have a stable fuel supply [13].

However, at the same time, the application of hydrogen in fuel cell vehicles also has its own limitations. Among them, the most restrictive factor for the development of hydrogen energy is hydrogen storage. As is well known, the chemical formula of hydrogen is H_2, which is one of the simplest known molecules. Although hydrogen has a good mass-energy density, its volume-energy density is only 0.0108 MJ/L, which is 3,000 times smaller than the volume-energy density of gasoline. In addition, hydrogen gas will undergo a violent explosion when it comes into contact with air, and this requires a more secure storage method. Therefore, in order for hydrogen fuel-powered vehicles to be put into mass production and use, there must be a compact, safe, reliable, inexpensive, and energy-saving way to store hydrogen gas [14].

The storage of hydrogen is extremely difficult, and because hydrogen is the small molecule, it has high diffusion and buoyancy. Currently, there are two common ways to store hydrogen (Figure 1.2). The first method is to use high-pressure cylinders to store hydrogen, which is the most commonly used method. A container made of pressure-resistant materials is used to inject compressed hydrogen into the container to achieve the purpose of hydrogen storage. Current on-board hydrogen storage technology relies on this high-pressure storage method, with the highest pressure reaching 700 bar [4, 15]. However, this storage method has great safety hazards. Hydrogen fuel cell vehicles that use high-pressure cylinders to store hydrogen are prone to explode in the event of a vehicle collision, and the cost of high-pressure cylinders is often expensive, which limits the practical application of hydrogen fuel cell vehicles. The second method is low-temperature storage, where hydrogen gas becomes a liquid at low temperatures. This storage method requires a temperature of 20 K, which increases the cost of hydrogen storage [15]. Both of the above hydrogen storage methods undoubtedly require us to find a suitable method to ensure the safe storage of hydrogen, so that the large-scale use of hydrogen fuel cell vehicles can be

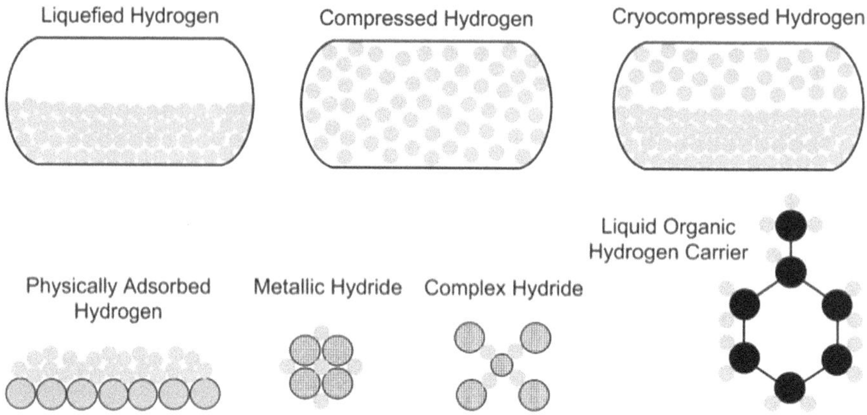

FIGURE 1.2 Summary of hydrogen storage mechanisms. Reproduced from Ref [16].

achieved. How to find a safe, reliable, and inexpensive method of hydrogen storage has been a problem that has plagued researchers for many years.

Chemical hydrogen storage uses reversible reactions between hydrogen and chemical substances, such as solid metal hydrides, borohydrides, ammonia borane, liquid alkenes, ammonia, and so forth. Taking the liquid organic materials as an example, the principle of hydrogen storage is to use the reversible and hydrogenation reactions of certain unsaturated liquid organic compounds such as olefins, acetylenes, or aromatics and hydrogen gas to achieve hydrogen storage (chemical bonding) and use dehydrogenation reactions to achieve hydrogen release. The hydrogen storage density of liquid organic materials is 5–10 percent, with a large hydrogen storage capacity, and the hydrogen storage material is a liquid organic material, which can be transported at room temperature and pressure, making it convenient and safe. In principle, every unsaturated compound (organic molecule with carbon-carbon double or triple bonds) can absorb hydrogen during the hydrogenation process. The organic liquid hydrogen storage technology uses certain hydrogen storage agents such as olefins, acetylenes, or aromatics, and hydrogen gas to produce reversible reactions that achieve hydrogenation and dehydrogenation. In this technology, hydrogen chemical bonds combine with organic hydrocarbon carrier molecules (hydrogenation) and can be released in the reverse process (dehydrogenation). However, liquid organic hydrogen storage also faces a series of technical problems that urgently need to be solved, such as strict technical requirements, complex hydrogenation and dehydrogenation equipment, high operating costs, high dehydrogenation reaction temperature, which is difficult and has low efficiency, consuming 30 percent energy of its energy storage; the reaction process requires the use of catalysts, and the catalyst activity is not stable enough, is easily poisoned, and high-temperature conditions can easily become deactivated; the hydrogen gas purity is not high, and there is a chance of producing impurities gases due to side reactions [17, 18].

In recent years, solid-state porous materials have gradually entered the field of view of scientists. These materials have a large specific surface area and porosity and can be used for hydrogen storage. Compared to traditional hydrogen storage methods, using solid-state porous materials for hydrogen storage is safer and more cost-effective and is expected to become the mainstream energy carrier for hydrogen fuel cell vehicles (FCVs). The US Department of Energy (DOE) has set system-level hydrogen storage goals to achieve storage targets of 5.5 wt percent and 40 g/L by 2025, and final targets of 6.0 wt percent and 50 g/L by 2030. These goals also indicate the storage capacity that needs to be achieved under different conditions of use. In the past few decades, solid-state porous materials have received increasing attention due to their unique properties, and scientists have conducted a lot of research on them. Representative materials include metal-organic frameworks (MOFs), covalent organic frameworks (COFs), porous organic polymers (POPs), carbon-based materials, and zeolites. In these years of research, BN has also shown corresponding potential in hydrogen storage applications. Next, we will introduce solid-state hydrogen storage materials based on boron nitride, including the physical and chemical properties of boron nitride materials, common applications, synthesis of porous boron nitride materials, and research progress on boron nitride materials in hydrogen storage in recent years. We will also discuss the advantages and difficulties faced by boron nitride materials when used for hydrogen storage.

1.2 BORON NITRIDE MATERIALS

1.2.1 Structures

BN is composed of the same number of boron (B) and nitrogen (N) atoms. BN has three crystal forms, and its crystal structure is similar to that of carbon analogues (Figure 1.3). They are hexagonal boron nitride (h-BN) similar to graphite, cubic boron nitride (c-BN) similar to diamond, and wurtzite boron nitride (w-BN) similar to zinc oxide fiber. Under standard conditions, h-BN is the most stable phase; c-BN is similar to diamond, has extremely high hardness, second only to diamond, but its thermal stability is far superior to diamond, and is used to manufacture abrasives with excellent performance [19]. w-BN is composed of sp3 hybridized boron and nitrogen atoms, which are similar to c-BN, but adjacent BN3 or NB3 tetrahedra are stacked at different angles. h-BN is a typical layered structure similar to graphite. In the 2D layer, N and boron B atoms are alternately arranged and connected by B-N covalent bonds, while weak van der Waals bonds connect 2D layers. Unlike graphite, in each 2D layer, boron atoms are precisely located above or below adjacent layer nitrogen

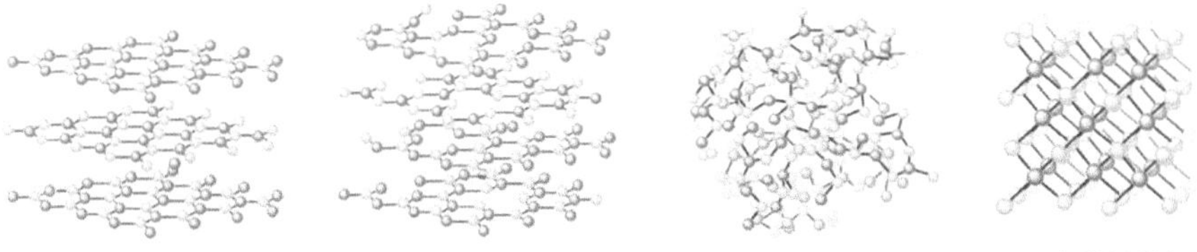

FIGURE 1.3 Structures of boron nitride materials. Reproduced from Ref [22].

atoms in each continuous BN layer, which is a characteristic of the atomic stacking of h-BN [20]. This stacking feature implies the strong polarity of the B-N bond, indicating that the B-N covalent bond carries some ionic bond characteristics. Since nitrogen atoms have higher electronegativity, the B-N covalent bond sigma bond is more localized to nitrogen atoms, and the lone pairs of electrons in the nitrogen atom 2p orbital are only partially delocalized with the boron atom 2p orbital. This electron distribution is different from the average distribution of the C-C bond in the graphite layer.

With a structure that is similar to graphite, and referred to as "white graphite", h-BN shares some properties with graphite, such as anisotropy along the basal and perpendicular planes, high mechanical strength, excellent thermal conductivity, and good lubrication. The unique structural features of h-BN give it distinct electrical, chemical, and optical properties. For example, the reduction of electron delocalization in the π-bonds of B-N bonds gives it a large band gap, leading to its electrical insulation properties and colorlessness. The unique properties of boron nitride make it important in areas such as insulation and thermal conductivity media, deep ultraviolet sources, dielectric layers, cosmetics, and transparent ceramics (Figure 1.4) [21].

1.2.2 Properties and Application

Due to its unique structure h-BN has many special properties. These properties make boron nitride contribute significantly in many fields. Next, we will introduce some representative applications and properties.

1.2.2.1 Porosity and Hydrogen Storage

Boron nitride materials can store hydrogen through either physical adsorption mechanisms or chemical adsorption mechanisms. The amount of hydrogen stored based on physical adsorption at low pressure is completely reversible, and the amount of hydrogen storage is closely related to the physical properties of the material, including specific surface area (SSA), pore width, and pore volume. Changes in these physical properties correspondingly improve or reduce the hydrogen storage performance of the material. Theoretically, boron nitride has greater potential in hydrogen storage applications than materials such as carbon-based materials and metal-organic frameworks. The adsorption of hydrogen molecules on the surface is due to the strong hydrogen bond and closed-shell electronic configuration and is essentially a form of physical adsorption, while the binding energy of hydrogen molecules on solid surfaces is usually very small. For example, the binding energy of hydrogen molecules on activated carbon surfaces is only 60 meV, which requires extreme conditions of high pressure (several hundred bar) and low temperature (77 K) to store hydrogen using activated carbon materials [23]. However, due to the partial ionic character of the B-N bond in boron nitride, an additional dipole moment is created, and the hydrogen molecule's induced dipole moment is highly sensitive to the local electric field, making boron nitride more capable of adsorbing hydrogen molecules and advantageous for its hydrogen storage performance [24].

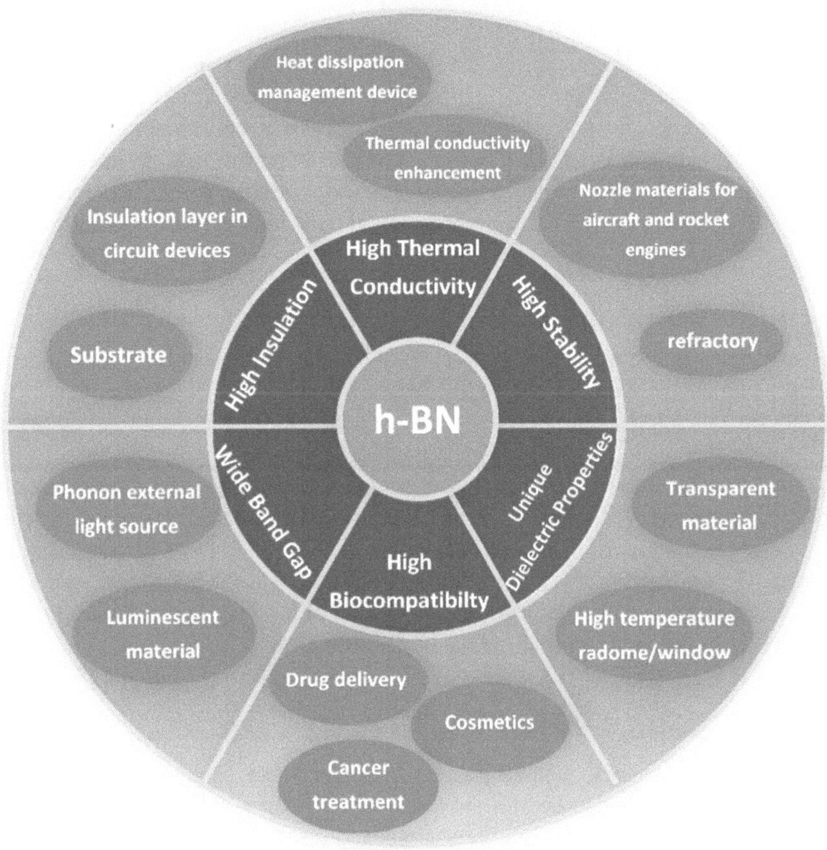

FIGURE 1.4 Summary of properties and applications for h-BN.

1.2.2.2 Biomedical Applications

Compared with carbon-based materials, boron nitride nanosheets or other boron nitride nano materials have better biocompatibility and lower toxicity [25]. They have shown good application prospects in fields such as drug delivery [26, 27], cell labeling [28], BNCT [29], and nanoenzymes [28]. However, boron nitride is a surface hydrophobic material with a relatively inert structure, making it difficult to chemically modify it effectively. Currently developed modification methods and selectable functional groups for modification are still limited. The application of boron nitride nanomaterials in this field will largely depend on whether these issues can be effectively addressed.

1.2.2.3 Insulation and Applications in Microelectronics

Up to now, there are mainly two types of applications for BN in the field of microelectronics. The first type is to use the electrical insulation properties of BN materials as a partition or barrier in high-conductive elements, such as potential barriers in field effect tunneling transistors and atomic-thin insulation for separating metal channels. The second type is due to the smooth surface of boron nitride, which is usually used as a substrate for high-performance devices. The surface of h-BN is extremely smooth, without charge and dangling bonds, and has excellent physical properties such as ultra-high in-plane mechanical strength, high chemical stability and thermal stability, and high thermal conductivity, and so forth. These properties make h-BN a unique 2D material in the design and manufacture of high-performance devices.

1.2.2.4 Mechanical, Thermal Properties and Composite Applications

BN nanosheets demonstrate mechanical strength and thermal conductivity comparable to graphene, and are considered to be the hardest and best thermal conducting material on Earth [30]. Therefore, using BN nanomaterials as composite fillers can enhance the mechanical strength of the matrix, increasing its thermal conductivity while maintaining its electrical insulation characteristics. However, the most critical problem encountered by researchers in the process of mechanical enhancement is how to effectively transfer the load to the nanofiller. As BN nanofillers are connected to the matrix used through weak van der Waals forces or p-p interactions, appropriate functional groups are used to chemically modify the surface of BN nanofillers to enhance this interaction, or to introduce other strong interactions.

1.3 SYNTHESIS OF POROUS H-BN

As mentioned above, h-BN can be used for hydrogen storage, but it requires a unique porous structure; h-BN can be customized into various porous structures and can be used for hydrogen storage, pollution treatment, drug delivery, and so forth. Similar to other porous materials, porous h-BN can be obtained under templated and non-templated conditions. This section will provide a detailed introduction to the synthesis of soft/hard templated porous h-BN and non-templated porous h-BN. For templated synthesis, various template materials have been used, such as silica dioxide, activated carbon, graphene aerogels, zeolite, cationic surfactants, block copolymers, and so forth. Porous BN materials obtained under templated conditions have surface areas ranging from 100 to 950 m^2/g. However, in recent years, the surface area of porous BN materials synthesized based on non-templated strategies has greatly increased, up to 1900 m^2/g [31]. The theoretical surface area of porous BN materials can reach up to 4800 m^2/g, indicating the enormous potential of porous BN materials in hydrogen storage [32, 33].

1.3.1 Template Method

So far, various porous templates have been used to prepare porous boron nitride. The first template to be use was activated carbon, with B_2O_3 and N_2 as the B and N sources, respectively, and reacted at high temperature to obtain the desired result. However, the specific surface area (SSA) of the resulting porous BN material is only 167.8 m^2/g, which is much smaller than the specific surface area of the activated carbon template (779 m^2/g) [34]. This phenomenon is due to the difficulty of filling the precursor in the template, which is a common problem with this hard template strategy. Afterward, various templates such as mesoporous carbon/mesoporous silica have been used as hard templates, such as SBA-15, CMK-3 carbon [35–38], and carbon aerogel [39]. Compared with carbon templates, silica templates require the use of HF to remove them. Moreover, due to the hydrophilicity of silica, the structural performance of the BN material obtained by this method is poor. In addition, because mesoporous carbon cannot be directly prepared, its synthesis itself also requires the use of mesoporous silica as a template, so it does not have a competitive advantage. It has been reported that using small-molecule ammonia borane (BH_3NH_3) as a BN precursor and infiltrating an SBA-15 template results in a porous boron nitride product with a specific surface area of 327 m^2/g and a pore volume of 0.50 cm^3/g, showing better filling properties of the small-molecule precursor [40]. In addition to hard templates, soft template methods can also be used to prepare porous boron nitride, wherein the advantages of the template agent used are eliminated simultaneously during the preparation process, thereby avoiding the need for additional template removal steps. For instance, using the aggregation properties of tri(methylamino)borazine (MAB), Meile et al. used CTAB micelles as a template to prepare a transparent gel, which, after high-temperature ceramization, obtained a porous BN with a specific surface area of up to 800 m^2/g and a pore size of 6 nm [41, 42]. Malenfant et al. used poly(ethylene oxide)-b-poly(10-phenylborasiloxane) hybrid organic-inorganic block copolymers to self-assemble in tetrahydrofuran to synthesize an ordered porous BN nanomaterial with a specific surface area of up to 950 m^2/g and a pore size of up to 20 nm [43].

1.3.2 Non-Template Method

Non-template method refers to the process of preparing porous BN by utilizing the self-reaction between BN precursors without template involvement. For instance, Rao's group reported the preparation of BN sheets using boric acid as the boron source and urea as the nitrogen source, with a specific surface area of up to 927 m^2/g [44]; Chen's group used boron oxide-guanidine hydrochloride as precursor to obtain BN nanosheets with a high specific surface area of 1425 m^2/g [45]. We proposed the use of boric acid-melamine (2B·M) as a raw material to synthesize micrometer-sized porous BN materials with ribbon-like morphology, which is the first BN material with SSAs exceeding 1000 m^2/g (reaching 1488 m^2/g) [46]. Afterward, 2B·M has been widely used as an efficient precursor for preparing porous BN [47–49]. In addition, various precursor systems have also been reported successively, and the porous BN materials prepared have high specific surface area. For example, BN prepared using boric acid-dicyandiamide as a precursor has a specific surface area of 1900 m^2/g [31], and BN prepared using boric acid-guanidine carbonate has a specific surface area of 1290 m^2/g, both exhibiting mainly micropore structures [50]. Furthermore, BN-based porous monoliths (BNPMs) were also prepared through a facile two-step template-free reaction including self-bubbling precursor solidification and high-temperature pyrolysis. The resulting BNPMs have an interconnected open-bubble structure and show the specific surface area up to 1406 m^2/g [51].

1.4 PROGRESS ON POROUS BNS FOR HYDROGEN STORAGE

In recent years, there has been an increasing amount of research on using porous BN materials for hydrogen storage, which suggests the enormous potential of BN in the field of hydrogen storage. This chapter will introduce relevant research achievements in this field, which include theoretical predictions and experimental progress.

1.4.1 Theoretical Progress

The affinity between H_2 molecules and the material surface can be tuned by modifying its surface composition. Several theoretical works have been conducted and stressed the importance of structural modifications in H_2 adsorption, for example, the adsorption behaviors of H_2 on BN nanotubes with different doping and structural defects haven been explored by pseudopotential density functional methods. It was concluded that the binding energy of H_2 molecule on BNNTs increased by 40 percent compared to that on CNTs, and the binding energy remarkably increased by ≈50 percent via substituting one B atom with a C dopant in the BNNTs [24]. In addition, Lei and coworkers conducted first-principles calculations and found that hydrogen adsorption energy on BN nanosheets doped with N sites was 20–80 percent higher than that of pure BN nanosheets, which is beneficial for hydrogen storage [52]. Similar conclusions were also seen in other theoretical studies associating with the H_2 adsorption on the O-doped BN nanostructures [53, 54]. Furthermore, C and O co-doping of the BN was also studied by first principle calculations, which showed stronger H_2 adsorption affinity up to 0.28 eV or 27 kJ /mol. These values are much higher than those doped only by C (0.16 eV) or O (0.18 eV) atoms [55].

Xue et al. explored Li or Ca-decorated carbon-doped BN monolayers (BC2N) for H_2 storage theoretically [56, 57]. The BC2N–Li may serve as high-capacity material for hydrogen storage with gravimetric hydrogen capacity of 9.94 wt percent at ambient conditions. It was estimated that the BC2N–Li could desorb H_2 at c.a. 373 K and at a pressure of 1.5 atm. These mild conditions are advantageous while in practical applications. More recent density functional theory and molecular dynamics simulations showed that the lithium- and oxygen-doped pillared BN structures exhibited improved gravimetric and volumetric hydrogen capacities at room temperature, with values on the order of 9.1–11.6 wt percent and 40–60 g/L [54].

Recent theoretical studies on hydrogen storage using boron nitride (BN) have shown that carbon-doped porous BN has a bandgap of 1.8 eV and is capable of storing 5.1 wt percent of hydrogen [58]. In this work, the authors propose a single layer of BN, specifically p-BN, using first-principles calculations. Compared to perfect h-BN, p-BN has a reduced bandgap of 0.7 eV, but this can be adjusted by adding carbon. When the carbon content is 50 percent, the bandgap of this material can also be adjusted to a moderate degree, namely 1.8 eV, and has a rich pore structure which can store 5.1 wt percent of hydrogen. There is also a cage-like BN material that can store 12 wt percent of hydrogen [59]. These examples reflect the huge potential for BN or BCN-based materials as hydrogen storage devices.

In addition to composition modifications, it is also highly interests to know the theoretical SSA upper limits of porous BNs. It is known that the highest SSA for

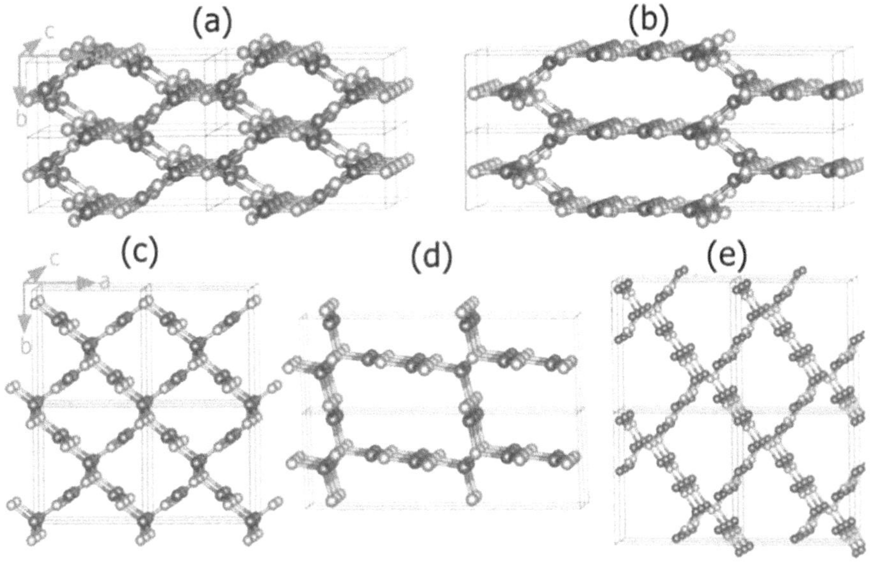

FIGURE 1.5 Proposed porous BN structures with high SSAs. The SSAs (m^2/g) for the BN models are: ~4800 (a), ~3800 (b), ~3700 (c), ~3400 (d), and ~3300 (e). It is reproduced from Ref. [33].

BN materials are still around 2000 m^2/g [31, 60, 61]. These values are still below the cutting-edge SSA values of porous carbons and MOFs. Because of the fundamental role of SSA and pore volume on H_2 physisorption, it is important to know the potential to further improve the SSAs of porous BNs.

Zeng et al. have proposed several theoretical BN models (Figure 1.5) with high porosity [32, 33]. A normal sp2 hybridized BN structure domain is connected by sp3-BN chains. They not only possess mechanical and thermal stability, but also have a high specific surface area of up to 4800 m^2/g. Additionally, the calculated hydrogen adsorption energy (−0.18 eV, i.e., −17.3 kJ /mol) is very close to the optimal adsorption energy (−0.15 eV), making it very favorable for room temperature reversible hydrogen storage. The increase of the specific surface area is favorable, even at room temperature, to the uptake of H_2. *Ab initio* simulations showed a H_2 uptake of about 8.7 wt percent for the pillared BN at 27°C and 100 bars with a slab distance of 29 Å and a specific surface area higher than 2200 m^2/g [62]. Studies have shown that porous BN is an ideal choice for hydrogen storage materials, and finding porous BN with higher specific surface areas and pore volumes is a feasible way to find porous BN products that meet hydrogen storage requirements.

1.4.2 Experimental Progress

In recent years, researchers have discovered numerous porous hexagonal boron nitride materials with high specific surface areas and porosity [31, 46, 50, 63]. Boron nitride materials can store hydrogen through physical adsorption or chemical adsorption mechanisms.

The study of hydrogen storage materials based on boron nitride nanostructures can be traced back to 2002, when the technology for macroscopic preparation of boron nitride nanotubes was solved. Ma et al. studied the hydrogen absorption performance of different forms of BN nanotubes. One is the ordinary boron nitride multi-wall nanotube, and the other is the bamboo-shaped nanotube. They found that at room temperature, bamboo-shaped and multi-wall BNNTs can absorb 2.6 wt percent and 1.8 wt percent of H_2 at 10 MPa (Figure 1.6). As pointed out by the authors, one issue with the use of BNNTs for hydrogen storage is that about 70 percent of the adsorbed hydrogen is retained in the material after the system is restored to ambient pressure. When the structure is heated to 573K, the retained hydrogen is released, indicating that most of the hydrogen may have been chemically absorbed under such high pressure [64].

Later, Tang et al. discovered that under the same testing conditions, the irreversible H_2 absorption capacity could increase to 4.2 wt percent (Figure 1.6) in a Pt-catalyzed activated BNNT. It can be understood that after the activation process, the SSA of BNNT can be increased from 254.2 m^2/g to 789.1 m^2/g, which helps to improve the H_2 adsorption performance of the sample. Similarly, most H_2 molecules are chemisorbed and need to be released at high temperatures [65]. Later studies found that such a high absorption capacity was partly due to the joint contribution of the introduced BNNT and residual Pt particles during the activation process. Because collapsed BNNTs activated through another pathway (using SO2 at high temperatures) showed very limited H_2 adsorption rate, at approximately 0.6 wt percent under 2 MPa and 77K [66].

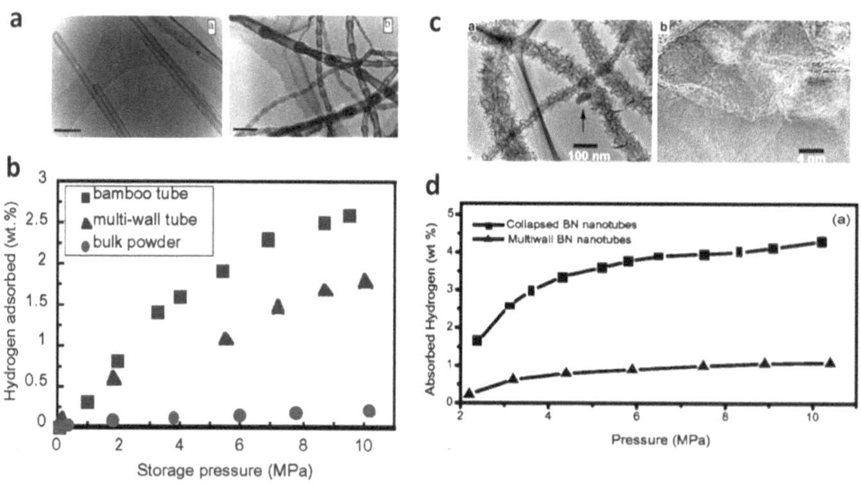

FIGURE 1.6 Hydrogen adsorption properties of different boron nitride nanotubes: (a) TEM images of multi-walled BNNTs and bamboo-like BNNTs; (b) H_2 adsorption isotherms of tube samples at room temperature; (c) TEM image of collapsed BNNTs; (d) Comparison of H_2 adsorption between folded BNNTs and multi-walled BNNTs at room temperature; (a, b) are reproduced from Ref. [64]; (c, d) are reproduced from Ref. [65].

Lin et al. used a boron-nickel catalyst to synthesize BNNT through ball milling in a nitrogen/hydrogen mixture at 1025ºC. Their research found that at 6 MPa and room temperature, BNNT can absorb a renewable H_2 capacity of 1.8-2.2 wt percent. They conducted temperature-programmed desorption analysis and found that 50 percent of the hydrogen adsorbed on BNNT was chemisorbed and could be released at a temperature of 250ºC [67].

Thereafter, the hydrogen storage research on boron nitride materials has mainly focused on the design of hollow and porous boron nitride materials. Portehault et al. developed a pyrolysis method for preparing porous BCN materials starting from a complex filling of borane (as B source) and mesoporous graphite nitride carbon (as N source). The prepared product has a very high surface area of up to 1560 m^2/g. This reactive hard template method ensures the adjustable composition and texture of the product. The results showed that the H_2 absorption rates of these compounds were 0.55-1.07 wt percent at 77 K and 100 KPa conditions [68]. Lian et al. reported the synthesis of thin-walled (1–3 nm) BN hollow spheres with no template sealed reaction using NH_4BF_4, NaN_3, and 0.64 g of sulfur (S) in a high-pressure reactor. The prepared boron nitride hollow spheres have a high specific surface area of up to 215 m^2/g and a large pore volume, and have a high hydrogen absorption capacity of up to 4.07 wt percent at 298 K and 10 MPa [69].

Lei et al. found that a 4 at percent oxygen-doped BN sheet with SSA of 536 m^2/g has a H_2 storage capacity of 5.7 wt percent at room temperature at 5 MPa. When the hydrogen pressure was lowered to ambient conditions, about 89 percent of the stored hydrogen was released. According to its SSA, a relatively high value of hydrogen uptake capacity was reported, which may be explained by the O-doping of the BN surfaces [52].

More recently, Muthu et al. decorated acid treated halloysite clay nanotubes (A-HNTs) with h-BN, and found that the hydrogen adsorption increased along with the increasing content of h-BN in A-HNT–h-BN nanoclay composites. A 2.88 wt percent of hydrogen storage capacity and 100 percent desorption were observed for the A-HNT–5wt percent h-BN nanoclay composite at 50°C [70].

1.4.3 Cases of Porous BNs for Hydrogen Storage

1.4.3.1 BN Porous Microbelts

Due to the low pressure, the main adsorption mechanism of hydrogen on the surfaces of C and BN is a physical mechanism, and the amount of hydrogen adsorption is largely affected by the specific surface area (SSA) and pore size of the adsorbent [71]. In order to obtain porous BN materials with high SSA and small pore width, the hard template method is used to produce porous BN. However, the SSA and pore width of the BN obtained by this strategy still cannot meet the needs of commercial hydrogen storage. In addition, this hard template method strategy also has other disadvantages, such as insufficient filling rate of precursors in the template, and difficulty in removing the template. In this regard, we have developed an easy-to-operate template-free method to prepare a new type of BN porous microstrip. This method can synthesize BN in one step at normal pressure. The prepared sample has the largest SSA and micropore volume, reaching 1488 m^2/g and 0.502 cm^3/g, and the maximum hydrogen adsorption capacity is 2.3 wt percent.

Using boric acid and melamine (2B·M) as precursors, synthesis was conducted at 900–1100°C, resulting in samples BNPB-900, BNPB-1000, and BNPB-1100. Compared with using organic amine borane precursors and organic borazine precursors, the advantages of using 2B·M precursors are simple production and low cost. However, in addition to providing N sources, melamine also provides C atoms. The key challenge in using 2B·M precursors is to remove the C element from the product. Prior to this work, a series of BN nano materials were prepared using 2B·M as precursor [72–74]. In order to remove the C element, ultra-high temperature (>1700–2000°C) is usually used to obtain pure BN composition [72]. However, it is difficult to retain the original nano morphology of BN at such high temperatures, and inevitable recrystallization of BN phase will occur. In contrast, we provide a simple method to prepare high-purity BN materials at low temperature (≥900°C) and atmospheric pressure (P = 1 atm) by reacting the 2B·M precursor with ammonia. The morphology of the product is controllable because the appearance of the precursor varies according to different crystallization conditions. In addition to O_2, the C component in BN can be removed by reacting with ammonia at 1000°C [38].

$$C + NH_3 \rightarrow CH_4 + N_2$$

The prepared BNPB has a belt-like structure. A large number of pores are randomly distributed on the surface of the BN belt (Figure 1.7). HRTEM characterization showed that the average distance between adjacent (0002) stripes is 0.38 nm, which is much larger than the interlayer spacing of (0002) face of h-BN and MWBNNTs (0.33-0.34 nm). It is worth noting that dislocation structures are commonly present in BN phases. After annealing at 1500°C for 5 hours, the ordering degree of BNPB-1100 did not improve significantly. This result can be attributed to the high thermal stability provided by the dislocation structure in the porous BN belt. These dislocations not only widen the average distance between BN (0002) layers, but also twist adjacent layers together, providing considerable strength to prevent further ordering of BN (0002) layers at high temperature.

The BET specific surface area of the sample was calculated through N_2 adsorption-desorption isotherms at 77 K. As shown in Figure 1.7d, the BET specific surface area test results for BNPB-900, BNPB-1000, and BNPB-1100 were 1161, 1488, 1144 m^2/g respectively. The isotherms and hysteresis loops of all BN porous microbelts obtained at different temperatures belong to typical type I and H4 type lines (according to the IUPAC classification), which demonstrate that they contain micropores and slit-shaped mesopores. At a pressure of P/P0=0.99, the total pore volume of BNPB-1100 is 0.880 cm^3/g. Although the total pore volume of the sample also increases with temperature, BNPB-1000 has the highest micropore volume ratio, which is 0.502 cm^3/g. These new BN porous microbelt materials have such high microporosity and SSAs due to the dislocation structure mentioned above.

Under the conditions of 77 K and 1 MPa, the adsorption and desorption isotherms of H_2 on BN porous microbelts measured by volumetric method are shown in Figure 1.7f. As the synthesis temperature increased from 900 to 1100°C, the sample exhibited an enhanced H_2 uptake capacity, from 1.6 to 2.3 wt percent. In comparison, the H_2 uptake capacity of bulk h-BN material measured under the same conditions

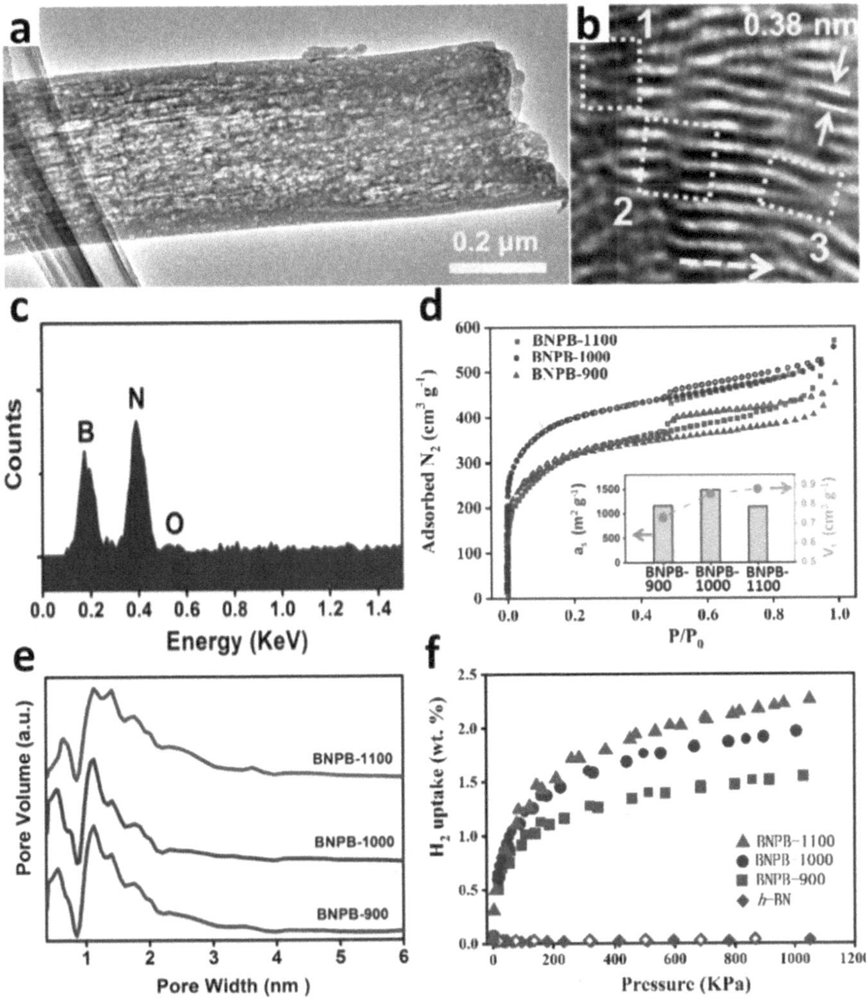

FIGURE 1.7 (a) TEM and (b) HRTEM image of BNPB-1100; (c) EDX spectrum of BNPB-1100; (d) N_2 adsorption-desorption isotherms of BNPB samples and corresponding (e) pore size distribution profiles; (f) H_2 adsorption-desorption isotherms at 77K. Reproduced from Ref. [46].

was very low, far less than 0.1 wt percent. However, further increasing the synthesis temperature did not increase the H_2 uptake capacity.

1.4.3.2 BCNO Porous Microbelts

Carbon (C) and boron nitride (BN) are iso-electronic materials, and both can form similar nano/micro structures. C and BN nanotubes and nanosheets exhibit excellent properties such as high mechanical strength and outstanding thermal conductivity [75]. However, nano-BN has very high chemical inertness and thermal stability, which is opposite to C analogs [76]. The B-N bond has partially ionic bond characteristics, which makes BN a wide bandgap semiconductor material [77]. By introducing C atoms into the BN lattice, the bandgap of BN or C can be controlled, or bandgap-controlled BCN hybrid compounds can be formed. Moreover, the polar B-N bond is also considered as the reason why the BN surface has higher H_2 affinity than the C surface [24]. In addition, C- and O-doping can introduce structural defects and modify the local electric field on the BN surface, providing more choices to screen suitable H_2 adsorbents in the BN system. Graphene oxide (GO)-carbon nanotubes (CNTs) composites have higher H_2 adsorption capacity than reduced graphene oxide (rGO)-CNTs composites [78], indicating the importance of the presence of O elements in improving the H_2 affinity of related adsorbents. Therefore, the study of quaternary BCNO materials is of great significance for developing hydrogen storage materials with superior properties.

As mentioned in the previous section, for hydrogen storage materials dominated by physical adsorption, increasing the SSAs and micropore volume of the material can effectively increase the adsorption of H_2 [71], although the exact correlation between structural parameters and H_2 adsorption is still unclear. Previous literature has reported that BNNTs exhibit partial chemical adsorption under high pressure of 10 MPa [64], but at 77 K and low pressure, the main hydrogen adsorption mechanism on BN or BCNO surfaces is reversible physical adsorption [68]. Regulating the pore width of the H_2 adsorbent is considered an effective method to improve the adsorption performance of the carbon-containing system.

In this work, we reported the preparation of a BCNO adsorbent with ultra-narrow and tunable pore widths, focusing on the detailed description of the relationship between their structural parameters and H_2 adsorption behavior. Dark brown BCNO porous samples (Figure 1.8) were prepared by heating the boron triazine (2B·M) precursor under argon atmosphere at different temperatures (900, 1100, and 1300°C, corresponding to BCNO-900, -1100, -1300) for 3 hours. The yield could reach 22-26 wt percent compared with the raw material 2B·M. The chemical composition of the newly prepared porous BN samples was analyzed in detail by energy dispersive spectroscopy (EDS) as shown in Figure 1.8d. The results showed that the composition of these samples was 30–40 at percent B atoms, 12–18 at percent C atoms, 34–35 at percent N atoms, and 13–18 at percent O atoms. However, X-ray photoelectron spectroscopy (XPS) characterization showed that as the synthesis temperature increased, the B-N structure domain increased, while the B-O and C-N structure domains decreased. It can be clearly observed from Figure 1.8b that many pores are randomly distributed on the banded surface of BCNO-1100 [63].

Through isothermal adsorption characterization, it was discovered that all BCNO microstrip samples showed similar specific surface areas (1040-1070 m^2/g) and pore volumes (0.39, 0.35, and 0.37 cm^3/g). However, with the increase of synthesis temperature, there was a significant change in the corresponding pore width. BCNO-900 and BCNO-1100 samples had a sharp and narrow distribution peak at 0.4 nm and 0.6 nm, respectively, while BCNO-1300 exhibited two major pore widths (0.4 nm and 1.1 nm). Therefore, this BCNO porous material is an ideal model for studying the effects of pore width and surface chemical environment on H_2 adsorption.

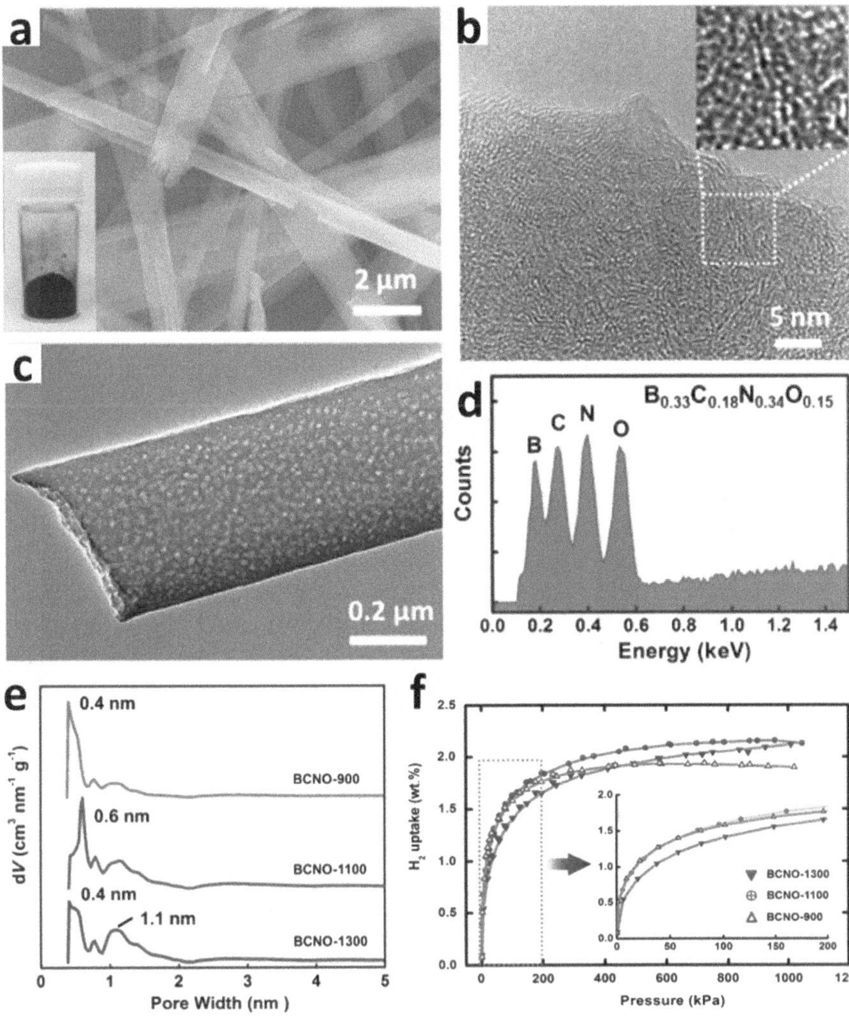

FIGURE 1.8 (a) SEM image of BCNO porous microbelts prepared at 1100 oC. (b) Low magnification TEM image of BCNO microstrip. The inset in (b) is an enlarged area after low-pass filtering. (c) HRTEM image of BCNO strip. (d) Energy-dispersive X-ray spectrum (EDS) of the sample. (e) Pore size distribution profiles. (f) H_2 adsorption-desorption isotherms at 77K. Reproduced from Ref. [63].

Figure 1.8f shows that at 1 MPa, BCNO-1100 (2.14 wt percent) exhibits a similar H_2 uptake to BCNO-1300 (2.12 wt%). However, the former clearly displays a higher H_2 uptake at low pressures of 0-1 MPa, indicating that smaller pore widths are favorable for H_2 adsorption. Interestingly, although BCNO-900 has a smaller pore width (0.4 nm) than the other samples, its H_2 uptake is far lower than that of BCNO-1100 and BCNO-1300 at pressures of 0.45-1 MPa. Only at very low pressures (<50 kPa) does the hydrogen absorption capacity of the sample exhibit a negative correlation with pore width. Previous studies have also reported a similar correlation between H_2 uptake and pore width for porous carbon materials [79].

In this work, we found that ultra-narrow pore widths are beneficial for H_2 adsorption at low pressures, and that micropores with pore widths as large as 1.1 nm can still effectively store hydrogen at ~1 MPa and 77 K. Therefore, increasing the volume of gaps with widths below 1 nm and increasing the proportion of B-N structural domains in the material should be an effective way to improve the hydrogen storage performance of the porous BCNO quaternary system.

1.4.3.3 BN Microsponges

Boron nitride materials have a higher H_2 adsorption potential than carbon analogues. We developed a porous BN microsponge (BNMSs) with extremely high SSAs, up to 1900 m^2/g, through a simple, one-step non-template reaction [31]. The SSAs and pore volumes of the prepared BNMSs can be easily tuned over a wide range (920-1900 m^2/g and 0.516-1.070 cm^3/g, respectively). Under conditions of 1 MPa and -196°C, the H_2 uptake of these BNMSs ranged from 1.65 to 2.57 wt percent, displaying good and fully reversible H_2 absorption.

Boric acid (BA) was used as the B source and dicyandiamide (Dcy) was used as the N source to remove the C element from the precursor under NH_3 atmosphere, obtaining relatively pure BN products. Elemental analysis (B, C, N, O) showed that the main components of the product were B and N (B+N > 93 at percent), with a stoichiometric ratio of B/N (mol/mol) of 1.00±0.02. Trace amounts of oxygen (5-7 at percent) were also detected in the sample, which may be due to adsorbed water and unreacted B_2O_3, and trace amounts of C element (<0.4 at percent) were also detected. No obvious mesopores or macropores were observed in the obtained BNMSs sample. The high magnification HRTEM characterization of these BNMSs materials showed the existence of dislocations in the obtained BN phase. Although the product did not change significantly with the increase of Dcy/BA ratio, the degree of order of the BN phase improved significantly with the increase of temperature, which to some extent hindered the formation of highly porous BN materials. Figure 1.9 shows a typical HRTEM image of the BNMS-Dcy-3-800 sample, revealing branching dislocation structures and two types of pores (P1 and P2) in the sample. It is worth noting that the dislocation structures in BNMSs introduce sp3 hybridization bonds between BN layers to ensure such a high porosity and avoid the ordered and collapsed pores during the synthesis process.

Through the N_2 isothermal adsorption-desorption study of BNMSs samples, the specific surface areas (SSAs) and pore characteristics were analyzed. As shown in Figure 1.10, as the Dcy/BA ratio increases from 1:1 to 3:1, the SSAs of BNMSs increase from 920 to 1840 m^2/g. However, as the Dcy/BA ratio further increases to 4:1, the SSA of the sample decreases to 1140 m^2/g. The corresponding total pore volume (P/P0=0.995) also showed a similar trend as a function of the Dcy/BA ratio (Figure 1.10b), indicating that an appropriate N precursor/BA ratio is necessary for optimizing SSAs and pore volume. In addition, the effect of temperature was also studied, and the BN microsponge material synthesized at 800°C with a Dcy/BA ratio of 3:1 showed the maximum SSAs and total pore volume (reaching 1900 m^2/g and 1.070 cm^3/g, respectively). Combined with the results of HRTEM, XRD, and

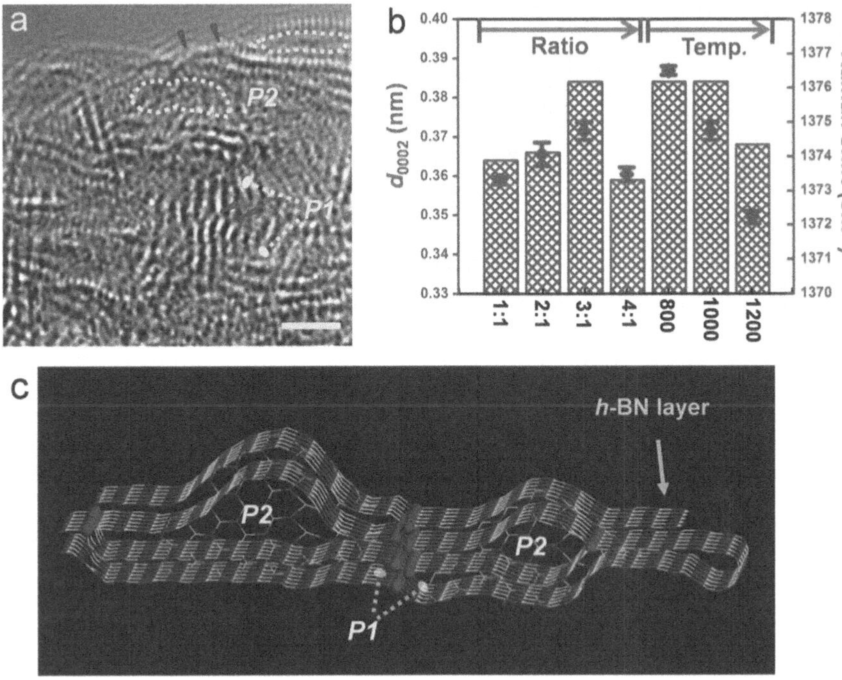

FIGURE 1.9 A typical HRTEM image of the BNMS-Dcy-3-800; (a) the porous structure of BN microsponge. Part (b) summarizes XRD (002) spacing and Raman E2g displacement for the microsponge samples. Part (c) illustrates dislocation and pore structure in the BN microsponge. It is reproduced from Ref. [31].

Raman spectroscopy, the decrease in performance values of these samples as temperature increases can be attributed to the ordering or crystallization of the BN phase in BNMSs. The PSD analysis of BNMSs shows that the main pore size is 1.0 nm, with two broad peaks in the ranges of 1.2-2.1 nm and 2.8-4.5 nm.

The H_2 adsorption amounts of BNMSs were obtained by measuring their adsorption-desorption isotherms at −196°C and 0-1 MPa (Figure 1.11). All BNMSs samples exhibited reversible physical adsorption. Among BNMSs with different Dcy/BA ratios, the highest H_2 adsorption amount was achieved by BNMSs with a Dcy/BA ratio of 3:1 at −196°C and relatively low pressure (1 MPa), which was 2.39 wt percent. Lower synthesis temperatures were also favorable for H_2 adsorption. For example, the BNMS-Dcy-3-800 sample showed the highest H_2 uptake among all BNMS samples, which was 2.57 wt percent (1 MPa) and 1.34 wt percent (0.1 MPa). Taking into account the compressed H_2 filled in free pore volume, the total H_2 adsorption amount of BNMS-Dcy-3-800 was calculated to be 2.60 wt percent (−196°C and 1 MPa), corresponding to a gravimetric energy capacity of 0.87 KWh/kg. However, these capacities were still lower than the target set by the US Department of Energy. By plotting the curves of H_2 adsorption amounts of BNMSs against pore volume and SSA, it can be clearly observed that H_2 adsorption increases with an increase in pore

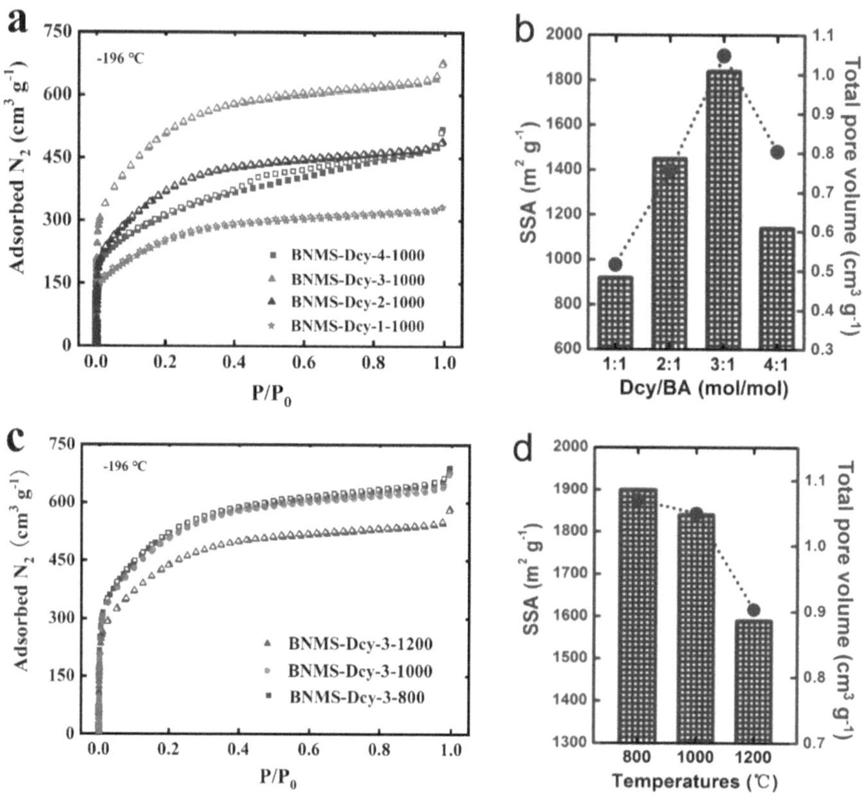

FIGURE 1.10 At a fixed synthesis temperature of 1000°C, the nitrogen adsorption-desorption isotherms (a), BET specific surface area (SSAs), and total pore volume (b) of BN micro-sponges with different Dcy/BA ratios were obtained. When Dcy/BA=3:1, the nitrogen adsorption-desorption isotherms (c), BET SSA, and total pore volume (d) of BN micro-sponges were obtained at different synthesis temperatures. Reproduced from Ref. [31].

volume and SSA. However, there are still some factors that may affect H_2 adsorption besides SSAs and pore volume, such as pore width, pore geometry, or surface chemistry state.

In summary, we have successfully synthesized BNMS materials through a simple, one-step non-template method, which has an increased (0002) spacing and a large number of dislocation structures. By combining an appropriate ratio of N precursor/boric acid and moderate synthesis temperature, we can produce microsponge BN products with ultra-high SSA (up to 1900 m^2/g), ultra-high porosity (1.070 cm^3/g) and high H_2 adsorption capacity (2.57 wt percent at −196°C and 1 MPa).

1.4.3.4 C and O Co-doped BN Microsponges

In addition to increasing the specific surface area of the adsorbent and designing appropriate pore sizes, the surface chemical state also has a great influence on

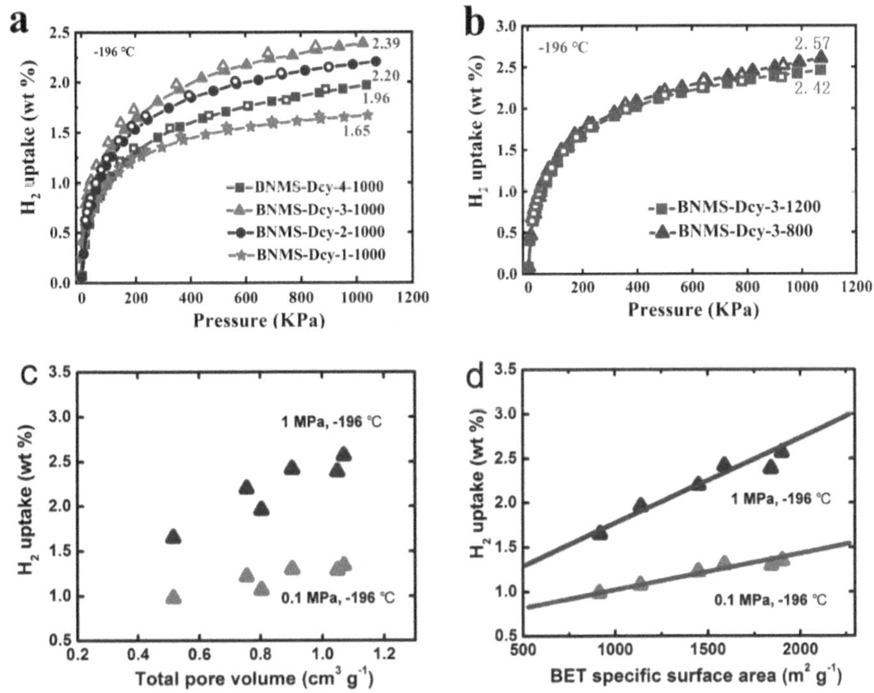

FIGURE 1.11 (a) Under the synthesis temperature of 1000 °C, at different Dcy/BA ratios, (b) when Dcy/BA=3:1, the hydrogen adsorption-desorption isotherms at different temperatures (−196 °C). The hydrogen adsorption capacity of BNMSs at 0.1 MPa and 1 MPa (−196 °C) varies with the total pore volume (c) and BET SSA (d). Reproduced from Ref. [31].

hydrogen storage capacity. Theoretical calculations predict that the adsorption energy of H_2 on the BN surface is 40 percent higher than that of carbon analogues. Further chemical modification of the BN surface will enhance its affinity for H_2 [24], for example, substituting a C atom for a B atom in BNNT or substituting an O atom for an N atom in BN nanosheets will increase the adsorption energy of H_2 molecules [80]. Theoretical simulations have shown that compared to a single BN structure, the adsorption energy of H_2 on O-doped sites is higher due to enhanced polarization effects [52–54]. Furthermore, Kumar et al. reported that co-doping of C and O in boron nitride using first-principles calculations is more effective for H_2 adsorption (adsorption energy = 0.28 eV or 27 kJ/mol) compared to C-(0.16 eV), O-(0.18 eV), and 2C-(0.17 eV) doping [55]. These advances provide a feasible pathway for designing high-efficiency adsorbents based on co-doped BN structures.

Recently, we reported on porous BN materials with different structures and chemical states, and experimentally demonstrated the importance of chemical states in their H_2 adsorption performance. We first synthesized sponge-like porous BN materials and systematically studied the effect of the N-/B- precursor molar ratio and reaction temperature on their structure and structural parameters. At 800°C, the H_2 uptake capacity of unit SSA co-doped BNMS can reach 5.5×10^{-3} wt percent g/m^2, which is 2.5-4.7 times that of undoped BN materials. This work provides solid experimental evidence that C/O co-doping significantly enhances the surface H_2 affinity of BN and highlights the importance of surface chemical states in designing advanced H2 adsorbents for high-density and low-pressure H2 storage.

We conducted an elemental analysis of the BNMS material. Different analytical methods were used to determine the content of B, C, N, and O in the samples. The results showed that the BNMS sample is mainly composed of B and N, and the B/N ratio is close to 1:1. The C content in samples prepared above 1000°C is very low, ≤0.1 at percent. However, in the BNMSs prepared at 700-900°C, a small amount of C content was detected, which was 2.2 at percent (BNMS-1.5-700) (Figure 1.12). In addition, these materials contain a certain amount of oxygen. It can be seen from Figure 1.12b that the O content generally decreases with the increase of the synthetic temperature. BNMS samples synthesized at lower temperatures have higher O content. For example, for BNMS-1.5-700, the O content is as high as 10.0 percent. XPS analysis provided similar trends in variation. XPS and IR spectroscopy analysis provided further structural information on bonding/chemical states. The results confirmed the existence of oxygen in materials synthesized at different temperatures. B1s at 190.6 eV and 192.0 eV correspond to B-N and B-O bonds in the material, respectively. C1s at 288.7 eV and 286.4 eV are components of C=N and C=O, respectively. The results showed that the C=N bond decreases with the increase of the synthesis temperature. The infrared spectrum in Figure 1.12d also confirmed that for the BNMS samples prepared at low temperatures, there are not only dominant B-N vibrations, but also

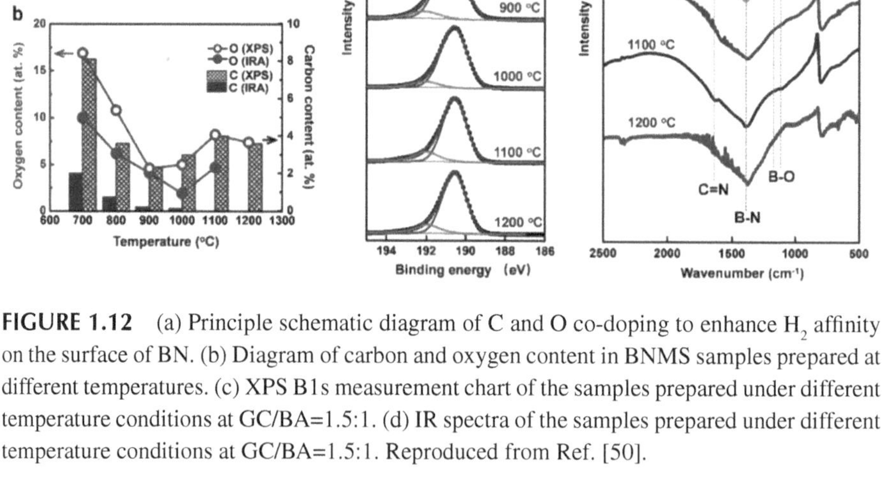

FIGURE 1.12 (a) Principle schematic diagram of C and O co-doping to enhance H_2 affinity on the surface of BN. (b) Diagram of carbon and oxygen content in BNMS samples prepared at different temperatures. (c) XPS B1s measurement chart of the samples prepared under different temperature conditions at GC/BA=1.5:1. (d) IR spectra of the samples prepared under different temperature conditions at GC/BA=1.5:1. Reproduced from Ref. [50].

B-O and C=N bonds [81–83]. These results all indicate that the chemical state of the BNMS samples prepared at 700-900°C deviates from the other samples.

The N_2 adsorption and desorption isotherms at −196°C were taken to determine the BET SSA and other properties of BNMS samples, as shown in Figure 1.13. According to the IUPAC classification, the recorded isotherms belong to the typical Type I and Type H4 loops, indicating the presence of micropores and narrow mesopores. When the GC/BA ratio changed from 1:1 to 3:1, the BNMS-1.5-1100 prepared under 1100°C had the highest BET SSA at P/P0=0.995, which was 1080 m^2/g, with a total pore volume of 1.01 cm^3/g (Figure 1.13). The optimal pore size distribution curve obtained by QSDFT showed that narrow mesopores with a width of 0.8~1.0 nm dominated. Figure 1.13d shows that as the temperature increased from 700 to 1000°C, both the SSA and total pore volume showed a valley at 800°C, which may be caused by the structural reorganization caused by the reaction of carbamide with BA, followed by

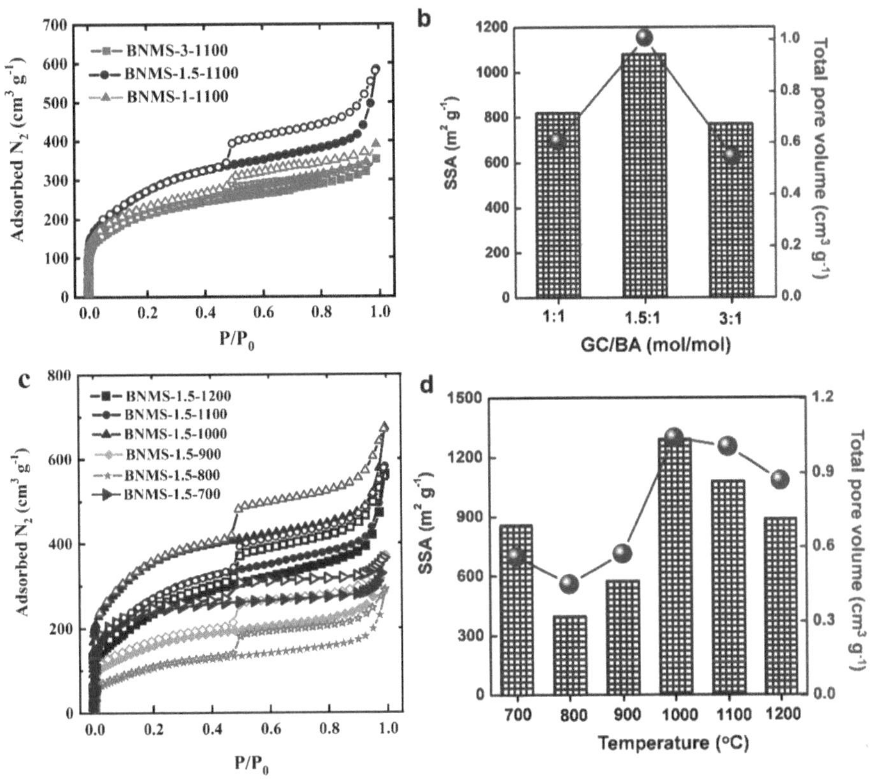

FIGURE 1.13 Isothermal N_2 adsorption-desorption curves (a) and BET specific surface area and total pore volume (b) of BNMSs prepared at 1000 °C under different GC/BA ratios. Isothermal N_2 adsorption-desorption curves (c) and BET specific surface area and total pore volume (d) of BNMSs prepared at different temperatures with Dcy/BA = 1.5:1. Reproduced from Ref. [50].

pore development. The BNMS-1.5-1000 sample had the highest SSA, which was 1290 m^2/g, with a total pore volume of 1.04 cm^3/g.

Isothermal adsorption and desorption method was used to investigate the adsorption performance of H_2 in the pressure range of −196°C to 0-1 MPa. Among the BNMSs prepared with different GC/BA ratios, the sample prepared with a GC/BA ratio of 1.5:1 had the highest H_2 adsorption capacity at relatively low pressure (1 MPa), which was 2.11 wt percent, consistent with the change in material SSA. Surprisingly, as the temperature increased from 700°C to 1200°C, the H_2 adsorption capacity decreased monotonically (from 2.33 wt% to 2.10 wt% at 1 MPa and from 1.35 wt% to 1.17 wt% at 0.1 MPa). This trend was not consistent with the change in material SSA or total pore volume (Figure 1.14). Figure 1.14c summarizes the H_2 adsorption capacity of the BNMSs prepared under different conditions. Figure 1.14d shows the relationship between the H_2 uptake capacity and the SSA value of the material at 1 MPa and 0.1 MPa. It can be seen that in the BNMS samples prepared with any GC/BA ratio above 1000°C, the H_2 adsorption capacity is basically linearly correlated with the SSA value, while the BNMSs prepared at 700-900°C deviate

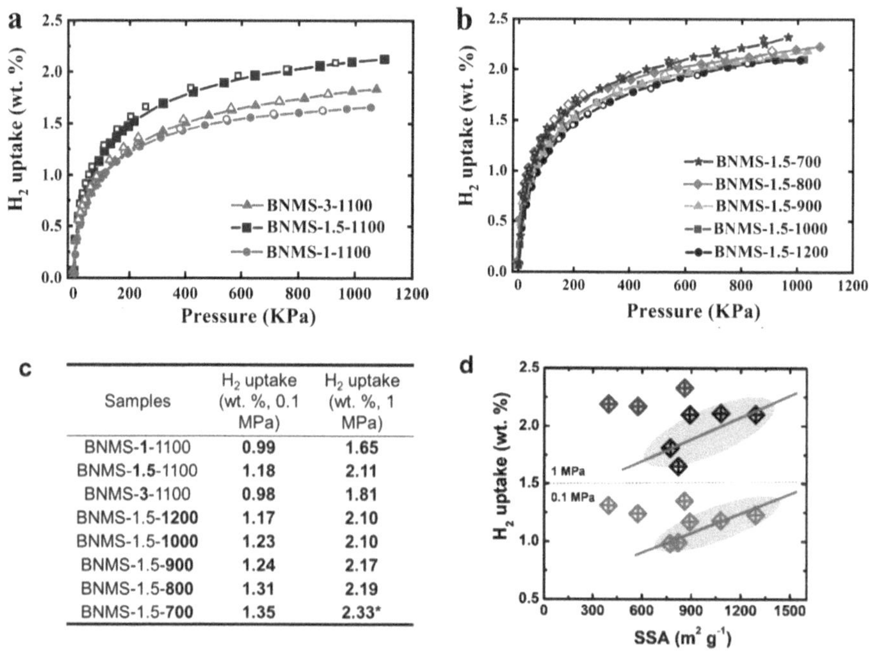

Samples	H_2 uptake (wt. %, 0.1 MPa)	H_2 uptake (wt. %, 1 MPa)
BNMS-**1**-1100	**0.99**	**1.65**
BNMS-**1.5**-1100	**1.18**	**2.11**
BNMS-**3**-1100	**0.98**	**1.81**
BNMS-1.5-**1200**	**1.17**	**2.10**
BNMS-1.5-**1000**	**1.23**	**2.10**
BNMS-1.5-**900**	**1.24**	**2.17**
BNMS-1.5-**800**	**1.31**	**2.19**
BNMS-1.5-**700**	**1.35**	**2.33***

FIGURE 1.14 (a) H_2 adsorption-desorption isotherms of C and O co-doped BNMS samples prepared at different GC/BA ratios at 1100°C. (b) H_2 adsorption-desorption isotherms of BNMS samples prepared at different temperatures with a GC/BA ratio of 1.5:1. The measurement temperature was −196°C. (c) Summary of hydrogen absorption capacity of BNMSs. (d) Relationship between H_2 adsorption capacity and SSA of the studied materials. The red symbols represent BNMSs synthesized at 700-900°C, and the black symbols represent BNMSs synthesized at ≥1000°C. Reproduced from Ref.[50].

significantly from this linear range. By calculation, the H_2 absorption capacity of BNMS-1.5-800 sample per SSA was 5.5×10^{-3} wt percent g/m^2, which was 2.5-3.5 times higher than that of the synthetic materials above 1000°C and 3.1-4.7 times higher than that of the BNMS samples prepared using BA and dicyanamide [31], and 2.7-4.1 times higher than that of reported BN porous microbelts [46].

Although the properties of material pores (such as SSA and total pore volume) play an important role in the physical adsorption of H_2, the surface chemical structure also has an important influence. The doping of C and O in the BN structure is considered a source of high affinity for H_2. Many theoretical works have shown the importance of structural modification on H_2 adsorption [24, 52]. In this work, BNMS materials doped with C and O have a higher specific surface area for hydrogen gas adsorption, which obviously provides strong experimental evidence.

1.4.4 Other Porous Materials for Physisorption Hydrogen Storage

In the previous sections, we primarily discussed BN-based hydrogen storage materials, including BNNTs, BNMSs, and porous BN microstrips. We also highlighted some key advancements in the field of hydrogen storage from our research group in recent years. However, it is important to acknowledge that alongside BN, there are several other porous materials that hold significant potential in the field of hydrogen storage. These materials include metal organic frameworks (MOFs), covalent organic frameworks (COFs), porous organic polymers (POPs), zeolite, and carbon-based materials, which primarily rely on physical adsorption. In this section, we will provide a brief overview of the pertinent information and research progress related to these materials.

1.4.4.1 MOFs

Metal-organic frameworks (MOFs) are a type of porous framework material consisting of inorganic units (metal ions or clusters) and organic ligands, connected through coordination bonds. Since Yaghi et al. first utilized MOFs for hydrogen storage in 2003, research on their application for hydrogen storage has continued for nearly two decades [84]. Over the years, there have been numerous significant advancements in the design and synthesis of MOFs. In this section, we will highlight some representative progress.

Highly porous MOFs have extremely high specific surface area and porosity, making them ideal materials for hydrogen storage at low temperatures (77 K). Gauldron et al. synthesized and employed theoretical calculations to study a series of highly stable zirconium based MOFs, including NU-1101, NU-1102, and NU-1103, for low-temperature hydrogen adsorption [85]. These MOF provided good samples to understand the relationship between gravimetric and volumetric capacity [86], Figure 1.15 (a-c) shows the structures of NU-1101, NU-1102, and NU-1103, respectively, with the ball in the middle representing the largest pore. Figure 1.15 (d) summarizes various measured and simulated textural and thermodynamic characteristics for these MOFs. Among them, NU-1103 has the highest specific surface area (6245 m^2/g) and porosity (2.72 cm^3/g), much higher than those of NU-1101 (4340 m^2/g, 1.72 cm^3/g)

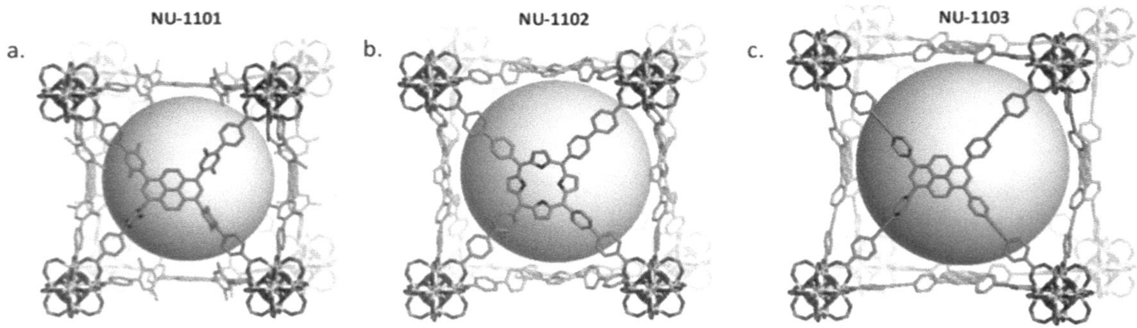

d.

property	NU-1101 exp.	NU-1101 sim.	NU-1102 exp.	NU-1102 sim.	NU-1103 exp.	NU-1103 sim.
pore volume (cc/g)	1.72	1.72	1.65	2.05	2.72	2.97
gravimetric BET area (m^2/g)	4340	4440	3720	4710	6245	6885
gravimetric NASA (m^2/g)		4420		4710		5645
volumetric BET area (m^2/cm^3)	1990	2030	1500	1900	1860	2050
volumetric NASA (m^2/cm^3)		2025		1900		1682
smallest pore diameter (Å)		9.5		11.1		13.5
largest pore diameter (Å)		17.2		20.5		24.2
heat of adsorption (kJ/mol)	5.5	5.5	4.5	4.9	3.8	4.5
H_2@100 bar/77 K (g/L)	48.7	52.9	45.3	54.2	44.9	52.6
ΔH_2@100 bar/77 K → 5 bar/77 K (g/L)	29.9	33.3	30.5	34.1	33.3	40.0
ΔH_2@100 bar/77 K →5 bar/160 K (g/L)	46.6	50.7	43.7	52.4	43.2	51.1
H_2@100 bar/77 K (wt %)	9.5	10.2	9.9	11.6	13.0	15.1
ΔH_2@100 bar/77 K →5 bar/77 K (wt %)	6.1	5.9	6.9	7.6	10.1	11.9
ΔH_2@100 bar/77 K →5 bar/160 K (wt %)	9.1	9.8	9.6	11.3	12.6	14.4

FIGURE 1.15 (a–c) MOF structures with largest pore in each structure represented by a sphere: (a) NU-1101, (b) NU-1102, and (c) NU-1103. (d) Measured and simulated properties for NU-series MOFs, including absolute H_2 adsorption loadings and H_2 deliverable capacity (ΔH_2). Reproduced from Ref. [85].

and NU-1102 (3720 m^2/g, 1.65 cm^3/g). When the hydrogen storage conditions change from 100 bar to 5 bar at 77 K, the hydrogen storage capacity of NU-1101, NU-1102, and NU-1103 decreases to the 39, 33, and 26 percent of the original values, respectively. Interestingly, under operating conditions, NU-1103 has the maximum gravimetric reversible hydrogen uptake of 12.6 wt percent, which is much higher than those of NU-1101 (9.1 wt%) and NU-1102 (9.6 wt%). However, the reversible volumetric uptake of NU-1103 is 43.2 g/L, which is lower than those of NU-1101 (46.6 g/L) and NU-1102 (43.7 g/L), respectively. All these MOFs demonstrate excellent hydrogen adsorption capacity, which not only can be used to study the relationship between reversible gravimetric uptake and volumetric uptake, but also show potentials as the candidates for hydrogen storage.

Recently, Chen et al. reported a porous MOF assembled from triangular prismatic ligands with metal trimer Al_3O and Fe_3O building units [87]. Based on a 6-connected edge-transfer network, NU-1500 is isotropic with a specific surface area of 3560 m^2/g and delivers a reversible H_2 capacity of 44.6 g/L or 8.2 wt percent under 77 K/100 bar → 160 K/5 bar. To explore the tradeoff between volumetric and gravimetric capacity, they performed molecular simulations based on a database of 2,800 MOFs. The results showed that the NU-1501, as an analogue of NU-1500 based on the same metal trimer and ligand with an additional phenyl group on each arm of the linker, is an ideal equilibrium position of reversible volumetric H_2 capacity and reversible gravimetric H_2 capacity. Accordingly, such targeted MOF materials were precisely designed and synthesized, including NU-1501-Al and NU-1501-Fe. Based on the N_2 adsorption-desorption isotherms, the BET specific surface area of NU-1501-Al reached to 7310 m^2/g [88]. More importantly, under the operating conditions of 77 K/100 bar → 160 K/5 bar, the prepared NU-1501-Al not only possesses a very high reversible gravimetric H_2 capacity of 14.0 wt percent but also maintains a relatively high reversible volumetric capacity of 46.2 g/L.At 296 K and 100 bar, NU-1501-Al can absorb approximately 2.9 wt percent (approximately 8.4 g/L) of H_2, a value significantly higher than the H_2 capacity of previously reported MOFs. Furthermore, the iron trimer-based isostructural NU-1501-Fe analogue also displays the same high hydrogen capacity, demonstrating the excellent hydrogen storage performance of this MOF system.

Similar to the above-mentioned BNs, MOF surfaces and hydrogen molecules only have weak interactions. This is because MOFs materials have low polarizations for hydrogen. For instance, in some highly porous MOFs materials, the isosteric heat of adsorption (Qst) for hydrogen is often less than −8 kJ/mol, indicating that despite highly porous, these MOFs often do not exhibit high hydrogen capacities. The ideal range for Qst in MOFs used for hydrogen storage is considered to be between −15 to −25 kJ/mol. Therefore, one strategy to enhance hydrogen storage in MOFs at room temperature is to increase the hydrogen binding energy while maintaining a high specific surface area and porosity. Increasing the charge density of MOFs has been proven to be an effective method to improve the hydrogen-MOF interaction. For example, Eddaoudi et al. investigated the adsorption of H_2 in M-soc-MOFs with different inorganic trinuclear clusters (In^{3+}, Fe^{3+}) and anions (NO^{3-}, Cl^-, and Br^-) [89, 90]. The study found that highly localized charge density, favorable pore size, and geometric shape resulted in a strong interaction between soc-MOFs and H_2, with a hydrogen Qst as high as −7.9 kJ/mol in Fe-soc-MOF-1a with NO^{3-} anions.

Long et al. studied a series of improved M2(dobdc)-type frameworks (dobdc4-=2,5-dioxido-1,4-benzenedicarboxylate) by replacing dobdc with m-dobdc (m-dobdc4-= 4,6-dioxido-1,3-benzenedicarboxylate). The resulting M2(m-dobdc4-) (M=Mg, Mn, Fe, Co, Ni) frameworks have exposed M2+ cation sites with a higher apparent charge density [91]. Figure 1.16(a) shows the H_2 isosteric heats of adsorption for M2(m-dobdc), where the Qst decreases with increasing hydrogen adsorption. This is because as hydrogen molecules gradually adsorb, the exposed metal cation sites become occupied, leaving only weaker adsorption sites available, resulting in a sharp decline in the enthalpy of hydrogen binding. Figures 1.16(c)

a. b.

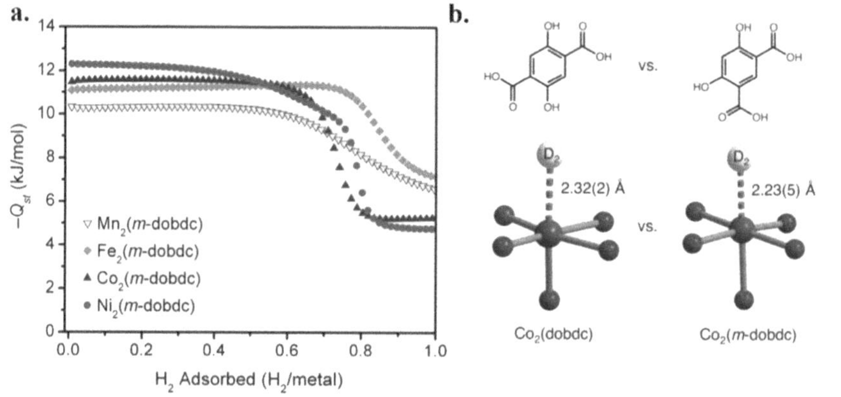

c.

		Mn[4b,9e]	Fe[4c]	Co[24a]	Ni[24b]
M_2(dobdc)	Langmuir (m^2/g)	1797	1535	1432	1574
	BET (m^2/g)	1102	1360	1341	
M_2(*m*-dobdc)	Langmuir (m^2/g)	1741	1624	1504	1592
	BET (m^2/g)	1349	1295	1264	1321

d.

	Mn	Fe	Co	Ni
M_2(dobdc)	−8.8	−9.7	−10.8	−11.9
M_2(*m*-dobdc)	−10.3	−11.1	−11.5	−12.3

FIGURE 1.16 (a) H_2 isosteric heat of adsorption curves for the M_2(m-dobdc) MOFs, as a function of the amount adsorbed. (b) The comparison between M-D2 distance for Co_2(dobdc) and Co_2(m-dobdc) revealed by neutron diffraction experiments. (c) Langmuir and BET surface areas of the M_2(dobdc) and M_2(m-dobdc) Frameworks. (d) Comparison of Low-Loading H_2 Isosteric Heats of Adsorption (Qst, kJ/mol) in M_2(dobdc) and M_2(m-dobdc). Reproduced from Ref. [91].

and (d) compare the BET surface areas and H_2 isosteric heats of adsorption at low hydrogen loadings between the M_2(dobdc) and M_2 (m-dobdc) frameworks. The results show no significant changes in BET surface area between the two framework types, but the H_2 adsorption enthalpy in M_2(m-dobdc) is generally higher by 0.4-1.5 kJ/mol compared to M_2(dobdc). For example, the H_2 adsorption enthalpy in Mn_2(m-dobdc) framework is −10.3 kJ/mol, while the corresponding Mn_2(dobdc) framework has an enthalpy of only −8.8 kJ/mol. In order to explore the relationship between structural modifications and enhanced hydrogen adsorption performance, various spectroscopic techniques were used along with simulations. Neutron diffraction experiments (Figure 1.16(b)) showed that the distance between the center of the D2 electron density and the main binding site in Co_2(m-dobdc) is 2.23(5) Å, shorter than that of Co_2(dobdc) (2.32(2) Å), indicating a stronger metal-hydrogen interaction in Co_2(m-dobdc) compared to Co_2(dobdc). Additionally, they conducted further high-pressure adsorption studies and found that these types of MOFs exhibited high volumetric hydrogen capacities at near-ambient temperatures. For example, at operating conditions of 298K and 100 → 5 bar, Ni_2(m-dobdc) has a volumetric hydrogen capacity of 11.0 g/L [91].

1.4.4.2 COFs

Covalent organic frames (COFs) are a class of porous crystals composed of organic building units that are connected by covalent bonds to form two-dimensional or three-dimensional frames. COFs usually have good chemical and thermal stabilities, and large surface area and pore volume. Since they are made of light atoms without using heavy metal atoms and ions, they can form low-density, metal-free materials, which have attracted wide attention as hydrogen adsorbents.

In 2009, Furukawa, Yaghi and coworkers studied the hydrogen adsorption of seven different porous organic frameworks (COFs) [92]. They categorized these COFs into three groups. The first group consisted of two-dimensional structures with one-dimensional small pores (COF-1, COF-6, pore size of 9 Å). The second group consisted of two-dimensional structures with one-dimensional large pores (COF-5, COF-8, COF-10, pore sizes of 27, 16, 32 Å, respectively). The third group had three-dimensional structures with medium-sized pores (COF-102, COF-103, pore size of 12 Å for both). By studying the pore structures of these COFs, it was discovered that the BET surface area of the third group of COFs was the largest. COF-102 and COF-103 had BET surface areas of 4650 and 4630 m_2/g, respectively, while COF-1, COF-5, COF-6, COF-8, COF-10 had the BET surface areas of 750, 1670, 750, 1350, 1760 m_2/g, respectively. In Figure 1.17, COF-102 and COF-103 exhibited the highest hydrogen adsorption capacity at 77K and high pressure with the values of 7.24 and 7.05 wt. percent, respectively. In contrast, the maximum hydrogen adsorption capacity of the first and second groups ranged from only 1.48 to 3.92 wt. The authors found that the hydrogen adsorption capacity of the 3D-COFs (COF-102, COF-103) was significantly higher than that of the 2D-COFs and comparable to the best-performing MOFs or other hydrogen storage materials. Under the operational conditions of 77K and 5 bar→85 bar, COF-103 exhibited a reversible hydrogen capacity of 6.4 wt percent (29.2 g/L) [92].

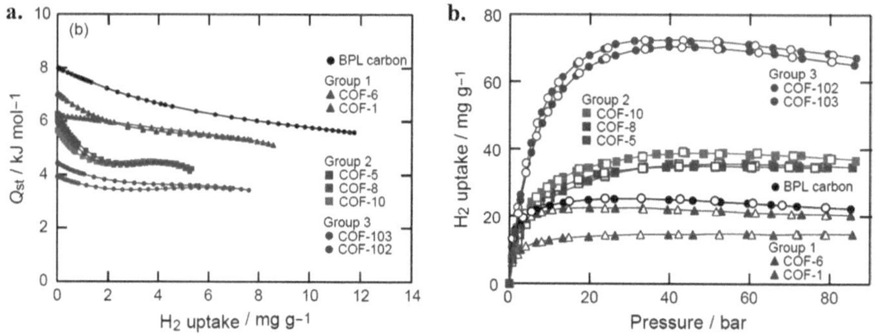

FIGURE 1.17 (a) Coverage dependency of adsorption enthalpies for COFs. (b) High-pressure H_2 isotherms for COFs measured at 77 K. Data for BPL carbon are shown for comparison. Reproduced from Ref. [92].

The authors also performed theoretical simulations of these materials, and the results were in good agreement with the experimental findings [93]. The authors further highlighted that the hydrogen storage capacity of the COFs in the third group was comparable to that of MOFs, indicating that the hydrogen capacity is correlated with the pore size rather than the structural backbone. This suggests a physical limitation for these materials, as high hydrogen adsorption capacity requires COFs with large pore volumes. However, designing COF materials with such large pores comes at the expense of microporosity and strong adsorption sites, which are prerequisites for enhancing the density of hydrogen adsorption in the material. The decreasing heat of adsorption (Qst) with increasing hydrogen adsorption in the third group of materials also supports this, as it indicates the saturation of strong adsorption sites as hydrogen molecules progressively adsorb into the material.

Due to the charge polarization mechanism, a large number of hydrogen molecules can form collective interactions with metal atoms at the molecular level. Compared to sacrificing the high pore volume of COF materials to achieve high-density strong binding sites, the method of incorporating metal atoms into COFs to introduce high-affinity sites appears to be more sensible. As early as 2009, Cao et al. calculated the hydrogen uptake of a series of lithium-doped COFs compared to undoped COFs. The results demonstrated that Li-doped COFs not only exhibited higher hydrogen affinity but also had higher hydrogen adsorption capacities compared to the original COFs. Under conditions of 298 K and 100 bar, Li-doped COF-105 and COF-108 respectively exhibited hydrogen mass uptake of 6.84 wt percent and 6.73 wt percent. The authors suggested that the high hydrogen affinity is due to the formation of a dative bond between the hydrogen molecule and the lithium atom [94].

Pramudya and coworkers incorporated a series of transition metal ions into COFs to explore their hydrogen adsorption capacity [95]. The d orbitals of transition metals strongly interact with hydrogen molecules, leading to an elongation of the H-H bond and an increase in hydrogen affinity. As shown in Figure 1.18, the hydrogen uptake of all hybrid species increased by 2-4 times. The average heat of adsorption (Qst) of COF-301-$CoCl_2$ reached 15.8 kJ/mol and under conditions of 298 K and 120 bar, its excess volumetric hydrogen uptake reached 22.2 g/L, significantly higher than

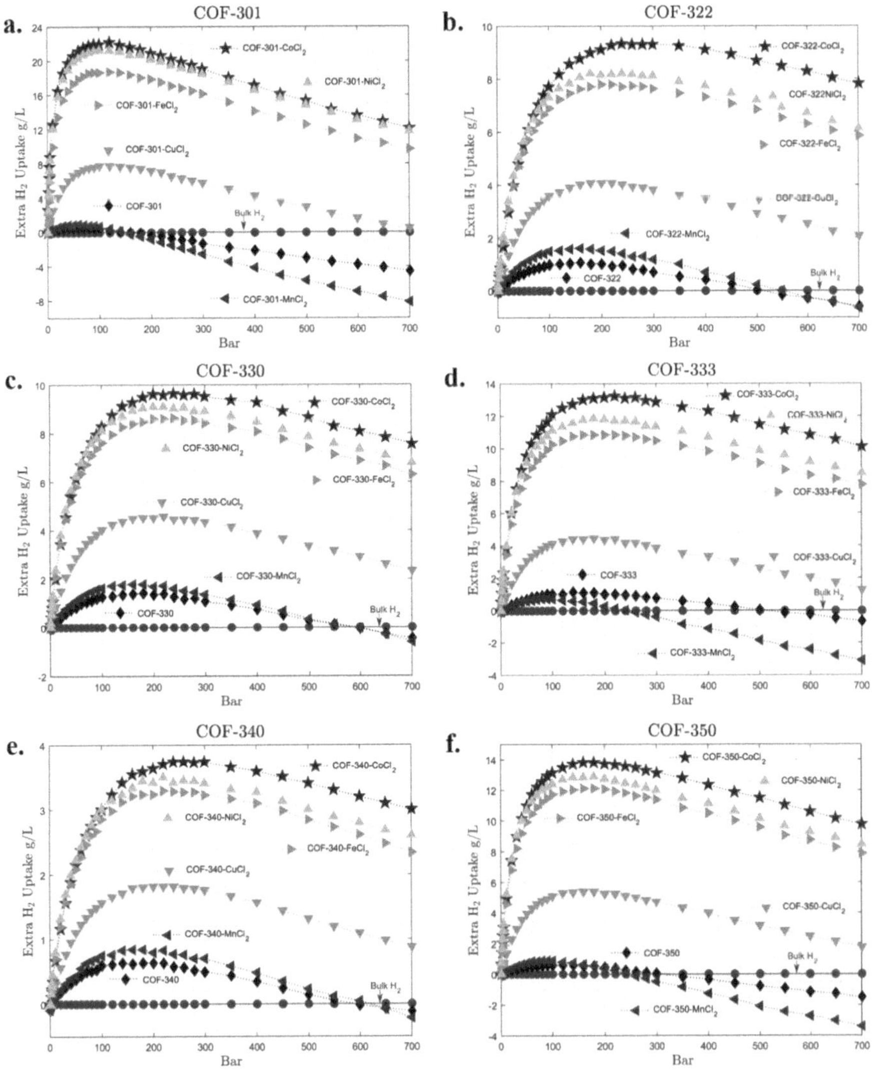

FIGURE 1.18 The relative H_2 uptake for (a) COF-301, (b) COF-322, (c) COF-330, (d) COF-333, (e) COF-340, and (f) COF-350 with different transition metals in 298K. It is reproduced from Ref. [95].

that of the undoped COF materials. Similar trends were observed in COF-322, COF-330, COF-333, COF-340, and COF-350. This work fully demonstrates the significant importance of metal ion doping in enhancing the hydrogen storage capacity of COFs.

1.4.4.3 POPs

Amorphous porous organic polymers (POPs) have also been explored as potential candidates for hydrogen storage due to their ease of processing and good mechanical stability. In 2009, Ben et al. synthesized PAF-1, which consists of tetrahedral tetraphenmethane structural units and possesses an ultra-high BET surface area of 5,600 m^2/g. PAF-1 exhibits excellent thermal and hydrothermal stability. The H_2 adsorption isotherm at 77K and 48 bar demonstrates that PAF-1 has a H_2 adsorption capacity of 7.0 wt percent [96]. Based on this work, Yuan et al. synthesized a series of PAF-1 analogues by replacing the central carbon atoms of the tetrahedral monomers with adamantane (PPN-3), silicon (PPN-4), and germanium (PPN-5). They showed that the BET area of PPN-4 was even higher (6461 m^2/g) than that of PAF-1, and its H_2 adsorption capacity reached 8.34 wt percent [97].

1.4.4.4 PCs

Carbon-based materials, such as porous carbon, carbon nanotubes, and fullerenes, are considered suitable adsorbents for gas storage due to their structural diversity. Furthermore, the porosity and structural composition of these materials can be tuned for hydrogen storage. Carbon-based materials are also highly appealing due to their abundance, workability, lightweight nature, and high chemical and thermal stability for H_2 storage. This significant interest led to progress in the field during the late 1990s and early 2000s. However, advancements stalled when the upper limit seemed to have been reached after the development of ultra-reactive porous carbon with high specific surface area. Initially, activated carbons were favored due to their low cost, but they often exhibited a large number of mesopores and a wide pore distribution. Similarly, although porous carbon has a high specific surface area, achieving a high H_2 adsorption capacity is challenging due to its low H_2 adsorption energy. Many efforts have been made to increase the H_2 adsorption capacity. In 2017, Blankenship et al. synthesized an oxygen-rich microporous carbon from cellulose acetate, which exhibited a specific surface area of up to 3800 m^2/g and a gravimetric hydrogen excess uptake of 7.0 percent at 77K and 20 bar [98]. The calculated enthalpy of adsorption was 10 kJ/mol, indicating strong physical adsorption of H_2 by the material. Additionally, metal doping has proved to enhance the storage capacity by providing stronger binding sites within the hydrogen adsorption framework [99].

1.4.4.5 Zeolites

Zeolite is a microporous adsorbent composed of aluminosilicate materials with a defined structure and pore size. Zeolites often contain counter ions that can influence the binding energy and interactions of guest molecules. However, the binding energy of H_2 in silica is typically similar to that in carbon. Moreover, zeolites usually have a composition containing heavier elements and lower pore fractions, which hinders their development as H_2 adsorbents. These types of materials have gained attentions in recent years. Nonetheless, despite extensive research on solid porous materials for hydrogen storage applications, the current hydrogen storage materials still fall short of the performance required by the US Department of Energy for high-performance hydrogen storage systems. This poses a significant challenge to the hydrogen storage economy.

1.5 SUMMARY AND PERSPECTIVE

In this chapter, we first introduce the relevant information of BN materials, including the types, synthesis methods, characteristics, and major applications of BN materials.

Next, we introduce solid-state hydrogen storage materials mainly based on BN substrates, including their preparation methods, chemical compositions, structural characteristics, and hydrogen storage performance. We also summarize our research results in recent years and provide detailed introductions to four kinds of porous BN hydrogen storage materials, including BN porous microstrip materials, BCNO porous microstrip materials, BN microsponge porous materials, and carbon-oxygen co-doped porous BN materials.

Hydrogen, as a clean energy, has been highly anticipated by scientists and can relieve the current energy crisis. Moreover, the proposed hydrogen fuel cell vehicle urgently needs a high-performance hydrogen storage device. In recent years, solid-state hydrogen storage materials have developed rapidly, but there are still significant challenges to meeting the requirements of high-performance hydrogen storage systems set by the US Department of Energy. To address this, we provide relevant ideas for designing high-performance hydrogen storage devices from the following aspects:

(1) improving the structural performance of materials. In the previous sections, we mentioned that solid-state porous materials are mostly dominated by physical adsorption mechanisms. The H_2 adsorption capacity of physical adsorption is usually related to the structural characteristics of hydrogen storage materials, such as specific surface area, total pore volume, and pore width. Therefore, preparing materials with high specific surface area and high porosity would be a good approach. At the same time, pore width also affects H_2 adsorption. From our previous work, we found that H_2 adsorption mainly depends on the existence of micropores. If the proportion of micropores in the total pore volume in high SSAs porous materials can be improved, it will significantly increase the H_2 storage capacity of the material.

(2) improving the H_2 binding energy on the surface of materials. In addition to the structural characteristics of materials, the binding energy between H_2 and the surface of materials is also a crucial factor affecting H_2 adsorption capacity. For example, although many porous carbon materials with high SSA and high porosity have been reported, the low H_2 adsorption energy on the carbon material surface often leads to low overall H_2 adsorption capacity. In the carbon-oxygen co-doped BN materials introduced in 4.4, we used C and O to replace part of the N or B atoms in the BN lattice to create stronger hydrogen binding sites. As a result, although the SSA of the prepared BN material is only 397 m^2/g, it has a hydrogen adsorption capacity of 2.19 wt percent at 1 MPa and −196°C, which is higher than the undoped BN sample with an SSA as high as 1290 m^2/g. Therefore, designing the chemical state of the material surface rationally is an important factor to be considered when designing high-performance hydrogen storage devices.

(3) reducing material costs. Once high-performance hydrogen adsorbents and related adsorption systems are fully developed, the next step will involve the scale-up and commercialization of this technology. Therefore, environmentally friendly, cheap, and safe synthesis methods are conducive to the large-scale

manufacture of porous materials. It is a wise approach to not only improve the hydrogen storage performance of solid-state hydrogen storage materials but also consider the synthesis costs of the materials. Ultimately, solid-state porous materials are expected to achieve the next generation of hydrogen storage systems that will be commercially viable and implementable at a large scale.

REFERENCES

[1] Chu, S.; Majumdar, A., Opportunities and challenges for a sustainable energy future. *Nature* **2012**, *488* (7411), 294–303.

[2] Khan, N.; Kalair, E.; Abas, N., et al., Energy transition from molecules to atoms and photons. *Eng Sci Technol* **2019**, *22* (1), 185–214.

[3] Momete, D. C., Analysis of the potential of clean energy deployment in the European Union. *IEEE Access* **2018**, *6*, 54811–54822.

[4] Durbin, D. J.; Malardier-Jugroot, C., Review of hydrogen storage techniques for on board vehicle applications. *Int J Hydrogen Energ* **2013**, *38* (34), 14595–14617.

[5] Jacobson, M. Z.; Colella, W. G.; Golden, D. M., Cleaning the air and improving health with hydrogen fuel-cell vehicles. *Science* **2005**, *308* (5730), 1901–1905.

[6] He, T.; Pachfule, P.; Wu, H., et al., Hydrogen carriers. *Nat Rev Mater* **2016**, *1* (12).

[7] Rosen, M.; Scott, D., Comparative efficiency assessments for a range of hydrogen production processes. *Int J Hydrogen Energ* **1998**, *23* (8), 653–659.

[8] Sdanghi, G.; Maranzana, G.; Celzard, A., et al., Review of the current technologies and performances of hydrogen compression for stationary and automotive applications. *Renew Sust Energ Rev* **2019**, *102*, 150–170.

[9] Manoharan, Y.; Hosseini, S. E.; Butler, B., et al., Hydrogen fuel cell vehicles; current status and future prospect. *Appl Sci-Basel* **2019**, *9* (11), 2296.

[10] Cullen, D. A.; Neyerlin, K. C.; Ahluwalia, R. K., et al., New roads and challenges for fuel cells in heavy-duty transportation. *Nat Energy* **2021**, *6* (5), 462–474.

[11] Alaswad, A.; Baroutaji, A.; Achour, H., et al., Developments in fuel cell technologies in the transport sector. *Int J Hydrogen Energ* **2016**, *41* (37), 16499–16508.

[12] Fan, L. X.; Tu, Z. K.; Chan, S. H., Recent development of hydrogen and fuel cell technologies: A review. *Energy Rep* **2021**, *7*, 8421–8446.

[13] Jorgensen, S. W., Hydrogen storage tanks for vehicles: Recent progress and current status. *Curr Opin Solid St M* **2011**, *15* (2), 39–43.

[14] Khan, M. T. I.; Monde, M.; Setoguchi, T., Hydrogen gas filling into an actual tank at high pressure and optimization of its thermal characteristics. *J Therm Sci* **2009**, *18* (3), 235–240.

[15] Allendorf, M. D.; Hulvey, Z.; Gennett, T., et al., An assessment of strategies for the development of solid-state adsorbents for vehicular hydrogen storage. *Energ Environ Sci* **2018**, *11* (10), 2784–2812.

[16] Usman, M. R., Hydrogen storage methods: Review and current status. *Renew Sust Energ Rev* **2022**, 167.

[17] Zhu, Q. L.; Xu, Q., Liquid organic and inorganic chemical hydrides for high-capacity hydrogen storage. *Energ Environ Sci* **2015**, *8* (2), 478–512.

[18] Wang, C. L.; Astruc, D., Recent developments of nanocatalyzed liquid-phase hydrogen generation. *Chem Soc Rev* **2021**, *50* (5), 3437–3484.

[19] Tian, Y. J.; Xu, B.; Yu, D. L., et al., Ultrahard nanotwinned cubic boron nitride. *Nature* **2013**, *493* (7432), 385–388.

[20] Weng, Q. H.; Wang, X. B.; Wang, X., et al., Functionalized hexagonal boron nitride nanomaterials: Emerging properties and applications. *Chem Soc Rev* **2016**, *45* (14), 3989–4012.

[21] Jiang, X. F.; Weng, Q. H.; Wang, X. B., et al., Recent progress on fabrications and applications of boron nitride nanomaterials: A review. *J Mater Sci Technol* **2015**, *31* (6), 589–598.

[22] Itskou, I.; L'Hermitte, A.; Marchesini, S., et al., How to tailor porous boron nitride properties for applications in interfacial processes. *Accounts Mater Res* **2023**, *4* (2), 143–155.

[23] Benard, P.; Chahine, R., Determination of the adsorption isotherms of hydrogen on activated carbons above the critical temperature of the adsorbate over wide temperature and pressure ranges. *Langmuir* **2001**, *17* (6), 1950–1955.

[24] Jhi, S. H.; Kwon, Y. K., Hydrogen adsorption on boron nitride nanotubes: A path to room-temperature hydrogen storage. *Phys Rev B* **2004**, *69* (24)5407.

[25] Ciofani, G.; Danti, S.; Genchi, G. G., et al., Boron nitride nanotubes: Biocompatibility and potential spill-over in nanomedicine. *Small* **2013**, *9* (9–10), 1672–1685.

[26] Weng, Q. H.; Wang, B. J.; Wang, X. B., et al., Highly water-soluble, porous, and biocompatible boron nitrides for anticancer drug delivery. *Acs Nano* **2014**, *8* (6), 6123–6130.

[27] Cheng, Y. M.; Han, Y. X.; Zhang, W., et al., Gram-scale synthesis of boron nitride nanosheets by salt-template method for anticancer drug delivery. *Chem Eng J* **2022**, *437* (1), 135304.

[28] Deng, Y. X.; Long, Y. Y.; Song, A. L., et al., Boron dopants in red-emitting B and N Co-doped carbon quantum dots enable targeted imaging of lysosomes. *Acs Appl Mater Inter* **2023**, *15* (13), 17045–17053.

[29] Li, L. P.; Li, J. Y.; Shi, Y. X., et al., On-demand biodegradable boron nitride nanoparticles for treating triple negative breast cancer with boron neutron capture therapy. *Acs Nano* **2019**, *13* (12), 13843–13852.

[30] Lee, C.; Wei, X. D.; Kysar, J. W., et al., Measurement of the elastic properties and intrinsic strength of monolayer graphene. *Science* **2008**, *321* (5887), 385–388.

[31] Weng, Q. H.; Wang, X. B.; Bando, Y., et al., One-step template-free synthesis of highly porous boron nitride microsponges for hydrogen storage. *Adv Energy Mater* **2014**, *4* (7), 1301523.

[32] Dai, J.; Wu, X. J.; Yang, J. L., et al., Unusual metallic microporous boron nitride networks. *J Phys Chem Lett* **2013**, *4* (20), 3484–3488.

[33] Dai, J.; Wu, X. J.; Yang, J. L., et al., Porous boron nitride with tunable pore size. *J Phys Chem Lett* **2014**, *5* (2), 393–398.

[34] Han, W. Q.; Brutchey, R.; Tilley, T. D., et al., Activated boron nitride derived from activated carbon. *Nano Lett* **2004**, *4* (1), 173–176.

[35] Dibandjo, P.; Chassagneux, F.; Bois, L., et al., Comparison between SBA-15 silica and CMK-3 carbon nanocasting for mesoporous boron nitride synthesis. *J Mater Chem* **2005**, *15* (19), 1917–1923.

[36] Vinu, A.; Terrones, M.; Golberg, D., et al., Synthesis of mesoporous BN and BCN exhibiting large surface areas via templating methods. *Chem Mater* **2005**, *17* (24), 5887–5890.

[37] Alauzun, J. G.; Ungureanu, S.; Brun, N., et al., Novel monolith-type boron nitride hierarchical foams obtained through integrative chemistry. *J Mater Chem* **2011**, *21* (36), 14025–14030.

[38] Schlienger, S.; Alauzun, J.; Michaux, F., et al., Micro-, mesoporous boron nitride-based materials templated from zeolites. *Chem Mater* **2012**, *24* (1), 88–96.

[39] Song, Y. X.; Li, B.; Yang, S. W., et al., Ultralight boron nitride aerogels via template-assisted chemical vapor deposition. *Sci Rep-Uk* **2015**, *5*, 10337.

[40] Rushton, B.; Mokaya, R., Mesoporous boron nitride and boron-nitride-carbon materials from mesoporous silica templates. *J Mater Chem* **2008**, *18* (2), 235–241.

[41] Dibandjo, P.; Bois, L.; Chassagneux, F., et al., Thermal stability of mesoporous boron nitride templated with a cationic surfactant. *J Eur Ceram Soc* **2007**, *27* (1), 313–317.

[42] Bernard, S.; Miele, P., Nanostructured and architectured boron nitride from boron, nitrogen and hydrogen-containing molecular and polymeric precursors. *Mater Today* **2014**, *17* (9), 443–450.

[43] Malenfant, P. R. L.; Wan, J. L.; Taylor, S. T., et al., Self-assembly of an organic-inorganic block copolymer for nano-ordered ceramics. *Nat Nanotechnol* **2007**, *2* (1), 43–46.

[44] Nag, A.; Raidongia, K.; Hembram, K. P. S. S., et al., Graphene analogues of BN: Novel synthesis and properties. *Acs Nano* **2010**, *4* (3), 1539–1544.

[45] Lei, W. W.; Portehault, D.; Liu, D., et al., Porous boron nitride nanosheets for effective water cleaning. *Nat Commun* **2013**, 4, 1777.

[46] Weng, Q. H.; Wang, X. B.; Zhi, C. Y., et al., Boron nitride porous microbeits for hydrogen storage. *Acs Nano* **2013**, *7* (2), 1558–1565.

[47] Li, J.; Lin, J.; Xu, X. W., et al., Porous boron nitride with a high surface area: hydrogen storage and water treatment. *Nanotechnology* **2013**, *24* (15), 155603.

[48] Li, J.; Huang, Y.; Liu, Z. Y., et al., Chemical activation of boron nitride fibers for improved cationic dye removal performance. *J Mater Chem A* **2015**, *3* (15), 8185–8193.

[49] Lin, J.; Xu, L. L.; Huang, Y., et al., Ultrafine porous boron nitride nanofibers synthesized via a freeze-drying and pyrolysis process and their adsorption properties. *Rsc Adv* **2016**, *6* (2), 1253–1259.

[50] Weng, Q. H.; Zeng, L. L.; Chen, Z. W., et al., Hydrogen storage in carbon and oxygen co-doped porous boron nitrides. *Adv Funct Mater* **2021**, *31* (4), 2007381.

[51] Xue, Y. M.; Dai, P. C.; Jiang, X. F., et al., Template-free synthesis of boron nitride foam-like porous monoliths and their high-end applications in water purification. *J Mater Chem A* **2016**, *4* (4), 1469–1478.

[52] Lei, W. W.; Zhang, H.; Wu, Y., et al., Oxygen-doped boron nitride nanosheets with excellent performance in hydrogen storage. *Nano Energy* **2014**, *6*, 219–224.

[53] Tokarev, A.; Kjeang, E.; Cannon, M., et al., Theoretical limit of reversible hydrogen storage capacity for pristine and oxygen-doped boron nitride. *Int J Hydrogen Energ* **2016**, *41* (38), 16984–16991.

[54] Shayeganfar, F.; Shahsavari, R., Oxygen- and lithium-doped hybrid boron-nitride/carbon networks for hydrogen storage. *Langmuir* **2016**, *32* (50), 13313–13321.

[55] Kumar, E. M.; Sinthika, S.; Thapa, R., First principles guide to tune h-BN nanostructures as superior light-element-based hydrogen storage materials: role of the bond exchange spillover mechanism. *J Mater Chem A* **2015**, *3* (1), 304–313.

[56] Qiu, N. X.; Zhang, C. H.; Xue, Y., Tuning hydrogen storage in lithium-functionalized BC2N sheets by doping with boron and carbon. *Chemphyschem* **2014**, *15* (14), 3015–3025.

[57] Qiu, N. X.; Tian, Z. Y.; Guo, Y., et al., A first-principle study of calcium-decorated BC2N sheet doped by boron or carbon for high hydrogen storage. *Int J Hydrogen Energ* **2014**, *39* (17), 9307–9320.

[58] Zhang, H.; Tong, C. J.; Zhang, Y. S., et al., Porous BN for hydrogen generation and storage. *J Mater Chem A* **2015**, *3* (18), 9632–9637.

[59] Kinal, A.; Sayhan, S., Accurate prediction of hydrogen storage capacity of small boron nitride nanocages by dispersion corrected semi-empirical PM6–DH2 method. *Int J Hydrogen Energ* **2016**, *41* (1), 392–400.
[60] Wu, P. W.; Zhu, W. S.; Chao, Y. H., et al., A template-free solvent-mediated synthesis of high surface area boron nitride nanosheets for aerobic oxidative desulfurization. *Chem Commun* **2016**, *52* (1), 144–147.
[61] Marchesini, S.; McGilvery, C. M.; Bailey, J., et al., Template-free synthesis of highly porous boron nitride: Insights into pore network design and impact on gas sorption. *Acs Nano* **2017**, *11* (10), 10003–10011.
[62] Shahsavari, R.; Zhao, S., Merger of energetic affinity and optimal geometry provides new class of boron nitride based sorbents with unprecedented hydrogen storage capacity. *Small* **2018**, *14* (15), 1702863.
[63] Weng, Q. H.; Wang, X. B.; Wang, X., et al., Preparation and hydrogen sorption performances of BCNO porous microbelts with ultra-narrow and tunable pore widths. *Chem-Asian J* **2013**, *8* (12), 2936–2939.
[64] Ma, R. Z.; Bando, Y.; Zhu, H. W., et al., Hydrogen uptake in boron nitride nanotubes at room temperature. *J Am Chem Soc* **2002**, *124* (26), 7672–7673.
[65] Tang, C. C.; Bando, Y.; Ding, X. X., et al., Catalyzed collapse and enhanced hydrogen storage of BN nanotubes. *J Am Chem Soc* **2002**, *124* (49), 14550–14551.
[66] Terao, T.; Bando, Y.; Mitome, M., et al., Effective synthesis of surface-modified boron nitride nanotubes and related nanostructures and their hydrogen uptake. *Physica E* **2008**, *40* (7), 2551–2555.
[67] Lim, S. H.; Luo, J. Z.; Ji, W., et al., Synthesis of boron nitride nanotubes and its hydrogen uptake. *Catal Today* **2007**, *120* (3–4), 346–350.
[68] Portehault, D.; Giordano, C.; Gervais, C., et al., High-surface-area nanoporous boron carbon nitrides for hydrogen storage. *Adv Funct Mater* **2010**, *20* (11), 1827–1833.
[69] Lian, G.; Zhang, X.; Zhang, S. J., et al., Controlled fabrication of ultrathin-shell BN hollow spheres with excellent performance in hydrogen storage and wastewater treatment. *Energ Environ Sci* **2012**, *5* (5), 7072–7080.
[70] Muthu, R. N.; Rajashabala, S.; Kannan, R., Synthesis, characterization of hexagonal boron nitride nanoparticles decorated halloysite nanoclay composite and its application as hydrogen storage medium. *Renew Energ* **2016**, *90*, 554–564.
[71] Yang, S. J.; Kim, T.; Im, J. H., et al., MOF-derived hierarchically porous carbon with exceptional porosity and hydrogen storage capacity. *Chem Mater* **2012**, *24* (3), 464–470.
[72] Ma, R.; Bando, Y.; Sato, T., CVD synthesis of boron nitride nanotubes without metal catalysts. *Chem Phys Lett* **2001**, *337* (1–3), 61–64.
[73] Yin, L. W.; Bando, Y.; Golberg, D., et al., Porous BCN nanotubular fibers: Growth and spatially resolved cathodoluminescence. *J Am Chem Soc* **2005**, *127* (47), 16354–16355.
[74] Zhu, Y. C.; Bando, Y.; Ma, R. Z., Aluminum borate-boron nitride nanocables. *Adv Mater* **2003**, *15* (16), 1377–+.
[75] Taha-Tijerina, J.; Narayanan, T. N.; Gao, G. H., et al., Electrically insulating thermal nano-oils using 2D fillers. *Acs Nano* **2012**, *6* (2), 1214–1220.
[76] Chen, Y.; Zou, J.; Campbell, S. J., et al., Boron nitride nanotubes: Pronounced resistance to oxidation. *Appl Phys Lett* **2004**, *84* (13), 2430–2432.
[77] Song, L.; Ci, L. J.; Lu, H., et al., Large scale growth and characterization of atomic hexagonal boron nitride layers. *Nano Lett* **2010**, *10* (8), 3209–3215.
[78] Aboutalebi, S. H.; Aminorroaya-Yamini, S.; Nevirkovets, I., et al., Enhanced hydrogen storage in graphene oxide-MWCNTs composite at room temperature. *Adv Energy Mater* **2012**, *2* (12), 1439–1446.

[79] Masika, E.; Mokaya, R., Hydrogen storage in high surface area carbons with identical surface areas but different pore sizes: Direct demonstration of the effects of pore size. *J Phys Chem C* **2012**, *116* (49), 25734–25740.
[80] Lale, A.; Bernard, S.; Demirci, U. B., Boron nitride for hydrogen storage. *Chempluschem* **2018**, *83* (10), 893–903.
[81] Weng, Q. H.; Ide, Y.; Wang, X. B., et al., Design of BN porous sheets with richly exposed (002) plane edges and their application as TiO2 visible light sensitizer. *Nano Energy* **2015**, *16*, 19–27.
[82] Feng, G. D.; Guan, M. M.; Lai, Q., et al., One-pot synthesis and application of novel amino-functionalized silica nanoparticles using guanidine as amino group. *New J Chem* **2016**, *40* (10), 8444–8450.
[83] Weng, Q. H.; Kvashnin, D. G.; Wang, X., et al., Tuning of the optical, electronic, and magnetic properties of boron nitride nanosheets with oxygen doping and functionalization. *Adv Mater* **2017**, *29* (28), 1700695.
[84] Rosi, N. L.; Eckert, J.; Eddaoudi, M., et al., Hydrogen storage in microporous metal-organic frameworks. *Science* **2003**, *300* (5622), 1127–1129.
[85] Gomez-Gualdron, D. A.; Wang, T. C.; Garcia-Holley, P., et al., Understanding volumetric and gravimetric hydrogen adsorption trade-off in metal-organic frameworks. *Acs Appl Mater Inter* **2017**, *9* (39), 33419–33428.
[86] Wang, T. C.; Bury, W.; Gomez-Gualdron, D. A., et al., Ultrahigh surface area zirconium MOFs and insights into the applicability of the BET theory. *J Am Chem Soc* **2015**, *137* (10), 3585–3591.
[87] Chen, Z. J.; Li, P. H.; Zhang, X., et al., Reticular access to highly porous acs-MOFs with rigid trigonal prismatic linkers for water sorption. *J Am Chem Soc* **2019**, *141* (7), 2900–2905.
[88] Gomez-Gualdron, D. A.; Moghadam, P. Z.; Hupp, J. T., et al., Application of consistency criteria to calculate BET areas of micro- and mesoporous metal-organic frameworks. *J Am Chem Soc* **2016**, *138* (1), 215–224.
[89] Liu, Y. L.; Eubank, J. F.; Cairns, A. J., et al., Assembly of metal-organic frameworks (MOFs) based on indium-trimer building blocks: A porous MOF with soc topology and high hydrogen storage. *Angew Chem Int Edit* **2007**, *46* (18), 3278–3283.
[90] Cairns, A. J.; Eckert, J.; Wojtas, L., et al., Gaining insights on the H-2–sorbent interactions: Robust soc-MOF platform as a case study. *Chem Mater* **2016**, *28* (20), 7353–7361.
[91] Kapelewski, M. T.; Geier, S. J.; Hudson, M. R., et al., M2 (m-dobdc) (M = Mg, Mn, Fe, Co, Ni) Metal–organic frameworks exhibiting increased charge density and enhanced H2 binding at the open metal sites. *J Am Chem Soc* **2014**, *136* (34), 12119–12129.
[92] Furukawa, H.; Yaghi, O. M., Storage of hydrogen, methane, and carbon dioxide in highly porous covalent organic frameworks for clean energy applications. *J Am Chem Soc* **2009**, *131* (25), 8875–8883.
[93] Klontzas, E.; Tylianakis, E.; Froudakis, G. E., Hydrogen storage in 3D covalent organic frameworks. A multiscale theoretical investigation. *J Phys Chem C* **2008**, *112* (24), 9095–9098.
[94] Cao, D.; Lan, J.; Wang, W., et al., Lithium-doped 3D covalent organic frameworks: high-capacity hydrogen storage materials. *Angewandte Chemie International Edition* **2009**, *48* (26), 4730–4733.
[95] Pramudya, Y.; Mendoza-Cortes, J. L., Design principles for high H2 storage using chelation of abundant transition metals in covalent organic frameworks for 0–700 bar at 298 K. *J Am Chem Soc* **2016**, *138* (46), 15204–15213.

[96] Ben, T.; Ren, H.; Ma, S. Q., et al., Targeted synthesis of a porous aromatic framework with high stability and exceptionally high surface area. *Angew Chem Int Edit* **2009**, *48* (50), 9457–9460.

[97] Yuan, D. Q.; Lu, W. G.; Zhao, D., et al., Highly stable porous polymer networks with exceptionally high gas-uptake capacities. *Adv Mater* **2011**, *23* (32), 3723–3725.

[98] Blankenship, T. S.; Blankenship, T. S.; Balahmar, N., et al., Oxygen-rich microporous carbons with exceptional hydrogen storage capacity. *Nat Commun* **2017**, *8*.

[99] Wang, L. F.; Yang, R. T., New sorbents for hydrogen storage by hydrogen spillover – a review. *Energ Environ Sci* **2008**, *1* (2), 268–279.

2 Brake Friction Materials for High-speed Trains

Yelong Xiao and Mingxue Shen

2.1 INTRODUCTION

Since the inauguration of the world's first high-speed railway, Japan's Tokaido Shinkansen, in 1964, global high-speed rail mileage has steadily grown over more than half a century, surpassing 60,000 kilometers in operation. Moreover, future demand for high-speed railways worldwide continues to grow. High-speed railways in China have developed rapidly and gone on to make remarkable achievements. Since the completion of the Beijing-Tianjin Intercity Railway, the first high-speed railway in China, in 2008, the total operating length of high-speed railways had surpassed 40,000 kilometers by the end of 2022, representing about 59.5 percent of the world's total. China has developed a comprehensive high-speed rail network, along with advanced high-speed railway systems and manufacturing platforms for high-speed trains.

High-speed trains in the world have been operating at or above 300 km/h and are moving towards higher speeds. The AGV-V150 train, manufactured by France, set a world rail speed record of 574.8 km/h; the 300X train manufactured by Japan and the ICE/V train manufactured by Germany reached maximum test speeds of 443 km/h and 406.9 km/h, respectively; and the CRH380BL train manufactured by China, set a Chinese record of 487.3 km/h on the Beijing-Shanghai high-speed railway in 2011. In June, 2017, Fuxing Hao, the latest China Standardized train, was designed with a maximum speed of 400 km/h, and made its debut on the Beijing-Shanghai high-speed railway, heralding a new era for China's railway equipment technology. China is currently conducting performance verification tests on the even faster CR450 train, which operates at speeds up to 400 km/h. It has successfully reached 403 km/h in a tunnel with a relative rendezvous speed of 806 km/h, and 435 km/h on an open line with a relative rendezvous speed of 870 km/h. These achievements have set new world records for rendezvous speeds of high-speed rail trains both in tunnels and on open lines.

As one of the nine core technologies of high-speed trains, the braking technology plays a crucial role in ensuring the safety, smoothness, and comfort of train operations. Brake pad is a critical component of the brake system for determining the emergency braking capability and ensuring the operational safety of high-speed trains. Given that high-speed trains operate at elevated speeds, the braking system's requirements become increasingly stringent with higher speeds. In case of emergencies, high-speed trains primarily depend on the mechanical braking. Brake pads and

DOI: 10.1201/9781003386223-2

brake discs come into contact under pressure, where friction between them rapidly converts kinetic energy into thermal energy, thereby achieving effective braking. Mechanical braking system is a crucial safety device for high-speed trains, serving as the final safeguard for their safe operation. According to the International Union of Railways (UIC), high-speed trains must be able to stop within a specified distance using only mechanical braking. As train speeds increase, the braking distance lengthens, and the temperature at the friction interface rises more rapidly, impacting the safety and reliability of the trains [1].

For instance, to ensure that a CRH380 train can stop within a 3.8 km safety distance from a speed of 300 km/h, the brake system must absorb substantial energy. The friction interface must dissipate a braking energy density of 450 J/mm², with instantaneous temperatures potentially exceeding 1000°C [2]. Under high temperatures and pressures, the friction coefficient of the brake material is prone to decrease and be unstable, leading to longer braking distances for high-speed trains. To ensure the safe operation of high-speed trains, brake materials must maintain the relatively stable friction coefficient under all conditions. It is particularly crucial to prevent the friction coefficient from dropping to excessively low levels under high temperatures and high pressures. Therefore, brake materials must possess a suitable and stable friction coefficient, excellent wear resistance, high heat resistance and thermal fatigue resistance, sufficient mechanical strength, good compatibility with the brake disc, good environmental adaptability, environmental friendliness, and other properties.

Copper metal matrix composites (Cu-MMCs), known for their excellent overall performance and cost-effectiveness, are currently the most ideal choice for high-speed train brake pads. These materials are employed in high-speed trains operating at speeds of 300 km/h and above. As Cu-MMCs are increasingly utilized for high-speed train brake pads, it is crucial for theoretical research to keep pace. In-depth studies are needed to understand the tribological behavior and frictional mechanisms of Cu-MMCs under complex service conditions. Additionally, ongoing research should focus on the material design and application testing of Cu-MMCs, and their tribological compatibility with various brake disc materials.

2.2 BRAKING TECHNOLOGY FOR HIGH-SPEED TRAINS

The braking of high-speed trains involves artificially halting the train's motion, slowing it down, preventing acceleration, or bringing it to a complete stop. It is primarily categorized into four types: service braking, emergency braking, parking braking, and downhill braking. The emergency braking distance is a crucial indicator of a train's braking ability and operational safety. According to "Technical Management Regulations for Railway" in China, there are established limits for the emergency braking distance of high-speed trains at various speeds under dry conditions on flat track (with a slope within ±4 mm/m and a curve radius of at least R3000 m), as shown in Table 2.1. Currently, high-speed trains primarily utilize electric braking and mechanical braking to generate stopping force. Additionally, other braking modes, such as eddy-current braking, magnetic rail braking, and wind-resistance braking, may be employed to enhance the braking ability of trains [3].

TABLE 2.1
Maximum Acceptable Braking Distances at Various Braking Speeds

Initial stop-braking speeds (km/h)	Maximum acceptable braking distances (m)
200	2000
250	3200
300	3800
350	6500

2.2.1 Electric Braking

There are two types of electric braking: rheostatic braking and regenerative braking. The electric traction motor is reconnected as a generator during rheostatic braking, converting the train's kinetic energy into the electrical energy, which is then dissipated through a resistor to achieve the braking effect. This mode was primarily used in early high-speed trains like the French TGV-B. In contrast, regenerative braking converts the train's kinetic energy into electrical energy, which is then fed back into the power supply network. This mode is more economical and environmentally friendly, and has been widely adopted in high-speed trains worldwide since the 1990s [4].

2.2.2 Eddy-current Braking

The eddy-current braking consists mainly of the track eddy-current braking and rotary eddy-current braking [5]. In the track eddy-current brake system, the electromagnetic coil is mounted on the bogie of high-speed trains and positioned a few millimeters above the rail surface. The relative motion between the coil and the rail induces eddy currents on the rail surface, which create a Lorentz force that produces braking force. This technology has been utilized in Germany's ICE3 high-speed train. Rotary eddy-current braking involves a metal disc mounted on the train's axle. During braking, this disc rotates within a magnetic field generated by the electromagnetic coil, inducing eddy currents on the disc's surface and producing a Lorentz force that generates braking force. This technology has been employed in Japan's Shinkansen 700, 300, and 100 series high-speed trains (Figure 2.1).

2.2.3 Magnetic Rail Braking

Magnetic rail braking is an auxiliary braking mode used in emergencies. The magnetic rail brake system comprises the electromagnets and lifting air cylinders mounted on the bogie of trains. The electromagnet is fitted with a segmented wear plate underneath. During braking, the lifting cylinder lowers the electromagnet, which, when energized, adheres to the rail surface with strong electromagnetic force. The

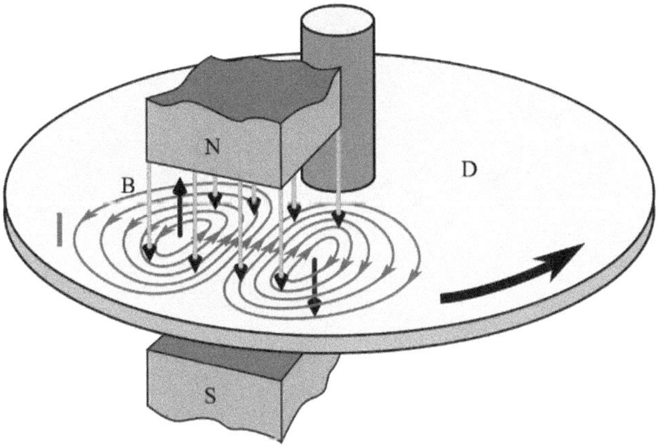

FIGURE 2.1 A schematic of the eddy-current braking system.

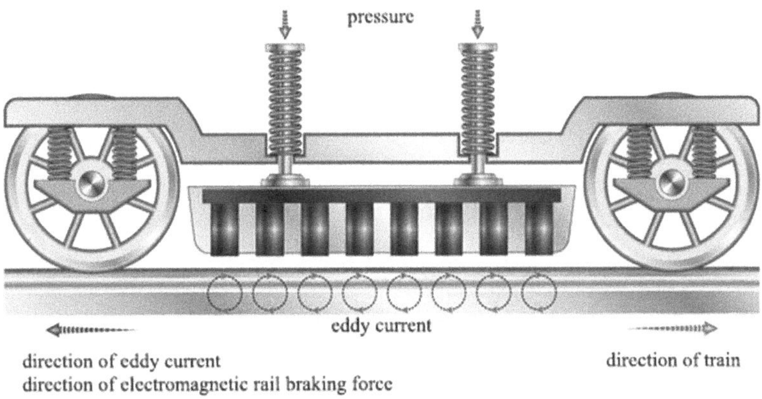

FIGURE 2.2 A schematic of the magnetic rail braking system.

friction between the wear plate and the rail generates a significant braking force. This technology has been implemented in Germany's ICE2 and ICE1 high-speed trains (Figure 2.2).

2.2.4 Wind-Resistance Braking

Wind-resistance braking is an effective braking mode for high-speed trains running at speeds of over 350 km/h during auxiliary or emergency braking. This braking mode relies on aerodynamic drag, a fundamental principle from aerodynamics, which originated in the aviation industry and jet technology. By utilizing the pressure difference between the front and back of a wind-resistance plate, the system generates braking force. The higher the speed of the train, the greater the braking ability of the wind-resistance braking. This braking mode is noted for reducing wheel-rail wear,

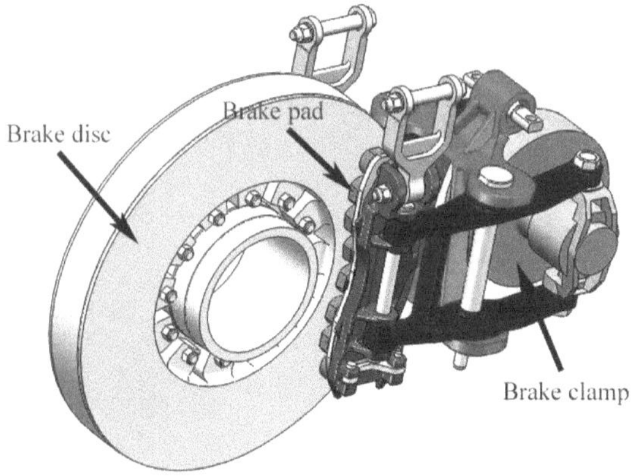

FIGURE 2.3 A schematic of the mechanical braking system.

featuring a simple structure, and being environmentally friendly. Japan's Shinkansen Fastech 360 high-speed train exemplifies the application of this innovative braking technology [6].

2.2.5 Mechanical Braking

In the mechanical brake system, hydraulic or compressed air pressure from the system is used to activate the brake calipers, which then press the brake pads against the brake discs (Figure 2.3). This action generates braking force through friction between the pads and the discs, converting the kinetic energy of high-speed trains into thermal energy that is dissipated into the atmosphere. Mechanical braking causes the wear of the pads, which manifests in two essential types: wheel braking and axle braking.

2.3 BRAKE MATERIALS UTILIZED IN TRAINS

If other braking modes fail, the mechanical braking must ensure that the train can stop within the specified distance to guarantee safe operation. Brake pad materials for trains typically include polymer matrix composites, metal matrix composites, as well as carbon matrix composites being an area of ongoing research (Figure 2.4).

2.3.1 Polymer Matrix Composites

Polymer matrix composites are made by mixing phenolic resin, reinforcing material, and friction modifier [7], which are then hot-pressed and solidified, as shown in Figure 2.5. These composites are lightweight, have a high friction coefficient, possess excellent wear resistance, and do not produce sparks during braking. They are widely used in passenger cars, freight locomotives, and subways operating at speeds up to 160 km/h. In China, the CRH1A/B/E trains, which operate at speeds of 200–250 km/h, also utilize polymer matrix composites. However, the material contains many

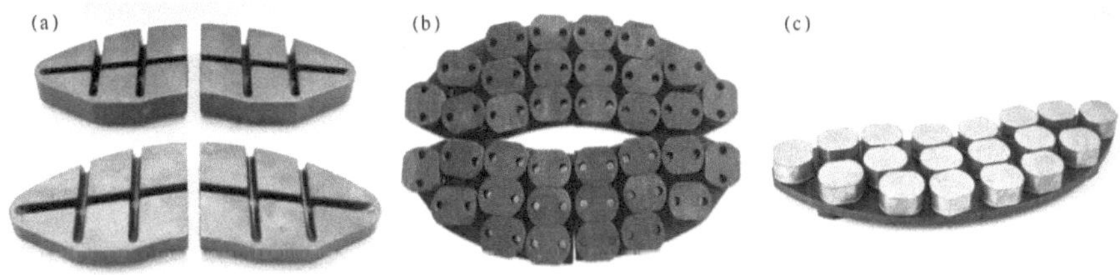

(a) Polymer matrix composites

(b) Carbon matrix composites

(c) Metal matrix composites

FIGURE 2.4 Various material for brake pads: (a) polymer matrix composites, (b) carbon matrix composites, (c) metal matrix composites.

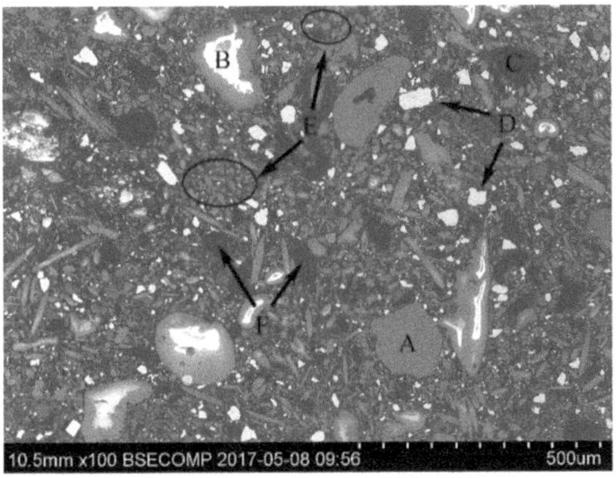

A: Aluminium reinforcing fibre
B: Calcium silicate
C: Graphite
D: Barium sulfate
E: Clumps of glass fibres
F: Resin binder (dark region)

FIGURE 2.5 Metallographic drawing of a polymer matrix composites [7].

organic compounds, and the covalent bonds of organic compounds are prone to form free radical under high temperatures. Additionally, polymer matrix composites have poor thermal conductivity, making it difficult to dissipate the heat generated during braking. This can lead to thermal cracking of the brake disc due to significant temperature rises. The braking performance of the material deteriorates significantly in high-temperature or humid environments, with a commonly used maximum heat-resistance temperature of around 400°C. This limitation means that the material is difficult to withstand the extreme high-temperature conditions [8]. Consequently, polymer matrix composites are inadequate for meeting the braking requirements of high-speed trains.

2.3.2 Carbon Matrix Composites

The carbon/carbon (C/C) composite consists of carbon fibers embedded in a carbon matrix. This advanced structural material features low density, high temperature resistance, high specific strength/modulus, and excellent friction and wear resistance, and has been extensively researched. Since the 1980s, Japan has explored

C/C composites for high-speed train braking, conducting on-board tests on the Shinkansen. France has also tested a C/C composite, "Sepcagh, SA3D," on the TGV high-speed train, with both showing excellent overall performance. However, the high-temperature mechanical properties of C/C composites are only effective in inert environments because they begin to oxidize in air at 450°C, significantly diminishing their mechanical properties [9]. Additionally, C/C composites have a low friction coefficient in humid environments, a lengthy manufacturing cycle, high costs, and are difficult to meet the demanding and variable conditions of high-speed train operation.

In the mid-to-late 1990s, C/C-SiC composites emerged as the latest generation of high-performance brake materials. These composites primarily consist of carbon fiber-reinforced carbon, a silicon carbide matrix, and silicon. They not only retain the key benefits of C/C composites—such as high temperature resistance, high specific strength, high thermal conductivity, and low density—but also possess additional advantages including corrosion resistance, oxidation resistance, and enhanced environmental adaptability (for instance, their friction coefficient remains stable even in humid conditions) [10]. Currently, C/C-SiC composites are being evaluated for use in Japan's Shinkansen and France's TGV-NG high-speed trains. However, their long production cycles, high costs, and low toughness indicate that further research is needed to optimize their application in high-speed trains.

2.3.3 Metal Matrix Composites

Metal matrix composites are produced by adding lubricants and friction components to metal or alloy powders through powder metallurgy technology. These composites are generally classified into two main types based on their matrix: iron metal matrix and copper metal matrix friction materials. Iron metal matrix materials are known for their excellent friction and wear resistance under high temperatures and heavy loads, and they are relatively inexpensive [11]. However, they have a tendency to adhere to steel brake discs, resulting in damage to brake discs and significant fluctuations in the friction coefficient. This leads to increased wear at low speeds and potential noise issues, which limits their suitability for use in high-speed train brake pads [12].

Cu-MMCs offer stable friction coefficients, excellent wear and heat resistance, good environmental adaptability, and low damage to brake discs, making them widely utilized in high-speed trains (Figure 2.6) [2, 13]. In Japan, Cu-MMCs for brake pads are manufactured from copper reinforced with Fe, Ni, and Sn alloying elements, and incorporate substantial amounts of ceramic phases and graphite, and brake pads can be used for trains at a speed of 300 km/h. In France, Cu-MMCs used for TGV high-speed trains primarily consist of copper, iron, molybdenum, tin, nickel, carbides, and silicates. In Europe, the BM40 and 664HD types of Cu-MMCs, developed by Becorit and Flertex, are engineered to meet the braking requirements of Germany's ICE3 high-speed trains. Additionally, the German company Knorr-Bremse has developed Cu-MMCs for high-speed train brake pads that offer superior wear resistance and thermal stability, along with a longer service life compared to conventional brake pads.

(a) Parallel to the friction surface (b) Perpendicular to the friction surface

FIGURE 2.6 Microstructure of Cu-MMCs: (a) parallel to the friction surface; (b) perpendicular to the friction surface [2].

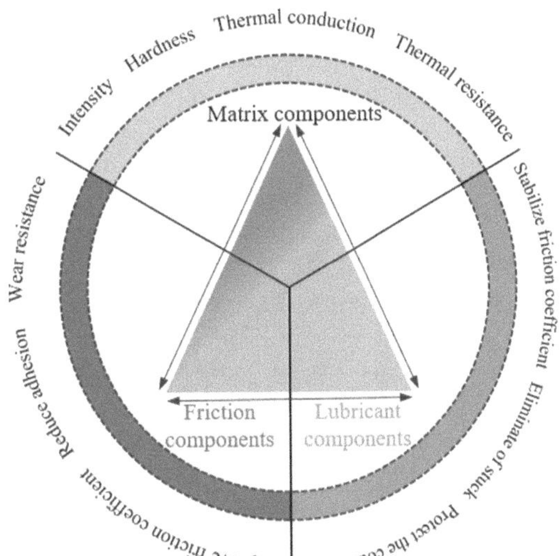

FIGURE 2.7 Ingredients of copper metal matrix composites.

2.4 MATERIALS DESIGN OF CU-MMCS FOR HIGH-SPEED TRAINS

Cu-MMCs are widely employed in high-speed trains due to their outstanding overall performance. These composites consist of a matrix made from pure copper or copper alloy, combined with various reinforcing materials such as fibers or particles. The resulting structure is a blend of independent substances embedded within the copper or copper alloy matrix. The copper matrix imparts excellent electrical and thermal conductivity, as well as superior friction and wear characteristics to Cu-MMCs. The components of Cu-MMCs can be broadly categorized into three types: matrix components, lubricant components, and friction components, as illustrated in Figure 2.7.

2.4.1 Metal Matrix

The metal matrix is the primary component of Cu-MMCs, serving both as the load-bearing structure and the medium that integrates various additives. It also acts as the main channel for heat dissipation. The strength, tribological properties, and heat resistance of Cu-MMCs are significantly influenced by the matrix's composition and structure. During braking, the matrix endures substantial heat load and plastic deformation, leading to softening, oxidation, and potential melting. Therefore, it is crucial for the matrix to possess robust anti-wear and heat-resistant properties to maintain effective friction and wear performance.

Many studies focus on the characteristics of copper powder, solid solution strengthening from alloying elements, and second phase strengthening to enhance the frictional properties of Cu-MMCs and improve the braking performance. Finer copper powder particles lead to a lower coefficient of friction and elevated temperatures, as the softened copper at the friction surface provides a lubricating effect, which can significantly reduce the coefficient of friction [14]. Additionally, using iron-cobalt-copper pre-alloyed copper powder helps prevent compositional segregation during powder mixing, resulting in a material that forms stable oxide films with consistent and low wear.

Alloying elements such as Fe, Ni, Mn, Ti, Sn, and W are commonly added to enhance the copper matrix's strength and anti-adhesive properties. Ni not only significantly improves the hardness and strength of the material but also increases the stability of the friction coefficient and reduces wear. The addition of Ti induces lattice distortion in the copper matrix, which boosts the material's hardness and strength, minimizes plowing effects, and enhances wear resistance. Rare earth elements, such as La, refine the grain structure of the copper matrix, contribute to solid solution and dispersion strengthening, and improve the material's microstructure, leading to better tribological and mechanical properties of Cu-MMCs. Fe is frequently added to Cu-MMCs for its strengthening effects and ability to tailor the friction coefficient and frictional stability. Fe is commonly used in Cu-MMCs for high-speed train brake pads. Fe, as a hard particle, enhances the material's mechanical properties and increases the friction force on the friction surface, resulting in a higher friction coefficient. At elevated speeds, the frictional heat generated causes Fe to react with atmospheric components, forming a protective oxide film. Additionally, Fe's good bonding ability with the copper substrate helps reduce wear [15].

The addition of ceramics such as Al_2O_3, TiB_2, and other second phase particles, enhances the copper matrix, improving its tribological properties and resistance to microcutting forces [16]. This results in greater wear resistance and a significantly lower wear rate. While materials with Al_2O_3 dispersion-reinforced copper powder exhibit stable friction coefficients, they tend to show higher wear amounts. On the other hand, reinforcing the copper matrix with nanoparticles, carbon fibers, or short carbon fibers can lead to substantial improvements in mechanical strength, fatigue resistance, corrosion resistance, and braking performance [17, 18]. Adding carbon fibers enhances the thermal conductivity and density of Cu-MMCs, but does not significantly improve tribological properties. Adding more than 2 percent graphene nanosheets into Cu-MMCs yields excellent mechanical and tribological properties [19]. Furthermore, the tribological performance is enhanced through the synergistic reinforcement of reduced graphene oxide and short carbon fibers, which help form a lubricating film on the friction surface of Cu-MMCs, thereby reducing the wear rate.

2.4.2 Lubricant Components

The lubricant component enhances lubrication in Cu-MMCs, improving wear resistance and anti-seize properties, which reduces adhesion between friction pairs and ensures smoother operation, thereby enhancing braking performance. The friction and wear characteristics of the material are influenced by the content of the lubricant component. Increased content of the lubricant component generally leads to better wear resistance and a lower friction coefficient. However, excessive lubricant component can reduce both the mechanical strength and the friction coefficient of the material. The lubricant components are primarily categorized into three types: low melting point metals, graphite and other non-metals (such as metal sulfides).

2.4.2.1 Low Melting Point Metals

Low melting point metals such as lead, bismuth, and antimony are commonly used as solid lubricants. Lead, in particular, is frequently employed due to its ability to form a lubricating film when the surface temperature exceeds its melting point. This film smooths sliding, enhances high-temperature performance, and improves the copper matrix's resistance to seizing and adhesion. Lead significantly affects the friction and wear characteristics of Cu-MMCs, contributing to better stability in frictional operations and reducing brake noise. However, due to health concerns, bismuth—another low melting point metal with similar properties—has become a more common substitute for lead in recent years. Replacing lead with bismuth, a non-toxic alternative with anti-adhesive properties, and incorporating graphite, which reduces friction and wear, can lead to improved frictional performance without the health risks associated with lead. Additionally, some researchers have experimented with aluminum as a substitute for lead in Cu-MMCs. Aluminum offers reduced density and hardness, an increased friction coefficient, and a smooth, flat friction surface, making it a viable option for environmentally friendly applications in Cu-MMCs.

2.4.2.2 Non-metal

Non-metals used as lubricants include graphite and boron nitride, among others. Graphite's crystal structure is composed of numerous planar hexagonal mesh layers, where each layer consists of carbon atoms arranged in a hexagonal ring. The bonds within a single plane are extremely strong, resulting in a high melting point and excellent chemical stability. However, the bonding between these layers is weak, held together only by van der Waals forces. This weak interlayer bonding allows the layers to slide easily over one another under shear force, which effectively reduces friction. Due to its low shear resistance, graphite is well-suited as a solid lubricant and is widely used for its excellent tribological properties. Despite its relatively poor mechanical strength, research has shown that a 4 percent graphite content in materials can provide a high and stable coefficient of friction along with a low wear rate [20].

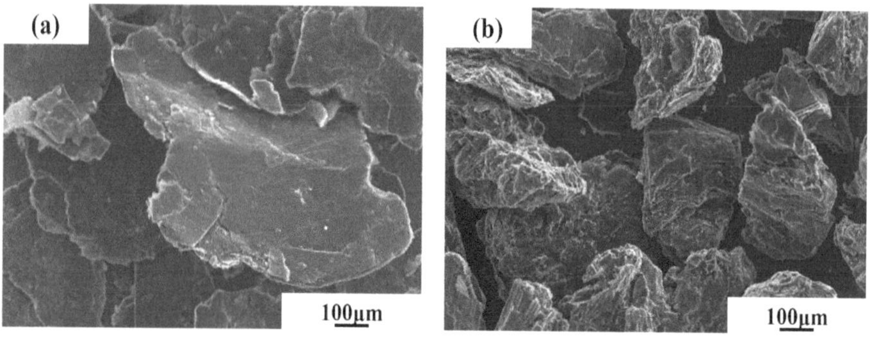

(a) Flake graphite (b) Granular graphite

FIGURE 2.8 SEM images of (a) flake graphite and (b) granular graphite [21].

When investigating the influence of different graphite types on the frictional properties of Cu-MMCs, it was found that Cu-MMCs with natural flake graphite exhibited superior wear resistance compared to those with artificial graphite. Cu-MMCs containing large-size flake graphite showed higher compressive strength than those with small-size flake graphite. Additionally, the combination of flake graphite and artificial graphite notably enhanced the material's strength. While granular graphite generally offers better tribological properties than flake graphite, its lower strength often leads to delamination. Optimizing the ratio of granular to flake graphite, with each type at 5 wt. percent, yielded the best tribological performance, as illustrated in Figure 2.8) [21].

The copper matrix and graphite show distinct physicochemical properties, which may lead to interface defects under high cyclic stresses during braking, adversely affecting frictional characteristics. To enhance the braking performance of Cu-MMCs, modifying the graphite surface to improve interfacial bonding with the copper matrix is beneficial. The addition of nickel plating graphite to copper–graphite composites enhances bonding, reduces cracking, and improves the wear resistance through diffusion, resulting in better braking performance [22]. BN can serve as an alternative solid lubricant to graphite and MoS_2, with practical applications involving BN contents of 6-12 percent, though it is sensitive to vacuum heat treatment. High-temperature calcination under vacuum conditions (5×10^{-4} mmHg) significantly increases the friction coefficient. Research comparing the tribological properties of h-BN and graphite reveals that while lower h-BN content can stabilize frictional characteristics, graphite offers superior lubricity and forms a denser friction film, enhancing tribological performance [23]. Adding a small amount of h-BN to graphite helps stabilize the friction coefficient and improve the wear resistance of Cu-MMCs.

2.4.2.3 Metal Sulfides

Metal sulfides, such as MoS_2 and WS_2, serve as effective lubricant components. Both MoS_2 and WS_2 are lamellar crystalline materials with low hardness, good corrosion resistance, thermal stability, and excellent lubrication properties at high temperatures. They perform well across a wide range of temperature, load, and speed variations. While MoS_2 and graphite are comparable in their ability to enhance frictional characteristics, they operate through different mechanisms. When comparing the effects of graphite and MoS_2 on the friction and wear properties of Cu-MMCs, it has been observed that Cu-MMCs containing MoS_2 exhibit a denser microstructure, higher density, increased hardness, and lower surface roughness. Research into MoS_2 as a lubricant component has revealed that sulfur, generated during the sintering process, reacts to form a variety of sulfides. The decomposition of MoS_2, reacting with iron, produces ferrous sulfide and iron sulfide, which contribute to lubrication and wear resistance. Consequently, some researchers have explored the use of FeS as a substitute for MoS_2 in Cu-MMCs. Experimental results demonstrated that the mechanical properties, including hardness, density, and frictional characteristics of Cu-MMCs containing FeS were comparable to those containing MoS_2. Both types of composites exhibited stable friction coefficients due to the formation of an FeS friction film [24, 25]. WS_2, as a lubricant component, plays a crucial role in enhancing lubrication. By adding larger WS_2 particles into Cu-MMCs, a smoother and more continuous transfer film is obtained, which helps to minimize cracks and improve wear resistance. Additionally, the tribological behavior of WS_2 was compared under two different morphologies: flake and spherical. The results indicated that spherical WS_2 demonstrated more stable wear resistance compared to the flaky WS_2, particularly under elevated loads. The flake WS_2 tends to move layer by layer through slip, whereas the spherical WS_2, due to its shape, facilitated the formation of a physical deposition film on the friction surface. This film effectively transformed sliding friction into rolling friction, thereby reducing the friction coefficient and enhancing the lubrication effect by impeding the relative motion of the frictional pair [26].

2.4.3 Friction Components

To counterbalance the reduced friction coefficient caused by lubricant components, friction-enhancing additives are typically added into Cu-MMCs. These additives play several crucial roles: they enhance the friction coefficient, improve the wear resistance, and prevent stalling. Additionally, they help cut through friction surface oxides and build-up, clean the friction interface, and prevent the transfer of brake material to the surface of the brake disc, which contributes to maintaining a stable friction coefficient. Furthermore, friction components can obstruct the flow of the matrix at elevated temperatures, thus enhancing both the wear resistance and temperature stability of Cu-MMCs. Hard ceramic particles or hard metal particles and their compounds are commonly used as friction components in Cu-MMCs.

2.4.3.1 Ceramic Particles

Ceramic particles are widely used as reinforcing phases in Cu-MMCs due to their excellent mechanical properties, high-temperature stability, and low production costs. Commonly employed ceramics include Al_2O_3, SiO_2, SiC, and ZrO_2. These particles enhance the strength of Cu-MMCs and significantly improve the friction and wear performance [27]. The effectiveness of these particles depends on their content and particle size distribution within the copper matrix. They contribute to friction and maintain surface roughness, which reduces adhesive wear at the friction interface and enhances the frictional stability and braking smoothness. Xiong et al. [15] demonstrated that SiO_2 markedly improved the friction coefficient and wear resistance of Cu-MMCs at high speeds compared to Fe. However, the content of ceramic particles in Cu-MMCs is limited by the strength of the particle-matrix interface. Kelly et al. [28] found that the wear resistance and mechanical properties are influenced by the volume fraction and spatial distribution of the ceramic particles. Zhan [29] observed that optimal interfacial bond strength allows for effective load transfer between the copper matrix and ceramic phase, thus increasing material strength. Surface modifications of ceramic particles with coatings like Ni, Cu, or Co-B have been studied to improve compatibility with the copper matrix, enhance interfacial bonding strength, and reduce particle dropout during braking, thereby improving wear resistance.

2.4.3.2 Hard Metal Particles

The fundamental properties of ceramic particles, such as Al_2O_3 and SiO_2, are markedly different from those of the copper matrix. These differences lead to poor wettability, resulting in a mechanically bonded interface that is prone to detachment. In contrast, hard metal particles and intermetallic compound particles exhibit better interfacing with the copper matrix. They enhance the hardness and strength of the matrix, making them among the best choices for friction components in Cu-MMCs.

Fe also plays a significant role as a friction component. It can improve the strength and friction wear performance of Cu-MMCs [30]. Fe powder is the second most abundant component in Cu-MMCs after copper, and it has a substantial effect on the friction and wear performance. By increasing the surface roughness of Cu-MMCs, Fe powder enhances the friction coefficient. Additionally, Fe accelerates the formation of the oxide layer on the friction surface, which reduces the damage to the friction surface. Peng et al. [31] conducted braking tests at speeds up to 380 km/h and found that Cu-MMCs with higher Fe content exhibited lower friction coefficients and wear rates, with a stable friction coefficient. CrFe, as shown in Figure 2.9, is another common hard metal friction component known for its high hardness, wear and corrosion resistance, and chemical stability [32]. When added to Cu-MMCs, CrFe improves the friction coefficient, frictional stability, and wear resistance more effectively compared to Al_2O_3 [33]. Mo-Fe further enhances the softening resistance of the copper matrix, reducing the degree of plowing and thus improving the wear resistance of Cu-MMCs.

2.5 FRICTION BEHAVIOR AND MECHANISMS OF CU-MMCS FOR HIGH-SPEED TRAINS

2.5.1 Effect of Braking Conditions on Tribological Properties of Cu-MMCs

The tribological performance, which is characterized by the friction coefficient and wear rate, is not the inherent properties of materials. It varies with the basic properties of materials as well as working conditions, including braking speed, applied pressure, and environmental conditions. During braking, the temperature of the worn surface

FIGURE 2.9 Topography of CrFe powder [32].

significantly influences the friction performance. This is because changes in surface temperature effect the rates of oxidation, diffusion, metal adhesion, and elastic-plastic deformation, among other material behaviors. The state of the worn surface temperature is affected by factors such as braking speed and applied pressure. To meet the design and functional requirements of friction pairs used in high-speed trains, laboratory simulation experiments are typically conducted using universal friction and wear test equipment under controlled conditions. These experiments help in investigating the influence of working conditions on wear mechanisms of Cu-MMCs.

2.5.1.1 Investigation on Speed-load Sensitivity to Tribological Properties of Cu-MMCs

The influence of braking speed on the friction and wear performance of materials primarily depends on surface temperature, while braking pressure affects tribological performance by altering the real contact area. In high-speed train operations, variations in braking parameters lead to changes in the tribological performance of Cu-MMCs under different speeds and loads [34]. Research indicates that increased braking speed generates significant frictional heat on the worn surfaces, causing the copper matrix to soften and the surface micro-convexities to deteriorate. This results in a decreased friction coefficient and an increased wear rate. Additionally, the elevated temperature of the friction surface accelerates oxidation wear, with substantial oxide detachment due to shear forces, further increasing the wear rate [35].

The friction coefficient exhibits distinct three-stage characteristics as the initial braking speed (IBS) increases [34]. In the low-speed stage (80–200 km/h), the friction coefficient decreases from 0.395 to 0.358 with rising IBS. During the medium-speed stage (250–320 km/h), the coefficient reaches its peak value of 0.398 at 320 km/h. In the high-speed stage (350–380 km/h), the friction coefficient drops to 0.379 as IBS continues to increase. The wear rate rises with increase in IBS, particularly increasing sharply once IBS surpasses 250 km/h. Beyond 380 km/h, the rate of increase slows down. The variation in the friction coefficient is closely linked to changes in the friction film. At low speeds, the formation of the friction film on the brake pad surface causes the coefficient to decrease. In the medium-speed range, improved contact between the disc and pad leads to exposure of hard-phase particles due to partial destruction of the friction film, which increases the friction coefficient. In the high-speed stage, the reduction in the contact platform and the high temperatures that cause the friction film to melt contribute to a decrease in the friction coefficient.

To prevent skidding during braking and ensure the safe operation of high-speed trains, it is essential to continuously adjust the braking load to manage the braking force between the brake pads and the discs, thus accommodating the adhesion limit between the wheel and rail. At a constant braking speed, increasing the braking load initially causes the friction coefficient of Cu-MMCs to rise, then fall, while the wear rate gradually increases and eventually stabilizes. When the braking pressure is 0.9 MPa, the wear rate of Cu-MMCs reaches the maximum value. However, increasing the braking load effectively reduces the adhesive wear of Cu-MMCs.

Based on the analysis of the worn surface changes in Cu-MMCs and the effects of braking load, several observations can be made: At low braking loads, the pressure stress on the material's surface micro-convex bodies is minimal, making plastic deformation unlikely. During braking, the mechanical interlocking of these micro-convex bodies increases, enhancing the friction resistance. However, the actual contact area between the friction surfaces remains small, which elevates the friction coefficient of the friction pair. At medium braking loads, the rise in surface temperature leads to high-temperature softening and significant plastic deformation of the micro-convex bodies. This reduces the mechanical interlocking effect between the micro-convex bodies. Consequently, the material of Cu-MMCs transfers to the brake disc, forming adhesive points. These adhesive points are sheared during the friction process, resulting in increased adhesive wear and wear rate. Under high braking loads, the frictional heat accumulates rapidly on the surface, causing a swift rise in temperature. This leads to severe oxidation, high-temperature softening, and plastic deformation of the micro-convex bodies, and in some cases, local melting that forms a molten surface film. The actual contact area significantly increases, and wear debris, oxides, and dislodged materials are compressed under high loads, forming a thick and dense friction film on the surface of Cu-MMCs. This film greatly reduces the surface roughness and significantly lowers the friction coefficient, thereby improving braking stability. In summary, the tribological behavior of the friction pair is influenced by factors such as braking load and surface temperature. The friction coefficient and wear rate of the material vary with different braking speeds and loads.

2.5.1.2 Impact of External Environments on Tribological Properties of Cu-MMCs

High-speed trains often operate in complex and variable environments, where environmental factors such as humidity, extreme cold, high temperatures, and other external conditions can significantly influence the performance of their mechanical braking systems. The friction coefficient and wear rate of brake friction materials are particularly influenced by air humidity. When the relative humidity is between 20 and 80 percent, its effect on the friction coefficient of Cu-MMCs is minimal.

However, when humidity exceeds 80 percent, the friction coefficient tends to increase slightly. In wet conditions, at low speeds and pressures, water acts as both an isolating and lubricating agent. The lubrication provided by the water film can lower the friction coefficient while increasing the wear rate. Conversely, at high speeds and pressures, water primarily serves as a coolant, which helps to reduce the softening of the copper matrix. The evaporation of water due to high temperatures and the centrifugal forces involved effectively disrupt the water film, causing the friction and wear performance of Cu-MMCs to revert to conditions similar to dry friction. As a result, while a humid environment can significantly lower the friction coefficient of Cu-MMCs at low speeds, its impact is less pronounced at higher speeds. Cu-MMCs exhibit strong adaptability to low temperatures and are relatively insensitive to humidity, ensuring stable braking performance across varying environmental conditions [2].

The relationship between the average friction coefficient and the initial stop-braking speed for Cu-MMCs tested under both dry and wet conditions is shown in Figure 2.10. The introduction of water spray to the sliding interface not only dissipated much of the heat generated during the friction process but also likely formed a water film, reducing solid–solid contact and resulting in a lower friction coefficient [2]. Additionally, as the braking speed or braking load increased, it became more challenging to maintain a sufficiently thick water film to support the braking force and prevent solid–solid contact, leading to a significant increase in the friction coefficient [35].

The low-temperature environment can lead to unstable friction performance, abnormal wear, damage to the contact surface, and even slippage. These issues

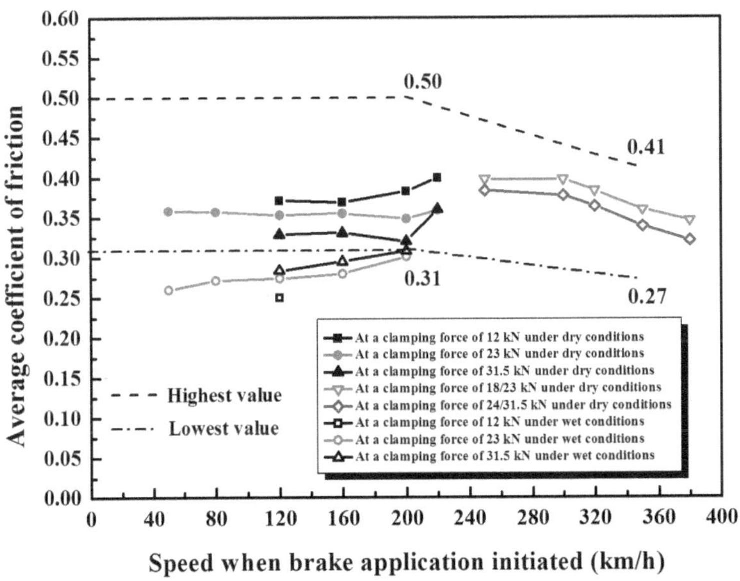

FIGURE 2.10 Results of the average coefficient of friction of Cu-MMCs [2].

result in a decreased friction coefficient, extended braking distances, and potentially serious problems such as the inability to stop the train, thereby posing significant safety risks to braking systems. In addition, low temperatures combined with rainy or snowy conditions can lead to water condensation and icing on the friction surfaces of the friction pair, altering the friction interface and further impacting tribological performance. Research on the tribological performance of Cu-MMCs in ice or snow conditions has shown that hard particles embedded in ice or snow can scrape the brake disc. This interaction causes the friction coefficient of the material to increase with higher braking pressure. When braking speeds are below 160 km/h, the friction coefficient increases with the increase of speed. However, at speeds above 160 km/h, the substantial frictional heat generated melts the ice and snow on the friction surface, making the friction coefficient less sensitive to speed changes. Under emergency braking conditions, the average friction coefficient of the material increases in a non-linear fashion with both braking speed and pressure. During continuous braking at low braking pressure, the average friction coefficient is influenced by an ice film on the contact surface, resulting in significant fluctuations in the friction coefficient.

Wind and sand may cause dust to accumulate on the surface of the friction pair, leading to unstable braking performance, corrosion, and oxidation of braking materials. This third media results in insufficient braking force and alterations in the materials' microstructure. Sand, consisting of various small solid particles with differing sizes, hardness, and chemical properties, can damage equipment through wear, erosion, corrosion, and penetration. These particles can infiltrate gaps and adhere to the surface of the friction pair, scratching it and eventually leading to thermal cracks under cyclic thermal stress, which can cause braking system failure. Additionally, the friction heat and thermal stress generated during braking create a strong thermal-mechanical coupling effect at the contact interface, complicating the working conditions.

2.5.2 Tribological Properties of Cu-MMCs at Elevated Temperatures

The high temperatures generated during the braking process of high-speed trains are a crucial factor that must not be overlooked, significantly affecting the service life and reliability of Cu-MMCs. The considerable kinetic energy dissipated during braking results in substantial frictional heat accumulation on the surface of the friction pair, causing a sharp rise in temperature. This elevated heat can lead to the softening of the copper matrix of Cu-MMCs, potentially causing severe damage and significantly shortening the lifespan of Cu-MMCs. Additionally, the softening of the copper matrix can create a self-lubricating effect, which markedly diminishes the reliability of the friction coefficient [36].

The variation in instantaneous friction coefficients of Cu-MMCs at temperatures from 25 to 800°C is shown in Figure 2.11(a). At 25°C, the friction curve exhibits a slight decrease during the initial running-in period, and then remains stable. As temperature increases, the friction curve exhibits a wider range of fluctuations, with more significant variations observed at 600°C. At 800°C, the friction coefficient is notably higher. However, with a further increase in temperature to 800°C, the friction

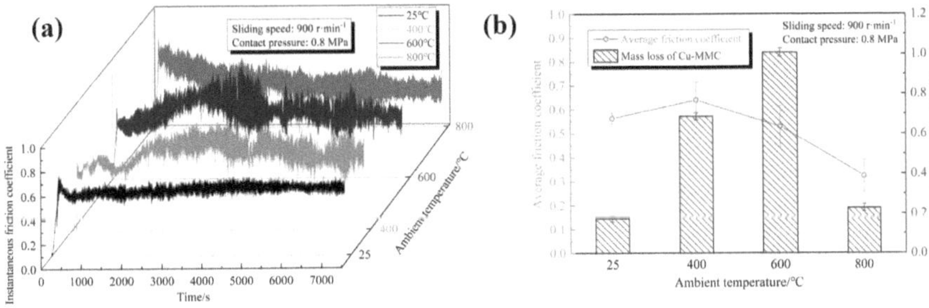

(a) Instantaneous friction coefficient (b) Average friction coefficient and mass loss of Cu-MMC

FIGURE 2.11 Evolution of the (a) instantaneous friction coefficient; (b) average friction coefficient and mass loss of Cu-MMC at various temperatures [37].

coefficient initially drops significantly before stabilizing. This trend in the coefficient of friction is referred to as "thermal decline," a phenomenon where frictional properties deteriorate at high temperatures due to changes in the microstructure of the Cu-MMC surface [37]. Figure 2.11 (b) presents the evolution of the average friction coefficient and wear amount of Cu-MMCs at temperatures ranging from 25 to 800°C. With the increase of temperature, the average friction coefficient initially increases slightly but then decreases sharply, reaching the maximum value at 400°C and the minimum value at 800°C. The amount of wear increases with the increase of temperature and reached the maximum value at 600°C, and then decreases significantly when the temperature reaches 800°C.

The tribological properties of Cu-MMCs are significantly influenced by high temperatures. At a brake interface temperature of 183°C, although the size of the peeling and contact plateau remains relatively small, a substantial amount of wear debris accumulates on the friction surface, compacting in interface defects or depressions. As the temperature rises to 305°C, the quantity of grinding chips markedly decreases, a continuous oxide film forms, and cracks become visible. Concurrently, the contact plateau expands and the surface smooths out. When the interface temperature reaches 456°C, severe spalling occurs on the friction surface. This is attributed to the softening of the copper matrix at high temperatures, which causes plastic flow beneath the surface. As a result, the contact platform size further increases and becomes smoother due to the effects of plastic flow [38].

By examining the friction and wear characteristics of Cu-MMCs with SiC additions at temperatures ranging from 400 to 800°C, Figure 2.12 presents BSE cross-sectional images of friction surfaces tested under various conditions. The findings indicate that the friction coefficient remains high and stable below 600°C. However, at 600°C, increasing the sliding speed accelerates the escape of substances, leading to the formation of a thick friction film. This film tends to break more readily due to increased hardening, resulting in an abnormal rise in the wear amount. At 800°C, the copper matrix softens and oxidizes, reducing its strength and preventing the formation of a continuous, stable, and dense oxide film. This results in a lower and less

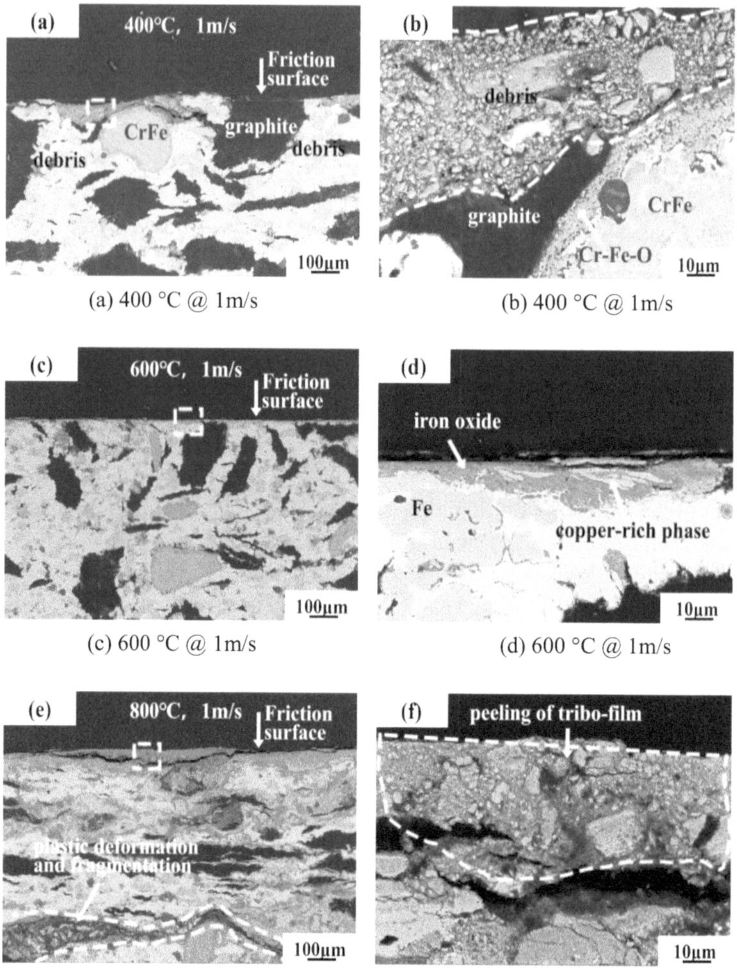

(a) 400 °C @ 1m/s (b) 400 °C @ 1m/s

(c) 600 °C @ 1m/s (d) 600 °C @ 1m/s

(e) 800 °C @ 1m/s (f) 800 °C @ 1m/s

FIGURE 2.12 BSE cross-sectional images of friction surface of Cu-MMCs tested at the conditions of (a, b) 400°C, (c, d) 600°C, (e, f) 800°C. Figures (b), (d), (f) and are the magnified BSE images of the region marked in Figures (a), (c), (e), respectively [39].

stable friction coefficient. The increased fluidity of the copper matrix causes material transfer and disrupts the formation of the friction layer. Consequently, the degree of seizure of grinding chips decreases, breaking the dynamic balance and leading to further intensification of the wear amount, and the production of significant particulate matter. In contrast, the tribological properties of the new material are similar to those of commercial material at temperatures ranging from 400 to 800°C, as they are less affected by these factors [39, 40].

2.5.3 Tribological Properties of Cu-MMCs Mated with Different Brake Discs

The performance of high-speed train braking systems depends not only on the fundamental properties of materials but also on the compatibility of the material structure and properties between the brake pad and brake disc. Different mating materials significantly influence the tribological performance of Cu-MMCs. Improper matching can result in uneven temperature and stress distribution on the brake discs, increased vibration and noise, and excessive material wear. This directly impacts the braking performance of the friction pair, as well as the reliability, stability, comfort, and service life of the braking system.

Currently, cast steel is commonly used for brake discs paired with Cu-MMCs. However, due to manufacturing processes and other factors, cast steel discs often have inherent defects such as inclusions, segregation, and porosity. Additionally, the significant thermal load generated during braking creates a substantial temperature gradient within the brake disc, leading to thermal stress. These defects, combined with the cyclic action of uneven temperature and stress fields, can lead to the formation of micro-cracks in the brake disc. As these micro-cracks propagate, they may eventually cause fatigue fractures in the brake discs [36, 37, 41]. While current research predominantly focuses on the wear mechanisms of brake materials, less attention has been given to the wear characteristics of steel brake discs, which are also crucial components. Railway standards specify that the friction surface of brake discs should be free from abnormal wear patterns such as groove wear, ploughing wear, wave wear, and metal inlay. Nonetheless, issues such as surface scratches and abnormal wear are observed in the steel brake discs currently in operation on high-speed trains.

As high-speed train technology continues to advance, the market for brake pads and brake discs is expanding, with an increasing array of new Cu-MMCs being introduced. An urgent challenge is ensuring compatibility between these newly developed brake pads and brake discs. As the braking energy handled by the brake discs increases, traditional steel discs are difficult to maintain performance due to excessive temperature rise and thermal cracking. This underscores the need for high-performance brake pairs to ensure the stability and efficiency of braking systems. Recent research has focused on the innovative mechanical brake system incorporating Cu-MMCs and carbon-based composites such as C/C and C/C-SiC materials. As depicted in Figure 2.13, self-mated C/C-SiC composites have a flatter friction surface compared to those paired with steel. These advanced materials are emerging as promising candidates for future high-speed train braking systems. Carbon matrix composites possess several advantages, including low density, high strength, and excellent high-temperature resistance. They maintain stable friction characteristics and superior wear resistance, especially under high-energy braking conditions, demonstrating exceptional overall performance. The adoption of C/C or C/C-SiC composites for brake discs can significantly reduce the weight of high-speed trains, thereby decreasing the energy consumption during train operation and parking. This improvement enhances the train operation efficiency and addresses issues such as thermal cracking, hot spots, and performance degradation seen in traditional steel discs [42].

When Cu-MMCs slide against C/C composites, an oxide film forms on the surfaces of Cu-MMCs, reducing their roughness. Cu-MMCs and C/C composites

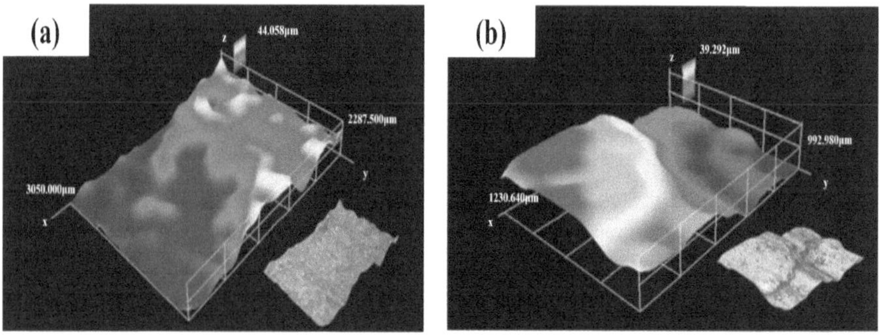

(a) Self counterpart (b) Steel counterpart

FIGURE 2.13 3D morphologies of friction surface of C/C–SiC mated with different counterparts at 6000 r/min. (a) self counterpart; (b) steel counterpart [45].

have distinct properties, with C/C composites exhibiting greater deformation ability, making them less prone to being plowed by Cu-MMCs. Despite the softening and plastic deformation of Cu-MMCs during friction, there is no bonding between the friction pairs, preventing material transfer films on C/C composites. During friction, significant oxidation and peeling of matrix carbon on C/C composites occur, leading to broken and pulled-out carbon fibers, which increases the wear rate, and decreases the friction coefficient and stability. In contrast, when C/C-SiC composites slide against Cu-MMCs, friction films form on the surfaces of C/C-SiC composites, blocking direct contact between the materials and providing better frictional stability and a lower wear rate. C/C-SiC composites, with their extended service life, have been successfully used in high-speed and high-energy load braking systems such as in engineering machinery, military aircraft, helicopters, and maglev trains [43–45].

2.5.4 Wear Mechanisms of Cu-MMCs

Cu-MMCs are influenced not only by variations in braking parameters, such as speed and load, but also by the service environment, including high temperatures, extreme cold, and humidity. These factors can significantly impact the performance of Cu-MMCs. Consequently, the wear mechanism of Cu-MMCs may change depending on the specific conditions they encounter.

To study the friction and wear mechanisms of Cu-MMCs, the concept of the "third body" is introduced. The third body refers to the friction film or transfer film formed from grinding debris on the friction surface. The braking stability of Cu-MMCs is closely linked to the generation and consumption of this third body on the surface. When the third body is reduced, the material surface lacks the protection of the friction film, leading to unstable friction coefficients and increased wear. Higher braking pressure promotes the formation of a dense third body [35]. At low initial test speeds, adhesive wear on the surface increases with the increase of pressure, while oxidative wear is minimal. At high initial test speeds, increased pressure results in a thicker oxide film, reducing adhesive wear but increasing oxidative wear [46].

According to the classical molecular-tribology theory [47], the friction force in the process of friction is the sum of the mechanical force (mechanical component) and molecular adhesion force (molecular component) between two contact surfaces. The tribological properties of materials are not the intrinsic or inherent properties of materials, but they have a great correlation with the characteristics of friction contact surfaces, which depend on the characteristics of friction pair materials and friction conditions. Braking speed and high-temperature environments primarily alter the surface temperature of Cu-MMCs, thereby affecting the friction and wear mechanism of Cu-MMCs. At low braking speeds, the surface temperature remains low, making it difficult for thermal effects to cause significant oxidation or softening of the friction material. Consequently, the material transfer is not obvious, and a third layer on the friction surface is hard to form. The stable friction surface remains similar to the original surface of Cu-MMCs, with wear mechanisms predominantly involving slight oxidation, adhesive wear, and abrasive wear [48]. As braking speed increases, the surface temperature of friction materials increases, causing significant oxidation of the friction material. This results in the surface material being extensively oxidized, and the contact surfaces are almost oxidized to high valence states. Oxides and wear debris are accumulated and compacted on the friction surface of Cu-MMCs. A mechanical mixing layer (MML) gradually forms, effectively isolating metal-to-metal contact and reducing adhesion between the contact surfaces. As the friction proceeds, the mechanical mixing layer is gradually peeled off, with tiny granular wear debris collecting in stripping pits to form new wear debris [49], as shown in Figure 2.14.

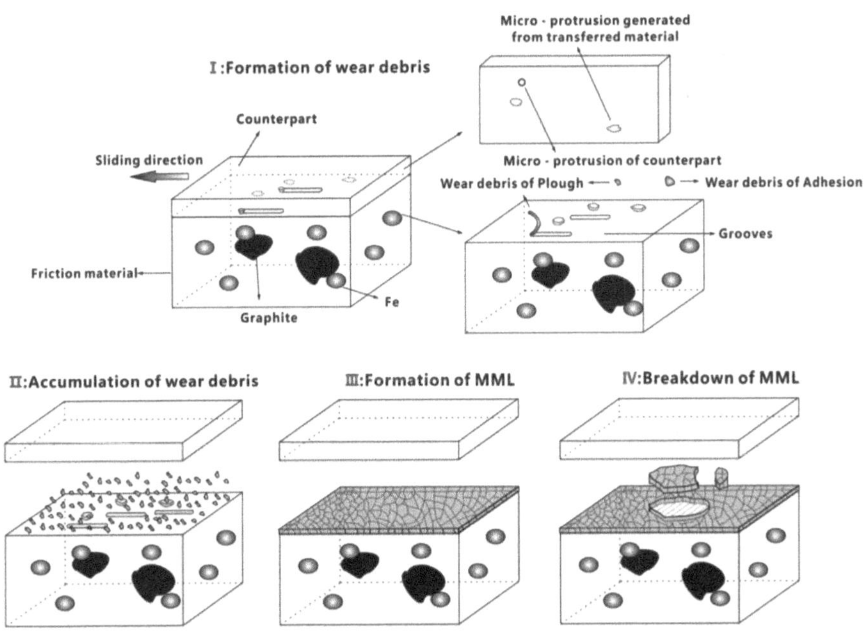

FIGURE 2.14 Schematic diagram of wear mechanism [49].

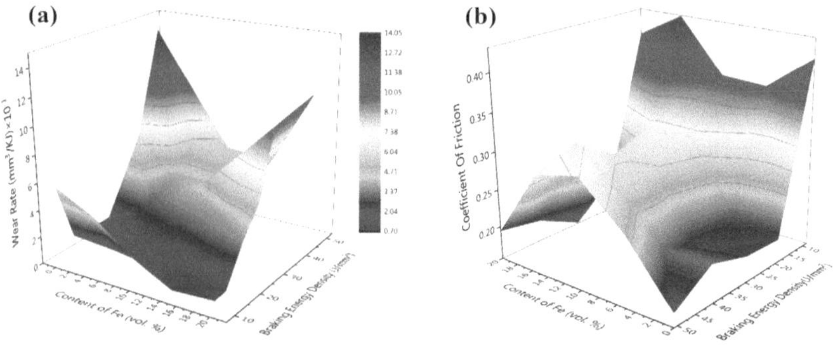

(a) Wear rate (b) Coefficient of friction

FIGURE 2.15 Three-dimensional maps of the wear rate (a) and mean friction coefficient (b) in terms of Fe content and braking energy [49].

The friction surface of Cu-MMCs is covered with a dense and continuous mechanical mixing layer, leading to a more stable friction curve. While this mechanical mixing layer helps to mitigate damage to Cu-MMCs, it also results in a lower friction coefficient. Consequently, Cu-MMCs exhibit the lower friction coefficient and wear rate.

The humidity in the environment also influences the wear mechanism of Cu-MMCs. Under wet friction conditions, at low speeds and pressures, water acts as a lubricant. The lubrication and flow of the water film can reduce the friction coefficient and increase the wear rate. However, at high speeds and pressures, water serves primarily as a coolant, reducing the softening of the copper matrix. The evaporation of water and the centrifugal effects of high temperatures can disrupt the water film, making the friction and wear performance of Cu-MMCs comparable to that in dry friction conditions [51]. High-speed trains often encounter rain and snow during operation. In cold, rainy, or snowy environments, condensation and ice can form on the surface of Cu-MMCs, altering the friction interface and impacting the wear mechanism [52]. When ice melts during braking, hard particles trapped in the ice can damage the friction surface, generating debris. As braking pressure and temperature rise, this debris accumulates. Newly generated debris adheres to existing debris and can become trapped between the brake pad and the brake disc, leading to increased wear of Cu-MMCs [53]. The primary wear mechanisms for Cu-MMCs in these conditions include spalling wear, adhesive wear, and abrasive wear. As the temperature decreases, spalling damage initially intensifies before eventually weakening, while adhesive damage shows a pattern of weakening followed by strengthening [54].

At various temperatures, the friction surface of Cu-MMCs displays distinct characteristics due to the complex physical and chemical changes caused by temperature and friction, leading to notable differences in tribological performance. At room temperature, the thermal effects generated during friction are insufficient to induce oxidation or softening in the friction material, and the material transfer is not obvious.

Consequently, a third body layer is rarely formed, and the friction surface remains largely similar to the original surface of Cu-MMCs. Many intact graphite particles are visible on the friction surface, which, due to their lubricating properties, results in a smooth friction curve and relatively low wear rate. At room temperature, the wear mechanisms of Cu-MMC are predominantly mild oxidation, adhesive wear, and abrasive wear [48]. However, when the temperature increases to 600°C, substantial heat energy accelerates the softening and oxidation of the friction material. The friction surface experiences partial oxidation, forming both low- and high-valence oxides. Graphite particle oxidation is exacerbated, leading to deeper and wider grooves on the friction surface. Adhesive point formation and rupture become more frequent, and the peeling phenomenon intensifies.

The material transfer from the friction surface of Cu-MMCs to the brake disc results in the formation of adhesive points that are subsequently sheared off during friction [50]. At 600°C, the third body layer on the friction surface of Cu-MMCs is not obvious, exposing a significant area of Cu-MMCs. This exposure leads to a rougher and more irregular friction surface, causing increased wear and reduced frictional stability. As the temperature rises to 800°C, a large amount of heat energy promotes severe oxidation of the friction material surface, converting it almost entirely to a high-valence state. This exacerbates softening, making the contacting materials more susceptible to shearing and rapid oxidation. Compared to brake discs, the friction surface of Cu-MMCs is softer, which facilitates the accumulation of oxides, wear debris, and other materials, forming a friction film. This film isolates the metal-to-metal contact, reducing adhesion between two contact surfaces. As the friction proceeds, the thickness of the third body layer gradually increases, covering the friction surface of Cu-MMCs with a dense, continuous mechanical mixing layer, leading to a more stable friction curve. However, while this layer decreases damage to Cu-MMC, it also reduces the friction coefficient, resulting in lower friction coefficients and wear rates of Cu-MMC.

Friction and wear maps can provide an overall framework for tribological behavior, offering a comprehensive elaboration of the effects of various conditions and material components on tribological properties of Cu-MMCs. For instance, variations in component content within Cu-MMCs lead to differing friction and wear properties. Figure 2.15 illustrates three-dimensional maps of the wear rate and average friction coefficient as functions of iron content and braking energy. Mapping friction and wear behavior as a function of material composition and operating conditions is a valuable tool, assisting in material selection for component design.

2.6 SUMMARY AND PERSPECTIVE

With the continuous advancement of rail transportation, high-speed trains have become increasingly popular, with expectations of speed, comfort, and safety deeply ingrained in public consciousness. Brake friction materials are vital to high-speed trains braking systems, as they convert the kinetic energy of trains into heat energy to facilitate braking. This chapter delves into the research on brake friction materials for high-speed trains, covering their classification, tribological properties, and practical applications. Cu-MMCs are extensively utilized in high-speed trains. This chapter discusses in detail the influence of matrix components, lubricant components, and friction components on the friction performance of Cu-MMCs, along with proposed optimizations. Additionally, the compatibility of Cu-MMCs with various brake disc materials is a key factor affecting braking performance. This chapter summarizes the evolution of friction performance in Cu-MMCs under various braking conditions and reviews techniques for predicting the service life of Cu-MMCs. Future research should focus on exploring new materials and technologies that enhance braking performance while reducing environmental impacts.

REFERENCES

[1] Xiao J K, Xiao S X, Chen J, et al. Wear mechanism of Cu-based brake pad for high-speed train braking at speed of 380 km/h. *Tribology International*, 2020, 150: 106357.
[2] Xiao Y, Zhang Z, Yao P, et al. Mechanical and tribological behaviors of copper metal matrix composites for brake pads used in high-speed trains. *Tribology International*, 2018, 119: 585–592.
[3] Sharma R C, Dhingra M, Pathak R K. Braking systems in railway vehicles. *International Journal of Engineering Research & Technology*, 2015, 4(1): 206–211.
[4] De La Torre S, Sánchez-Racero A J, Aguado J A, et al. Optimal sizing of energy storage for regenerative braking in electric railway systems. *IEEE Transactions on Power Systems*, 2015, 30(3): 1492–1500.
[5] Wang P J, Chiueh S J. Analysis of eddy-current brakes for high speed railway. *IEEE Transactions on Magnetics*, 1998, 34(4): 1237–1239.
[6] Bijwe J, Nidhi, Majumdar N, et al. Influence of modified phenolic resins on the fade and recovery behavior of friction materials. *Wear*, 2005, 259(7–12): 1068–1078.
[7] Cai P, Wang Y, Wang T, et al. Improving tribological behaviors of friction material by mullite. *Tribology International*, 2015, 93: 282–288.
[8] Windhorst T, Blount G. Carbon-carbon composites: A summary of recent developments and applications. *Materials and Design*, 1997, 18(1): 11–15.
[9] Farhan S, Li K, Guo L, et al. Effect of density and fibre orientation on the ablation behaviour of carbon–carbon composites. *New Carbon Materials*, 2010, 25(3): 161–167.
[10] Li G, Yan Q, Xi J, et al. The stability of the coefficient of friction and wear behavior of C/C–SiC. *Tribology Letters*, 2015, 58: 13.
[11] Ram Prabhu T, Varma V K, Vedantam S. Effect of reinforcement type, size, and volume fraction on the tribological behavior of Fe matrix composites at high sliding speed conditions. *Wear*, 2014, 309(1–2): 247–255.
[12] Shen M, Li H, Du J, et al. New insights into reducing airborne particle emissions from brake materials: Grooved textures on brake disc surface. *Tribology International*, 2022, 174: 107721.
[13] Su L, Gao F, Han X, et al. Effect of copper powder third body on tribological property of copper-based friction materials. *Tribology International*, 2015, 90: 420–425.
[14] Rodrigues A C P, Österle W, Gradt T, et al. Impact of copper nanoparticles on tribofilm formation determined by pin-on-disc tests with powder supply: Addition of artificial third body consisting of Fe_3O_4, Cu and graphite. *Tribology International*, 2017, 110: 103–112.

[15] Xiong X, Chen J, Yao P, et al. Friction and wear behaviors and mechanisms of Fe and SiO_2 in Cu-based P/M friction materials. *Wear*, 2007, 262(9–10): 1182–1186.

[16] Tjong S C, Lau K C. Abrasive wear behavior of TiB_2 particle-reinforced copper matrix composites. *Materials Science and Engineering A*, 2000, 282(1–2): 183–186.

[17] Shirvanimoghaddam K, Hamim S U, Karbalaei Akbari M, et al. Carbon fiber reinforced metal matrix composites: Fabrication processes and properties. *Composites Part A: Applied Science and Manufacturing*, 2017, 92: 70–96.

[18] Trinh P Van, Trung T B, Thang N B, et al. Calculation of the friction coefficient of Cu matrix composite reinforced by carbon nanotubes. *Computational Materials Science*, 2010, 49(4): S239–S241.

[19] Chen F, Ying J, Wang Y, et al. Effects of graphene content on the microstructure and properties of copper matrix composites. *Carbon*, 2016, 96: 836–842.

[20] Pérez B, Echeberria J. Influence of abrasives and graphite on processing and properties of sintered metallic friction materials. *Heliyon*, 2019, 5(8): e02311.

[21] Zhang P, Zhang L, Wei D, et al. Effect of graphite type on the contact plateaus and friction properties of copper-based friction material for high-speed railway train. *Wear*, 2019, 432–433: 202927.

[22] Zhang P, Zhang L, Fu K, et al. Effects of Ni-coated graphite flake on braking behavior of Cu-based brake pads applied in high-speed railway trains. *Journal of Tribology*, 2019, 141(8): 081301.

[23] Chen B, Bi Q, Yang J, et al. Tribological properties of solid lubricants (graphite, h-BN) for Cu-based P/M friction composites. *Tribology International*, 2008, 41(12): 1145–1152.

[24] Jin Z, Chao M A, Xiao K, et al. Effect of WS_2 particle size on mechanical properties and tribological behaviors of Cu-WS_2 composites sintered by SPS. *Transactions of Nonferrous Metals Society of China*, 2018, 28(6): 1176–1185.

[25] Peng T, Yan Q, Zhang Y, et al. Low-cost solid FeS lubricant as a possible alternative to MoS_2 for producing Fe-based friction materials. *International Journal of Minerals, Metallurgy and Materials*, 2017, 24: 115–121.

[26] Hu N, Zhang X, Wang X, et al. Study on tribological properties and mechanisms of different morphology WS_2 as lubricant additives. *Materials*, 2020, 13(7): 1522.

[27] Zhang P, Zhang L, Fu K, et al. The effect of Al_2O_3 fiber additive on braking performance of copper-based brake pads utilized in high-speed railway train. *Tribology International*, 2019, 135: 444–456.

[28] Markov D, Kelly D. Mechanisms of adhesion-initiated catastrophic wear: Pure sliding. *Wear*, 2000, 239(2): 189–210.

[29] Zhan Y, Zhang G. The effect of interfacial modifying on the mechanical and wear properties of SiC_p/Cu composites. *Materials Letters*, 2003, 57(29): 4583–4591.

[30] Zhang P, Zhang L, Fu K, et al. Effects of different forms of Fe powder additives on the simulated braking performance of Cu-based friction materials for high-speed railway trains. *Wear*, 2018, 414–415: 317–326.

[31] Peng T, Yan Q, Li G, et al. The influence of Cu/Fe ratio on the tribological behavior of brake friction materials. *Tribology Letters*, 2018, 66: 18.

[32] Cheng Y, Xiao Y, Du J, et al. Size effect of CrFe particles on tribological behavior and airborne particle emissions of copper metal matrix composites. *Tribology International*, 2023, 183: 108376.

[33] Fan J, Zhang C, Wu S, et al. Effect of Cr-Fe on friction and wear properties of Cu-based friction material. *Materials Science and Technology*, 2018, 34(7–8) : 869–875.

[34] Peng T, Yan Q, Li G, et al. The braking behaviors of Cu-based metallic brake pad for high-speed train under different initial braking speed. *Tribology Letters*, 2017, 65: 135.
[35] Zhang P, Zhang L, Fu K X, et al. Fade behaviour of copper-based brake pad during cyclic emergency braking at high speed and overload condition. *Wear*, 2019, 428–429: 10–23.
[36] Wang Z, Han J, Domblesky J P, et al. Crack propagation and microstructural transformation on the friction surface of a high-speed railway brake disc. *Wear*, 2019, 428–429: 45–54.
[37] Xiao Y, Cheng Y, Shen M, et al. Friction and wear behavior of copper metal matrix composites at temperatures up to 800°C. *Journal of Materials Research and Technology*, 2022, 19: 2050–2062.
[38] Lu C, Jiang X, Chen X, et al. Experimental study on the evolution of friction and wear behaviours of railway friction block during temperature rise under extreme braking conditions. *Engineering Failure Analysis*, 2022, 141: 106621.
[39] Zhang P, Zhang L, Wei D, et al. A high-performance copper-based brake pad for high-speed railway trains and its surface substance evolution and wear mechanism at high temperature. *Wear*, 2020, 444–445: 203182.
[40] Zhang P, Zhang L, Wei D, et al. Substance evolution and wear mechanism on friction contact area of brake disc for high-speed railway trains at high temperature. *Engineering Failure Analysis*, 2020, 111: 104472.
[41] Wu S C, Zhang S Q, Xu Z W. Thermal crack growth-based fatigue life prediction due to braking for a high-speed railway brake disc. *International Journal of Fatigue*, 2016, 87: 359–369.
[42] Chen F, Li Z, Luo Y, et al. Braking behaviors of Cu-based PM brake pads mating with C/C–SiC and 30CrMnSi steel discs under high-energy braking. *Wear*, 2021, 486–487: 204019.
[43] Xiao P, Li Z, Zhu Z, et al. Preparation, properties and application of C/C-SiC composites fabricated by warm compacted-in situ reaction. *Journal of Materials Science and Technology*, 2010, 26(3): 283–288.
[44] Li Z, Xiao P, Xiong X, et al. Preparation and tribological properties of C fibre reinforced C/SiC dual matrix composites fabrication by liquid silicon infiltration. *Solid State Sciences*, 2013, 16: 6–12.
[45] Li Z, Liu Y, Zhang B, et al. Microstructure and tribological characteristics of needled C/C-SiC brake composites fabricated by simultaneous in filtration of molten Si and Cu. *Tribiology International*, 2016, 93: 220–228.
[46] Zhang X, Zhang Y, Du S, et al. Influence of braking conditions on tribological performance of copper-based powder metallurgical braking material. *Journal of Materials Engineering and Performance*, 2018, 27: 4473–4480.
[47] Goryacheva I G, Dobychin M N. Some results concerning the development of the molecular-mechanical theory of friction. *Journal of Friction and Wear*, 2008, 29: 243–250.
[48] Khruschov M M. Principles of abrasive wear. *Wear*, 1974, 28(1): 69–88.
[49] Zhou H, Yao P, Xiao Y, et al. Friction and wear maps of copper metal matrix composites with different iron volume content. *Tribology International*, 2019, 132: 199–210.
[50] Wang H, Ma G, Xu B, et al. Microstructure and vacuum tribological properties of 1Cr18Ni9Ti steel with combined surface treatments. *Surface and Coatings Technology*, 2011, 205(11): 3546–3552.

[51] Eriksson M, Lundqvist A, Jacobson S. A study of the influence of humidity on the friction and squeal generation of automotive brake pads. *Proceedings of the Institution of Mechanical Engineers, Part D: Journal of Automobile Engineering*, 2001, 215(3): 329.

[52] Olofsson U, Sundh J, Bik U, et al. The influence of snow on the tread braking performance of a train: A pin-on-disc simulation performed in a climate chamber. *Proceedings of the Institution of Mechanical Engineers, Part F: Journal of Rail and Rapid Transit*, 2016, 230(6): 1521–1530.

[53] Wu Y, Liu Y, Chen H, et al. An investigation into the failure mechanism of severe abrasion of high-speed train brake discs on snowy days. *Engineering Failure Analysis*, 2019, 101: 121–134.

[54] Ma L, Ding S, Zhang C, et al. Study on the wear performance of high-speed railway brake materials at low temperatures under continuous braking conditions. *Wear*, 2023, 512–513: 204556.

3 Tribological Behavior of Wheel–Rail Contact under Extreme Service Conditions

Mingxue Shen and Yelong Xiao

3.1 INTRODUCTION

Wheel and rail achieve the guidance, support, traction, and braking of trains through rolling friction contact, which is a key core component for safe train operation. The safety, punctual operation, and precise control of train motion are closely related to the adhesion effect and contact state of the wheel–rail contact interface. The quality of the train operation is closely related to the rolling contact state of the wheel and rail. When rolling contact occurs, the shape of the contact spot is approximately oval, and its area is equivalent to the size of the nail clipper [1], as shown in Figure 3.1. The contact spot with such a small area needs to provide sufficient normal force to support several tons or even tens of tons of axle load during train operation, as well as the tangential force required to realize train traction and braking. Therefore, wheel–rail rolling friction contact is a very complicated problem.

In China, where high-speed railways are extensively developed, the characteristics of cross-regional and long-distance operation determines the complexity and particularity of the high-speed railway operating environment, and also brings considerable challenges to the service performance of high-speed railway wheels and rails [2]. Taking the Qinghai–Tibet Railway and the Southern Xinjiang Railway in China as examples, the high-speed railways often operate under extreme cold climates (the lowest temperature can reach −45°C) or dry and hot climates (the surface temperature of the rails for the Southern Xinjiang Railway can be 60°C) [3]. Besides, the climate inside tunnels (for example, along the Sichuan–Tibet Railway) is an extremely hot and humid condition, while the ambient temperature outside can be extremely low. This means that within a short time the trains will undergo alterations between severe cold and extreme heat during operation. Significantly, the interface friction and damage failure for the wheel–rail rolling-sliding contact, which is a complicated open-system tribology, are strongly disturbed by the external environment [4]. To make the wheel–rail contact system more reliable, the impact of service temperatures on the tribology behavior of wheel–rail needs to be systematically understood.

 DOI: 10.1201/9781003386223-3

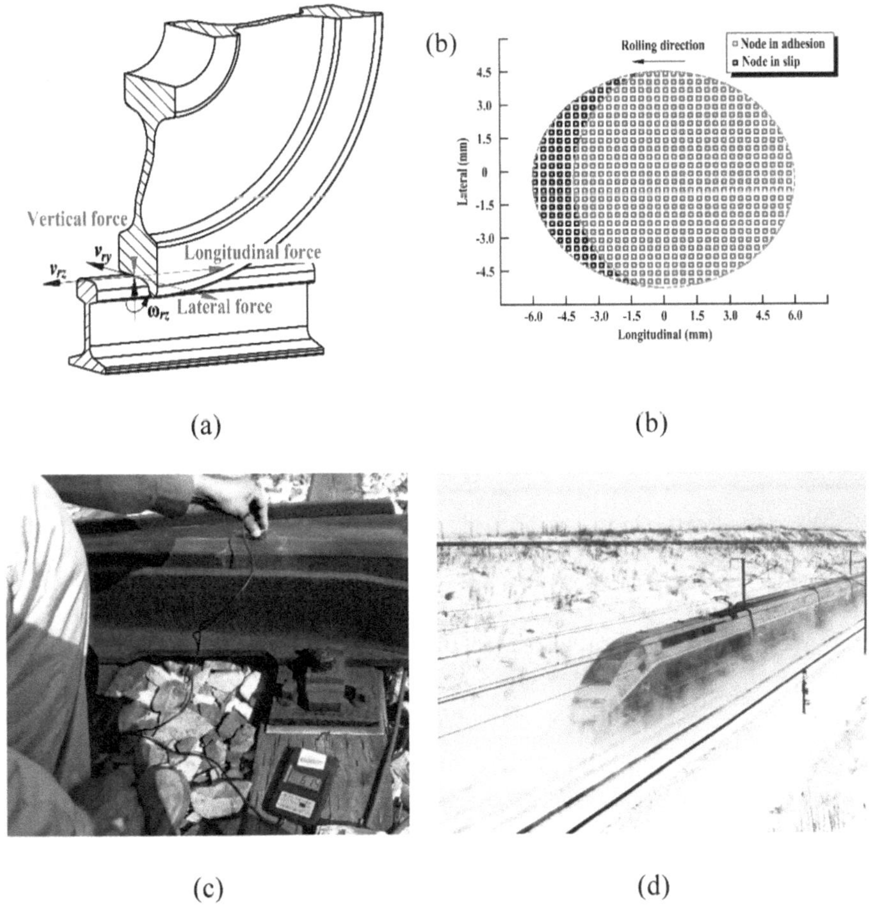

FIGURE 3.1 Schematic diagram of wheel–rail interface, and workers measure the temperature of the rail surface in summer: (a) schematic diagram of wheel–rail contact, (b) sticking-slip region at the contact location, (c) the workers measure the temperature of the rail surface in summer, and (d) a train on snow-covered tracks.

The wheel–rail system is always in an open environment, and the contact and service behavior of wheel–rail are strongly interfered with by the external environment. At present, China has built the world's fastest high-cold railways – the Harbin-Dalian High-speed Railway, the highest altitude and longest plateau railway; the Qinghai-Tibet Railway and the Lanzhou-Xinjiang High-speed Railway that cross desert sandstorm areas; and the tropical high-temperature Hainan high-speed railway that operates around the island; and the Sichuan-Tibet Railway, which is under construction. The environment along above lines are extremely harsh, which makes the service of wheel and rail need to overcome complex environmental conditions such as high temperature, high humidity, high cold, acid rain corrosion, and sandstorms. The climate environment will have a serious impact on the service performance of wheel–rail materials. Temperature, humidity, and corrosive environment are important factors affecting the adhesion characteristics of wheel–rail contact interface, and their changes will have an important impact on the wear of wheel–rail materials and the traction/braking performance of trains, greatly threatening the safety of train operation.

3.2 WHEEL–RAIL INTERACTION AND ITS INTERFACIAL BEHAVIOR

3.2.1 Wheel–Rail Interaction

The research of the wheel–rail relationship involves the knowledge of rigid body motion geometry, mechanical design, vehicle dynamics, tribology, solid mechanics (rolling contact mechanics, elastic-plastic theory, fracture mechanics and finite element, etc.), vibration and noise, thermodynamics and materials science, and so forth.

In the process of railway transportation, the traction force is transferred by the adhesion creep between the wheel and rail contact spots. Therefore, wheel–rail interface adhesion characteristics are the most direct and major factors affecting train traction and braking. Interface adhesion behavior directly affects many aspects, such as the utilization of locomotive traction power, wheel and rail surface damage, train running stability and train braking. Wheel-rail interaction directly causes vibration, impact, and damage to both vehicles and rail systems. It is also the primary source of deterioration and functional loss in wheel-rail systems. The creep force between wheel and rail is transferred upward to the locomotive and acts downwards on the track structure, resulting in relatively complex vibration forms. These vibrations are coupled with each other, which can cause scratches on the wheel–rail contact surface and, in severe cases, may become an inducing factor for rail breakage. This is shown in Figure 3.2. Abrasions on wheels and rails increase maintenance costs, but also lead to elastic deformation of wheels and rails, as well as plastic deformation in the wheel–rail contact area, which ultimately endangers the safe operation of trains [5].

FIGURE 3.2 Rail surface damage.

3.2.2 Adhesion of Wheel–Rail Interface

3.2.2.1 Adhesion Coefficient between Wheel and Rail

Adhesion coefficient is a crucial indicator of the adhesion condition between wheel and rail interface, which represents the possible degree of traction or braking force transmitted by the vehicle to the rail. Specifically, it is the maximum ratio of the tangential force in the circumference of the wheel to the vertical load of the wheel. Wheel–rail interface adhesion is a dynamic random variable that constantly changes with different service environments, times and locations. The adhesion coefficient between the wheel and the rail is defined as the ratio of the tangential force to the normal force of the wheel, as shown in Formula 3.1. Under normal working conditions, the adhesion coefficient should be less than or equal to the friction coefficient between the wheel and the rail.

$$\mu = \frac{F_T}{F_N} = \frac{T_N}{F_N \cdot r} \tag{3.1}$$

Where: μ is the adhesion coefficient between wheel and rail; F_T is the shear force; F_N is the normal load; F_N is torque; r is the radius of the rail sample.

The maximum possible adhesion in the wheel–rail contact area is called available adhesion. The maximum possible adhesive force F_{max} corresponds to the peak value of wheel traction force F in Figure 3.3. Therefore, the ratio of F_{max} to static load F_N is defined as the available adhesion coefficient μ_{max}.

$$\mu = \frac{F_{max}}{F_N} \tag{3.2}$$

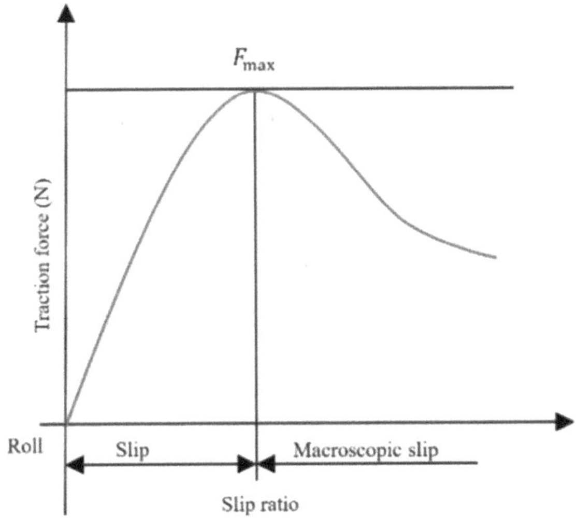

FIGURE 3.3 Relationship between traction force of wheel circumference and creep rate.

3.2.2.2 Low Adhesion of Wheel and Rail and Its Hazards

Wheel–rail rolling contact is not only a dynamic process, but also a very complex tribological problem, involving the influence of many factors such as physics and chemistry. The main factors affecting wheel–rail adhesion are: (1) surface condition and environmental factors; (2) the quality of ballast bed; (3) wheel creep; (4) traction motor power, power supply and control system; (5) axle load; (6) speed; (7) the geometric shape of the ballast bed, wheel, and rail; (8) metallography, materials, and hardness; (9) vehicle dynamic effect.

An extremely harsh service environment can easily lead to the low adhesion between wheel and rail. This problem can result in insufficient traction and longer braking distance of the train while, in severe cases, it can lead to serious safety accidents such as trains running into signals, rushing out of platforms, or colliding with each other. Additionally, idling, skidding, or sliding due to low adhesion may cause significant damage to the wheel tread or rail head. It is reported that the annual loss of British railways due to low adhesion is about 50 million pounds sterling, while Swedish railways is about 100 million Swedish kronor; in the Netherlands, the low adhesion of wheels and rails caused by rainwater once paralyzed most of the lines for a day.

3.2.3 Wheel–Rail Tribology in Service Environment

Olofsson et al. [2] pointed out that the cause of low adhesion of the local wheel–rail in the month of October was fallen leaves along the railway. Under the repeated rolling of the wheel and rail, fallen leaves form a black oxide film with a certain thickness, which is the fundamental reason for the low adhesion of the wheel and rail. Chen et al. [6] confirmed in their recent studies that the formation of black oxide film depends on environmental humidity (such as condensation, frost, etc.), and that the influence of fallen leaves on wheel–rail adhesion is relatively weak in dry conditions, this is shown in Figure 3.4 (a) and (b); in turn, the alternating change of cold seasonal temperatures leads to the adsorption of water on the rail surface, which can dissolve the acid in fallen leaves and chemically react with the rail to form oxides. In addition, it is pointed out that how to restrain the formation of black oxide film and in a timely way remove it is an effective countermeasure to prevent railway wheel–rail skid, especially in mountainous areas. Environmental humidity and oxidation products are the main culprits of low adhesion of wheel and rail, and which have a profound influence on their service behavior. Numerous studies have shown that various pollutants on wheel–rail contact surfaces will have varying degrees of impact on wheel–rail adhesion. The water and oil media on the rail surface can significantly reduce the adhesion coefficient. Especially small amounts of water and oil can form a very thin paste, resulting in extremely low adhesion coefficient. British scholar White et al. pointed out through the data analysis of actual operating lines that low adhesion accidents caused by non-leaf conditions often occurred in the morning (06:00 ~ 09:00), when dewdrops are prone to form on the rail heads, as shown in Figure 3.4 (c), but the specific critical moisture level could not be verified at first. Japanese scholar Nakahara et al. reported on the relationship between adhesion characteristics of trains

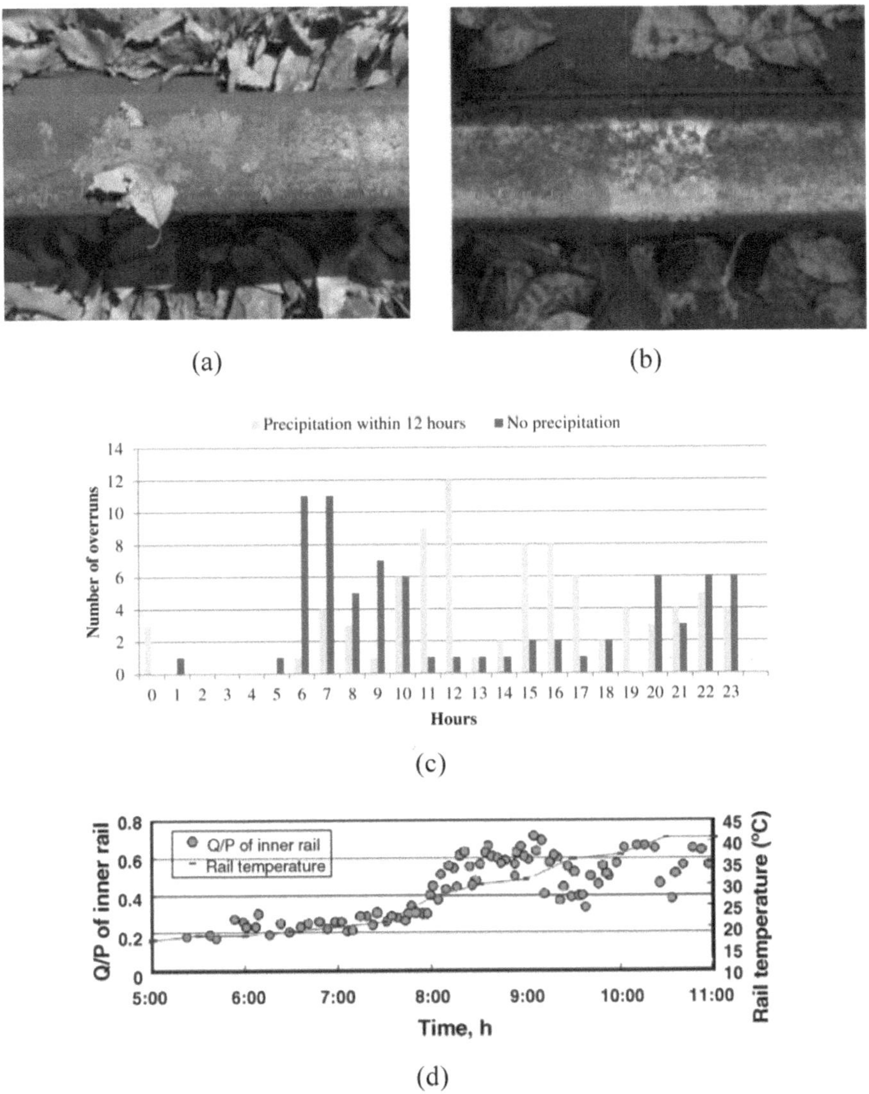

FIGURE 3.4 Influence of different ambient humidity on wheel–rail adhesion [11], which is closely related to ambient temperature and humidity: (a) dry environment; (b) humid environment; (c) the number of accidents caused by low wheel–rail adhesion in different time periods; (d) wheel–rail adhesion coefficient and rail temperature at different time periods.

during different running periods, pointing out that the adhesion coefficient in morning period was close to 0.2, as shown in Figure 3.4(d), and believing that low adhesion was caused by more residual oxides on rail surface in morning hours. In addition, environmental humidity is also the main factor affecting wheel–rail low adhesion, which may have a greater impact than rainwater on wheel–rail adhesion. Moreover, friction oxidation products at wheel–rail interface and humidity have a synergistic effect on reducing wheel–rail adhesion performance. Therefore, studying the adhesion and damage at the wheel rail interface requires a comprehensive consideration of the effects of environmental humidity and oxidation products.

In on-site vehicle testing and research, it was found that sanding or using other hard particles when low adhesion occurs is an effective measure to improve the wheel–rail adhesion coefficient and avoid wheel skid [7, 8]. In addition, other studies have found that a large number of sand particles entering the wheel–rail contact interface in dry conditions will produce a solid lubrication effect, resulting in a decrease of wheel–rail adhesion coefficient. However, the addition of sand inevitably affects the tribological behavior of wheel–rail interface [9–11].

3.2.4 Common Research Methods in Wheel–Rail Tribology

Currently, research methods in wheel–rail tribology include experimental methods and numerical simulation methods. Among them, the experimental methods include small-sized wheel and rail pin-disk test, small-sized double-disk test, full-sized simulation test, and line test. However, wheel–rail line tests and full-scaled experiments are difficult to be popularized in practice because of their long testing cycles and high costs. In contrast, the small-sized pin-disk test and small-sized double-disk test have been favored by domestic and foreign researchers in wheel/rail tribology because of their advantages of better control of variables and lower costs.

The wheel–rail pin-disk testing machine has a simple structure, convenient operation, and can easily add a third body medium to the test process, so it is often used to study the influence of the third body medium on wheel–rail adhesion in tribology, as shown in Figure 3.5(a). The wheel–rail double-disc testing machine is another small-size experimental machine commonly used in the research of wheel–rail tribology. It simulates the rolling sliding contact of the wheel–rail through two small diameter rollers, which are directly processed from the on-site wheel tread and rail head. Compared with the full-size experiments, the double-disc experiment not only reduces the sample production costs, but also facilitates the micro-characterization analysis of the wheel–rail sample after the experiment. In addition, the double-disc testing machine simulates wheel–rail tests under different conditions by adding environmentally controllable modules (such as external equipment such as refrigeration, heating, and humidity generator), as shown in Figure 3.5(b). A full-size wheel–rail experiment is a wheel–rail tribology test with the same size as the on-site wheel–rail in the laboratory, one which has good simulation performance and can simulate the wheel–rail contact relationship 1:1. This method is generally used for dynamic performance tests and wheel–rail adhesion tests and other dynamic data testing, as shown in Figure 3.5(c).

In 2010, the China Academy of Railway Sciences and RENK Test System jointly developed a 500 km/h wheel-straight rail test bench, with a track wheel diameter of 3 m, which is closer to the real wheel–rail relationship. This testing machine is composed of a track wheel system, a test wheelset system, a hydraulic vibration excitation system, a track contact interface environment simulation system, a track wheel profile numerical control repair device, a high-pressure hydraulic supply unit,

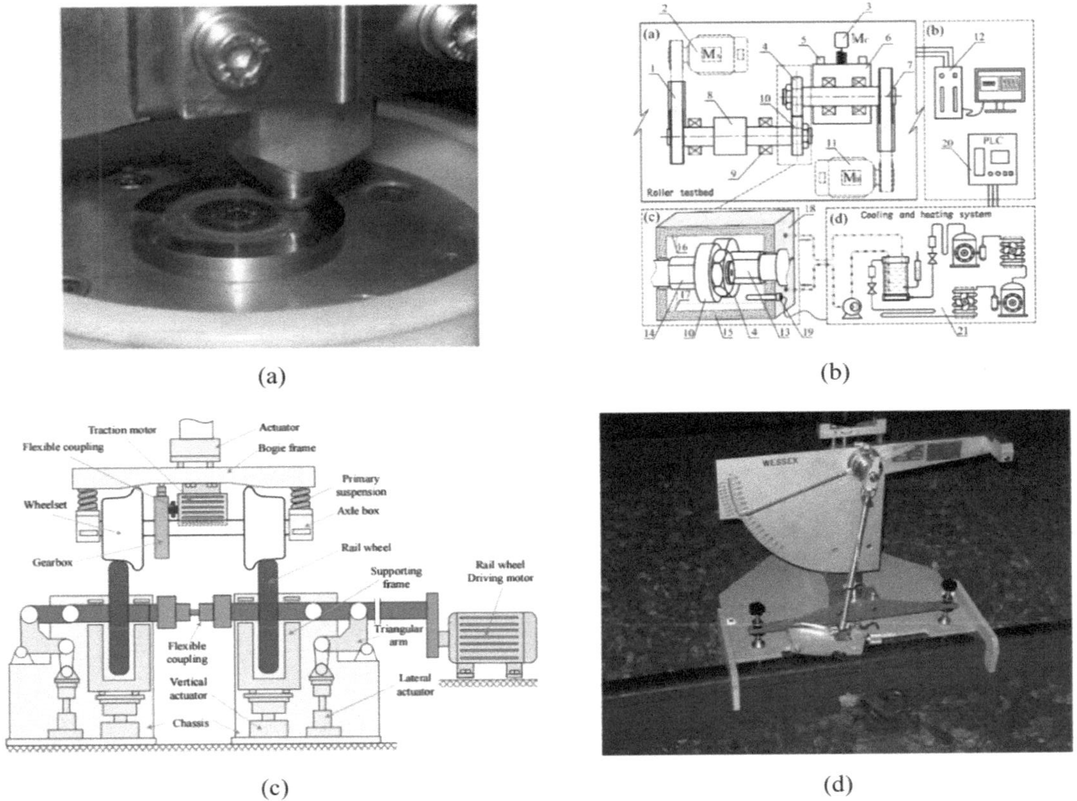

FIGURE 3.5 (a) Pin-disc testing machine [12]; (b) Sub-sized double-disc tester with cooling and heating system [13]; (c) 1:1 wheel-rail rolling test device [14]; (d) Field investigation [15].

a lubrication unit, electrical equipment, a measurement and data acquisition system, control system and other parts. Tests on adhesion, creep, motion stability, wear, contact fatigue, derailment, noise of high-speed, and heavy-duty wheel–rail relationships can be systematically carried out.

In addition, a pendulum friction tester is also a common method to test the rail adhesion of the line, which calculates the friction coefficient of the tested plane by measuring energy loss. The tester has a simple structure and strong applicability, making it one of the methods for quickly evaluating adhesion. Its structure is similar to that of a pendulum impact tester, but only a short distance (127 mm) of the rail can be measured each time, and the measured data is the friction coefficient in the pure sliding state, which cannot completely simulate the wheel–rail contact state, as shown in Figure 3.5(d).

3.3 WHEEL–RAIL TRIBOLOGY IN A WIDE TEMPERATURE RANGE ENVIRONMENT

3.3.1 Response to Interface Adhesion

With the expansion and extension of the railway network, more and more railways are being laid in hot and cold environments. For example, the minimum service temperature of the Harbin-Dalian High-speed Railway in winter is lower than −45°C, while in summer, when the outdoor temperature is 35°C, the maximum temperature of the rail surface can reach 65°C [16]. The Lanzhou-Xinjiang Line has a continental climate, which is dry, with little precipitation throughout the year, hot in summer and cold in winter. The seasonal temperature difference and the temperature difference between day and night are large, which makes the adhesion of the train during operation quite different.

The variation curve of adhesion coefficient of wheel–rail rolling-sliding contact under different ambient temperatures (−50 ~ 60°C) [17], as shown in Figure 3.6. It is found from the figure that as the temperature decreases, the adhesion coefficient value shows an upward trend. The main reason is that atmospheric humidity (including absolute humidity and relative humidity) decreases significantly with the decrease of temperature. The reduction of moisture in the operating environment can significantly inhibit the friction oxidation behavior between wheel and rail interfaces, further leading to the lack of lubrication between metal oxides and water at the contact interface, resulting in the increase of adhesion coefficient. On the contrary, with the increase of temperature and ambient humidity, the oxidation reaction at the friction interface becomes more intense, and the third-body lubricant generated on the contact surface will reduce the adhesion coefficient. In addition, the investigation found [18] that the adhesion coefficient is related to the oxides generated at the wheel–rail contact interface at different temperatures. At low temperature, due to the reduction of oxidation, the adhesion coefficient of the wheel–rail is higher than that at room temperature.

In addition, through the test of different temperatures had found that [19], the adhesion coefficient values will change with the temperature change during the test.

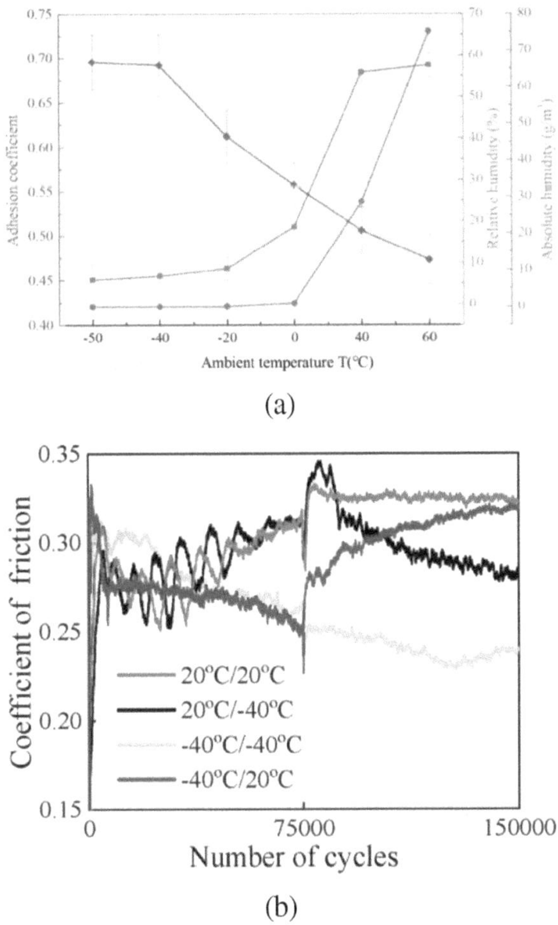

FIGURE 3.6 (a) Evolution of mean adhesion coefficient of wheel/rail with ambient temperature [17]; (b) COF evolution curves as a function of number of cycles [19].

In Figure 3.6(b), the changing curve of adhesion coefficient have been exhibited, the four changing curves of the adhesion coefficient represent the constant temperature of 20°C, the condition of adhesion coefficient reduced from 20°C to −40°C at 75,000 cycles, the constant temperature of −40°C and the condition of adhesion coefficient increased from −40°C to 20°C at 75,000 cycles, respectively. At a constant room temperature of 20°C, the adhesion coefficient showed an upward trend and stabilized at about 0.31 after 75,000 cycles. When the test temperature decreased from 20°C to −40°C, the adhesion coefficient gradually decreased and finally stabilized at about 0.27. Under the constant temperature condition of −40°C, the adhesion coefficient continues to decline, and it can be seen from the figure that the adhesion coefficient does not tend to be stable under this condition. However, as the temperature rises from −40°C to 20°C, the adhesion coefficient increases sharply, finally reached 0.32. By comparing the adhesion coefficient after reaching the stable state (after 75,000 cycles), it can be found that the adhesion coefficient under the service condition of 20°C is higher than that of the service condition, −40°C.

3.3.2 Response to Surface Damage

Ambient temperature has a significant influence on fatigue fracture, crack initiation and propagation, stress-strain, and microstructure transformation of wheel–rail materials [20–23]. The ductile-brittle transition temperature of wheel–rail materials used for high-speed trains is about −20°C [20]. Research has shown that the properties of materials before and after the tough-brittle transition will show different results [21–24]. Fang et al. [25] found through tensile tests that the ductile-brittle transition temperature of high-speed rail wheel steel ER8C in China is above -20°C. When the temperature is lower than the ductile-brittle transition temperature of the material, the yield strength of the material approaches the tensile strength, and the brittleness increases. The rolling contact fatigue cracks will accelerate the initiation and propagation. In addition, some scholars [26, 27] have demonstrated that the wheel–rail damage behavior presents seasonal characteristics under extreme service temperatures, mainly manifested in the phenomenon of rail spalling more serious in winter compared to summer, and rail fracture accidents generally occur in winter, usually from November to March of the following year, accounting for 60 ~ 80 percent of the annual number of broken rails, as shown in Figure 3.7(a, b).

The surface damage of wheels under different service environments is shown in Figure 3.7(c–h). It can be obviously seen from the figure that the decrease of ambient temperature increases the wheel surface wear, and the main wear forms at low temperatures (T≤0°C) are fatigue wear and lamination wear, among which lamination wear will produce widely distributed rough pits on the surface. In addition, delamination wear exists on all wheel wear surfaces at low temperature, but the delamination characteristics are different at different temperatures. Taking −50°C as an example, serious delamination wear appears on the surface, as shown in Figure 3.7(c). At 0°C, small and dense pits appear on the worn surface, as shown in Figure 3.7(f). Therefore, with the increase of temperature, the phenomenon of delamination wear is reduced, and the worn surface gradually becomes smooth. Under high temperature conditions (T = 40 and 60°C), slight ploughing phenomena and brown adhesive layer appear locally on the worn surface. The main wear mechanisms are abrasive wear and adhesive wear, as shown in Figure 3.7(f) and (g).

The microstructure of the worn surface for rail at different ambient temperatures is shown in Figure 3.7(i–l). It can be seen from Figure 3.7 that there is multiple ploughing on the worn surface at 20°C, and surface ploughing is one of the main characteristics of abrasive wear: that is, the main wear form of rail at room temperature is abrasive wear. At the temperature of −15°C, a large area of flaky adhesion is attached to the rail wear surface. By contrast, the adhesive size and area are smaller when at −30°C condition, but there are a few surface cracks in this condition. When the temperature

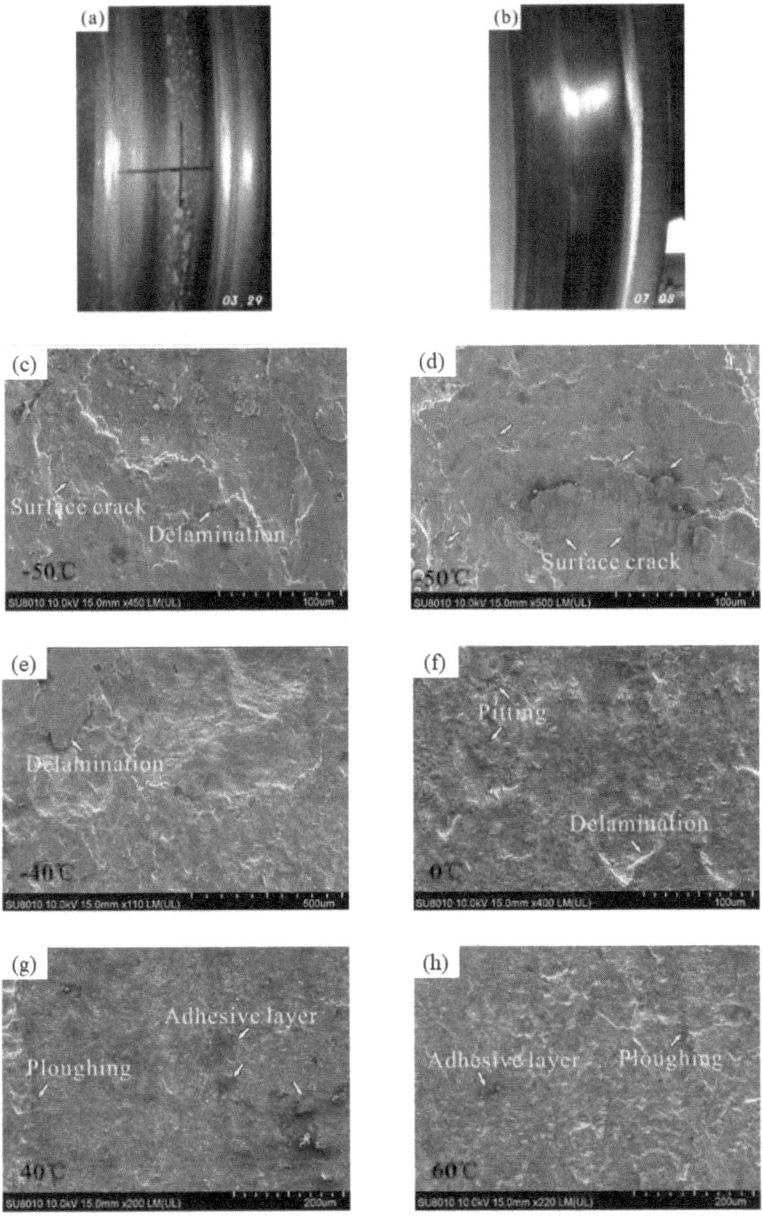

FIGURE 3.7 (a,b) Typical damage characteristics of wheels of Harbin-Dalian high-speed rail in different seasons [17], (c–h) wear surface of the wheel at different service temperatures [13], (i–l) micrographs of worn surface under different temperature conditions: (i)20°C; (j) -15°C; (k) -30°C; (l) -40°C, (m,n) SEM micrographs of microstructure at different temperatures, (m) -40°C; (n) 20°C [28].

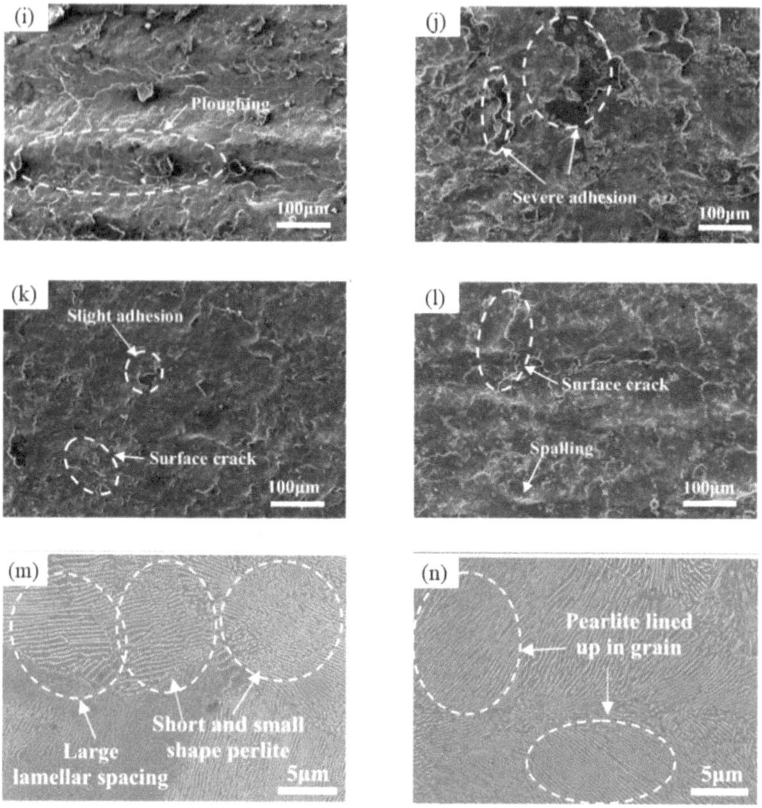

FIGURE 3.7 (Continued)

is further reduced to –40°C, cracks and slight spalling occurs on the worn surface, as shown in Figure 3.7(l). Therefore, when the temperature drops from −15 to –40°C, adhesion wear gradually alleviated and long cracks appear on the surface, and the main wear form changes to fatigue wear. Due to the decrease of temperature, the brittleness of the rail material increases and the fracture toughness decreases, so it is difficult to produce adhesive wear at low temperatures. As mentioned above, abrasive wear occurs at room temperature, while adhesive wear is predominant at –20°C. When the temperature drops further to –40°C, the rail surface eventually experiences wear predominantly characterized by fatigue. In conclusion, the rail surface will be affected by temperature, and the various temperatures will contribute to different wear and damage mechanisms.

From the perspective of material microstructure [28], at low temperatures (–40°C), most of the pearlite in Figure 3.7(m) is arranged in different directions, with large laminar spacing and short rod-like shape. When the temperature rises to room temperature, the number of short and small pearlite gradually decreases, while the number of arranged pearlite gradually increases. The spacing between the pearlite layers is more uniform, which makes it difficult for cracks to initiate and propagate. When the temperature is lower than the ductile brittle transition temperature (DBTT) of U71Mn rail material, the local pearlite becomes more disordered, and its size gradually decreases. The disordered ferrite easily slides under shear stress, so subsurface cracks are more likely to form at the boundary between ferrite and pearlite. Compared with room temperature, the subsurface damage of track materials is more serious at low temperatures.

3.3.3 Fatigue Crack Behavior

The fatigue crack behavior of wheel–rail contact interface is significantly affected by service temperature. With the decrease of temperature, the fatigue crack becomes more and more serious, as shown in Figure 3.8(a–f). This is due to the decrease in temperature leading to a decrease in humidity in the environment, which increases the adhesion coefficient between wheels and rails. As a result, the surface shear stress increases, leading to an increase in crack propagation rate. In the low temperature environment, the mechanical properties of wheel–rail materials also decline, which also has a certain impact on the service performance of wheel–rail [29, 30].

Before reaching the ductile brittle transition temperature, the crack propagation rate, depth, and length gradually increase with the decrease of temperature and reach the maximum at the ductile brittle transition temperature. When the temperature is lower than the ductile brittle transition temperature, the depth and length of the crack propagation will decrease gradually. This is because when the wheel–rail system operates at room temperature, the pearlite is arranged linearly, making it difficult to generate cracks. However, when the material is operated at the ductile brittle transition temperature [31], the arrangement characteristics of some pearlite will become more disordered, and the size will become smaller and more prone to slip. When the wheel–rail system operates at −30°C and −40°C, the arrangement of pearlite is mostly irregular short strip and multi-channel arrangement, which can effectively prevent crack propagation compared with that at −15°C (ductile brittle transition temperature) [28].

At the same time, the type of rail material will be different under different temperature conditions. Zhou et al. [27] found that the fatigue crack length of eutectoid steel at room temperature (20°C) is smaller than that of eutectoid steel at low temperature (−40°C), while that of eutectoid steel at −40°C is longer than that of eutectoid steel at 20°C, as shown in Figure 3.8 (g, h). The reason may be that eutectoid steel has better wear resistance and rolling contact fatigue (RCF) than eutectoid steel at room temperature. At low temperatures, eutectoid steel still has lower wear rate and hardening layer thickness than eutectoid steel. However, at low temperature, the brittleness of rail material plays a dominant role in crack propagation [28]. Due to its low ductility, eutectoid steel has longer crack lengths than eutectoid steel. Therefore, compared with eutectoid steel, hypereutectoid rail steel has better wear resistance at low temperature, but poor RCF resistance.

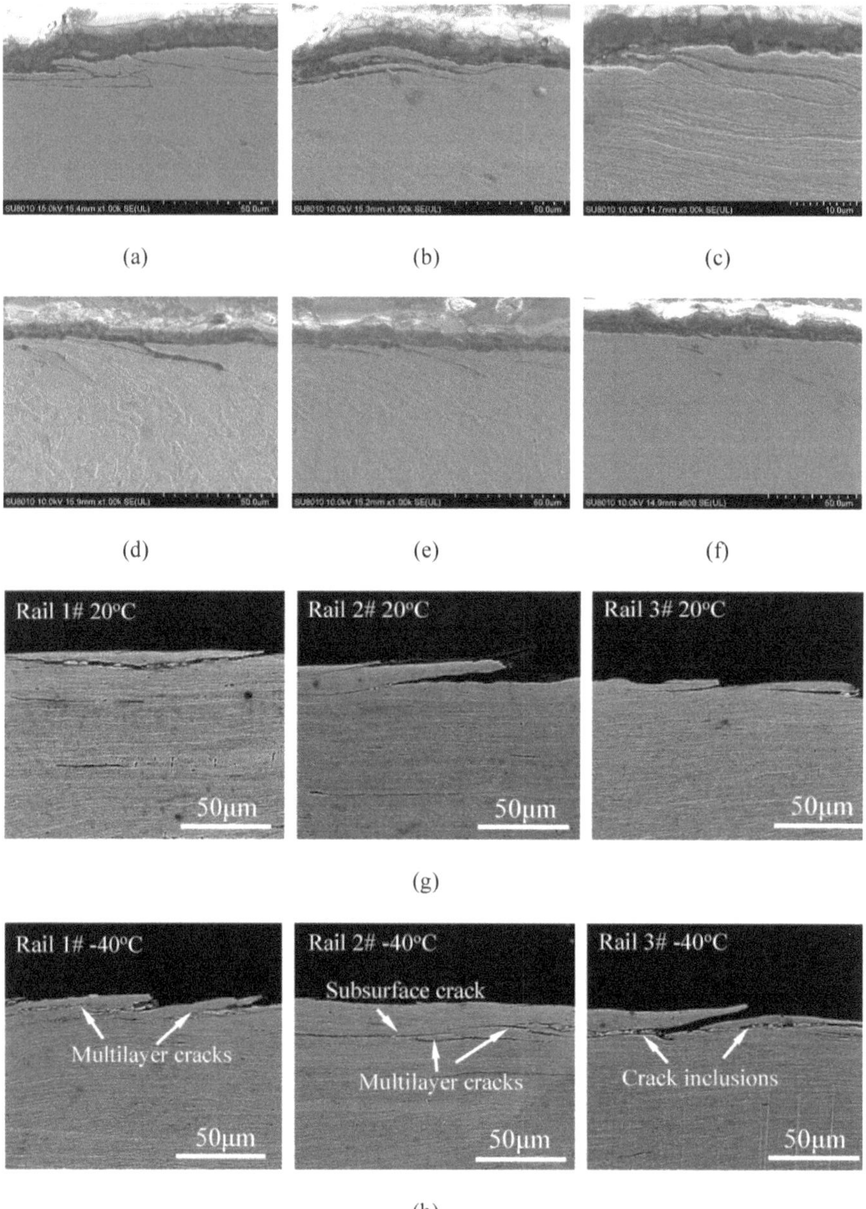

FIGURE 3.8 (a–f) SEM micrographs of subsurface cracks at different temperatures, (a) −50°C; (b) −40°C; (c) −20°C; (d) 0°C; (e) 40°C; (f) 50°C; (g, h) SEM observations of subsurface cracks of rails: (g) at room and (h) at low temperatures [27].

3.3.4 Tribo-oxidation Effect

As an open tribological system, the surface of the rail cannot avoid local oxidation, and the oxidation products formed by the rail itself (such as oxidation debris) always accompany the rolling contact of the wheel and rail. The friction oxidation behavior of wheel and rail is more pronounced at high temperature. The moisture in the air at room temperature oxidizes iron ions on the surface of the rail into reddish-brown oxides (i.e., rust). However, at low temperatures, the strong bonding between the worn debris particles and the wheel and rail surface will form a dense friction film. It is precisely because of the existence of many friction films currently that wear and friction oxidation effects of the wheel and rail in low temperature environments are reduced. The decrease of temperature also reduces the content of O and Fe^{3+} in the friction film at this time [32], as shown in Figure 3.9.

With the increase of temperature, the friction oxidation effect between wheel and rail materials is significantly enhanced. A substantial increase in O element content and a significant decrease in Fe element content can be observed and analyzed on the surface of some wheel–rail materials. At this time, the friction oxidation between the wheel–rail gradually forms a rubber layer composed of iron oxide in a high-humidity environment, and these black oxides play a wear-resistant role in the wheel–rail

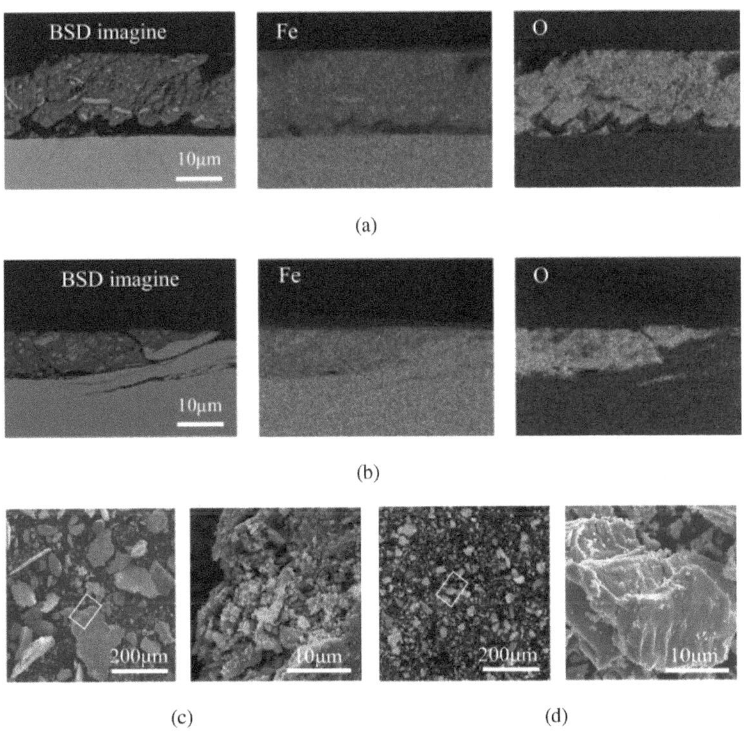

FIGURE 3.9 (a,b) EDS detection on rail subsurface at (a) 20°C and (b) −40°C, (c, d) SEM observation of the morphologies of ejected wear debris at (c) 20°C and (d) −40°C [32].

operation [13]. Below room temperature, the wear mechanism between wheel and rail gradually changes from fatigue wear to abrasive wear as humidity increases. At this point, the friction oxide produced on the surface of wheel and rail materials acts as abrasive particles to accelerate the process of wheel and rail. However, when the humidity reaches a certain level, an anti-friction layer will be generated between wheel and rail to avoid excessive wear [33]. In addition, the size of wear debris produced by friction oxidation between wheel and rail at different temperatures also varies. Compared with room temperature, the wear debris produced at low temperature in a thin and irregular shape, which are not only reduced in size and quantity, but also has a smoother surface [32], as shown in Figure 3.9(c, d). It also indicates that the friction oxidation between wheel and rail materials decreases with the decrease of temperature.

3.4 WHEEL–RAIL ADHESION BEHAVIOR IN THE THIRD-BODY MEDIUM

3.4.1 Influence of Water Medium

When water is added to the wheel–rail operating condition, the adhesion coefficient of wheel–rail will decrease relative to the dry conditions. This is because, compared with the dry state, the addition of water medium forms a lubricant (water film) between the debris generated by wheel–rail friction and water [34], as shown in Figure 3.10(a).

However, when the water medium reaches a certain level, the water film generated between the water medium and the wheel–rail friction pair reaches a certain thickness, and the adhesion coefficient between the wheel and rail will also decrease to the minimum value and remain stable (see Figure 3. 10(b)). In addition, further increasing the amount of water medium added will actually cause a decrease in the water film thickness [35].

Moreover, the different temperature of water medium may also affect the different adhesion characteristics of wheel–rail operation. The effect of increasing the temperature of the added water medium on the adhesion characteristics between the wheel and rail decreases, as shown in Figure 3. 10(c), that is, the wheel–rail adhesion coefficient increases with the increase of the temperature of the water medium [35]. This may be due to the increase in temperature of the added water medium, which leads to a decrease in the viscosity of the water film generated between the water medium and the wheel–rail friction pair, resulting in an increase in the adhesion coefficient between the wheel and rail [36]. Japanese scholar Chen et al. [36–38] evaluated the adhesion characteristics of wheel–rail rolling contact under the condition of water medium with different temperatures (0 to 80°C) through a double-disc wheel–rail test bench. The results showed that water temperature had an important influence on wheel–rail adhesion, and the increase of water temperature lead to higher adhesion coefficient. Even during the process of water temperature dropping from 50°C to 5°C at different rolling speeds, the adhesion coefficient can be reduced from 0.2 to 0.02. Therefore, they put forward the method of increasing water temperature to improve the adhesion coefficient of wheel and rail surface in wet and cold environments, to improve the safety of train operation in rainy weather.

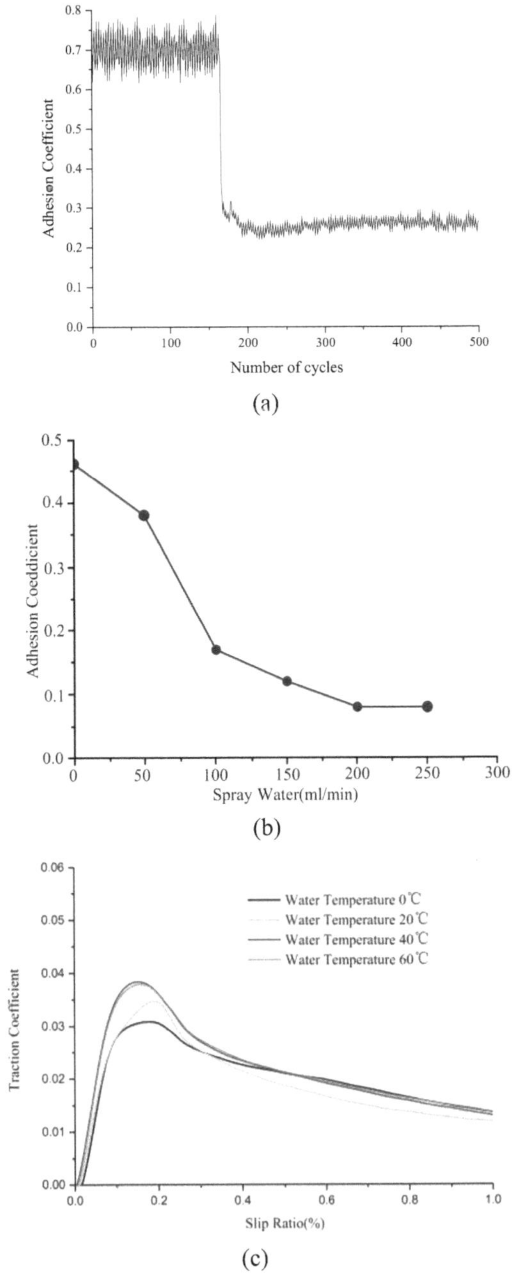

FIGURE 3.10 (a) Adhesion coefficient of wheel/rail under dry and water added conditions; (b) Adhesion coefficient versus amount of water spray at the speed of 250km/h, and (c) Adhesion coefficient versus water temperature.

3.4.2 Reduce Wear between Wheel–Rail Interface

Studying strategies to reduce wheel and rail wear has important practical significance. Usually, wheel–rail wear occurs between the tread and the top of the rail, as well as the rim and the side of the rail. For tread and rail top, tread damage mainly includes peeling, falling block and abrasion. The top surface of the rail will be spalling, plastic deformation, crushing and wave grinding. At the same time, the wear of the rim and the rail side is also relatively serious, especially when the train passes through a section with small curvature radius: the damage forms of the rim mainly include peeling, wear and fatigue damage. The rail side is prone to be damaged by peeling, crushing, wave wear and side wear. Friction and wear will inevitably shorten the life of wheel and rail, resulting in unstable train operation, even serious derailment accidents, which endanger people's lives and property.

At present, measures taken by relevant railway departments to improve friction and wear include wheel and rail lubrication, correcting of rail profile, enhancing wheel and rail material strength, reasonable setting and adjustment of track parameters (gauge, rail bottom slope, and outer rail elevation), and so forth. At present, some scholars' studies show that the textured design of wheel and rail interface also has the potential to improve friction and wear.

3.4.2.1 Wheel and Rail Lubrication

Wheel–rail lubrication is an effective method with less investment and quick effect, which can rapidly reduce the wear of the rim and rail side, and prolong the service life of the rail. The method of rim rail side lubrication has been used on railway curves in foreign countries since the 1940s, and it has also been applied on many lines in China in recent years. This method mainly uses lubricating devices to treat the rim or rail side (see Figure 3.11(a, b)). Because of the particularity of wheel–rail system, the lubrication form can only be boundary lubrication. When the third medium acts on the metal surface, a boundary lubrication film will be formed, and the boundary film has small shear strength. The friction process of metal during friction mainly occurs in the boundary film layer to effectively reduce the sliding resistance, protect the wheel–rail contact surface, and effectively reduce the wear of wheel–rail materials.

(a) (b)

FIGURE 3.11 (a) Rim lubrication. (b) Rail side lubrication.

The effect of boundary lubrication is largely dependent on the generation and normal functioning, and the type of lubricant determines the structure, form, and performance of boundary lubrication film, so how to select lubricants for a specific working condition determines the quality of lubrication effect. At present, the commonly used lubricants for on-site workers include lubricating oil, lithium molybdenum disulfide grease, vegetable oil, and graphite calcium-based grease.

3.4.2.2 Correct Rail Profile

The matching problem of geometric parameters of wheel–rail contact directly or indirectly affects the performance of wheels in straight line operation, curve passing, wheel and rail damage, and train running stability. In recent years, with the development of railway transportation in China towards high-speed and heavy-haul directions, the contact fatigue damage, surface peeling, surface plastic crushing, rail corrugation and other problems for wheels and rails has become increasingly serious, hindering the improvement of railway transportation ability and economic benefits. The reason is mainly caused by excessive wheel–rail contact stress. In order to reduce the wheel–rail contact stress and wear damage, selecting the optimal wheel–rail geometric contact parameter matching under the given wheel–rail system structure may be the most economical and effective method. For example, by using advanced on-site rail grinding technology to modify the rail profile and optimize geometric matching, the rim does not make contact with the side of the rail or reduce the contact force, thereby reducing or fundamentally preventing rail side wear and rim wear. The application of wheel–rail grinding turning tools can effectively repair milling defects in steel rails. On this basis, Liu et al. [39] proposed a new model to realize personalized rail grinding, which establishes an effective grinding profile according to the area difference between the wear profile and the target profile, controls the speed of the grinding wheel, grinding pressure, feed speed and other variables to establish an analytical model, and finally designs an algorithm to generate personalized grinding mode. The analysis data shows that the area difference between the predicted profile and the target profile is 13 mm^2, while the actual profile error is 12.2 mm^2. Experimental verification shows that the algorithm can effectively output the grinding mode and realize the purpose of rail profile correction. In addition, regular grinding before and after the rail service and before a certain depth crack, to maintain the rail head profile in the best geometric shape, which can optimize the wheel–rail interface contact relationship, thereby extending the rail life, controlling side wear and fatigue, reducing wave wear and reducing wheel–rail impact force, and so forth.

3.4.2.3 Texture Design of Wheel–Rail Interface

The related research and practice of tribology and bionics have shown that the smoother the surface of the friction pair, the more wear-resistant it is. Machining a texture with a certain morphology on the surface of the friction pair can capture wear particles, form a micro-oil reservoir to store lubricating medium, and produce dynamic lubrication effects, thus improving the load-bearing capacity and reducing surface wear.

The research has shown that when locomotives turn over a small-curvature curve, the inertia will make abnormal contact between the rail and the rim, resulting in "rail side wear" and exacerbating wear. Some studies have shown that it is theoretically feasible to achieve secondary lubrication of wheel–rail by designing laser textures on the wheel–rail wear contact surface, and the coupling effect of laser texture and lubricants can reduce the wear more effectively. At the same time, texture can also reduce the grinding and refinement of debris particles, reduce the repeated friction and shear effects of debris at the friction interface, and effectively reduce surface wear and prolong service life. Studies have shown that surface texturing treatment can reduce ploughing phenomenon and significantly reduce wear loss caused by wear [40]. Concave texture can also store lubricant, and extrude the texture to release lubricant during load contact, thus improving lubrication condition. When hydrodynamic lubrication is formed on the contact surface, tribological properties can be effectively improved, friction and wear can be reduced, and life can be prolonged [41]. And the textured surface design can improve the surface hardness of the wheel and rail to a certain extent, making the wheel–rail surface less prone to wear debris and more wear-resistant. There are many types of laser microtexturing designs, including but not limited to circles, squares, rectangles, and diamonds, and some irregular shapes (Figure 3.12(a)). Under varying working conditions, different textures are engineered by adjusting parameters like morphology, depth, and surface density, with the aim of reducing wear at the wheel and rail interfaces.

Shinya Fukagai et al. [41] found that the surface texture can optimize the traction coefficient during running-in, thereby reducing the risk of climbing and derailment caused by the high traction coefficient of the wheel and rail at sharp turns. In order to prevent this phenomenon, it is found that the surface texture can appropriately reduce the traction coefficient of the wheel tread/rail top interface. During the experiment, they designed a glossy surface and two different pitch textures, 0.5mm/rev and 1.0mm/rev, respectively, to simulate the actual machining of the wheel. The experiment shows that the design of the texture on the surface of the wheel will increase the strength of the wheel surface, and the running-in curve of the traction coefficient shows an instantaneous upward trend, mainly due to the plastic deformation of the surface texture, and the peak value of the traction coefficient is different due to the depth of the plastic flow. At the same time, it was found that the work-hardened layer of the texture surface was thicker than the smooth surface. In short, the risk of wheel flange climbing and derailment during the running-in process can be reduced by appropriate texture design or surface treatment.

3.4.3 Role of Iron Oxides

As an open tribological system, rail surface cannot avoid local oxidation. The oxidation products formed by itself and rolling contact wear of wheel and rail always accompany the service process. Iron oxide is an important factor causing low adhesion between wheel and rail, but it is not a sufficient condition. Although little attention has been paid to the phenomenon of low adhesion between wheels and rails caused by surface oxides of steel rails, and relevant research is limited. However, the existing

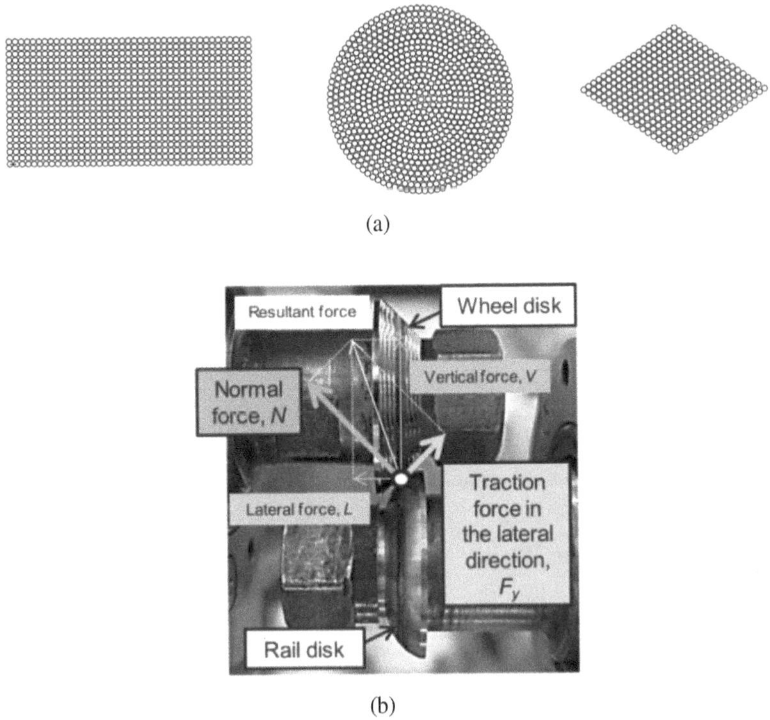

FIGURE 3.12 (a) Texture map of texture surface, (b) A set of wheel disk and rail disk after attaching on the experimental apparatus.

studies has fully shown that the surface oxides of wheel and rail have little influence on the adhesion coefficient under dry conditions, and even play a role in increasing the adhesion. Zhu et al. [42–44] studied the influence of temperature and humidity on the tribological properties of surface corroded rail through the pin-disk test, and the results showed that when the temperature was low and there were oxides on the surface, the increase of humidity would significantly reduce the friction coefficient. He also pointed out that moisture is a necessary condition for surface oxides to cause low adhesion between wheels and rails, and improving the surface roughness is beneficial to the removal of oxide surface and the recovery of adhesion coefficient during wheel–rail rolling.

Another necessary condition for wheel–rail low adhesion phenomenon caused by iron oxides is to maintain a certain humidity at wheel–rail interface. White et al. [45] found that rail samples with prefabricated iron oxides and a small amount of water exhibit an ultra-low adhesion effect with an adhesion coefficient lower than 0.02 through a double-disc testing machine. Environmental humidity, as one of the main factors affecting wheel–rail low adhesion, may have a more significant effect on wheel–rail adhesion than does rain. The friction oxidation products and humidity at wheel–rail interface have a synergistic effect on reducing wheel–rail adhesion. Therefore, studying the adhesion and damage of the wheel rail interface requires

FIGURE 3.13 Common iron oxides in wheel rail contact.

comprehensive consideration of the effects of environmental humidity and oxidation products.

It is generally considered that, α-Fe_2O_3, α-FeOOH, β-FeOOH, and γ-FeOOH are the four common types of iron oxides in wheel–rail contact, as shown in Figure 3.13. Among them, α-Fe_2O_3 and β-FeOOH have significant effects on adhesion characteristics and wear of wheel–rail contact. α-Fe_2O_3 is the most common iron oxide on rail surface, which usually produces in wet environments. When combined with the intermediate water layer, it will affect the adhesion of wheel and rail; β-FeOOH mainly exists on rail surface in coastal areas with high chloride ion content, and has anti-friction effect. When the vehicle is in a free rolling condition, the wheel–rail adhesion is significantly reduced.

3.4.4 Response Measures for Wheel–Rail Low Adhesion

The existence of the third body medium can induce low adhesion of wheel and rail, leading to adverse consequences such as wheel–rail slipping, idling, scratches, extended braking distance, and low operating efficiency, thereby posing challenges to the normal operation of trains. Therefore, effective control low adhesion between wheel and rail is a prerequisite to ensure the normal traction and braking during trains operation. Currently, sprinkling sand, installing grinding wheel tread cleaning devices, installing anti-slip systems and other adhesion control devices, using electromagnetic viscosification, and chemical methods (spraying sodium silicate) are commonly used measures to slow down low adhesion of wheel and rail.

3.4.4.1 Sprinkle Hard Particles

Sprinkling sand/alumina particles at the wheel–rail interface is a common way to increase the viscosity of the wheel rail, and its viscosification effect is related to factors such as particle crushing strength, particle size, and feed speed. Literature points out that the higher the crushing strength of particles, the better the viscosification effect of wheel and rail [46].

Sprinkling sand is an ancient method of thickening and is still in use today (Figure 3.14). The vehicle is equipped with a sand storage box, which is used to discharge dry sand onto the track near the driving wheel through a throat. Through grinding, the pollutants on the wheel–rail contact surface are removed, the surface roughness between the wheel–rail is improved, and the pollution film is scraped to expose the inner metal material, increasing traction adhesion and shortening the braking distance during braking.

Sprinkling sand can effectively improve the low adhesion effect of wheel–rail interface caused by the presence of third body medium. Within a certain particle size range, the larger the sand particle size, the better the viscosification effect of wheel and rail. When sand particles exceed a certain size, the viscosification of wheel–rail is not obvious. However, when the sand particle size is too small, the adhesion coefficient of wheel and rail will be further reduced under the originally low adhesion condition. Within a certain range, the viscosification effect with the increase of the sand feed rate, and the change in viscosification effect does not significant when the feed rate exceeds a certain level. The larger the particle size of gravel, the better the effect of viscosification on wheel and rail, but the greater the degree of damage on wheel and rail surface [46].

The adhesion coefficient of wheel rail under low adhesion effect can also be rapidly improved by scattering alumina particles. The research shows that [46], compared to sanding alone, sanding with alumina particles has a more excellent viscosification effect. Contrary to gravel, in a certain range, the smaller the size of alumina particles, the more significant the viscosification effect of wheel and rail surface. The particle size of the currently used viscosified alumina is 0.3mm. Although the particle size is small, the ultimate particle size is large, and the mechanical strength is high. It is easy to insert into the wheel–rail contact surface to form a micro convex body. With the increase of the feed rate, the viscosification effect of scattered alumina particles first increases and then decreases [48]. This is because when the feed rate is too high, excessive alumina particles cannot be embedded in the wheel–rail interface, but stay between the wheel–rail interface and play a ball-like effect.

Sanding and alumina particles can improve the low adhesion effect caused by the presence of a third body medium between the wheel and rail, but as hard particles are sprayed onto the wheel rail contact interface, a large number of hard particles are embedded on the wheel rail surface. During the operation of the wheel, these hard particles embedded on the surface of the wheel and the rail are repeatedly squeezed under extreme contact stress, which will seriously exacerbate the fatigue wear of the

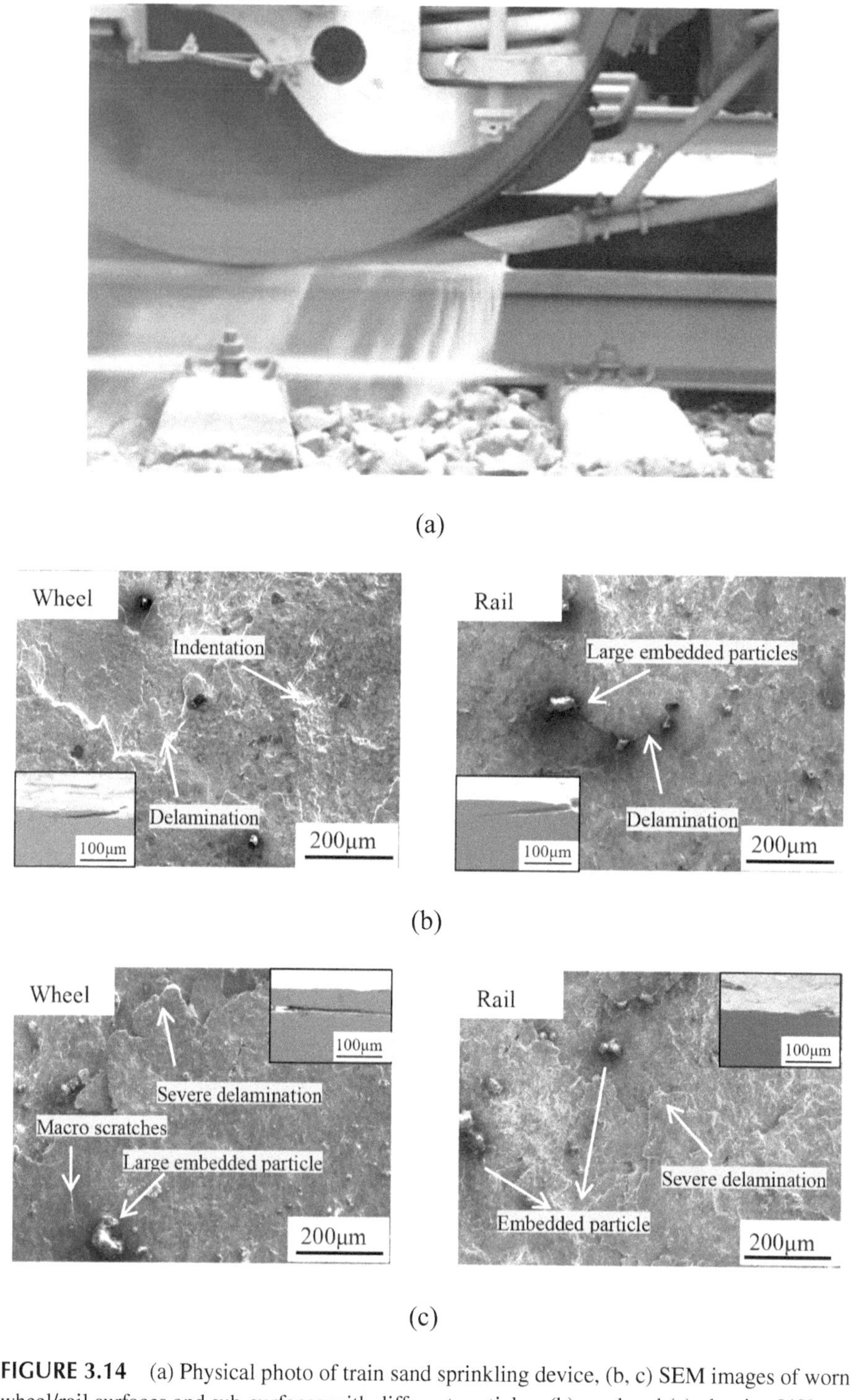

FIGURE 3.14 (a) Physical photo of train sand sprinkling device, (b, c) SEM images of worn wheel/rail surfaces and sub-surfaces with different particles: (b) sand and (c) alumina [48].

wheel and rail surfaces. Some embedded particles are crushed into smaller particles under this high stress, and then repeatedly squeezed as the surface cracks are rolled deep into the cracks, exacerbating the propagation of fatigue cracks. More small particles are still embedded on the wheel surface, increasing the adhesion between the wheel and rail surface and increasing the wear of the wheel rail surface. Compared to sanding aluminum oxide particles, sanding aluminum oxide particles can provide a better optimization coefficient for increasing the adhesion of the wheel rail contact surface, but at the same time, it also causes more severe damage to the wheel rail contact surface. Wang et al. [46] found that the pattern of sanding aluminum oxide particles on the wheel rail contact interface has more wheel rail wear, more severe surface peeling, and deeper fatigue contact cracks. In this way, sanding has a better economic effect on increasing the adhesion of the wheel and rail.

3.4.4.2 Grinder

The high-speed train grinder is a major component on the wheel tread cleaning device, which has the function of increasing adhesion, and is an important part to ensure that the high-speed train can shorten the braking distance and prevent vehicle sliding (Figure 3.15). The grinder is usually installed on the bogie of the high-speed train unit, which can remove impurities such as rainwater, dirty oil, rust and dust attached to the wheel tread during vehicle operation. When the vehicle brakes, it is pressed on the wheel tread to clean, ensuring that the wheel tread has a good surface condition.

In the presence of a third body medium at the wheel–rail interface, the adhesion coefficient of wheel–rail can be effectively increased by using the grinder tread cleaning device, and the viscosification effect increases with the increase of the pressure exerted by the grinder on the wheel tread. Compared to scattering hard particles, the effect of grinding is not so significant, but the damage to the wheel and rail surface is milder.

3.4.4.3 Other Viscosification Measures

In addition to scattering hard particles onto the wheel–rail interface and installing grinding tread cleaning devices, methods such as installing anti slip control systems,

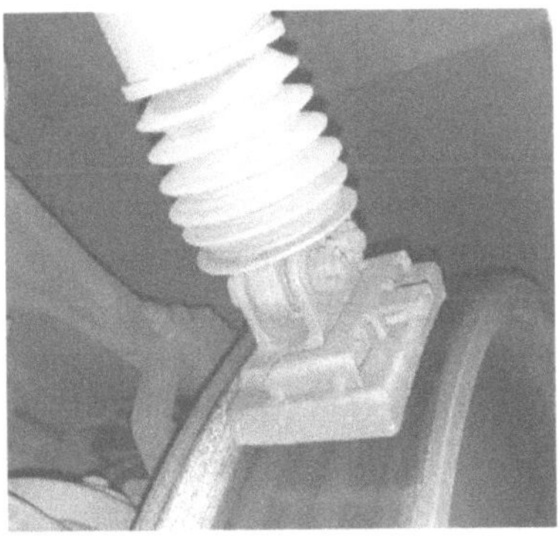

FIGURE 3.15 Physical drawing of grinder.

chemical methods, and installing electromagnetic adhesion devices can also be used to improve wheel rail adhesion. Anti-slip control system is based on the variation law of the adhesion creep characteristic curve–within a certain range, wheel–rail adhesion will first increase and then decrease with the increase of slip, controlling the wheel–rail slip to improve wheel–rail adhesion within a certain range. The chemical method is to dissolve the oxide layer and oil stains on the surface of the wheel rail with chemical substances such as silicates to maintain good surface conditions of the wheel rail, thereby improving the adhesion coefficient during wheel rail contact. The electromagnetic adhesion device is a device that installs a strong magnet on the wheel or winds a metal coil on the wheel to add a magnetic field, and improves the contact stress between the wheel and the track in a certain way to achieve the viscosification effect of the wheel and rail. This method is currently in the experimental stage and has not been applied to trains.

3.5 SUMMARY AND PERSPECTIVE

With the increasing diversity of railway operating environments, more and more environmental factors affect the friction and wear behavior of wheel–rail system. In coastal area, high temperature and humidity will greatly accelerate the oxidation rate of the wheel and rail surface, thus accelerating the oxidation wear of the wheel and rail surface. In high cold area, low temperature will make the wheel/rail brittle, thereby accelerating the initiation and propagation of rolling contact fatigue cracks, especially when the ambient temperature is at the ductile-brittle transition temperature of the wheel–rail material, the initiation and propagation of fatigue cracks are the most intense.

The safe and punctual operation of trains, as well as the precise control of their movement status, are closely related to the adhesion effect of wheel–rail contact interface. With the significant increase in train operating speed, the requirements for traction and braking force during acceleration and braking are also increased, making it increasingly important to solve the problem of low adhesion between high-speed wheel–rail. It is necessary to systematically study the various influencing parameters of wheel–rail adhesion to comprehensively understand the mechanism of wheel–rail adhesion.

At present, the main measures to deal with low adhesion are to sprinkle hard particles such as sand and aluminum oxide to the wheel–rail contact interface, and to install abrasive tread devices to increase the adhesion. The adhesion coefficient of wheel–rail under low adhesion effect can be improved in a very short time by spreading hard particles into the wheel–rail interface. Abrasive tread cleaning devices can help clean the impurities on the wheel tread, so that the wheel tread has a good surface condition, and then improve the wheel–rail surface adhesion. In addition, methods such as installing anti slip control systems, chemical methods, and installing electromagnetic viscosification devices can also be used to improve the adhesion of wheel–rail surfaces.

REFERENCES

[1] Telliskivi T, Olofsson U. Contact mechanics analysis of measured wheel–rail profiles using the finite element method. *Proceedings of the Institution of Mechanical Engineers, Part F: Journal of Rail and Rapid Transit*, 2001, 215(2): 65–72.
[2] Lyu Y, Bergseth E, Olofsson U. Open system tribology and influence of weather condition. *Scientific Reports*, 2016, 6: 32455.
[3] Liang B, Zhao N Y. A study on temperature distribution of surrounding rock and mechanical characteristics of lining of Monglian tunnel under high geothermal. *Advanced Materials Research*, 2011, 255–260: 2594–2600.
[4] Olofsson U, Lyu Y. Open system tribology in the wheel–rail contact-a literature review. *Applied Mechanics Reviews*, 2017, 69(6): 060803.
[5] Jin X. Research progress of high-speed wheel–rail relationship. *Lubricants*, 2022, 10(10): 248.
[6] Chen H, Furuya T, Fukagai S, et al. Wheel slip/Slide and low adhesion caused by fallen leaves. *Wear*, 2020, 446–447: 203187.
[7] Turley J. A successful approach to rail operations in low adhesion conditions. World Congress on Railway Research (WCRR), Tokyo, 1999.
[8] Arias-Cuevas O, Li Z. Field investigations into the adhesion recovery in leaf-contaminated wheel–rail contacts with locomotive sanders. *Proceedings of the Institution of Mechanical Engineers, Part F: Journal of Rail and Rapid Transit*, 2011, 225(5): 443–456.
[9] Arias-Cuevas O, Li Z, Lewis R. Investigating the lubricity and electrical insulation caused by sanding in dry wheel–rail contacts. *Tribology Letters*, 2010, 37(3): 623–635.
[10] Skipper W A, Chalisey A, Lewis R. A review of railway sanding system research: Wheel/rail isolation, damage, and particle application. *Proceedings of the Institution of Mechanical Engineers, Part F: Journal of Rail and Rapid Transit*, 2020, 234(6): 567–583.
[11] Lewis R, Dwyer-Joyce R S. Wear at the wheel/rail interface when sanding is used to increase adhesion. *Proceedings of the Institution of Mechanical Engineers, Part F: Journal of Rail and Rapid Transit*, 2006, 220(1): 29–41.
[12] Lewis S R, Lewis R, Olofsson U, et al. Effect of humidity, temperature and railhead contamination on the performance of friction modifiers: Pin-on-disk study. *Proceedings of the Institution of Mechanical Engineers, Part F: Journal of Rail and Rapid Transit*, 2013, 227(2): 115–127.
[13] Shen M, Qin Y, Ji D, et al. Role of ambient temperature in the adhesion and damage characteristics of wheel/rail interface during rolling-sliding contact. *Wear*, 2022, 506–507: 204458.
[14] Liu P, Yang S, Liu Y. Full-scale test and numerical simulation of wheelset-gear box vibration excited by wheel polygon wear and track irregularity. *Mechanical Systems and Signal Processing*, 2022, 167: 108515.
[15] Lewis S R, Lewis R, Olofsson U. An alternative method for the assessment of railhead traction. *Wear*, 2011, 271(1–2): 62–70.
[16] Sanchis I V, Franco R I, Fernández P M, et al. Risk of increasing temperature due to climate change on high-speed rail network in Spain. *Transportation Research Part D: Transport and Environment*, 2020, 82: 102312.
[17] Shen M X, Qin T, Li S X, et al. High-speed wheel–rail interfacial adhesion and surface damage behavior of wheel in wide temperature range. *Journal of Traffic and Transportation Engineering*, 2021, 21(3): 269–278.

[18] Shi L B, Ma L, Guo J, et al. Influence of low temperature environment on the adhesion characteristics of wheel–rail contact. *Tribology International*, 2018, 127: 59–68.

[19] Zhou L, Hu Y, Ding H H, et al. Experimental study on the wear and damage of wheel–rail steels under alternating temperature conditions. *Wear*, 2021, 477: 203829.

[20] Fang X Y, Cai Z B, Wang J G, et al. Evaluation of temperature-sensitive fatigue crack propagation of a high-speed railway wheel rim material. *Fatigue & Fracture of Engineering Materials & Structures*, 2019, 42(8): 1815–1825.

[21] Fang X Y, Zhao Y X, Liu H W. Study on fatigue failure mechanism at various temperatures of a high-speed railway wheel steel. *Materials Science and Engineering: A*, 2017, 696: 299–314.

[22] Ma L, Guo J, Liu Q Y, et al. Fatigue crack growth and damage characteristics of high-speed rail at low ambient temperature. *Engineering Failure Analysis*, 2017, 82: 802–815.

[23] Zhu Z, Li G, Dai G, et al. Mechanisms and new parameter attribute reduction of high-speed railway wheel rim steel subjected to low temperature fatigue. *Materials Science and Engineering: A*, 2016, 673: 476–491.

[24] Shaikh K, Hussain I, Chowdhry B S. Wheel defect detection using a hybrid deep learning approach. *Sensors* , 2023, 23(14): 6248.

[25] Fang X Y, Huang W, Yang X F, et al. Effects of temperature on fatigue cracks initiation and propagation for a high-speed railway wheel rim steel. *Engineering Failure Analysis*, 2020, 109: 104376.

[26] Lee C W, Kwon S J. Evaluation of surface defects of wheel and rail for Korean high-speed railway. *Materials Science Forum*, 2010, 654–656: 2499–2502.

[27] Zhou L, Wang W J, Hu Y, et al. Study on the wear and damage behaviors of hypereutectoid rail steel in low temperature environment. *Wear*, 2020, 456–457: 203365.

[28] Ma L, Shi L B, Guo J, et al. On the wear and damage characteristics of rail material under low temperature environment condition. *Wear*, 2018, 394–395: 149–158.

[29] ASCE. Committee on fatigue and fracture reliability of the committee on structural safety and reliability of the structural division. Fatigue reliability 1–4. *Journal of the Structural Division*, 1982, 108(ST1): 3–88.

[30] Terashita Y, Tatsumi M. Analysis of damaged rail weld. *Quarterly Report of RTRI*, 2003, 44(2): 59–64.

[31] Wang Y, Zhou H, Shi Y, et al. Mechanical properties and fracture toughness of rail steels and thermite welds at low temperature. *International Journal of Minerals, Metallurgy, and Materials*, 2012, 19: 409–420.

[32] Zhou L, Ding H, Han Z, et al. Study of rolling-sliding contact damage and tribo-chemical behaviour of wheel–rail materials at low temperatures. *Engineering Failure Analysis*, 2022, 134: 106077.

[33] Rong K, Xiao Y, Shen M, et al. Influence of ambient humidity on the adhesion and damage behavior of wheel–rail interface under hot weather condition. *Wear*, 2021, 486–487: 204091.

[34] Wang W J, Wang H, Wang H Y, et al. Sub-scale simulation and measurement of railroad wheel/rail adhesion under dry and wet conditions. *Wear*, 2013, 302(1–2): 1461–1467.

[35] Chang C, Chen B, Cai Y, et al. An experimental study of high speed wheel–rail adhesion characteristics in wet condition on full scale roller rig. *Wear*, 2019, 440–441: 203092.

[36] Chen H, Ishida M, Namura A, et al. Estimation of wheel/rail adhesion coefficient under wet condition with measured boundary friction coefficient and real contact area. *Wear*, 2011, 271(1–2): 32–39.

[37] Chen H, Ban T, Ishida M, et al. Experimental investigation of influential factors on adhesion between wheel and rail under wet conditions. *Wear*, 2008, 265(9–10): 1504–1511.

[38] Chen H, Tanimoto H. Experimental observation of temperature and surface roughness effects on wheel/rail adhesion in wet conditions. *International Journal of Rail Transportation*, 2018, 6(2): 101–112.

[39] Liu Y M, Yang T Y, He Z, et al. Analytical modeling of grinding process in rail profile correction considering grinding pattern. *Archives of Civil and Mechanical Engineering*, 2018, 18: 669–678.

[40] Shen M, Li H, Du J, et al. New insights into reducing airborne particle emissions from brake materials: Grooved textures on brake disc surface. *Tribology International*, 2022, 174: 107721.

[41] Fukagai S, Ma L, Lewis R. Tribological aspects to optimize traction coefficient during running-in period using surface texture. *Wear*, 2019, 424–425: 223–232.

[42] Zhu Y, Chen X, Wang W, et al. A study on iron oxides and surface roughness in dry and wet wheel– rail contacts. *Wear*, 2015, 328–329: 241–248.

[43] Zhu Y, Olofsson U, Chen H. Friction between wheel and rail: A pin-on-disc study of environmental conditions and iron oxides. *Tribology Letters*, 2013, 52: 327–339.

[44] Lyu Y, Zhu Y, Olofsson U. Wear between wheel and rail: A pin-on-disc study of environmental conditions and iron oxides. *Wear*, 2015, 328–329: 277–285.

[45] White B, Lewis R. Simulation and understanding the wet-rail phenomenon using twin disc testing. *Tribology International*, 2019, 136: 475–486.

[46] Wang C, Shi L B, Ding H H, et al.. Adhesion and damage characteristics of wheel/rail using different mineral particles as adhesion enhancers. *Wear*, 2021, 477: 203796.

[47] Arias-Cuevas O, Li Z, Lewis R. A laboratory investigation on the influence of the particle size and slip during sanding on the adhesion and wear in the wheel–rail contact. *Wear*, 2011, 271(1–2): 14–24.

[48] Cao X, Huang W L, He C G, et al. The effect of alumina particle on improving adhesion and wear damage of wheel/rail under wet conditions. *Wear*, 2016, 348–349: 98–115.

4 Surface Integrity in Robotic Compliant Grinding of Nickel-based Superalloys

Xiaokang Huang, Wenxing Guo, Ze Chai, Huabin Chen, and Xiaoqi Chen

4.1 INTRODUCTION

Nickel-based superalloys are essential materials widely used in the aerospace and energy sectors, as well as for other critical applications in modern society. These materials exhibit exceptional properties at high temperatures, including superior mechanical strength, excellent corrosion and creep resistance, and high stability. However, they also have a propensity for work-hardening during machining and poor thermal conductivity, making them typical difficult-to-machine materials. To overcome these challenges, robotic belt grinding has become a popular and effective method for machining superalloys, offering advantages such as integration with multiple sensors, high material removal capacity, low heat input, less workpiece damage, long tool life, and high flexibility.

The continuous improvement of robots' positioning and repetition accuracy, along with the emergence of artificial intelligence algorithms, have driven the application and development of robotic compliant grinding. Theoretical studies and industrial practices have shown that the surface integrity of ground workpieces exerts significant influences on workpiece properties, such as mechanical property and corrosion resistance. As robotic compliant grinding becomes more widely used for machining superalloy workpieces, a comprehensive understanding of the evolution of surface integrity and its effects on workpiece properties is crucial and remains a primary focus of both academia and industry.

This chapter is centered around modeling approaches related to surface integrity in robot compliant grinding. The effects of grinding parameters on surface integrity are also briefly introduced where necessary.

 DOI: 10.1201/9781003386223-4

4.2 OVERVIEW OF THE ROBOTIC COMPLIANT GRINDING

4.2.1 Grinding Surface Integrity

The surface integrity is an essential indicator to evaluate the quality of surface and sub-surface layers of machined workpieces, which has significant and comprehensive influences on the service performance of workpieces. As early as 1964, Field et al. [1] defined surface integrity as the inherent or enhanced condition of a surface, which is finished by machining processes or other surface construction operations. This definition does not refer to specific machining methods and is applicative widely.

In the grinding field, surface integrity is defined using more specific methods and can be divided into three aspects: surface conditions, residual stresses, and microstructures. Surface conditions consist of surface profiles (roughness, waveness and figure) and surface defects (burn, crack, scratch, notch, ditch). The residual stresses refer to the tensile residual stress (TRS) and compressive residual stress (CRS), both the values and directions. Microstructures include grain features (size, distribution, and shape), crystal defects (dislocations, stacking faults, etc.), phase transformations. Table 4.1 summarizes the surface integrity indices of ground workpieces.

Achieving high-quality surface conditions is important to ensure the accuracy of workpiece dimensions and reduce the likelihood of failure due to surface defects. Tensile residual stress is considered to have a negative effect on the fatigue strength of workpieces, whereas compressive residual stress can improve fatigue life. Workpieces with finer grains exhibit higher microhardness and tensile strength, impeding crack propagation and thus enhancing fatigue life. High-quality surface integrity enables workpieces to perform stably in challenging environments, such as high temperature, high pressure, corrosion, and vibration. This is particularly critical for superalloy workpieces used in aerospace and energy applications. Therefore, surface integrity is always the study focus in material machining and grinding.

TABLE 4.1
Surface Integrity Indices of Ground Workpieces

	Surface conditions	Residual stress	Microstructures
Indices	1. Surface profiles *roughness* *waveness* *figure* 2. Surface defects *burncrack* *scratch* *notch* *ditch*	1. Tensile residual stress (TRS) *value* *direction* 2. Compressive residual stress (CRS) *value* *direction*	1. Grain features *size* *distribution* *shape* 2. Crystal defects *dislocation* *stacking faults* 3. Phase transformations

4.2.2 Grinding Features

Compliant grinding refers to material grinding methods using one or more tools being compliant with the workpiece surface in a controllable and reversible manner [2], which is different from conventional rigid wheel grinding where the deformation of the grinding tool is negligible. Currently, the popular compliant grinding tools include abrasive belt, shape adaptive grinding (SAG) tool, rubber-based bonnet tools, grinding discs, flap wheel tool, cylindrical tools, rubber matrix tool, brush tool and so on [2], as shown in Figure 4.1.

Compliant grinding tools exhibit noticeable elastic deformation during grinding and conform well to the workpiece surfaces. Furthermore, the proportion of heat flow transmitted to the workpieces during grinding is low, resulting in minimal damage to the workpieces. In contrast, in rigid grinding, the offset rigid tools have severe friction with workpiece surfaces and force large abrasive grains to cut into workpiece materials, resulting in rapid increase of grinding forces and heat generation, which may overheat the grinding areas and induce damages in surfaces and sub-surfaces of workpieces. In addition, the stiffness requirement of the compliant grinding facilities is significantly lower than that of the rigid grinding facilities, due to that the elastic deformation of grinding tools can 'absorb' the disturbance of the workpiece's irregular surface to the machining process.

However, the tools' elastic deformation also complicates the controlling of the grinding process and the machining quality of workpieces. During grinding, a single abrasive grain will have straight motion perpendicular to the grinding direction and rotation motion parallel to the grinding direction (presented in Figure 4.2), and the motion extend varies with grinding forces, workpiece shapes, workpiece materials, abrasive grain positions as well as grain wear condition. Therefore, the grain-workpiece interactive mechanisms are distinct from conventional rigid grinding, which requires new and more complex methods in building models of grinding tool wear, material removal rate (MRR) prediction, grinding force-thermal fields, and material deformations during grinding. On the other hand, the tools' elastic deformation introduces an extra DOF (degree of freedom) for the grinding system, and the material removal rate (MRR) cannot be determined by the offset of the grinding tool. Therefore, precisely controlling grinding tool gestures and MRR is a great challenge for compliant grinding.

The tool gestures and material removal rate (MRR) play a critical role in shaping the surface and determining its quality and accuracy, as well as affecting the

FIGURE 4.1 Compliant grinding tools: (a) abrasive belt [3], (b) rubber-based bonnet tools [4], (c) rubber-based bonnet tools [5], (d) grinding discs.

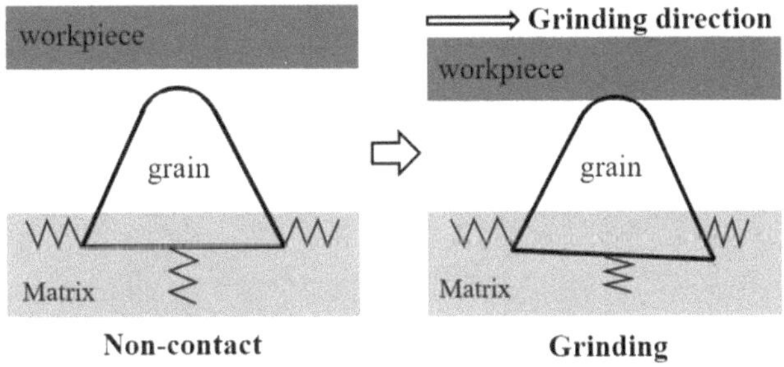

FIGURE 4.2 The elastic deformation of single abrasive grain in compliant grinding.

force-thermal fields and microstructures of the surface layers. Furthermore, when using a compliant tool with a certain offset, the elastic deformation reduces gradually from the center to the edge of the deformed area, and the grinding effects follow a sequence of cutting, plowing, and rubbing [6]. Such a continuous and force-related distribution of grinding effects further increases the difficulty of the grinding process control.

Industrial robots are currently widely used as flexible manufacturing units. They are capable of easily integrating various sensors and intelligent control algorithms, enabling real-time control of machining processes based on multi-information fusion. Therefore, the integration of compliant grinding tools and robots can enable precise control of the processes of compliant grinding in an efficient way, leading to intelligent compliant grinding and improving the grinding quality of complex superalloy workpieces.

4.2.3 Grinding Systems

The robot belt grinding system consists of two components: hardwares and intelligent control models/algorithms. Hardwares refer to actuators which execute the machining instructions and sensors which collect multi signals during grinding, while the control models process and analyze the signals and generate control instructions based on signal-analyzing results and preset parameters, achieving closed-loop and intelligent control of grinding.

4.2.3.1 Hardwares

The hardware component can be divided into two subsystems, that is, the machining subsystem including robots, grinding tools and PLC (Programmable Logic Controller), and the monitoring subsystem including multi sensors, data collection and conversion equipment, data analyzing and feedback facilities. Figure 4.3 presents an example related hardwares of a typical robot belt grinding system.

In the machining subsystem, the joint motion of the robot is controlled by the robot controller. The workpiece is grabbed from the feeding table by a specific clamp mounted at the end of the robot joint (or the force sensor, if applicable), and then is

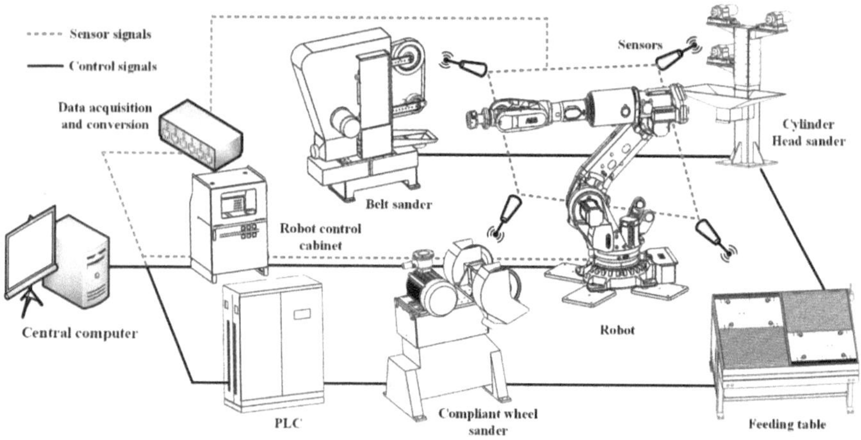

FIGURE 4.3 A robotic intelligent belt grinding system.

ground by different sanders to remove a certain number of materials and meet the requirement of surface integrity. The robot, sanders and feeding table work cooperatively under the control of the PLC.

In the monitoring subsystem, the six-axis force sensor is usually mounted at the end of the robot joint and integrated with the robot, which can not only record the forces and torques in three directions, but also enable constant-force grinding, which means that that the feed rate of the workpiece is automatically determined during grinding. Around the grinding area, sensors such as the infrared camera, the thermocouple, the microphone, the acoustic emission sensor, the high-speed cameras, and the laser three-dimension profilometer are used to monitor and measure workpiece temperatures, machining sounds and vibrations, grinding tool wear, material removal amount and ground surface conditions of workpieces. The measured data will be sent to the storage unit in the central computer after necessary processing such as data conditioning and data conversion.

To better describe the grinding process, in this chapter, the data obtained by sensors is defined as the characteristic parameters, while the indices of surface integrity are defined as the target parameters, and the parameters related to specific grinding actions such as normal forces, belt speeds, belt types, grinding times, grinding paths, workpiece gestures and coolants are defined as grinding parameters. Target parameters are directly and closely related to characteristic parameters, and characteristic parameters are determined by grinding parameters. In other words, characteristic parameters are the bridges connecting the grinding parameters and target parameters. For every grinding scenario, the main work is to choose proper grinding parameters to obtain the expected target parameters in an efficient way. Therefore, characteristic parameters are the key for intelligent and close-loop control of robot compliant grinding.

Figure 4.4 presents the basic structure of the intelligent control flow of robot belt compliant grinding. The established intelligent control models and algorithms in the

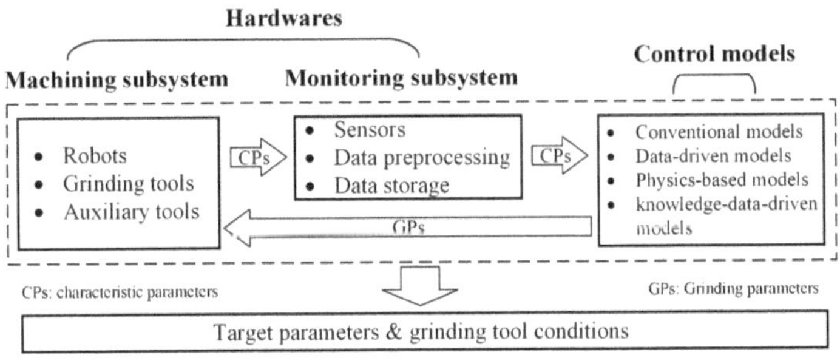

FIGURE 4.4 A structure of the intelligent control flow of robot belt compliant grinding.

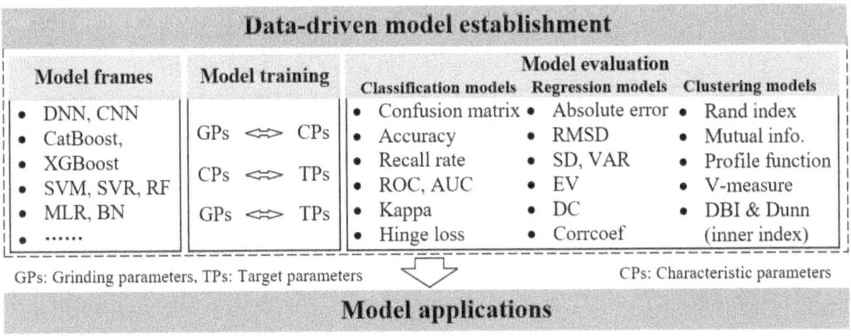

FIGURE 4.5 Basic work logic of supervised data-driven control models.

central computer will analyze characteristic parameters, to judge whether the required target parameters are already gained (if so, the grinding process will be terminated) and determine the following grinding process. Based on analysis results, the central computer sends instructions to the PLC, which controls (keep or adjust) the grinding parameters to ensure that the target parameters will be gained. Meanwhile, the control models will evaluate the condition of grinding tools and provide their current cutting capacity and remaining life.

4.2.3.2 Intelligent Control Models

Control models are the fundamentals and keys to achieve intelligent robot compliant grinding. Conventional typical control models include classical feedback control, PID (Proportion Integration Differentiation) control and fuzzy control. At present, as the rapid development of computer technologies, the artificial intelligent-based (AI-based) control models draw much attention from both academia and industries, which are efficient in solving high-nonlinear and strong coupling problems.

The supervised data-driven models are currently the most popular AI-based models, and their basic work logic is shown in Figure 4.5. To build a data-driven model, the first step is to select a model frame considering the problems to be solved, such as DNN (Deep Neural Network), CNN (Convolutional Neural Network),

CatBoost (Gradient Boosting+Categorical Features), XGBoost (Extreme Gradient Boosting), SVM (Support Vector Machine), SVR (Support Vector Regression), MLR (Multiple Linear Regression), RF (Random Forest), BN (Bayesian Network) and so on, and sometimes structure adjustments are required based on the original model frame and more than one frame needs to selected and fused. Then, a large amount of data is used to train the model frame, during which the grinding or characteristic parameters are regarded as the input date while the target parameters are output data. After specific iterative training processes, once the frame's performance is good enough, which is evaluated from many aspects by different indicators, the hyper-parameters of the model frame are determined, and the model is established. Then the model is tested in real grinding processes and finally is applied. Zhang et al. [7] recorded the sound signals during belt grinding Inconel 718, and the belt-condition-related sound signals were separated by Fast Fourier Transform and Discrete Wavelet Decomposition. Then a data-driven model was built using Optimally Pruned Extreme Learning Machine (OP-ELM) to accurately monitor the belt condition. Based on the grinding sound signals, Chen et al. [8] combined the RF and MLR to monitor the belt wear. Gao et al. [9] established a XGBoost model to predict the material removal during grinding. Wang et al. [10] extracted belt condition features from belt images, which were used to train a CatBoost-based model. This model was directly related to belt condition and not influenced by grinding parameters and performed well on predicting the material removal rate.

However, purely data-driven models completely ignore the physical principles of the grinding process, and accuracy highly depends on the scales of the training data. In practice, grinding conditions and workpiece features are various, and obtaining a large amount of data is challenging and sometimes hard to realize. These determine that the data-driven models have low generalization performance, poor robustness, and poor interpretability of the results. Such intrinsic defects significantly limit the widespread applications of purely data-driven models in robot compliant grinding. A few researchers tried to monitor grinding processes using indicators that are directly related to the physical process of grinding. Taking the belt condition (BC) as an example, Huang et al. [11] proposed a partitioned BC monitoring method based on a unified coefficient and images processing. The unified coefficient was directly calculated using two features extracted from belt images and highly related to BC. This method did not use conventional data-driven models, and was quite simple, less-computation, and has high adaptability.

However, grinding processes exhibit high nonlinearity and strong coupling, rendering knowledge-based methods extremely complex or unsolvable. Consequently, purely knowledge-based methods have limited applicability. To address this, knowledge-data-driven models have been proposed, incorporating the physics principles of machining processes to overcome the limitations of purely data-driven models. These models retain their superior capability in solving high-nonlinear and strongly coupled problems.

Currently, a typical representative of knowledge-data-driven models is the physics-informed neural networks (PINN). In a PINN model, the physical principles and data-driven models are integrated and promote each other, significantly reducing the scale

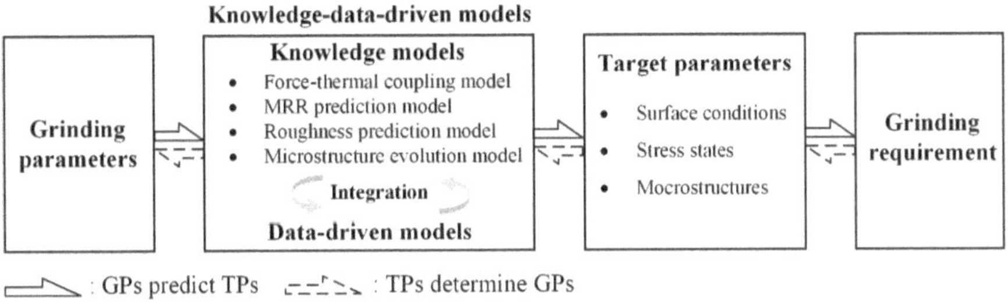

FIGURE 4.6 A structure of the knowledge-data-driven model for robot grinding.

of data set, improving the robustness and accuracy of the model. Thus, they have broad prospects in many fields [12]. Xie et al. [13] established a PINN model to predict the temperature fields during direct energy deposition (DED) by integrating the partial differential equation for heat conduction into a CNN. This model can achieve the same prediction accuracy as DNN and XGBoost by using only 20 percent of labeled data. Wang et al. [14] achieved high prediction accuracy of the machining tool wear by a PINN model, which explored enough information from physics and data domains, and eliminated the physical inconsistency in conventional data driven models.

Figure 4.6 presents a structure of the knowledge-data-driven model for robot grinding. The surface integrity of ground workpieces (target parameters) can be predicted by the grinding parameters; on the other hand, the optimized grinding parameters can also be obtained according to the requirements of target parameters. These will significantly promote the intelligent control level and machining efficiency of robot compliant grinding. Obviously, accurately establishing knowledge models are the key for intelligent control of robot complaint grinding.

4.3 FORCE AND THERMAL FIELDS DURING GRINDING

The grinding force field, thermal field and their coupling effects are the fundamental and direct causes of the microstructure alterations of the ground surface layer. Elucidating the mechanisms of grinding force-thermal coupling is the key to understanding surface integrity.

Due to the elastic deformation of grinding tools, the force field and thermal field of compliant grinding is much more complicated than that of conventional rigid wheel grinding. A study on abrasive grain-workpiece interaction [15] simplified abrasive grains as balls and analyzed the non-uniform elastic deformation caused by large deformation of the contact wheel in robot compliant grinding. Based on this, the study transformed the three-dimensional contact problem into a one-dimensional linear spring model to determine the contact pressure distribution at the interface between the grinding tool and the workpiece. Li et al. [16] established microscopic models of sliding force, plowing force, and cutting force of robot belt grinding Inconel 718, in which the statistical distribution of abrasive height was comprehensively considered. Following this, the grinding mechanisms were analyzed to predict the material removal rate.

However, the one-dimensional 'spring' model ignores the rotation motion parallel to the grinding direction of abrasive grains. Anthony Beaucamp et al. [6] theoretically and systematically analyzed the grain-workpiece interaction effects and forces of shape adaptive grinding (SAG) and built a two-dimensional 'spring' model of a single abrasive grain. According to this model, considering the distribution rules of abrasive size, they achieved to compensate and correct the cutting depth of abrasive grains in terms of the characteristics of sliding, plowing, and cutting, and proposed a prediction method of dynamic cutting depth at distinct positions.

Heat input and heat flow distribution are the key points of grinding thermal fields. Kim et al. [17] used inverse heat discrete algorithm to determine the total heat input for creep feed grinding, and established three heat source models, namely rectangle, right triangle, and irregular triangle. Jingliang et al. [18] built a model of the interaction between abrasive grains and workpiece materials in the grinding contact area of grinding wheel, and applied Rowe's heat distribution law to discrete grinding areas to obtain a novel heat input model. However, there are few reports on solving thermal field based on the interaction mechanism between abrasive grain and workpiece in robot compliant grinding. By using an infrared camera and finite element simulation, Ren et al. [19] proposed a dynamic heat input prediction model of robot belt grinding Inconel 718 based on BADs-LSSVM. The average accuracy of the model reached 96.7 percent, which provided an important basis for in-depth analysis of the thermal field. However, these models did not fully consider the changes of material mechanical and thermal parameters during grinding, and the accuracy, generalization and robustness of models are limited.

In robot belt grinding Inconel 718, Ren et al. [20] proposed a dynamic energy partition model based on grinding effects and temperature dependent mechanical properties. The main procedures are shown in Figure 4.7, which include four steps. Prior to commencing the calculation, various preparatory measures must be taken. These measures consist of determining the critical parameters, creating geometry and heat transfer models, defining input values for temperature-independent parameters, initializing the room temperature, and setting the first interval temperature to initiate the calculation. Next, the calculation proceeds with determining the temperature-dependent scratch hardness, the MRR dependent on grinding force, and the dissipated energy resulting from both air and chips. The third step involves determining the number of contact grains and their respective grinding effects based on force balance, height distribution of grains, and the revised scratch hardness, which allows for the calculation of the energy resulting from both plowing and cutting effects. Finally, an optimization algorithm is utilized to ensure that the energy absorbed by both grains and workpieces is optimized, using the principle that the maximum temperatures of both grains and workpiece are equal. It should be noted that the abrasive grains are heated and cooled in a circular fashion, while the maximum temperatures and temperature fields of both grains and workpieces are continuously recorded and updated in the entire grinding process.

A more complete model of the dynamic force-thermal coupled field for grinding turbine blades is shown in Figure 4.8, which consists of three parts: the theoretical model, the solution model, and the prediction model.

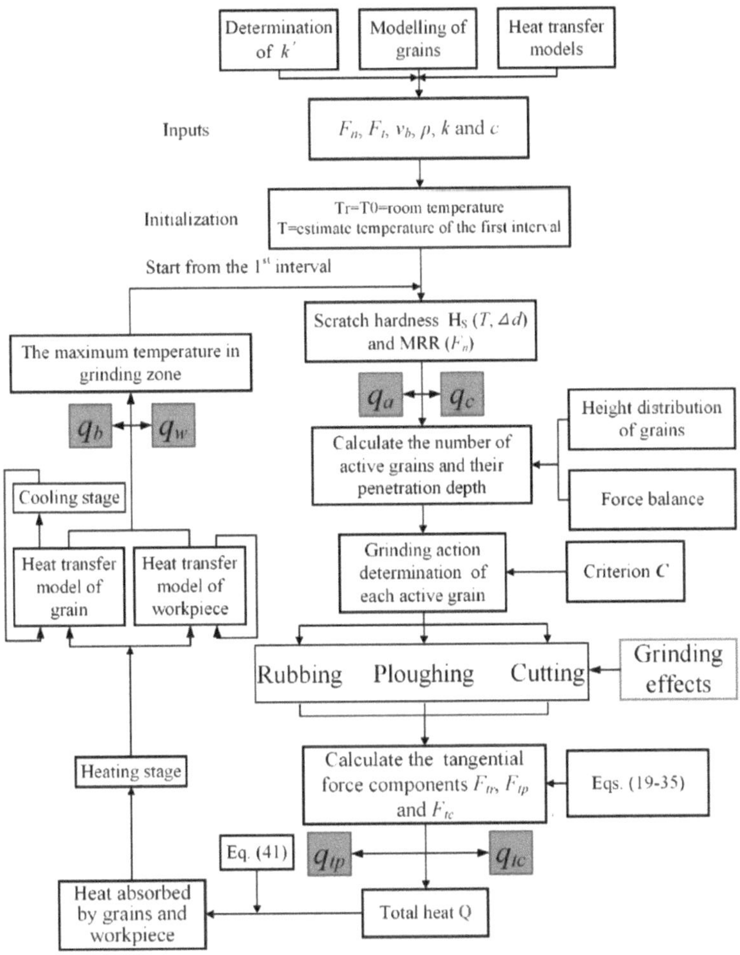

FIGURE 4.7 Procedures for energy partition calculation [21].

Theoretical model: Based on the two-dimension 'spring' single grain model and Hertz elastic contact theory, the micro and macro equation set of blade materials and abrasive grains can be obtained. Further, the real and dynamic cut depth of grains can be solved, and the multi grain-blade material elastic-plastic interactive model is established. Then, according to the interactive model, the total heat generated during grinding is gained, following which the heat input into blades can be solved. Considering the features of the blades' curved surfaces, the heat distribution model, and the differential equation of heat conduction for grinding blade can be obtained. Combining the interactive model, dynamic heat transfer model and force-thermal fields iterative methods, and considering the contact feature between blades and grinding tools as well as the alterations of blade material's mechanical and thermal properties, the theoretical model of dynamic force-thermal coupled field is finally built.

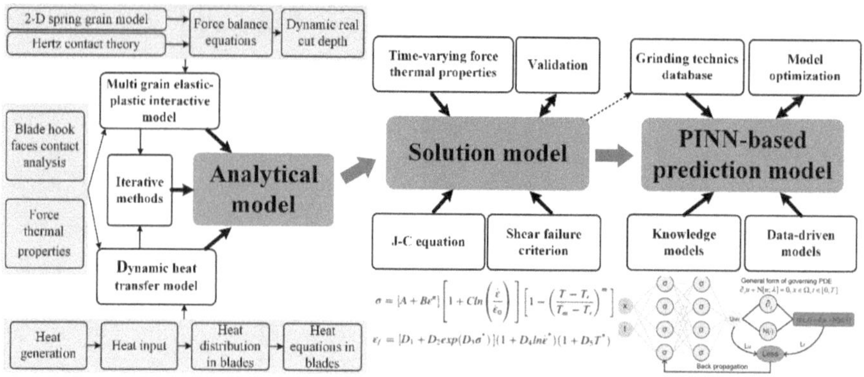

FIGURE 4.8 Modeling of the dynamic force-thermal coupled field.

Solution model: Efficient resolution of the established theoretical model is imperative, and the finite element simulation presents a viable approach. By combining the classic Johnson-Cook constitutive equation (J-C equation) with the shear failure criterion for blade materials, a finite element model is constructed using business softwares (e.g., ABAQUS, ANSYS). Then the stress, deformation, and thermal fields of the ground layer materials during the grinding process could be gained. The simulation results are validated by the measured data.

Prediction model: The objective of modeling the dynamic force-thermal coupled field is to forecast force and temperature-related parameters. Hence, a precise and efficient prediction model is indispensable. By virtue of the FE model, a considerable amount of data and the database can be obtained. Thereafter, a PINN-based prediction model can be constructed by integrating data-driven models and thermal physical models. Finally, Finally, using optimization theory to optimize the PINN model from three aspects: accuracy, speed, and stability.

4.4 SURFACE CONDITIONS OF GROUND WORKPIECES

The surface conditions of ground workpieces are primarily influenced by the elastic deformation of grinding tools and grinding parameters, which significantly determine the shape accuracy and service life of workpieces. In this section, three surface condition indicators, that is, surface roughness, surface burn and microcracks are discussed.

4.4.1 Surface Roughness

The evolution principles of surface roughness and the theoretical models are the research focuses. Wang et al. [21] established a model of material removal rate based on the Hertz elastic contact theory, and then constructed the surface removal contour of abrasive belt grinding (shown in Table 4.2). Anthony et al. [22] explored the changes of SAG grain size and shape during grinding, and put forward a concept of

TABLE 4.2
Theoretical Models of Surface Integrity

Author	Model purpose	Model description
Wang et al. [21]	Calculate surface removal contour	$h'h'(xi) = 2\int_{0}^{Tc(xi)/2} C_g P_t(xi,t)^{b_1} V_s^{b_2} dt$ T_c: contact time, C_g: contact constant, P_t: contact pressure, b_1, b_2: constants, V_s: belt speed
Wang et al. [24]	Calculate the evolution rule of fatigue damage	$\Delta D_i^{fat} = \sum_{\alpha=1}^{12}\left[\frac{\lvert\tau_{\max}^{\alpha}\rvert}{S_{oct}}\right]^{mfat}\left[\frac{\gamma_{\max}^{\alpha}}{\gamma_{oct}}\right]^{nfat}$ $\tau_{\max}^{\alpha}$: maximum shear stress of the slip system α, $\gamma_{\max}^{\alpha}$: maximum slit strain rate, S_{oct}: reference stress rate, γ_{oct}: reference strain rate.
Wang et al. [25]	Calculate critical conditions for crack propagation	$\Omega(d,h) = \left[2H\Pi\left(\frac{h-\tan\psi d}{\cos\phi}\right)^{\frac{1}{2}} - \pi^{\frac{1}{2}}K_{IC}N^{\frac{1}{m}}\right]$ $d - 2H\Pi\pi\left(\frac{h-\tan\psi d}{\cos\phi}\right)^{\frac{3}{2}}$ H: material hardness, Π: stress coefficient, h: crack depth, ψ half angle of cone, d: cut depth of grain, KIC: fracture toughness, N: density of cracks, m: material constant
Bouktib et al. [26]	Calculate the grain cut depth and analyze residual stress with a FE model	$h_{res} = h_{k+1}\left\{1-\left[\frac{h_{k+1}}{h_y}\left(\frac{RY_{k+1}^2}{0.045h_{k+1}E^2}\right)^n\right]^{-0.3}\right\}$ $\left\{1-\left[\frac{h_{k+1}}{h_y}\left(\frac{RY_{k+1}^2}{0.045h_{k+1}E^2}\right)^n\right]^{-0.65}\right\}$ $Y_{k+1} = Y_k\left(\frac{0.3E}{Y_k}\sqrt{\frac{h_k}{2R}}\right)^n$ h_{k+1}: real cut depth, h_y: cut depth at plastic deformation, Y_{k+1}: real yield stress at current grinding, Y_k: yield stress at previous grinding, R: radius of grain

(continued)

TABLE 4.2 (Continued)
Theoretical Models of Surface Integrity

Author	Model purpose	Model description
Ding et al. [27]	Calculate Phase transition fraction	$f = f^{eq}\left\{1-\exp\left[-k(T,\dot{T})(t)^{m(T,\dot{T})}\right]\right\}$ (diffusion type) $f'_\alpha = f_\gamma\left\{1-\exp\left[-\beta(M_s - T)\right]\right\}g(T,\dot{T})$ (non-diffusion type) $k(T,\dot{T}) = C_{k1}(T)^{C_{k2}}(\dot{T})^{C_{k3}}$, $m(T,\dot{T}) = C_{m1}(T)^{C_{m2}}(\dot{T})^{C_{m3}}$ $g(T,\dot{T}) = C_{g1}(T)^{C_{g2}}(\dot{T})^{C_{g3}}$ f_{eq}: phase equilibrium fraction, $k(T,\dot{T})$ $m(T,\dot{T})$ and $g(T,\dot{T})$: function of temperature and heating rate, f_y: maximum phase fraction of austenite, β: 0.01, material constant
Sugui et al. [28] and Xu et al. [29]	Calculate the driving force elements diffusion	$F = \frac{2A}{3\alpha_0}\left[1-\frac{E}{(E+\sigma_a)}\right] + \lvert\Omega dS\rvert + \left(\frac{B}{2E}\right)(\sigma_a - \sigma_{mis})^2$ First term: interatomic potential energy; Second term: interfacial energy; Third term: external stress. α_0: average lattice parameter of γ/γ' phases, A and B are the constant, σ_0 : applied stress, σ_{mis} : misfit stress between the γ/γ' interfaces, Ω: interface energy per unit area of γ/γ' phases, dS: change in total interface area of γ/γ' phases

equivalent thickness of grinding ships in compliant grinding, which provided a reference for controlling the grinding surface roughness. Following this, a research team proposed the ductile-brittle transition criteria in SAG from the perspective of microscopic cut depth of abrasive grains and material response features, thereby controlling the surface roughness of the ground workpieces [23].

For compliant grinding, the grinding tool deforms elastically, resulting in the difference of contact pressures in various positions in the grinding area, and thus the changes of surface roughness. Discretization could be an efficient method to deal with this problem. Zhao et al. [30] used a discretization method to simplify and segment the entire belt-workpiece-contact curved surface during grinding process into several local planes with different contact pressures, and calculated the cutting depth of abrasive grains on each plane at different times. According to the cutting depth, they derived the envelope curves of cutting edges, and the surface roughness simulation based on the grain-workpiece interaction mechanisms was realized.

The roughness of ground surfaces is closely related to the interactions between abrasive grains and workpiece materials. Therefore, taking deep insight into the interaction mechanisms can promote the modeling of surface roughness. Zhu et al. [6] systematically investigated the grain-workpiece interaction mechanisms of SAG. Based on 2-D 'spring' grain model and the distribution model of grain sizes, meanwhile considering the dynamic compensation for grains' elastic deformation caused by sliding, plowing, and cutting during the grinding process, they proposed a method to predict the dynamic cutting depth of grains at different grinding areas. Following this, the relationship of 'grinding normal force-cutting depth-grain wear' is clarified, and the prediction of cutting depth and width is realized. Under a given offset of grinding tool, as the tool wears and grains become blunt, the number of grains that have cutting effect is significantly reduced, while the number of grains that have sliding and plowing effects is greatly increased, leading to a decrease in the cutting depth of the wear tool, and an increase in the cutting width, which in turn leads to more obvious surface scratches. Therefore, based on the established model, the influences of tool wear can be compensated in real time, and the prediction and control of grinding surface roughness could be achieved.

Automatically controlling the robot pose and feed rate is the key step for the applications of theoretical studies about grinding surface roughness. From theoretical models to robot control algorithms has already become the focus of researchers. A study by Song et al. [31] considered the elastic deformation of abrasive belt during grinding and the time-varying features of the polishing contact force, and then the relationship between the grinding force and the cutting depth in the robotic belt grinding is comprehensively analyzed. Following this, the study established a robot pose error model considering belt elastic deformation, and finally gained the mapping relationship between the grinding force and the cutting depth.

4.4.2 Surface Burn

Surface burn is caused by the high temperature in the contact area during grinding. Mild burn will cause the formation of oxide films on workpiece surfaces. However, serious burn will introduce tensile residual stress and cause microstructure transformation in the ground surface layers, resulting in metamorphic layers and even white layers, which may deteriorate the surface quality and are usually detrimental to service performance of machined workpieces. In addition, burn could sometimes induce microcracks, leading to the workpiece being scrapped. Excessive grinding time, high contact pressure, poor heat dissipation conditions and other inappropriate grinding parameters will cause surface burn. Nickel-based superalloys have poor thermal conductivity, and the grinding heat is readily accumulated in the workpiece during grinding. Therefore, there is a pressing need for reliable methods and models to monitor temperatures in compliant grinding nickel-based superalloys so to avoid surface burn.

Wang et al. [32] studied the relationship between grinding force, belt speed and temperature of robot belt grinding Inconel 718, and investigated the surface burn

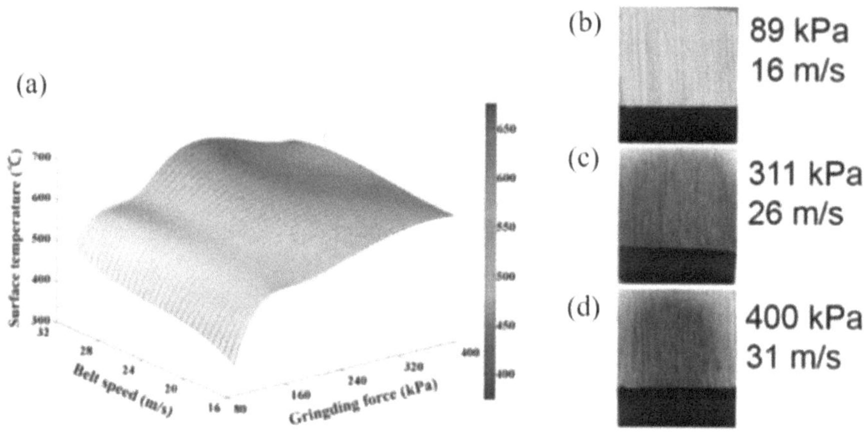

FIGURE 4.9 Grinding temperatures and surface burn: (a) Relationship between temperatures and grinding parameters, (b) normal grinding surface, (c), (d) surface burn [32].

during grinding. As shown in Figure 4.9(a), an increase in grinding force or belt speed leads to a higher temperature. However, as the grinding force or the belt speed exceeds a certain threshold, the temperature rises slowly. It can be found from Figure 4.9(b) that the grinding burn does not occur when the grinding temperature is below 580°C. With increasing the grinding force and belt speed, more grinding heat is generated in a short time and accumulate in the grinding area, leading to an increase of temperature. Therefore, the surface burn appears, as presented in Figure 4.9(c) and (d).

The key to avoid grinding surface burn is to accurately monitor the temperature and control the maximum temperature of the workpiece below a threshold value. At present, thermocouples, and infrared radiation (IR) cameras are two common methods to monitor the grinding temperature. Thermocouples, despite their ability to provide accurate grinding temperatures, have several limitations, such as damages to workpieces, wear-caused ineffectiveness, long response time and difficulties in obtaining a large-scale and continuous temperature field. IR camera, as a non-contact temperature measurement method, is widely used to measure grinding temperatures, especially temperature distributions. However, the accuracy of IR camera measurement is significantly influenced by the surface emissivity. At present, accurately obtaining the temperature-dependent emissivity for surfaces with different conditions in a fast way is still a challenge. Due to the elastic deformation of compliant tools during grinding, temperatures on the grinding surface are distinct at distinct positions. Hence, it is very necessary to measure the temperature distribution of grinding surfaces, which still needs more studies and efforts.

Using theoretical models is an efficient way to monitor and predict temperatures during grinding. Li et al. [33] established an analytical thermal field model for disc grinding, based on which, the structure of the disc can be designed according to the requirement, such as temperatures and surface roughness. Nevertheless, such theoretical models are rare, and their applications are very limited.

TABLE 4.3
Real-time and Nondestructive Detection Methods for Grinding Surface Burn

Signals	Introduction
Visual signal	**Principles**: Extract features such as hue, saturation HSV and so on from images of ground surfaces, and then establish the mapping relationship between features and the degree of burn/thickness of burn layer. **Limitations**: Image processing may take a long time.
Eddy Current	**Principles**: Monitor the alterations of microstructures and residual stress in the surface layer by analyzing the feature changes of eddy current. The signal processing time is very short. Limitations: Readily to be interfered by other factors.
Barkhausen Noise	**Principles**: Combine the Barkhausen noise and magnetoelastic acoustic emission to monitor the change of microstructures in the grinding surface layer. **Limitations**: Only applicable for magnetic materials; the facility is complex.
Acoustic emission and sound	**Principles**: Extract features from grinding sound signals and vibration signals, based on which to monitor and predict potential surface burn during grinding. **Limitations**: Poor interpretability of results; Readily to be interfered by other factors.

Other signals during grinding, such as visual signals and acoustical signals, are closely related with the surface conditions, and have been used to realize real-time and nondestructive detection for grinding surface burn. Table 4.3 summarizes four typical methods.

4.4.3 Microcracks

During grinding, alterations of microstructures in the surface layer could cause stress concentration and induce microcracks. Microcracks are serious grinding defects and will reduce the fatigue strength and fatigue life of workpieces. Exploring and clarifying formation mechanisms of microcracks are important for grinding quality improvement.

Some researchers established theoretical models to study the microcracks of superalloys during grinding. Zhu et al. [34] investigated the damage mechanism according to the initiation, propagation, and suppression of microcracks, using a single-grit scratching simulation in terms of an improved chip-thickness model considering elastic deformation of grinding tool and workpiece. And they proposed a new method to quantitatively describe the state of microcrack initiation using undeformed chip thickness. Wang et al. [24] systematically studied the rule and mechanism of influences of surface roughness on stress and fatigue property of DD6 nickel-based single crystal superalloy under grinding. Stress concentration has a positive relationship with surface roughness. In a certain range, the large surface roughness will intensify the stress concentration and reduce the fatigue life. In terms of the crystal plasticity theory and fatigue damage model, and considering resolved shear stress and slip rate of slip systems, they established a finite element model to predict surface morphologies. Following this, a geometrical equivalent simplification model to predict the fatigue life of ground workpieces (shown in Table 4.2). Overall, there is still a lack of systematic modeling research on the initiation, propagation, and suppression of microcracks in robot compliant grinding of superalloys.

In conclusion, the quality of ground workpieces is influenced by surface roughness, surface burn, and microstructures. Accurate theoretical models and monitoring methods are vital for achieving efficient and intelligent control of robot compliant grinding. Nevertheless, more research is necessary to establish systematic theoretical models for surface conditions in compliant grinding of superalloys.

4.5 RESIDUAL STRESSES OF GROUND SURFACE LAYERS

The machining process introduces or changes the residual stress of the surface layer via complex interaction of tool-workpiece materials. Compressive residual stress (CRS) can benefit the workpiece quality and improve the fatigue strength, while tensile residual stress (TRS) will promote the generation of microcracks and is usually detrimental to workpiece service performance.

In the grinding process, the residual stress is induced by three effects in workpiece materials: thermal expansion and contraction, thermomechanical plastic deformation (deformation caused by the abrasive grain), local phase transformation, as listed in Table 4.4. Thermal expansion and contraction tend to introduce TRS, while the thermomechanical plastic deformation is commonly the cause of CRS. Local phase transformation also exerts significant influences on residual stresses. When the density of the new-formed phase is larger than the previous phase, CRS is prone to appear. Otherwise, the TRS tends to form. During the grinding process, these three effects simultaneously exist, and the final type of residual stress depends on the balance of them. Tonshoff et al. [35] established the relationship between specific grinding power and residual stress, as depicted in Figure 4.10, which is a reference to comprehensively consider these three effects and to estimate the types of residual stresses.

Grinding-induced residual stresses present gradient distribution features along the grinding depth, which could exert special effects on mechanical properties, especially the tensile behavior of workpieces [36], and deserves much attention. Wang et al. [32] investigated the residual stress distribution under different grinding parameters

TABLE 4.4
Effects of Grinding Process on Workpiece Residual Stress

Types	Thermal expansion and contraction	Thermomechanical plastic deformation	Local phase transformation
CRS	Rare	Common	$\text{Density}_{\text{-new}} > \text{Density}_{\text{-previous}}$
TRS	Common	Rare	$\text{Density}_{\text{-new}} < \text{Density}_{\text{-previous}}$

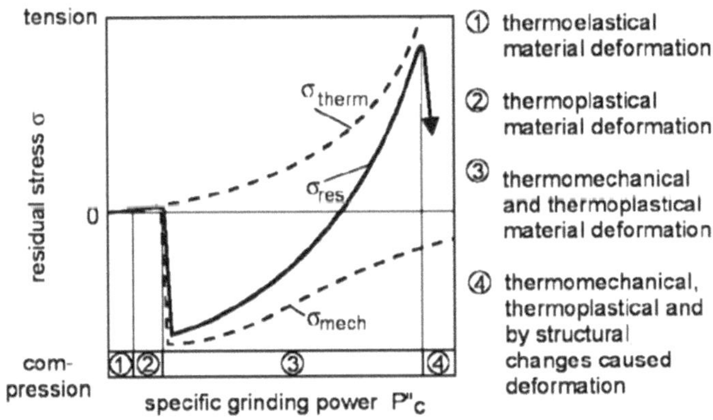

FIGURE 4.10 Relationship between specific grinding power and residual stresses [35].

along the grinding depth direction. The residual stresses for untreated specimens are near 0 and their changes are negligible. Under contact force 178 kPa and belt speed 31 m/s, significant CRS is induced in the 60 μm thick surface layer along both X and Y directions. However, as the contact force increases, obvious TRS appears and the maximum value of CRS is quite small, but the layer thickness increases to 75 μm. The residual stress is determined by the coupling effects of workpiece material deformation and temperature. When a specimen is treated with a larger contact force, the cutting depth of abrasive grain is deeper, and thus the material deformation is more severe. Therefore, the deformed layer is thicker. On the other hand, as the contact force increases, the temperature goes up, which relaxes the residual stress and induce thermal expansion, resulting in small CRS and even TRS.

In the grinding surface layer, the maximum CRS usually appears in the subsurface, that is, a certain distance from the machined surface. This could be attributed to the accumulation effect in the subsurface of material deformation. This phenomenon is common in other machining processes such as cutting, milling, and shot peening. Such a gradient structure could be used to strengthen the yield strength and improve the ductility of workpieces. Yang et al. [37] studied the residual stress in the surface layer of surface mechanical attrition treatment (SMAT) using synchrotron-based X-ray diffraction in an in-situ tension experiment. The results show that maximum CRS appears in the subsurface, where the strongest strain hardening exists. During tension, the layer with max. CRS remains elastic while adjacent layers start plastic deformation. Hence, higher yield strength is achieved through direct strain hardening within this layer, as well as significant dislocation hardening and hetero deformation-induced (HDI) hardening in adjacent tensile layers, and the ductility is simultaneously improved by the compatible deformation.

There are two ways to improve the stress states of ground surfaces.

(i) To avoid or relax the TRS by heating the surface layer. Feng [38] adjusted the temperature gradient and effectively controlled the residual stress of ground Inconel 718 surfaces. Simulated and experimental results demonstrated that this approach can greatly reduce TRS. Post processing such as heat treatment, shot peening and laser treatment is often used to relax the TRS. However, these methods may introduce extra microstructure damage and accuracy errors.

(ii) To introduce CRS by adjusting grinding parameters. Ding et al. [39] reviewed the residual stress in grinding from the aspects of generation mechanisms, influencing factors, modeling and simulation, measurement, and monitoring. The intelligent real-time monitoring of residual stresses, the distribution of residual stresses, the prediction of residual stresses of workpieces with complex shapes and the reliable theoretical models are concerns that need to be paid more attention in the future. Boukib et al. [26] established a theoretical model (shown in Table 4.2) to calculate the grain-cutting depth and its variation change rules, and studied the influences of initial surface roughness, grinding forces, grinding times and friction coefficient on the residual stress. Combined with the theoretical model and the FE model, the evolution rules of residual stresses in abrasive belt grinding were obtained. Wang et al. [40] explored the variation rules of residual stresses in abrasive belt grinding based on the established theoretical model and a two-dimensional finite element model. The results showed that increasing the grinding force and the number of abrasive grains could increase the amplitude and distribution depth of residual stresses. However, when the grinding force was too large, the MRR increased and thus reduced the amplitude and distribution depth of residual stresses. How to find the optimal parameters was the key. In addition, recently some scholars also attempted to use a molecular dynamics model to analyze the abrasive belt grinding-induced residual stresses at the molecular level [41].

In general, the number of theoretical models available for residual stresses in ground surfaces is limited. As the demand for grinding efficiency continues to rise, the grinding process has transitioned from traditional subtractive manufacturing to the integration of "subtractive manufacturing + surface strengthening". In such cases, achieving rapid measurement of residual stresses during grinding and accurately controlling them to the desired states, especially through theoretical models, deserves comprehensive and extensive research.

4.6 MICROSTRUCTURES OF GROUND SURFACE LAYERS

During grinding, the abrasive grains interact with workpiece materials with very high speeds. The coupling force-thermal effect and severe material deformation will introduce alterations of microstructures in ground surface layers, including grain feature and crystal defects, phase transformation, and white layers. These alterations are gradient and significantly affect the grinding quality. Understanding the evolution mechanisms and modeling the evolution process are essential.

4.6.1 Crystal Grain Features and Crystal Defects

Crystal grain refinement is common in the ground surface layer, which is mainly attributed to two severe-plastic-deformation-related causes: crystal grain crushing, and dislocation tangle. Under the cutting effect of abrasive grains, the crystal grains directly interacting with abrasive grains are readily crushed. During grinding, the accumulation of plastic deformation causes the entanglement of numerous dislocations, which gradually form a connected network of three-dimensional tangles, resulting in the formation of dislocation cells. Generally, the refined domains can also be observed in the surface layer materials. The domain refers to a region within a material where the atoms have a similar orientation or arrangement. The dislocation cells and domains are different from crystal grains, and they usually will not induce obvious alterations of grain boundaries. The dislocation cells and domains can be observed by a transmission electron microscope (TEM) and X-ray diffraction (XRD), respectively.

Wang et al. [32] observed the ground surface microstructures of Inconel 718 and revealed that the plastic deformation resulted in the introduction of a considerable number of dislocations in the initial equiaxed grains of Inconel 718. Moreover, the subgrain boundaries and new grain boundaries are formed in the grain, indicating the occurrence of dynamic recovery. The formation of new grains is facilitated by the increasing misorientation of subgrain boundaries. The grinding process induces plastic deformation due to shear force, leading to lattice rotation and grain fragmentation. Meanwhile, the high temperature generated during grinding will result in a homogeneous microstructure comprising equiaxed grains due to boundary splitting. Although the process of dynamic recrystallization is continuous, the grinding duration is short, leading to the presence of a certain number of dislocations in the equiaxed grains. Therefore, grain refinement during grinding is due to continuous partial dynamic recrystallization related to plastic deformation and thermal treatment.

Huang et al. [42] characterized domain size and microstrain of the surface layer of belt-ground Inconel 718 using XRD. As shown in Figures 4.11(a) and (b), comparing to the matrix material, the profiles of diffraction peaks (BG) of the ground surface-layer material present remarkable widening, indicating that significant domain refinement and large microstrain occur after grinding. The calculated domain size in the outmost surface is 12 nm, decreasing 78.2 percent relative to the matrix material. The maximum microstrain is 6.55×10^{-3}, about 8.4 times larger than that of the matrix material. The domain grain refinement and large microstrain contribute to the severe plastic deformation of surface materials during grinding. observed the dislocation cells in the ground surface layer of Inconel 718 using TEM. Figure 4.11(c) is the grain boundary distribution along the grinding depth direction in the surface layer, which is obtained by electron back scattering diffraction (EBSD). Along the grinding depth direction, a gradient grain refinement is observed, which could be attributed to the severe plastic deformation of materials and mechanical crushing of crystal grains under the impact and shear of abrasive grains.

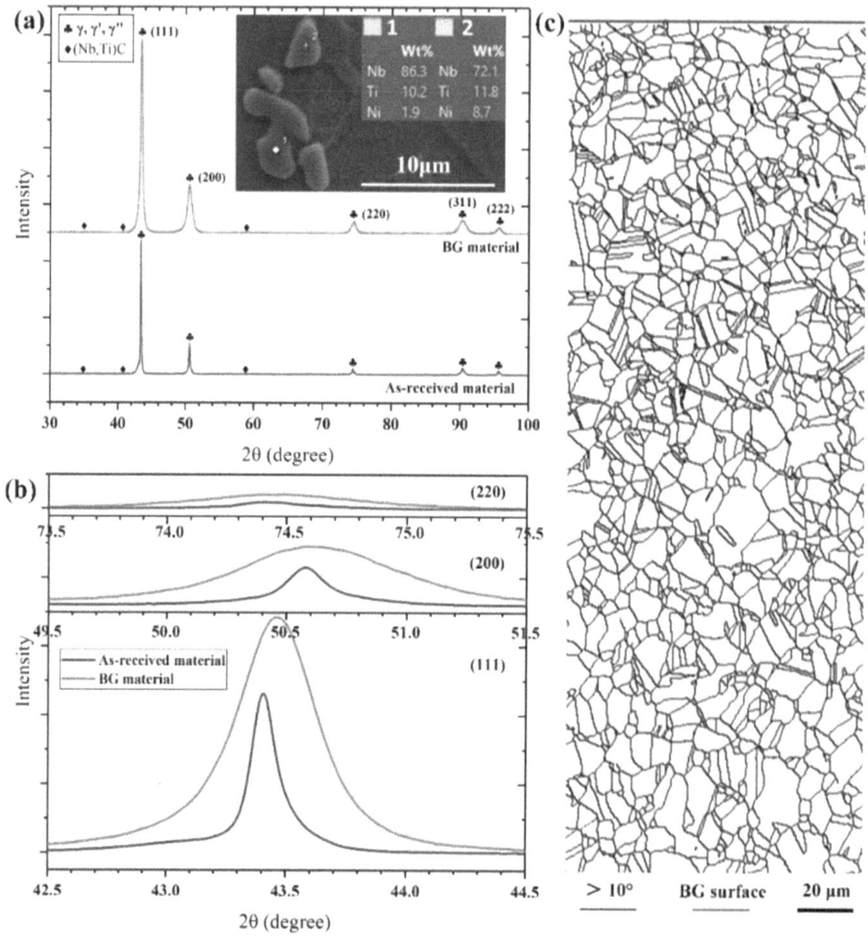

FIGURE 4.11 Microstructures of matrix Inconel 718 and ground surface layer: (a) full range XRD profile, (b) amplified XRD patterns of (111), (200), and (220) peaks, (c) the grain boundaries along the grinding depth direction in the surface layer [42].

Du et al. [43] investigated the gradient microstructures ground surface layer of GH4169 alloy using (TEM) and high-resolution transmission electron microscopy (HRTEM). Along the grinding depth direction, the surface severe deformed layer can be divided into three layers: the surface amorphous layer (SAL), the micro-shear band (MSB) and the nanocrystalline layer (NCL). The SAL has amorphous microstructures and contains plenty of vacancies, with some ordered structures with diameters of 1–3 nm still existing in the SAL. MSB is completely amorphous and is parallel to the grinding surface with width of 2 nm. The NCL comprises many nano grains with diameter of 10–150 nm, and many crystal defects including amorphizations, dislocations, stacking fault and twins are found in these grains. Gong et al. [44] conducted deep investigations into phase changes of nickel-based single crystal superalloy. Under the coupling effect of grinding forces and temperatures, a grinding metamorphic layer is formed, in which the γ phase and γ' phase are severely distorted along the grinding direction, and they tangle with each other to form a hardened layer. The phase distortion has gradient distributions along the grinding depth direction, which is ascribed to the mechanical drag effect of abrasive grains and the interaction between dislocations and phases.

During grinding, the mechanical drag effect of abrasive grains causes the single-direction severe plastic deformation of surface layer materials, which can induce alterations of crystallographic textures of surface layer materials. Huang et al. [42] reported that after belt grinding Inconel 718, the random orientation of surface layer materials is changed and a strong <100> texture is observed, which has a gradient distribution along the grinding depth direction. Feng et al. [45] investigated the orientation mapping of the ground surface nanostructure of Ni-based superalloy using a novel precession electron diffraction (PED) technique coupled with transmission electron microscopy (TEM). A strong shear texture and a minor copper texture are identified. Fergani et al. [46] proposed a systematic method to predict the texture evolution in machined workpiece by combining the FE model and Visco Plastic Self Consistent method. This model can predict shear texture for the main BCC phases, and the relative slip activity between the two phases, which aids in the planning of machining processes to realize required machining properties.

It should be pointed out that the subtractive grinding is different from the surface mechanical grinding treatment (SMGT) [47]. In SMGT, the machining tool (for example, a ball tip) repeatedly slides under a certain pressure on the workpiece surface, and usually in a very low temperature (for example, liquid nitrogen temperature), during which the materials will not be removed at all, but only deform severely. After SMGT, a gradient nano-grained (GNG) surface layer (several decades micrometers thick) will be synthesized on the substrate coarse-grain materials, enabling the entire workpiece to simultaneously have both tensile strength and plasticity. For the subtractive grinding, the material deformation is much weaker than that for SMGT. After grinding, a surface layer with gradient microstructures will form, and sometimes a very thin nano-grain layer will appear on the outmost surface. The gradient architecture of microstructures in the surface layer will introduce special mechanical behaviors and have unique deform mechanisms, which exert effects on the service performance of machined workpieces, and thus are worth in-depth study.

Current studies mainly focus on three aspects: (i) characterizing the alteration of crystal grain features and crystal defects of the ground surface layer; (ii) investigating the influence rules of different parameters; (iii) analyzing and revealing corresponding change mechanisms. Theoretical models are very rare and should be the emphasis of future works.

4.6.2 Phase Transformation

The maximum temperature of nickel-based superalloys workpiece during compliant grinding can reach near 1000°C. Although the lasting time of grinding is usually short, the force-thermal coupling effects can cause the phase transformation of the surface layer material during grinding.

Wu et al. [48] studied the thermal effects on phase transformation in grinding of Fe-Ni superalloy. The findings indicate that the grain boundary wetting phase transformation occurs with varying grinding temperature and thermal burn. As the grinding temperature increases, the volume fractions of martensite and austenite increase, while that of ferrite decreases. And a small feed speed of workpiece and grinding depth can reduce the phase transformation, thus improving the quality of ground surface.

The FE (finite element) model is an efficient method to analyze and predict the phase transformation under different grinding parameters. Duscha et al. [49] established a FE model to analyze the phase transformation kinetics of materials during deep, pendulum and speed stroke grinding. In the model, the temperature-dependent and phase history-dependent material properties were used to ensure more accurate simulated results. The FE model was evaluated using micrographs of ground workpieces, and a significant correlation was observed for various grinding parameters. Based on this model, the dynamic temperature distribution in workpiece during grinding can be determined, relying on which phase transformation can be calculated.

Theoretically models of material phase transformation during grinding are essential. Ding et al. [50] studied the influences of grinding parameters on phase transformation during grinding by proposed theoretical relationships. First, they established a link between the quantitative material phase and the diffusion-controlled kinetics and diffusion less transformations, meanwhile considering the strain rate and the grinding temperature under varying temperatures. Then, a mechanical physics-based model is built to predict the volume fraction of phase transformation in different temperatures under anisothermal conditions, based on which the relationship between phase volume fractions in sequential segmented isothermal processes and grinding parameters are clarified. Finally, the model was validated experimentally. Via correlation coefficients, linear correlations between grinding parameters and phase transformations are established. The influences of grinding parameters, temperatures, and strain rates were determined by the partial correlation coefficient analysis. This model enables to introduce expected phase transformations and to modify local properties of material, by adjusting grinding parameters. Ding et al. [51] further improved the thermotical perdition model and realize phase transformation at high heating rate during grinding. The flow chart of the model is shown in Figure 4.12.

The force- and thermal-related submodels are calculated by indirect inputs, including workpiece material properties, grinding tool topographic parameters, interaction geometry configuration and process parameters. The temperatures and heating rate are obtained by submodels, and are directly input into the phase transformation model. Finally, the phase transformation volume fractions under varying grinding parameters are gained by the Johnson-Mehl-Avrami (JMA) model and the Koistinen-Marburger (KM) model, taking influences of temperature and heating rate into account. The proposed physics-based model can predict the phase transformation volume fractions in continuous grinding process under anisothermal conditions. Moreover, experimental evidence confirms the feasibility of using this model to predict the extent and occurrence of phase transformation with respect to material properties, wheel topography, and thermal-mechanical loading during grinding.

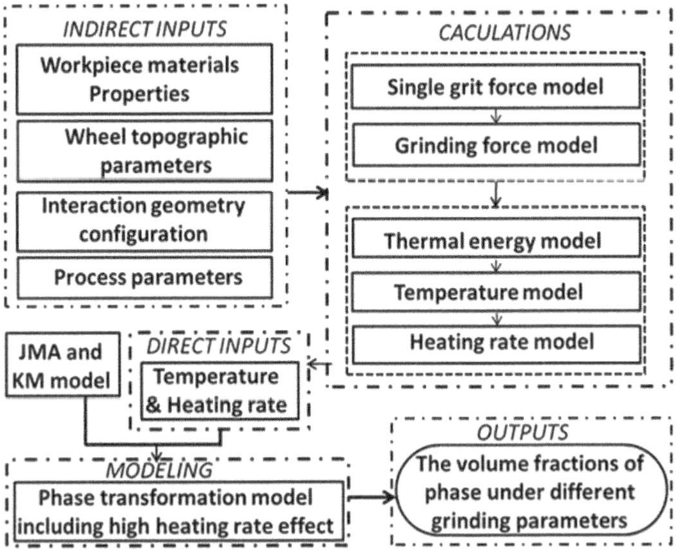

FIGURE 4.12 Flow chart of a phase transformation model [51].

Nevertheless, theoretical models about grinding nickel-based superalloys are scanty, and existing models usually ignored the influence of grinding tools' elastic deformation.

4.6.3 White Layers

During grinding, when the coupling effect of the severe plastic deformation and high temperatures is significant in the surface materials, a thin layer having quite different microstructures and mechanical properties with the matrix material will form, with a thickness ranging from a few micrometers to several hundred micrometers. This layer presents a white contrast under an optical microscope or a scanning electron microscope, and thus is defined as the white layer. This layer is usually composed of complicated crystals and various phases. Generally, a white layer is characterized by high hardness, poor ductility, high residual stresses, and altered corrosion resistance, which is detrimental and can result in premature failure of the workpiece. Hence, understanding the formation mechanisms and properties of the white layer is of great significance for designing and optimizing the grinding process and parameters for nickel-based superalloys.

Generally, the white layer is caused by three mechanisms: (i) rapid heating and quenching leading to transformation products; (ii) surface reaction with the environment; and (iii) plastic deformation generating a homogeneous structure or fine-grained structure [52]. For steels, the phase transformation of austenite-martensite caused by rapid heating and cooling is the main reason for the white layer formation. However, the mechanisms for nickel-based superalloys are quite different because the austenite structure is stable from room temperature to high temperature, even in melting temperature.

The dynamic recrystallization (DRX) is one probable reason for the formation of white layers. Zhang et al. [53] studied white layer formation during turning superalloy Inconel 718. After machining, a white layer is formed in the surface materials, and crushed and refined δ phase is observed therein, but no new phase is observed. Compared with hardened steel, there is no "dark layer", meaning that the corrosion resistance of the surface materials has no obvious alteration. Moreover, some defects including cracks, craters, and flaking are shown in the white layer. Few oxides, carbides, nitrides, or carbonitrides are generated. Some elements of Nb and Mo are diffused into the γ phase. They also found that the white layer shows typical crystal features of DRX, with refined and equiaxial grains. From the view of evolution of grain boundaries, they consider that continuous dynamic recrystallization is the main formation mechanism for the white layer. Initially, severe plastic deformation results in plenty of dislocations, which are subsequently converted into low-angle grain boundaries through dynamic recrystallization, leading to the formation of subgrains. As the subgrains gradually rotated under high temperature and further deformation, equiaxed and fine recrystallized grains were generated. This process continues until the strain and temperature reach a limit.

Some researchers pointed out that dynamic recrystallization plays a significant role in the white layer generation and have established some theoretical models. Xu et al. [29] investigated the grinding-induced white layer of Ni-based single-crystal superalloy. After grinding, a several micrometers thick whiter layer with dense structures is formed, and its thickness is significantly affected by the grinding parameters. Beneath the white layer, a severe deformed layer (SDL) with severe shear deformation flow usually exists, where slip lines on the γ' phase, the elongated and rotated γ'-cube phase, as well as the distorted and narrowed γ-channel phase are presented. They used the element diffusion theory to analyze the formation of the white layer. The EDS (Energy Disperse Spectroscopy) analysis indicates that elements Al and Ta in the white layer have great increases of about 107 percent and 127 percent, respectively, while the content of Cr and Co show obvious decrease. The results indicate that under high strain rate and high temperature during grinding, Al and Ta generating the γ' phase diffuse into the γ phase, while Cr and Co in the γ matrix phase diffuse into the γ' phase. Such diffusion is uphill because the Al and Ta in the γ' phase is large and Cr and Co in the γ phase is small. As shown in Table 4.2, Sugui et al. [28] and Xu et al. [29] pointed out that as grinding speed and feed speed increases, the severe plastic deformation and dislocation movement are greater and the temperature are higher, which will promote the diffusion of these elements, and thus resulting in thicker white layers.

Relying on more advanced characterization and test methods, such as TEM, transmission Kikuchi diffraction (TKD), atom probe tomography (APT) and compression tests of micro-pillars, more microstructure features and formation mechanisms of white layers for machined nickel-based superalloys are revealed.

In the white layer of milled P/M IN100, the outermost layer consists of a partially amorphous phase, while a double-phased nanocrystalline layer follows it. This variation is ascribed to the high strains, strain rates, and temperature gradients [54]. Kruk et al. [55] reported that the white layer of milled Inconel 718 consists of nanocrystalline γ phase without γ', γ'' or δ phase precipitates, and the DRX-induced grain refinement caused by heat accumulations, high strain rate and severe plastic deformation are considered as the principal mechanism of white layer formation. Chen et al. [56] investigated the white layer of broached Inconel 718 and revealed that it is produced as an adiabatic shear band. The white layer is composed of a nanocrystalline (20–50 nm) structure with a strong shear texture, which is generated by adiabatic shear localization where the rotational dynamic recrystallization based on mechanically-driven subgrain rotations is the principal metallurgical process. Moreover, Al-rich γ' and Nb-rich γ'' nanoclusters are observed in the white layer, both of which are elongated and aligned caused by the severe shear deformation.

Liao et al. [57] studied the white layer of S135H nickel-based superalloy subjected to severe plastic deformation by dry plunge turning, and the microstructure, formation mechanism and micro mechanics of the white layer are shown in Figure 4.13. The white layer consists of a recrystallisation layer with ultrafine grains, a dynamic recovery layer with low angle grain boundaries and subgrains, and a deformed layer closely next to the recovery layer. In the white layer, the main crystal defects are dislocation tangles, dislocation walls and mechanical twins. The formation mechanisms are revealed by the phase and grain structure evolution, which is composed of three steps (T-I to T-III), determined by the temperature and strain rate during machining. In this study, the recrystallization grains in the white layer do not form through classic nucleation mechanisms but rather develop from pre-existing dislocation cells and subgrains that are generated by dynamic recovery. On the other hand, the γ' precipitation and the growth of the recrystallized grains are suppressed by the high cooling rate during machining, leading to the white layer microstructures consisting of fine grains containing supersaturated γ phases along with γ' forming elements, or ultrafine precipitates that exhibit minimal intergranular defects. In addition, the deformation mechanisms of white layer materials are completely different from these of the bulk materials. The pilar of the bulk material shows clear slip bands while it of the whiter layer only bulks. Moreover, the materials of the white layer start plastic deformation without owing clear yield point, and they almost have no elastic recovery even at low stresses, indicating the deterioration of the material mechanical performance. The significance of plasticity of the white layer may be attributed to two reasons: one is the Bauschinger behaviour of the white layer; the other is the dissolution of the γ' phase in recrystallisation layer of the white layer.

4.6.4 Monitoring and Predicting of Microstructure Alterations

It is worthwhile to monitor and predict the microstructure changes during grinding. Duscha et al. [49] developed a thermo-mechanical-metallurgical FE model to predict phase transformation and residual stresses, and the model structure is shown in Figure 4.14. The model takes thermal and mechanical loads during grinding as well as the change of workpiece material properties into account. Meanwhile, the coupling effects of three physical fields, that is, thermal, metallurgical, and mechanical, are considered in the model, as represented by the solid arrows in Figure 4.14. The dotted arrows represent the neglected week couplings. Thermal stresses are induced by temperature gradients in the workpiece, while phase transformation leads to deformations and impacts stress evaluation. These transformations are exothermic or endothermic

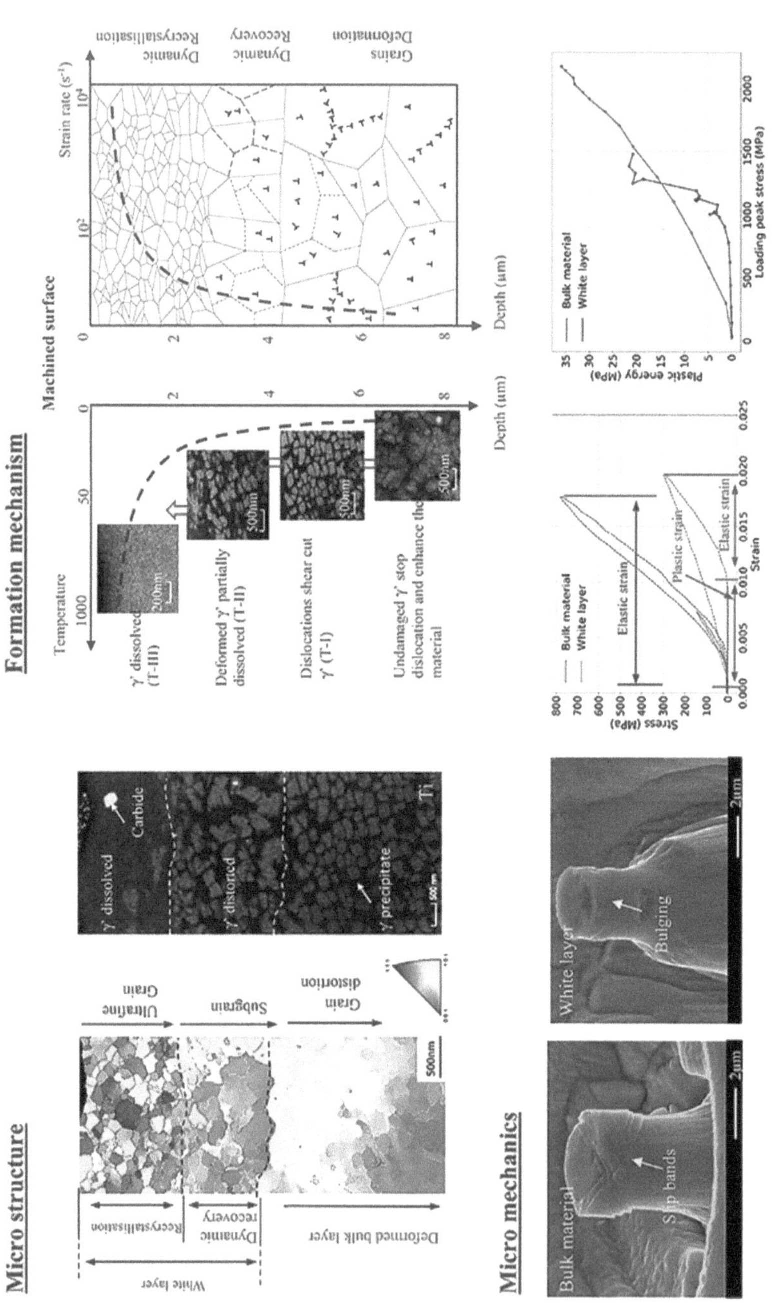

FIGURE 4.13 The microstructure, formation mechanism and micro mechanics of the white layer [57].

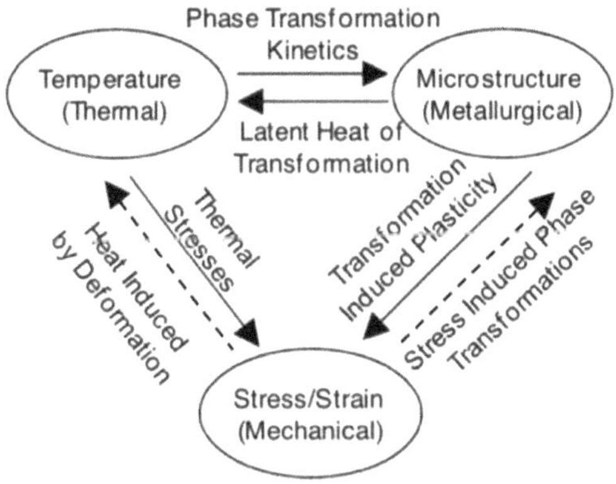

FIGURE 4.14 Thermo-mechanical-metallurgical Model [49].

reactions that affect the temperature field through latent heat and are influenced by stresses and strains. However, these coupling effects are not considered in this work.

Foeckere et al. [58] established a three-dimensional theoretical model to predict the thermo-metallurgical effects during grinding and grind-hardening. This model considers the grinding fluid effect and can obtain the 3D temperature field caused by grinding and grind-hardening. Based on this, the phase transformation in the surface layer and the thickness of the caused hardened layer will be gained by the model. The theoretical model is solved by a FE model, which uses a time-dependent deactivation of the finite elements to represent the transient material removal and consider the effects of the heat flow dissipated by the chips. Moreover, the material data of the FE model are homogeneous, isotropic, temperature-dependent, and phase-dependent. Finally, the theoretical calculation results are compared with experimental results, and the high accuracy of the model is confirmed.

In general, the microstructures present in ground surface layers are critical for controlling grinding quality. Future research efforts should focus on better understanding the relationship between grinding parameters and the rules governing microstructure evolution. Additionally, it is essential to develop efficient methods for monitoring and predicting microstructure alterations.

4.7 INFLUENCES OF SURFACE INTEGRITY ON WORKPIECE PROPERTIES

Surface integrity determines the shape accuracy of ground surfaces and workpieces and exerts significant influences on workpiece properties. This section summarizes the influences of surface integrity on microhardness, fatigue strength and corrosion behavior.

4.7.1 Microhardness

The occurrence of increased microhardness in the machined surface layer is a common phenomenon in grinding and has been widely reported. This increase in microhardness can enhance the mechanical properties of the workpiece, thereby prolonging its working life. Therefore, understanding the mechanism underlying this alteration and developing methods to predict microhardness are crucial areas of focus for research.

Microhardness is influenced by microstructures and residual stresses of the material. Huang et al. [42] studied the features of the robot-belt-grinding Inconel 718 surface layer and found that the microhardness significantly increases and has a gradient distribution in the ground surface layer. The thickness of the hardened layer is approximately 150 μm. The maximum microhardness on the outmost is about 702 HV, presenting an increase of 59.5 percent relative to that of the matrix material. They considered that the great increase of microhardness is attributed to the strain hardening induced by the severe plastic deformation under the force-thermal coupling field during grinding. The accumulation of plastic deformation increases the density of dislocations, leading to dislocation tangles, which will impede the movement of dislocations. Hence, the plastic deformation of materials is more difficult and the microhardness increases. Moreover, the generated crystallographic texture may also exert influences on the increase of microhardness. The gradient distribution of microhardness along the grinding depth direction is consistent well with the gradient microstructures in the surface layer. Wang et al. [32] pointed out that besides the strain hardening, the residual stress and the refined grains can also increase the microhardness of belt-ground Inconel 718. Additionally, due to the high strain and temperature during grinding, the deformation stored energy increases, being the driving force for recrystallization and dynamic recovery. The refined grains caused by recrystallization introduce plenty of grain boundaries, consequently increasing the microhardness. Ji et al. [36] reported that the hardness increased to 2.62 GPa from 1.96 GPa after machining. By comparing the hardness of the machined specimen and machined+heat treatment specimen, they concluded that the compressive residual stress contributed about 30 percent and the grain refinement contributed about 70 percent to the hardness increase.

The phases exert significant effects on microhardness. Wu et al. [48] revealed that morphology and distribution of the second phase significantly affect the microhardness of Fe-Ni superalloys. As the volume fraction of martensite and austenite increases and the ferrite volume fraction decreases, the microhardness increases. After grinding, the volume fraction of austenite also increases. Therefore, the influence of martensite and ferrite on microhardness is higher than that of austenite. Xu et al. [29] deeply investigated the effects of phases of DD5 nickel-based single-crystal superalloy on microhardness. With increasing the cutting speed and feed speed, the hardness of the outmost surface becomes large. During grinding, the phase-forming elements of Al and Ta diffuse into the surface layer, leading to the increased precipitation of γ' phase. High cutting speed and feed speed will increase the diffusion amount and depth of Al and Ta, and thus the amount of γ' phase precipitated and the microhardness. Therefore, the increment of γ' phase induced by segregation of Al and Ta may be a reason for the microhardness increase.

Researchers have proposed some prediction models for microhardness of the surface layer using FE models or theoretical models or their combination. The thickness of microhardness has attracted attention. Thai et al. [59] attributes the hardened effect into the phase transformation, which is determined by machining temperatures. Therefore, a temperature-dependent heat transfer FE model combining a triangular moving heat source was established to calculate the temperature field, therefore obtaining the thickness of the microhardness layer. This model is validated by experiments and has high accuracy. Following the similar idea, Ortega et al. [60] proposed an improved model, which considered the heat dissipated by grinding chips and the heat consumed by the fast austenitization process. Meanwhile, a model is established to evaluate surface softening during grinding. Then, a methodology is proposed to predict the hardness of the ground surface layer under non-isothermal tempering. Finally, the desired hardened depth of ground surface is achieved, and the control of the surface softening is realized. Experimental results show that the error of the prediction is below 6 %. Alonso et al. [61] proposed a parameter "area-based specific energy" (E) to on-line control the microhardness depth and material softening during grinding. The E has a linear relationship with the hardened depth and is independent of the grinding parameters, and the relationship slope can be obtained from the elements of the material. However, if the maximum chip thickness is larger than a threshold, the E will show a relationship with the tempering level.

The gradient values of microhardness along the grinding depth direction in the surface layer are important. Some prediction models based on machine learning algorithms have been built. These models are data-driven models and the main building procedures are similar and follows Figure 4.5. First, one or more model frames need to be selected. Then, plenty of experiments are carried out to obtain adequate mapping date between grinding parameters and the gradient microhardness, to train the model. Finally, when the accuracy and robustness of the trained model is high enough and reaches the requirement, the model is built and applied to predict the values of microhardness under different grinding parameters. However, the data-driven models have intrinsic limitations. Some theoretical models have been proposed for specific grinding scenarios. Shi et al. [62] built a model based on the hardening mechanism to predict the hardness of workpiece and help to select proper parameters. The basic structure of the model is "grinding parameters-equivalent strain-hardness", where the equivalent strain is the key intermediate parameter for the prediction model. The equivalent strain considers the cutting depth, grinding speed, feeding speed, and cooling liquid during grinding, while the hardness is calculated by a power equation of the equivalent strain. Sun et al. [63] established a hybrid model to study the dynamic hardness distribution, namely the hardness distribution changing rules with temperatures. Firstly, the theoretical model of dynamic forces considering three grinding effects and the dynamic characteristics of the grinding tool, and the workpiece was built and validated. Then, the dynamic temperature distribution during grinding is calculated via a finite difference method by coupling dynamic grinding forces. Following this, the prediction model of dynamic hardness is built by the superposition of two types of hardness, that is, the martensite and the residual parts. The dynamic temperature results in the inconsistent formation and distribution of the hardened layer in temporal and spatial field, and for a large cutting depth, with increasing the distance along the feeding direction, the inconsistence of hardness distribution becomes large.

4.7.2 Fatigue Strength

Microcracks usually initiate and propagate from the surface of the workpiece, leading to fatigue failure. Hence, the grinding surface integrity determines the fatigue strength to a great extent. Quan et al. [64] pointed out that for grinding GH4169 superalloy, omissions, scratches, texture disorders, and knock marks on the ground surface are the main reasons of the unstable high-cycle fatigue life of ground workpiece. During grinding, the squeezing effect causes a shallow compressive residual stress in the surface layer, thus improving the surface condition. Zhang et al. [65] studied the effects of surface integrity on high-cycle fatigue life of creep feed grinding of IC10. They found that surface roughness has a close relationship with the fatigue life, and large roughness is detrimental to the fatigue strength of ground workpiece. When the surface roughness is large, stress will concentrate in the deep grooves on the surface, where the microcracks are readily to generate. Apparently, this will reduce the fatigue life of workpieces. In contrast, microhardness depth and microhardness rate have opposite influences on fatigue life with surface roughness. Large microhardness depth and microhardness rate will improve the fatigue life. The hardening effect is induced by the plastic deformation during grinding, which prevents the protruding of dislocation lines. Therefore, the fatigue cracks have more difficulty in initiating and propagating. Moreover, the crack tends to initiate in the subsurface with low hardness rather than the hardened layer. Thus, the propagation of cracks needs larger tensile stress or more cycles under larger microhardness depth and microhardness rate.

Some theoretical models based on material science theories and finite element analysis are established to understand and predict the influencing rules of surface integrity on fatigue strength. Wang et al. [66] revealed the effect mechanism of surface roughness on fatigue behavior of DD6 nickel-based single crystal superalloy. Larger surface roughness will cause micro defects, which induce the concentration of stress, plastic strain, and fatigue damage under fatigue load, and promote the initiation of fatigue crack. Then, using the measured surface morphology, a FE model was established employing the rate-dependent crystal plasticity theory and the fatigue damage model based on the resolved shear stress and slip rate of slip systems. Finally, a surface-roughness-parameter determined equivalent model is built to predict the fatigue life, where the surface defects are simplified via sinusoidal shaped notches. By considering the surface topography and surface roughness parameters, a model of actual surface topography was built. Then, using sinusoidal shaped notches, the plate model is gained. Following this, the semi-empirical formula of the stress concentration factor K_t and surface roughness parameters is obtained. Finally, by introducing the K_t and equivalent fatigue damage parameters, the fatigue life prediction model considering surface roughness is proposed. The errors of the predicted results of fatigue life are within 1.5 times the error band.

4.7.3 Corrosion Behavior

In recent times, there has been a growing interest in the mechanisms and regulations governing the impact of grinding surface integrity on the corrosion resistance of ground workpieces. Wang et al. [67] studied the surface corrosion behavior of belt-grinding Inconel 718 in 3.5 wt % NaCl solution at room temperature. They studied the electrochemical dissolution behavior of the ground surface materials, and summarized four dissolution steps: (i) the dissolution starts at the defective pores of the oxide film; (ii) the Nb-rich areas generate around defective pores; (iii) the dissolution occurs at the boundary between the Nb-rich areas and matrix materials; (iv) material continues to dissolve into electrolyte, and undissolved materials is gradually peeled off, generating etch pits. The analyzing results of potentiodynamic polarization curves and electrochemical impedance spectroscopy illustrate that as the abrasive grain size decrease, the corrosion resistance of ground specimen becomes high, which is due to the influence of surface roughness and residual stresses. Large surface roughness is adverse to the passive film's formation and promotes the electrolyte's diffusion, thus increasing the corrosion rate. Experimental results show that as grain size decreases, the compressive residual stress increases, which is an advantage to the formation of the passive film, improving the corrosion resistance. Montero et al. [68] investigated the effects crystal orientation on hot corrosion of TMS-138 superalloy. The severe plastic deformation caused by surface treatment will induce the alterations of crystal orientation and γ-γ' coherent two-phase structure, which result in a high corrosion rate. Moreover, the type I hot corrosion is more sensitive to the crystal orientation that type II, and for type I, high atomic density orientations help to improve the hot corrosion resistance. Sudbrack et al. [69] studied the influences of grinding treatment on 815°C Oxidation of LDS-1101+Hf single-crystal nickel-based superalloys. The results exhibited that a thin protective Al_2O_3 layer is formed on the ground surface, which could improve the corrosion resistance of workpieces. The formation of this defective layer is attributed to the grinding-induced defects pathways. At the (001) orientation, the initiative and deformation-caused dislocations acting as one type of defect pathways could facilitate Al diffusion to generate the Al_2O_3 layer. Moreover, grinding-induced recrystallization grain boundaries may also be extra and faster defect pathways for Al cations from the matrix metal, and thus facilitate the formation of the Al_2O_3 layer. High surface roughness means more severe plastic deformation during grinding, which introduces more defective pathways. Hence, grinding treatment with a 9-μm or rougher surface facilitate the generation of the Al_2O_3 layer.

4.8 APPLICATIONS AND RESEARCH TRENDS

In this section, several typical applications of the grinding quality improvement of superalloy workpieces by controlling the surface integrity are introduced.

Li et al. [70] investigated the grinding surface integrity of GH4169 and the application in aero-engine blades. They explored the optimal grinding parameters, under which the blade surface meets several criteria, including profile errors within ± 0.01 mm, roughness less than 0.4μm, proper residual compressive stresses and hardening rate, absence of burrs and smearing caused by grinding chips, and smooth connection between the leading/trailing edge and suction/pressure surface. Zhao et al. [71] studied the surface condition improvement of belt grinding integrally bladed rotors of an aeroengine. Using single-factor experiment, the optimal range of grinding parameters (abrasive grain size, contact force, belt speed, and feed speed) was determined. Based on the central composite design experiments and quadratic

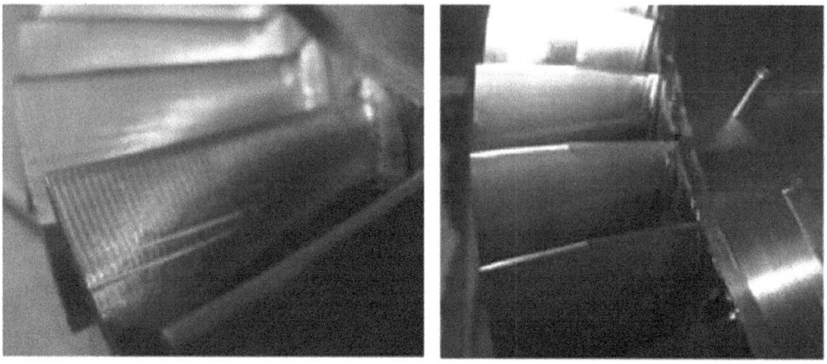

FIGURE 4.15 The integrally bladed rotors before and after grinding [71].

regression design technology, a surface roughness prediction model was established. Finally, the optimal grinding parameters were obtained. The roughness of ground surface under optimal parameters was reduced by 25 percent. As shown in Figure 4.15, the surface quality of integrally bladed rotors was greatly improved, increasing the fatigue life of the ground workpiece. Yao et al. [72] studied the surface integrity of Inconel 718. Under the grinding parameters, the surface integrity was significantly improved, with the softened layer and the plastic deformation layer respectively being 30–40 μm and 10–15 μm, the surface roughness being 0.112 μm, the surface residual stress being 700 MPa and 40–60 μm deep, and the hardness being 440 HV. Miao et al. [73] comprehensively investigated the surface integrity of grinding of nickel-based turbine blade fir-tree roots. After grinding, a deformed layer was formed, and the work hardening, and compressive residual stresses were measured. When the infeed speed was high, the grinding temperature would excessively large and thus resulting in surface burn. In this case, various surface defects (microcracks, oxidation, smearing) and subsurface defects (material fracture, white layer, material peeling, microcracks, recrystallization) were induced, and severe work hardening as well as tensile residual stresses were observed. They pointed out that the thermo-mechanical-structural coupling effects significantly influence the profile grinding of turbine blade fir-tree roots. The collaboration of surface integrity and blade root structure shape are the key for realizing the high-quality grinding.

The grinding-induced strengthening effects are beneficial to workpiece property and service performance, which however faces two problems: (i) the ground surface is usually with high roughness and needs to be polished; (ii) the polishing process will weaken or even remove the strengthened layer. Aiming at these problems, Huang et al. [42] proposed a method to improve surface strengthening and quality of belt ground (BG) Inconel 718 based on isotropic etching polishing (IEP). IEP is characterized by the formation of isotropic etching holes during the etching process rather than preferential corrosion attack, as opposed to conventional electrochemical polishing. IEP not only has a unique polishing mechanism, but also consumes less energy and utilizes readily available electrolytes. The BG surface was treated by IEP at proper parameters, with the electrolyte being H_2SO_4 and CH_3OH, and the evolution of BG surface consists of four stages: formation of isotropic etching holes, mergence of etching holes, new surface formation and smoothening and formation of the smooth surface. The material removal and surface roughness thickness could be controlled precisely during IEP. The IEP retains a maximum of 86.1 percent of the BG-induced <100> texture and 75.1 percent of the strengthened layer. Moreover, the comprehensive residual stress of the IEP surface is larger than that of the original BG surface, which is reasonable because the maximum value of the comprehensive residual stress is usually in the subsurface of the ground layer. In addition, the surface conditions are significantly improved by IEP. A mirror-like, smooth (and without defects or contaminants) surface is obtained after the IEP. Khan et al. [74] applied IEP to treat the 3D-printed Inconel 718 turbine blade and nozzle. The treated components not only have high-quality surfaces, but the corrosion resistance is improved by a compact and uniform passive film formed during the IEP.

4.9 SUMMARY AND PERSPECTIVE

In this chapter, the surface integrity in robotic compliant grinding of nickel-based superalloys is introduced and discussed, especially modeling-related works. The definition of grinding surface integrity, grinding features and grinding systems is introduced first. The force and thermal fields, surface conditions of ground workpieces, alterations of residual stresses and microstructures, influences of surface integrity on workpiece properties are discussed in detail. The typical applications of the grinding quality improvement of superalloy workpieces by controlling the surface integrity are introduced. Important research trends are proposed from four perspectives, namely, force-thermal coupling models, knowledge-driven models, monitoring and controlling systems, robot controlling algorithms. This chapter provides an in-depth understanding of evolution and modeling for surface integrity in robotic compliant grinding of nickel-based superalloys.

Based on the current research status, future research needs to focus on the following four aspects:

(1) *The force-thermal coupling models*: Initially, comprehensive analyses from a multiscale viewpoint of the effects of tool elastic deformation on the grinding process are imperative. Subsequently, the interaction mechanisms between abrasive grains and workpieces in compliant grinding of superalloys must be elucidated. Lastly, a systematic force-heat coupling field model can be established by integrating temperature-depended material force and thermal parameters.

(2) *Knowledge-driven model for grinding processes*: This consists of three primary aspects. The first is to explore the evolutionary law of workpiece surface integrity under force-heat coupling fields during grinding. Then, the quantitative relationships between surface integrities and properties need to be investigated. On this basis, knowledge-driven and knowledge-data-driven models for controlling workpiece surface integrity and properties in robot compliant grinding can be constructed.

(3) *Monitoring and controlling systems of the grinding process*: This means to optimize the monitoring and controlling methods of surface integrity, and to establish a more efficient and comprehensive system by integrating various measuring sensors/controlling facilities and robots, achieving accurate and efficient real-time monitoring and feedback controlling of compliant grinding.

(4) *Efficient robot controlling algorithms*: This requires developing accurate and robust algorithms to achieve intelligent and efficient grinding. For instance, (i) feed compensation algorithms of grinding tool wear and grinding tool elastic deformation; (ii) grinding path planning algorithms for complex curved surfaces and different compliant grinding tools; and (iii) automotive selection and optimization algorithms for grinding parameters.

REFERENCES

[1] FIELD M, KAHLES J F. The surface integrity of machined and ground high strength steels. *DMIC Report*, 1964, 210: 54–77.

[2] ZHU W-L, BEAUCAMP A. Compliant grinding and polishing: A review. *International Journal of Machine Tools and Manufacture*, 2020, 158: 103634.

[3] REN X, HUANG X, FENG H, et al. A novel energy partition model for belt grinding of Inconel 718. *Journal of Manufacturing Processes*, 2021, 64: 1296–306.

[4] BEAUCAMP A, NAMBA Y, COMBRINCK H, et al. Shape adaptive grinding of CVD silicon carbide. *CIRP Annals - Manufacturing Technology*, 2014, 63(1): 317–20.

[5] ZHU W L, BEAUCAMP A. Ultra-precision finishing of low expansion ceramics by compliant abrasive technologies: A comparative study. *Ceramics International*, 2019, 45(9): 11527–38.

[6] ZHU W-L, YANG Y, LI H N, et al. Theoretical and experimental investigation of material removal mechanism in compliant shape adaptive grinding process. *International Journal of Machine Tools and Manufacture*, 2019, 142: 76–97.

[7] ZHANG X, CHEN H, XU J, et al. A novel sound-based belt condition monitoring method for robotic grinding using optimally pruned extreme learning machine. *Journal of Materials Processing Technology*, 2018, 260: 9–19.

[8] CHEN J, CHEN H, XU J, et al. Acoustic signal-based tool condition monitoring in belt grinding of nickel-based superalloys using RF classifier and MLR algorithm. *International Journal of Advanced Manufacturing Technology*, 2018, 98(1): 859–72.

[9] GAO K, CHEN H, ZHANG X, et al. A novel material removal prediction method based on acoustic sensing and ensemble XGBoost learning algorithm for robotic belt grinding of Inconel 718. *International Journal of Advanced Manufacturing Technology*, 2019, 105(1-4): 217–32.

[10] WANG Y, HUANG X, REN X, et al. In-process belt-image-based material removal rate monitoring for abrasive belt grinding using CatBoost algorithm. *International Journal of Advanced Manufacturing Technology*, 2022, 123(7): 2575–91.

[11] HUANG X, REN X, YU H, et al. Partitioned abrasive belt condition monitoring based on a unified coefficient and image processing. *Journal of Intelligent Manufacturing*, 2023, 35: 905–923.

[12] RAISSI M, PERDIKARIS P, KARNIADAKIS G E. Physics-informed neural networks: A deep learning framework for solving forward and inverse problems

involving nonlinear partial differential equations. *Journal of Computational Physics*, 2019, 378: 686–707.

[13] XIE J, CHAI Z, XU L, et al. 3D temperature field prediction in direct energy deposition of metals using physics informed neural network. *International Journal of Advanced Manufacturing Technology*, 2022, 119: 3449–68.

[14] WANG J, LI Y, ZHAO R, et al. Physics guided neural network for machining tool wear prediction. *Journal of Manufacturing Systems*, 2020, 57: 298–310.

[15] YANG Z, CHU Y, XU X, et al. Prediction and analysis of material removal characteristics for robotic belt grinding based on single spherical abrasive grain model. *International Journal of Mechanical Sciences*, 2021, 190: 106005.

[16] LI L, REN X, FENG H, et al. A novel material removal rate model based on single grain force for robotic belt grinding. *Journal of Manufacturing Processes*, 2021, 68: 1–12.

[17] KIM H-J, KIM N-K, KWAK J-S. Heat flux distribution model by sequential algorithm of inverse heat transfer for determining workpiece temperature in creep feed grinding. *International Journal of Machine Tools and Manufacture*, 2006, 46(15): 2086–93.

[18] JIANG J, GE P, SUN S, et al. From the microscopic interaction mechanism to the grinding temperature field: An integrated modelling on the grinding process. *International Journal of Machine Tools and Manufacture*, 2016, 110: 27–42.

[19] REN X, CHAI Z, XU J, et al. A new method to achieve dynamic heat input monitoring in robotic belt grinding of Inconel 718. *Journal of Manufacturing Processes*, 2020, 57: 575–88.

[20] REN X, HUANG X, CHAI Z, et al. A study of dynamic energy partition in belt grinding based on grinding effects and temperature dependent mechanical properties. Journal of Materials Processing Technology, 2021, 294: 117112.

[21] WANG Y J, HUANG Y, CHEN Y X, et al. Model of an abrasive belt grinding surface removal contour and its application. *International Journal of Advanced Manufacturing Technology*, 2016, 82(9–2): 2113–22.

[22] BEAUCAMP A, NAMBA Y, CHARLTON P. Process mechanism in shape adaptive grinding (SAG). *CIRP Annals*, 2015, 64(1): 305–8.

[23] BEAUCAMP A, SIMON P, CHARLTON P, et al. Brittle-ductile transition in shape adaptive grinding (SAG) of SiC aspheric optics. *International Journal of Machine Tools and Manufacture*, 2017, 115: 29–37.

[24] WANG J J, WEN Z X, ZHANG X H, et al. Effect mechanism and equivalent model of surface roughness on fatigue behavior of nickel-based single crystal superalloy. International Journal of Fatigue, 2019, 125: 101–11.

[25] WANG W, YAO P, WANG J, et al. Controlled material removal mode and depth of micro cracks in precision grinding of fused silica - A theoretical model and experimental verification. *Ceramics International*, 2017, 43(15): 11596–609.

[26] BOUKTIB N, KHELLOUKI A. Investigation of belt-finishing effect on the residual stress field through repeated scratching on rough hard-turned surface. *Tribology International*, 2021, 153: 106644.

[27] DING, ZISHAN, LI, et al. Material phase transformation at high heating rate during grinding. *Machining Science & Technology*, 2016, 20(2): 290–311.

[28] SUGUI T, SHU Z, FUSHUN L, et al. Microstructure evolution and analysis of a single crystal nickel-based superalloy during compressive creep. *Materials Science and Engineering: A*, 2011, 528(15): 4988–93.

[29] XU Y, GONG Y, WANG Z, et al. Experimental study of Ni-based single-crystal superalloy: Microstructure evolution and work hardening of ground subsurface. *Archives of Civil and Mechanical Engineering*, 2021, 21: 1–11.

[30] ZHAO C, LI J, WANG W. Forming mechanisms based simulation and prediction of grinding surface roughness for abrasive belt rail grinding. *Procedia CIRP*, 2020, 87: 503–8.

[31] SONG K, XIAO G, CHEN S, et al. A new force-depth model for robotic abrasive belt grinding and confirmation by grinding of the Inconel 718 alloy. *Robotics and Computer-Integrated Manufacturing*, 2023, 80: 102483.

[32] WANG J, XU J, WANG X, et al. A comprehensive study on surface integrity of nickel-based superalloy Inconel 718 under robotic belt grinding. *Materials and Manufacturing Processes*, 2019, 34(1): 61–9.

[33] LI H N, AXINTE D. On the inverse design of discontinuous abrasive surface to lower friction-induced temperature in grinding: An example of engineered abrasive tools. *International Journal of Machine Tools and Manufacture*, 2018, 132: 50–63.

[34] ZHU D, SONG S, QU C, et al. Numerical investigation of crack initiation, propagation and suppression in robot-assisted abrasive belt grinding of zirconia ceramics via an improved chip-thickness model. *Ceramics International*, 2020, 46(14): 22030–9.

[35] TöNSHOFF H K, FRIEMUTH T, BECKER J C. Process monitoring in grinding. *CIRP Annals*, 2002, 51(2): 551–71.

[36] MOON J, BAEK S M, LEE S, et al. Effects of residual stress on the mechanical properties of copper processed using ultrasonic-nanocrystalline surface modification. *Materials Research Letters*, 2019, 7: 97–102.

[37] YANG M X, LI R G, JIANG P, et al. Residual stress provides significant strengthening and ductility in gradient structured materials. *Materials Research Letters*, 2019, 7(11): 433–8.

[38] LI F. Active control of the residual stress in Inconel718 grinding assisted by the strengthen induction heating. Journal of Mechanical Engineering, 2018, 54: 216.

[39] DING W, ZHANG L, LI Z, et al. Review on grinding-induced residual stresses in metallic materials. *International Journal of Advanced Manufacturing Technology*, 2016, 88(9-12): 2939–68.

[40] WANG W, SALVATORE F, RECH J. Characteristic assessment and analysis of residual stresses generated by dry belt finishing on hard turned AISI52100. *Journal of Manufacturing Processes*, 2020, 59: 11–8.

[41] YUN H, SHUAI L, GUIJIAN X, et al. Prediction of surface residual stress on titanium alloy generated by belt grinding using molecular system dynamics. *Procedia CIRP*, 2020, 87: 480–4.

[42] HUANG X, CHAI Z, CAO F, et al. Isotropic etching polishing of belt ground Inconel 718 to improve surface strengthening and quality. Surface and Coatings Technology, 2022, 436: 128292.

[43] DU S, JIANG Z, ZHANG D, et al. Microstructure of plastic deformation layer on grinding surface of GH4169 alloy. Journal of Mechanical Engineering, 2015, 51(12): 63–8.

[44] GONG Y-D, ZHANG WEI J, CAI M, et al. Experimental study on the grinding metamorphic layer of nickel-based single crystal superalloy. Journal of Northeastern University (Natural Science) *(China)*, 2020, 41(6): 846–51.

[45] FENG Z Q, CHEN Y X, WU G L, et al. Precession electron diffraction assisted orientation mapping of gradient nanostructure in a Ni-based superalloy [C]. *IOP Conference Series: Materials Science and Engineering. IOP Publishing*, 2015, 89(1): 012023.

[46] FERGANI O, PAN Z, LIANG S Y, et al. Microstructure texture prediction in machining processes. *Procedia CIRP*, 2016, 46: 595–8.

[47] FANG T, LI W, TAO N, et al. Revealing extraordinary intrinsic tensile plasticity in gradient nano-grained copper. *Science*, 2011, 331(6024): 1587–90.

[48] WU C, GUO W, LI R, et al. Thermal effect on oxidation layer evolution and phase transformation in grinding of Fe-Ni super alloy. Materials Letters, 2020, 275: 128072.
[49] DUSCHA M, ESER A, KLOCKE F, et al. Modeling and simulation of phase transformation during grinding. *Advanced Materials Research*, 2011, 223: 743–53.
[50] DING Z S, LI B Z, ZOU P, et al. Material phase transformation during grinding. *Advanced Materials Research*, 2014, 1052: 503–8.
[51] ZISHAN D, BEIZHI L, STEVEN L Y. Material phase transformation at high heating rate during grinding. *Machining Science and Technology*, 2016, 20(2): 290–311.
[52] GRIFFITHS B. Mechanisms of white layer generation with reference to machining and deformation processes. *Journal of Tribology*, 1987, 109(3): 525–30.
[53] ZHANG S, LIU Z, WANG B, et al. Phase transition and dynamic recrystallization mechanisms of white layer formation during turning superalloy Inconel 718. *Journal of Materials Research and Technology*, 2021, 15: 5288–96.
[54] WUSATOWSKA-SARNEK A M, DUBIEL B, CZYRSKA-FILEMONOWICZ A, et al. Microstructural characterization of the white etching layer in nickel-based superalloy. *Metallurgical and Materials Transactions A*, 2011, 42(12): 3813–25.
[55] KRUK A, WUSATOWSKA-SARNEK A M, ZIETARA M, et al. Characterization on white etching layer formed during ceramic milling of Inconel 718. *Metals and Materials International*, 2018, 24(5): 1036–45.
[56] CHEN Z, COLLIANDER M H, SUNDELL G, et al. Nano-scale characterization of white layer in broached Inconel 718. *Materials Science and Engineering: A*, 2017, 684: 373–84.
[57] LIAO Z, POLYAKOV M, DIAZ O G, et al. Grain refinement mechanism of nickel-based superalloy by severe plastic deformation - Mechanical machining case. *Acta Materialia*, 2019, 180: 2–14.
[58] FOECKERER T, ZAEH M F, ZHANG O B. A three-dimensional analytical model to predict the thermo-metallurgical effects within the surface layer during grinding and grind-hardening. International Journal of Heat and Mass Transfer, 2013, 56(1): 223–37.
[59] NGUYEN T, ZHANG L C. Realisation of grinding-hardening in workpieces of curved surfaces-Part 1: Plunge cylindrical grinding. *International Journal of Machine Tools and Manufacture*, 2011, 51(4): 309–19.
[60] ORTEGA N, ALONSO U, SaNCHEZ J A, et al. Modelling of the hardening and finishing stages of grind-hardened workpieces. *International Journal of Advanced Manufacturing Technology*, 2016, 82(1): 435–49.
[61] ALONSO U, ORTEGA N, SANCHEZ J A, et al. Hardness control of grind-hardening and finishing grinding by means of area-based specific energy. *International Journal of Machine Tools and Manufacture*, 2015, 88: 24–33.
[62] SHI X, XIU S. Study on the hardness model of grinding for structural steel. *International Journal of Advanced Manufacturing Technology*, 2020, 106(7): 3563–73.
[63] SUN C, WANG J, KONG X, et al. Research on dynamic hardening effect on workpiece's hardness distribution with depth of cut. *International Journal of Advanced Manufacturing Technology*, 2019, 103(9): 4013–28.
[64] QUAN F, CHEN Z, LI Q, et al. Effects of process combinations of milling, grinding, and polishing on the surface integrity and fatigue life of GH4169 components. *Proceedings of the Institution of Mechanical Engineers Part B Journal of Engineering Manufacture*, 2019, 234(3): 095440541986805.
[65] ZHANG S, YANG Z, JIANG R, et al. Effect of creep feed grinding on surface integrity and fatigue life of Ni3Al based superalloy IC10. *Chinese Journal of Aeronautics*, 2021, 34(1): 438–48.

[66] WANG J J, WEN Z X, ZHANG Y M, et al. Experimental characterization and computational modelling for fatigue behavior of a Ni-based single crystal alloy considering surface roughness. *Metals and Materials International*, 2021, 27: 4383–96.
[67] WANG J, XU J, ZHANG X, et al. An investigation of surface corrosion behavior of Inconel 718 after robotic belt grinding. *Materials*, 2018, 11(12): 2440.
[68] MONTERO X, ISHIDA A, MEIBNER T M, et al. Effect of surface treatment and crystal orientation on hot corrosion of a Ni-based single-crystal superalloy. Corrosion Science, 2020, 166: 108472.
[69] SUDBRACK C K, BECKETT D L, MACKAY R A. Effect of surface preparation on the 815°C oxidation of single-crystal nickel-based superalloys. *JOM*, 2015, 67(11): 2589--98.
[70] LI X, MA S, MENG F. Surface integrity of GH4169 affected by cantilever finish grinding and the application in aero-engine blades. *Chinese Journal of Aeronautics*, 2015, 28(5): 1539--45.
[71] ZHAO T, SHI Y, LIN X, et al. Surface roughness prediction and parameters optimization in grinding and polishing process for IBR of aero-engine. *International Journal of Advanced Manufacturing Technology*, 2014, 74(5): 653–63.
[72] YAO C F, JIN Q C, HUANG X C, et al. Research on surface integrity of grinding Inconel718. *International Journal of Advanced Manufacturing Technology*, 2013, 65(5): 1019–30.
[73] MIAO Q, LU M, DING W, et al. Creep-feed grinding of single crystal nickel-base turbine blade fir-tree roots: tool wear, grinding force, temperature, and surface integrity. *International Journal of Advanced Manufacturing Technology*, 2023, 126(3): 1453–1470.
[74] AJMAL K M, YI R, ZHAN Z, et al. A novel finishing approach for 3D printed Inconel 718 by utilizing isotropic electrochemical etching. Journal of Materials Processing Technology, 2022, 299: 117356.

5 Ceramic Matrix Composites for Commercial Aircraft Engines

Laifei Cheng, Chaokun Song, and Fang Ye

5.1 INTRODUCTION

In the long history, pottery has been the most important means of human life. China created a glorious ceramic culture as early as feudal society. In the late stage of the second industrial revolution, due to its excellent wear resistance, corrosion resistance and other characteristics, ceramics began to be gradually applied in machinery, metallurgy, petroleum, chemical industry, power and electronics and other fields.

Owing to the unique covalent bond and ionic bond structure, ceramics have excellent properties such as high temperature resistance, high specific modulus, high hardness, corrosion resistance and wear resistance. At the same time, ceramic materials have strong defect sensitivity and the fatal weakness of high brittleness and poor reliability because of the high bond energy, which seriously restricts their application as high temperature structural materials. It is shown that the strength and uniformity of ceramic materials can be improved by suppressing the volume effect, and the toughness cannot be improved by increasing the surface energy. Therefore, the development of ceramic matrix composite (CMC) is an effective way to improve the strength and toughness of ceramic materials.

This chapter introduces the basic knowledge, the basic characteristics and design principles of CMC used for commercial aircraft engines.

5.2 OVERVIEW OF CERAMIC MATRIX COMPOSITES

According to the reinforcement length-diameter ratio, the CMC can be grouped into particle toughening CMC, whisker toughening CMC, and continuous fiber toughening CMC. The particle-reinforcement length-diameter ratio is less than 5. The whisker-reinforcement length-diameter ratio is between 5 and 100. The continuous fiber-reinforcement length-diameter ratio is greater than 100. The diameter size of the reinforcement is the order of micrometer. Different length-diameter ratio give rise to different effect on toughening. Particle toughening is the simplest and easiest way to toughening the materials. But the particle toughening CMC is normally with

DOI: 10.1201/9781003386223-5

fracture toughness less than 6 MPa·m$^{1/2}$. The whisker is relatively small in the size of diameter, but has high strength. The whisker improves both the toughness and strength of the composite. The whisker toughening CMC is normally with fracture toughness less than 10 MPa·m$^{1/2}$. The particle toughening and whisker toughening can not really improve the toughness and the reliability of the ceramic thermal structural component. Continuous fiber-reinforced ceramic matrix composite (CFCC) can inhibit the volume effect of the defects in ceramics via dissipating fracture energy by crack deflection and fiber pull-out. Therefore, fiber-reinforcement will improve both fracture toughness and strength of the composite. Designing the structure of fiber preform can control the toughness and strength of the composite. Adding fibers into the composite hugely improves the toughness of the ceramic materials. This makes CFCC overcome the weakness of brittleness and poor reliability of ceramics, and it has similar fracture behavior to metal. CFCC has been the new thermal structural materials with high temperature resistance and low density resulting from the excellent strength and toughness resulting from it is insensitive to cracks and it does not occur catastrophic damage.

5.2.1 Preparation Method of CMC-SiC

Continuous fiber-reinforced silicon carbide ceramic matrix composite (CMC-SiC) includes carbon fiber and silicon carbide fiber toughened silicon carbide composites, which is the most studied, the most successful and the most widely used continuous fiber toughened ceramic matrix composite.

There are four different ways to synthesis CMC-SiC: chemical vapor infiltration (CVI), precursor infiltration pyrolysis (PIP), reactive melt infiltration (RMI) and slurry impregnation and hot pressing (SIHP). These synthesis methods will be discussed later.

5.2.1.1 CVI

Prof. Naslain et al. in the University of Bordeaux in France developed CVI in the 1970s. CVI equipment schematic diagram is shown in Figure 5.1. First, put the fiber preform into the reaction chamber. The gaseous precursor in a certain proportion diffused into porous preform with uniform temperature and no obvious forced gas flow. Then SiC is deposited in situ by chemical reactions on the fiber surface. At last, by-products are released from the reaction wall and transferred into the main air stream by means of diffusion, and then they are discharged from the deposition furnace.

The diffusion of the gas precursor produces a gradient in concentration of the precursor as a function of depth through the fiber preform. It results in a higher gas concentration and deposition rate at the entrance of the preform than in the interior of the preform. During the CVI process, the holes on the surface of the preform are closed prematurely and the gas transmission channel is cut off, which leads to density gradient (bottleneck effect) of the composite and affects the deposition quality. To prevent excessive surface deposition, the isothermal chemical vapor infiltration (ICVI), which lower the pressure and temperature, is used instead. Although this makes the

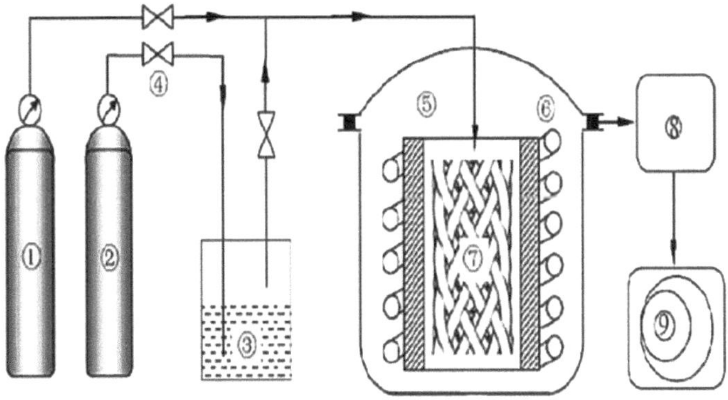

1-H_2 2-He 3-MTS 4-Control valve 5-CVD furnace 6-Induction coil
7-Perform 8-Scrubber 9-Mechanical valve

FIGURE 5.1 The schematic diagram of chemical vapor infiltration equipment.

whole process very long and with a low densification rate. ICVI suits for the densification for any component with complex structural components as long as the component wall is not too thick, and multiple components can be prepared in one furnace at the same time. Therefore, the ICVI process is widely used for commercial products.

5.2.1.2 PIP

a) The PIP process: first soak the porous fiber preform into the organic precursor solution. Then it is cross-linked and cured in inert atmosphere. Finally, the organic precursor is transformed into inorganic ceramics after pyrolysis [11]. During the heating process, some complicated chemical and physical changes, including molecular bond breaking, forming free radical fragment, evaporation of the small molecules, forming matrix, and change in volume and density. Due to the low yield of the product obtained by the PIP method, it is necessary to repeat the precursor impregnation and pyrolysis process in order to obtain the composite with high densification degree.

The main characteristics of the PIP process are as follows: (1) the composition and structure of precursor can be designable; thus the design of ceramics and ceramic matrix composites is realized. (2) The ideal composite consisted of reinforcement combined with matrix can be realized: in the converting process of organic polymer to ceramics, its structure has undergone a transformation from organic linear structure to three-dimensional network structure, and then to ceramic nanocrystalline. Therefore, it is possible to obtain the ideal composite by changing the pyrolysis conditions to detect and control different transformation stages. (3) Good processability: organic polymers have the general properties of resin materials, such as soluble, fusible, cross-linkable, cured, and so forth. Therefore the initial steps of the process of the preparation of ceramic matrix composite can use the steps of molding process technology of some plastic and resin matrix composites. It is convenient to prepare composites reinforced by unidirectional, two-dimensional or three-dimensional fiber preforms. Preform impregnated with organic polymer are machinable before pyrolysis, that its shape and size can be easily trimmed by turning, milling, drilling and other machining technologies. (4) Low pyrolysis temperature: the pyrolysis temperature of organic polymers is much lower than the sintering temperature of ceramic powder with the same composition; thus the degree of thermal damage to the fibers is smaller.

This method also has some disadvantages: during the drying (or crosslinking and curing process) and pyrolysis process of the precursor, there are many pores and cracks formed in the matrix. These defects are caused by the shrinkage of the matrix owing to the volatilization of solvent, low molecular components and the increase of density. Limited by the conversion rate of precursor, it is necessary to undergo repeated impregnation and pyrolysis process in order to obtain ceramic matrix composites with high densification degree, and the cost is high. In addition, high purity and stoichiometric ceramic substrates are difficult to be obtained.

5.2.1.3 RMI

RMI, or called molten silicon infiltration/liquid silicon infiltration (MSI/LSI), is developed by Fitzer E since the 1980s. RMI's process is as follows: first carbon matrix is introduced into preforms by CVI or PIP process. Then the preforms were embedded with silicon powder (with specific particle size) and placed in the RMI furnace. At the setting temperature, the molten silicon penetrates into the porous preform by capillary force, and it reacts with the carbon matrix to form the SiC ceramic matrix.

RMI process shows advantages such as short period, low cost and low porosity. Even for the complex structure component, the CMC with high densification degree after a single infiltration treatment. In addition, the change in the dimensions of the preform before and after RMI process is small, and the shape of the fiber skeleton and fiber strength can be basically maintained. Therefore, this method is very competitive in industrial production. However, when Si react with C in the matrix, it will react with the carbon fiber at the same time, which results in fiber damage and performance degradation. At the same time, there exists a certain amount of residual Si in the composite, which will lead to the decrease of the high temperature mechanical properties of the composite, especially the creep resistance.

5.2.1.4 SIHP

SIHP is a traditional way to synthesis CMC [2]. SIHP process: First, the fiber bundle is desized at high temperature, and it is passed through the slurry tank equipped with ceramic slurry, so that the ceramic slurry is evenly coated on the surface of each monofilament fiber. Then, the fibers soaked with the slurry are wound on the hub to obtained prepreg, and the prepreg is dried and cut into prefabricated sheets. Finally, the prefabricated sheets are placed in a graphite mold to be laminated to the required thickness, and then they are hot pressed and sintered to fabricate CMC. The composition and properties of the slurry directly affect the properties of the composite. The slurry is usually made up of solvent (water or alcohols), ceramic powder and organic binders. The surfactant is added to improve the wettability between the solvent and the ceramic powder and fiber. The shape of the ceramic powder should be spherical and the size should be as small as possible.

SIHP process shows advantages such as short sintering time and low cost. For example, the densification time required for SIHP is only one hour. Furthermore, liquid phase promotes the sintering of the composite at high temperature, which reduce the residual holes inside the composite and improve the densification. SIHP is also suitable for fiber reinforced glass or ceramic matrix composites, such as C/SiO_2, SiC/LAS, SiC/BAS, and so forth, and also suitable for the manufacture of ceramic matrix composites with enough liquid phase during the sintering process, such as C/Si_3N_4. In a word, the SIHP method can be used when enough liquid phase sintering additives are added.

SIHP process also shows some disadvantages: (1) The structure and shape of composites are limited: since the fiber preforms are obtained by laminated method, only simple composites with obvious anisotropy can be manufactured. (2) The height and size of the composites is limited: the mechanical load of 20~30 MPa must be applied to the preform, and the strength of the graphite mold limits the size of the component; at the same time, the pressure gradient exists along the height direction of preform due to the effect of friction. In order to ensure the uniformity of density inside the composite, the height to diameter ratio of the component should be less than 0.45. (3) It is not suitable for the material system which is fabricated by solid phase sintering: the mechanical load acting on solid particles during hot pressing sintering will cause serious damage to the fiber, which limits the application of this preparation method in the field of ceramic matrix composites.

5.2.2 Application of CMC-SiC

Ceramic matrix composites is a new type of thermal structural material, which shows characteristics such as low density, high thermostability, high specific strength, high specific modulus, oxidation resistance, creep resistance, insensitive to cracks, and so forth. It has wide application fields and even been introduced in high technology civil and military fields. The composites, which work as thermal structural/functional materials, have a life span of hundreds or thousands hours below 1650°C, tens of minute to hours below 2000°C, and seconds to minutes below 2800°C as shown in Figure 5.2. CMC-SiC composite which currently is the most mature and widely used material, and can be applied in high thrust-to-weight aero engine, high performance aerospace engine, thermal protection system of aerospace vehicle, braking system of aircraft/high-speed trains, nuclear station, gas power station, deep space detectors and other fields.

The limit service temperature of superalloy is 1000°C, which cannot satisfy the requirements for high thrust-weight ratio aero engine. It is urgent to developing new structure materials which can withstand high-temperature. CMC-SiC shows great potentials in the field of aero engine due to a series of excellent properties. It can be used in, such as turbine outer ring, tail nozzle, combustion chamber, turbine disk and turbine blade. It becomes one of the key thermal structural materials for high thrust-weight ratio aero engine and hence can increase the operating temperature by 350~500°C and reduce the weight by 30~70 percent. The CMC-SiC material has

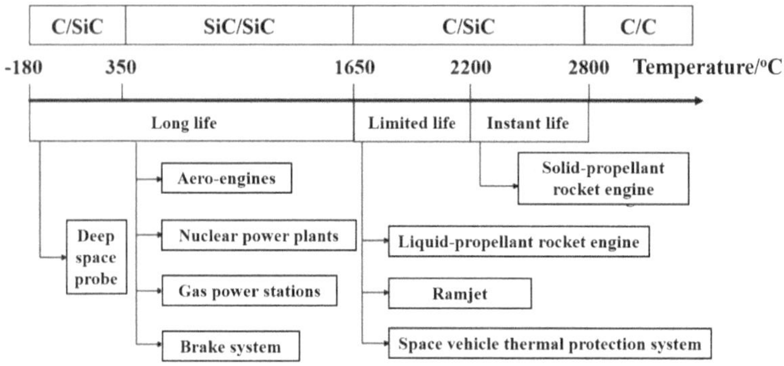

FIGURE 5.2 The applied field and temperature range of CFCC.

been successfully demonstrated and verified on the medium-load stationary parts of various military and civil engines, such as the combustion chamber, turbine outer ring, and the adjusting plate of the vector nozzle.

The technical requirements of aero engine play an important role in the development of CMC. The European Power Association (SEP), University of Bordeaux, Karlsruhe University in Germany and Oak Ridge National Laboratory in the United States had started the research on C/SiC composites since 1970s. The nozzle made by C/SiC composite material has been used on the M55 engine of Mirage 2000 and the M88 aero-engine of Rafale fighters. In the 1990s, France's Snecma Company developed the CERASEP series of SiC/SiC composite and successfully applied the composite to the nozzle regulator of the M-88 engine. This marked that the application of SiC/SiC composite materials in aerospace had begun, and the upgradation of the combustion chamber and other engine components had been completed. In 2015, GE Aviation successfully tested the world's first lightweight ceramic matrix composite of low-pressure turbine blades of the F414 turbofan engine validation machine. At the same time, the combustion chamber and high-pressure turbine components made by the ceramic matrix composites were tested in the GEnx engine, which promoted the development of this technology.

Northwest Polytechnical University in China, National University of Defense Technology-China, Shenyang Institute of Metals-China, Chinese Academy of Sciences, and the 43rd Research Institute of the Fourth Research Institute of Aerospace Science and Technology Corporation have all carried out relevant research in the past 20 years. The Science and technology on Thermostructure Composite Materials Laboratory of Northwestern Polytechnical University has developed some products, including sealing vane adjusting element, inner cone, flame tube, turbine outer ring, guide vane and turbine rotor.

CMC-SiC faces serious corrosion problems (it includes heat corrosion and water corrosion) and oxidation problems, significantly reducing its service life. The environment barrier coatings (EBCs) have been widely applied to improve the corrosion resistance property. There are many studies involving the corrosion of CMC-SiC, which is not within the scope of this chapter. During the period of service, oxygen

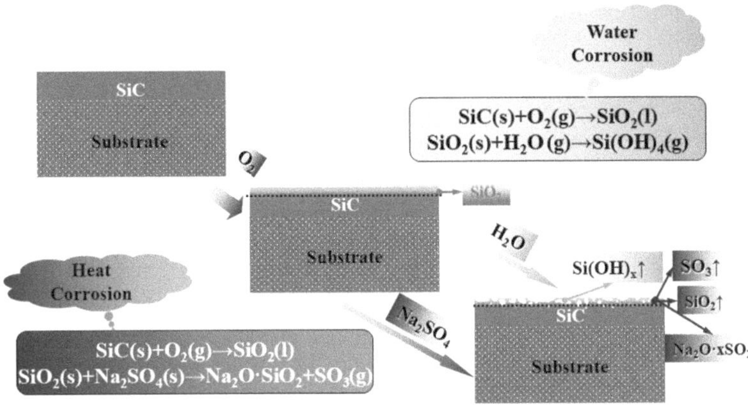

FIGURE 5.3 The faced corrosion problems of CMC-SiC.

diffuses rapidly along the holes or cracks, resulting in oxidation of the interphase and fiber, hence, leading to a decline in the mechanical properties of the material and even early failure. Therefore, the oxidation of the composite limits the application of CMC-SiC. It is imperative to improve the oxidation resistance of CMC-SiC and thereby prolong its service life.

In this chapter, measures are proposed and discussed to obtain long-term CMC-SiC. High degree of densification, high σ_{mc} and introduction the self-healing components are three significant strategies to prolong the duration of CMC-SiC, as shown in Figure 5.3. The corresponding manufacturing technologies to obtain CMC-SiC with high density are elaborated. The design principles of fiber preform architecture, interface shear strength, CTE match and matrix strengthening are summarized to enhance the σ_{mc}. The types of self-healing component and the effective self-healing temperature range are first distinguished and concluded.

The self-healing mechanism is deeply expounded upon. The CMC-SiC with excellent service performance can be obtained by microstructure and composition design combined with proper process technique. This chapter could help readers to quickly familiarize themselves with the progress and major scientific issues of long-term ceramic matrix composite (CMC). The principles of process control, microstructure design and composition choice are further provided for researchers to fabricate CMC with excellent characteristics (for example, mechanical property, self-healing performance et al.). The development tendency of thermal-structure composites for aero-engine is also predicted, and the valuable thinking space is built for subsequent researchers.

5.3 SIC/SIC WITH HIGH DENSITY

For components working under low stress, although no stress damage such as matrix microcracks will initiate when they are loaded, the holes derived from the preparation will still provide a channel for oxygen to diffuse inward, resulting in the oxidation of the interphase and fibers during long-term service, which will lead to rapid deterioration of material properties and failure. Meanwhile, high porosity has adverse effects on mechanical properties and thermal conductivity of SiC/SiC. When it is loaded, the holes acting as the crack source will reduce the σ_{mc}. In addition, the holes will increase the interface thermal resistance, which is not conducive to heat flow transfer and reduce the thermal conductivity. In this case, increasing the densification degree can effectively improve the environmental performance of the composites. When SiC/SiC with high density works, oxidation only occurs on the surface of the material. The compact oxide layer generated on the surface of SiC/SiC can effectively block the oxygen and prolong the service life. The methods of preparing CMC with high density are summarized in this chapter.

The porosity and volume density of the material can be obtained by Archimedean drainage method to evaluate the densification degree of the material. The porosity of SiC/SiC with high density should be less than or equal to 2 percent.

$$P_C = \frac{m_1 - m_0}{m_1 - m_2} \times 100\% \tag{5.1}$$

$$\rho_C = \frac{m_0}{m_1 - m_2} \times \rho_{H_2O} \tag{5.2}$$

m_0, m_1 and m_2 are the dry weight, floating weight and wet weight of the sample, respectively. ρ_{H_2O} is the density of water. P_C and ρ_C are the open porosity and density of the sample, respectively.

5.3.1 Melt Infiltration Technology

Melt infiltration (MI) is one kind of method for preparing SiC/SiC with high density. The process steps are as follows: firstly, the interphase and a certain volume fraction of SiC matrix are introduced into the SiC fiber preform successively by CVI process to obtain porous SiC/SiC. Then, a certain volume fraction of SiC particles are introduced into the porous SiC/SiC by slurry impregnation (SI) technology. And the large inter-bundle pores of porous SiC/SiC are divided into small pores by SiC particles. Finally, liquid silicon is introduced into the composites under capillary force to fill the remaining pores and obtain SiC/SiC with high density. Compared with RMI process, there is no carbon-silicon reaction in melt infiltration process, which is just a physical process. Therefore, under higher capillary force, liquid silicon can theoretically reach and fill all the open pores, so that the composite can achieve a higher densification degree.

Due to the high content of residual silicon, the working temperature of MI SiC/SiC is limited. On one hand, when the working temperature of the rotating components exceed 1400°C, the sweating of Si is harmful to the surrounding metal sub-components. On the other hand, when the working temperature is above 1250°C, the mechanical property decreases owing to the degradation of SiC fiber and interphase.

It is important to correctly predict the service life of MI SiC/SiC that can avoid the catastrophic fracture during the service.

5.3.2 Nano-infiltration and Transient Eutectic-phase Technology

Nano-infiltration and transient eutectic-phase (NITE) is another method for preparing SiC/SiC with high density. It is a composite technology for infiltration SiC particles combined with hot pressing and sintering. Akira Kohyama from Kyoto University in Japan have conducted extensive research on this method. In this method, TyrannoTM–SA SiC fiber is usually selected as the reinforcement, PyC or multilayer (PyC/SiC) n as the interphase, nano-sized β-SiC particles and sintering additives (Al_2O_3, Y_2O_3 and SiO_2) are used to obtain dense SiC matrix through transient liquid phase sintering at high temperature. Al_2O_3, Y_2O_3 and SiO_2 can form transient eutectic phase at low temperature, which can reduce the sintering temperature of SiC to a certain extent. The temperature of hot pressing sintering is 1720~1780°C and the pressure is 15~20 MPa. The final density SiC/SiC can reach is 2.96g/cm^3, and the porosity is reduced to 3.7 percent. Studies have shown that the increase of density is also conducive to improving the σ_{mc} of CMC. When the SiC/SiC prepared by this method is oxidized at 1000°C, the PyC interphase is still intact, indicating that the dense matrix can effectively prevent the diffusion of oxygen and greatly improve the environmental performance. However, because of the sintering temperature exceeding the limit temperature of most fibers, Tyrannotm-SA SiC fiber with good thermal stability can only be used as the reinforcement at present.

5.4 SiC/SiC with High Matrix Cracking Stress

Matrix cracking stress (σ_{mc}) is corresponding to matrix crack initiation in CMC. It can be obtained by tensile stress-strain curves or acoustic emission (AE) techniques. According to the tensile stress-strain curve, there are usually three methods to obtain the value of σ_{mc}, as shown in Figure 5.4(a). The first is the stress corresponding to the inflection point of the linear segment and the nonlinear segment. The second is the stress corresponding to 0.005 percent residual strain. The specific method is to cross the point of 0.005 percent strain on the horizontal axis and make a straight line parallel to the linear segment of the stress-strain curve, the stress corresponding to the vertical coordinate of the intersection of the line and the stress-strain curve is the σ_{mc}. The third is the stress corresponding to 0.05 percent strain. Acoustic emission technology can be used to in-situ monitor the damage process of materials. Two sensors are installed on the surface of the specimen during the test of the mechanical properties of materials, and the acoustic emission activities accompanied by rapid release of strain energy caused by the formation and expansion of matrix cracks inside the composites can be monitored in real time. Two parameters can be obtained by acoustic emission energy-stress curve (Figure 5.4(b)): σ_{min} and σ_{onset}. σ_{min} is the stress corresponding to the increase of acoustic emission energy from 0, that is, the first acoustic emission event, which represents matrix crack initiation. σ_{onset} is the stress corresponding to the rapid increase of acoustic emission energy, representing the propagation of crack.

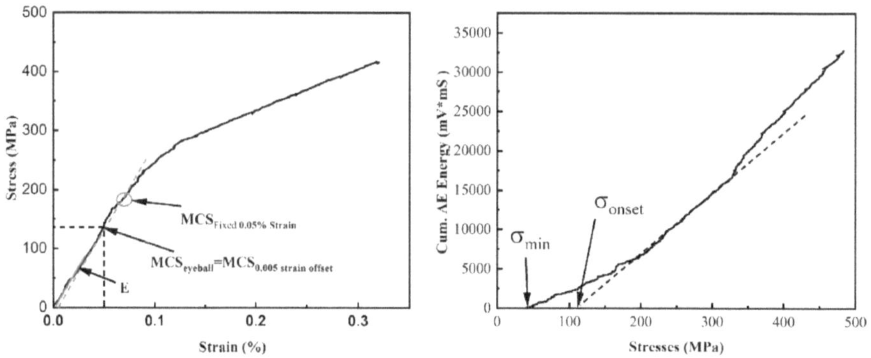

FIGURE 5.4 (a)SiC/SiC tensile stress-strain curve; (b) SiC/SiC acoustic emission energy-stress curve.

σ_{mc} is a very important parameter, which determines the time of damage generation in CMC. When the working stress exceeds the σ_{mc}, microcracks will occur in the matrix, and oxygen will quickly diffuse inwards along the cracks and oxidize the interphase and fibers, which significantly shortens the service life of CMC. It is an effective measure to prolong the service life of CMC by improving the σ_{mc} to ensure it serves without stress damage.

S. Sigl and A.G. Evans studied the crack growth resistance and matrix cracking stress in CMC. Matrix cracking stress can be expressed as:

$$\frac{\sigma_{mc}}{E}+\frac{\sigma_{r}^{m}}{E_{m}}=\left(\frac{6\mu V_{f}^{2}E_{f}G_{mc}}{\left(1-V_{f}\right)E_{m}Ea}\right)^{1/3}\left(\frac{q}{E_{m}}\right)^{1/3} \tag{5.3}$$

σ_r^m is the axial residual stress, *G*mc is the matrix toughness, *E*, E_m and E_f is Young's modulus of composites, matrix and fiber, respectively. μ is the friction coefficient, v_f is the fiber volume fraction, a is the fiber diameter, q is the normal compressive stress at interface. It can be seen from formula (5.3) that σ_{mc} is proportional to v_f and q, which also means that the matrix cracking stress will increase with the increase of fiber volume fraction and normal compressive stress at interface. When σ_r^m is negative, that is, when there is residual compressive stress in the matrix, σ_{mc} rises. Therefore, the matching of thermal expansion coefficient between fiber and matrix is very important. The axial thermal expansion coefficient of fiber should be greater than that of the matrix to ensure the axial compressive residual stress in the matrix. The radial thermal expansion coefficient of the fiber should be smaller than that of the matrix, so that the normal residual compressive stress exists at the interface, which increases the interfacial shear strength, improves the load transfer capacity, and enhances the σ_{mc}.

Morscher's research shows that for 2D and 3D composites with 0°/90° characteristics, matrix cracks originate from the 90° mini composites, which are

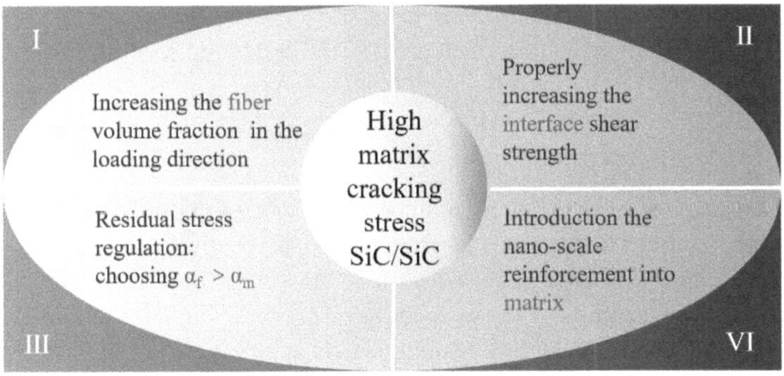

FIGURE 5.5 Several methods to improve the cracking stress of SiC/SiC composites matrix.

composed of fiber bundles, interphase, and matrix perpendicular to the 0° fibers. The cracking behavior of the matrix depends on the stress distribution of the matrix region in the 90° mini composites, which can be calculated by a simple rule-of-mixture theory:

$$\sigma_{\text{mini-matrix}} = \frac{(\sigma_T + \sigma_{th})}{E}\left(\frac{E - V_{\text{fmini}} E_{\text{mini}}}{1 - V_{\text{fmini}}}\right) \tag{5.4}$$

σ_T is the external stress, σ_{th} is the residual compressive stress in the matrix, V_{fmini} is the volume fraction of the 0° mini composites, and E_{mini} is the elastic modulus of the 0° mini composite, which can be calculated through the elastic modulus and volume fraction of each unit in the 0° mini composite using the rule-of-mixture. It can be seen from formula (5.4) that the cracking stress of the matrix is closely related to the residual stress in the matrix, the fiber volume fraction in the loading direction and the elastic modulus of the 0° mini composites. In summary, the σ_{mc} is closely related to the following four factors: the fiber preform architecture (fiber volume fraction in the loading direction), the residual stress (axial and radial), the interfacial shear strength and the elastic modulus of the 0° mini composites (matrix strengthening), as shown in Figure 5.5.

5.4.1 Fiber Preform Architecture

The fiber volume fraction in CMC-SiC is not less than 20 percent, so the fiber perform structure has a significant effect on the mechanical behavior of CMC and the σ_{mc}. Morscher studied the matrix cracking behavior of SiC/SiC with different preform structures, acoustic emission technology was employed to monitor the matrix cracking phenomenon. The results show that with the increase of fiber volume fraction in loading direction and the decrease of fiber volume fraction perpendicular to the loading direction, the turning point of linear segment in stress-strain curve moves to higher stress, which means that the increase of σ_{mc}. In the linear segment, the tensile strength of the composite can be calculated by the following formula:

$$\sigma_c = \sigma_f \upsilon_f + \sigma_m \upsilon_m \tag{5.5}$$

σ_c, σ_f and σ_m refer to the strength of composites, fiber and matrix, respectively. υ_f and υ_m refer to the volume fraction of fiber and matrix, respectively. As can be seen from formula (5.5), as the fiber volume fraction increases, σ_c improves. When the fiber volume fraction perpendicular to the loading direction is higher, the matrix crack in the thickness direction is easier to form under lower stress.

5.4.2 Interface Shear Strength

The volume fraction of interphase in CMC is not high, but it has a profound effect on the properties of composites. The interphase plays an important role in load transfer, mechanical fusing, protecting fibers from chemical damage and relieving residual stress. The interface shear strength, as a key parameter of interphase, has an important effect on σ_{mc}.

The interface shear strength of CMC can be obtained by push-out or push-in test. As shown in Figure 5.6(a), the push-out test applies a load to the single fibers in the CMC wafer sample to completely break the interphase and push the fibers out of the wafer sample. The load-displacement curve obtained by the push-out test can be divided into three stages. The initial stage is the elastic bending segment of CMC wafer samples, followed by the elastic deformation segment of fibers until the interface suddenly breaks and fibers slip, at which time the load reaches the maximum. The interface shear strength can be calculated by the following formula:

$$\tau = \frac{P}{2\pi re} \tag{5.6}$$

P is the maximum load, r is the fiber radius, and e is the thickness of CMC wafer sample. Push-in test applies a load to the single fiber until the interface breaks. As shown in Figure 5.6(b), the load-displacement curve obtained by the push-in test is S-shaped. The first stage is the elastic deformation stage of the fiber and matrix, and the slope of the load-displacement curve is S_0. As the load increases, the interface breaks and the load-displacement curve enters into nonlinear segment. The interface shear strength can be obtained by the following formula:

$$\tau = \frac{nP_c}{2\pi r^2} \tag{5.7}$$

$$n = \frac{S_0}{\pi E_f} \tag{5.8}$$

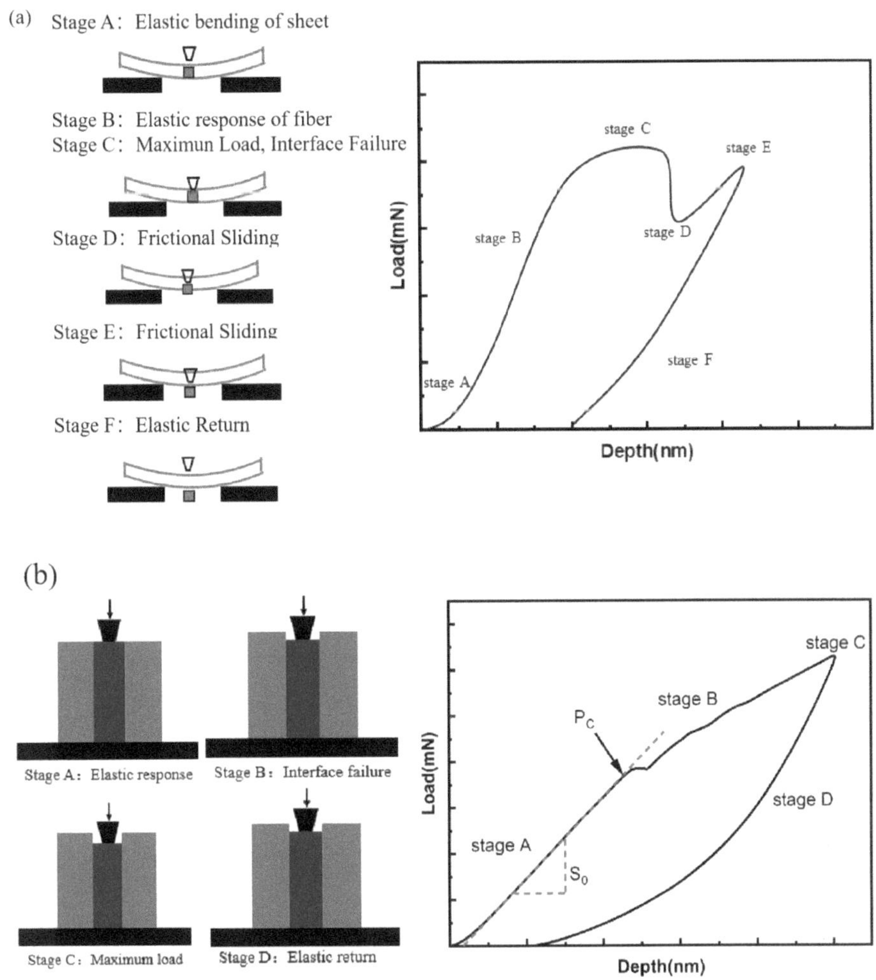

FIGURE 5.6 (a) Push-out test diagram and load-displacement curve; (b) push-in test diagram and load-displacement curve [3].

n is the parameter related to the elastic properties of the fiber and matrix and the binding of the surrounding fibers, which can be calculated by formula (5.8). P_c is the critical load capable of breaking the interface.

The increase of interface shear strength can make the stress transfer from the matrix to the fiber more quickly, effectively delaying the cracking of the matrix. At the same time, the interface debonding can still occur, which ensures the good mechanical properties of the composites. As shown in Figure 5.7, under the action of applied tensile stress (σ_T), the load is transferred from the matrix to the fiber through the interphase, redistributing the stress in the composite material. In this case, the stress (σ_m) in the matrix is equal to the external stress (σ_T) minus the stress on the fiber (σ). When the interface shear strength is high, $\sigma_m' = \sigma_T - \sigma'$; when the interface shear

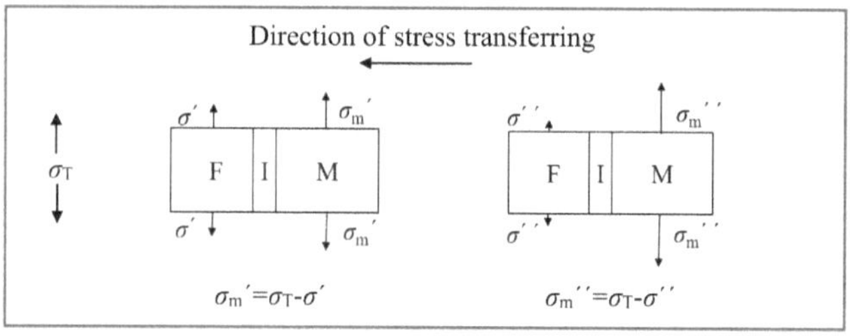

FIGURE 5.7 Diagram of stress transfer in composite materials, σ_T is external stress, $\sigma_m{}'$ and σ' are the stress in matrix and fiber at high interfacial shear strength, and $\sigma_m{}''$ and σ'' are the stress in matrix and fiber at low interfacial shear strength, respectively.

strength is low, $\sigma_m{}'' = \sigma_T - \sigma''$. When the interface shear strength is higher, the load can be transferred from the matrix to the fiber more quickly, resulting in $\sigma' > \sigma''$, so that $\sigma_m{}' < \sigma_m{}''$, matrix cracks will initiation under higher applied load.

5.4.3 Thermal Expansion Coefficient Match

At the preparation temperature, the thermal expansion between the fiber and the matrix is always matched. At temperature above or below the preparation temperature, the thermal expansion is always mismatch between the fiber and the matrix. Thermal stress occurs in CMC when temperature changes due to different thermal expansion coefficients between fiber and matrix. When the thermal stress is high, part of the thermal stress will be released. The interface thermal stress that is not released is called residual thermal stress. The residual thermal stress in CMC can be calculated by the following formulas:

$$\sigma_{th} = \frac{1}{(1+\upsilon_m)/E_m^*} \times (\alpha_f^r - \alpha_m)(T_o - T_p) \tag{5.9}$$

$$E_m^* = E_m \frac{1-\theta}{1+2.5\theta} \tag{5.10}$$

$$\theta = \frac{P}{1-\lambda V_f - P} \tag{5.11}$$

$$\sigma_r^m = E_m^* \frac{\lambda E_f V_f}{\lambda E_f V_f + E_m^* V_m} (\alpha_f - \alpha_m)(T_o - T_P) \tag{5.12}$$

σ_{th} and σ_r^m are the radial residual stress at the interface and the axial in the matrix, respectively. υ_m and υ_f are Poisson's ratio of the matrix and the fiber, respectively. α_f^r,

α_f and α_m are the radial, axial thermal expansion coefficients of the fiber and axial of matrix, respectively. T_O and T_P are room temperature and preparation temperature respectively. E_m^* is the effective Young's modulus of the matrix. P and θ are the porosity and relative porosity respectively. λ is the effective coefficient of fiber volume fraction in the loading direction. For the two-dimensional fiber preform, λ is 0.5.

The design of thermal expansion coefficient between fiber and matrix can effectively improve the σ_mc. The axial thermal expansion coefficient of the fiber should be greater than that of the matrix, so that residual compressive stress (σ_R) exits in matrix, and the tensile stress of the matrix when CMC is loaded is $\sigma_m = \sigma_T - \sigma_R$, and part of the external stress is offset, thus delaying the initiation of cracks in the matrix. The radial thermal expansion coefficient of the fiber should be less than that of the matrix, so that the radial residual stress is compressive stress, and the interfacial shear strength and the σ_{mc} can be improved to a certain extent.

5.4.4 Matrix Strengthening

One-dimensional nanomaterials (SiC nanowires, C nanotubes, BN nanotubes, etc.) have very high modulus and strength, and are an important reinforcement similar to fiber. At present, some studies introduce one-dimensional nanomaterials into the matrix of traditional fiber reinforced ceramic matrix composites to strengthen the matrix, so as to ensure that the load can be transferred rapidly from the matrix to the one-dimensional nano reinforcement when CMC is loaded, and significantly improve the modulus of the composite and the σ_{mc}.

Pavia extended the traditional Aveston-Cooper-Kelly shear-lag model to mixed brittle matrix composites containing micron-scale fibers and nano-scale reinforcement, taking the performance of traditional Nicalon SiC fiber reinforced CVI SiC composites and carbon nanotubes as reference, predicting the σ_{mc} of CMC with mixed reinforcement. The σ_{mc} increases significantly with the increase of the volume fraction, modulus of the nano-scale reinforcement, the interfacial friction between the nano-scale reinforcement and the matrix and the decrease of the radius of the nano-scale reinforcement. At the same time, they prepared the nanofiber reinforced composite with the same reinforcement volume fraction as the hybrid composite, and the σ_{mc} was significantly reduced. It shows that micron Nicalon SiC fibers also have an important effect on the σ_{mc}. The "hybrid effect" existed in this kind of micro/nanohybrid composites. Kim prepared SiC nanowires modified SiC/SiC using CVI process. With the introduction of SiC nanowires, the deposition sites of CVI SiC increase, the impingements of the growth fronts of SiC grains increase, and the size of SiC grains decreases. At the same time, rod-like SiC grains in SiC/SiC grow along the direction perpendicular to the fibers, and after the introduction of SiC nanowires, SiC grains grow randomly on the surface of the fibers without obvious growth direction. The bending strength-displacement curve shows that the bending strength, bending modulus and stiffness of the composites are significantly improved after the introduction of SiC nanowires, and the ability of SiC nanowires modified SiC/SiC to resist the initiation of cracks in the matrix is enhanced.

5.5 CMC-SIC MODIFIED WITH SELF-HEALING MATRIX

When CMC works in long term under moderate load, both matrix crack initiation and material damage will inevitably occur. Oxygen diffuses rapidly into material interior along the microcracks, which results in oxidation of interphase and fiber, causing rapid material failure. At present, self-healing of cracks is an effective way to prevent oxidation atmosphere from further spreading and the oxidation of interphase and fiber.

Considering the thermodynamical condition, the conditions of crack healing should satisfy the inequation that $G < \Delta G_{heal}$, where G and ΔG_{heal} respectively represent the released elastic strain energy in per unit area of crack growth (the driving force of crack growth) and the Gibbs free energy associated with crack healing (the driving force of crack healing). As the formula 13 shown below, the ΔG_{heal} is made up of three parts, among them; ΔG_{surf} is the released energy when surface converses into grain boundary. ΔG_{el} represents the elastic strain energy of materials under external compressive stress; ΔG_{chem} is identified as the chemical energy produced by chemical reactions. Therefore, correspondingly, there are three kinds of healing mechanisms for ceramic materials: the diffusion mechanism related with ΔG_{surf}, which means the sintering of ceramics; the thermal expansion mechanism depending on ΔG_{el}, that is selecting the fiber whose axial thermal expansion coefficient is higher than the matrix's to make the matrix bear the compressive stress; the oxidation mechanism corresponding to ΔG_{chem} is relied on the glass phase resulted from oxidation reactions of the self-healing components at lower activation energies to heal cracks. Compared with the diffusion mechanism requiring high temperature and the thermal expansion mechanism with limited choice of material system, the oxidation mechanism shows higher healing efficiency. By introducing B-containing self-healing components (B_xC, SiBC, SiBCN, etc.) into the composite, during the course of oxidation, the volume expansion of the self-healing components and the viscous flow of B_2O_3 or borosilicate glass can rapidly seal the cracks and achieve self-healing target, effectively preventing oxygen from diffusing into the material and extending the service life of the composite.

$$\Delta G_{heal} = \Delta G_{surf} + \Delta G_{el} + \Delta G_{chem} \tag{5.13}$$

5.5.1 CMC-SiC Modified with Borosilicate Glass

The process of self-healing is fairly complicated, including the oxidation of self-healing components, crack healing and the volatilization of glass phase. When introducing borosilicate glass directly into CMC-SiC, the crack healing phenomenon and the stability of glass phase can be studied separately, and the effective operating temperature range of glass phase with different components can be analyzed, which is of great importance for researching self-healing behavior of self-healing components.

Cheng et al. prepared borosilicate glass coating on the C/SiC surface (named as C/SiC+Sealant) to improve the environmental performance of the composite. Borosilicate glass coating with target SiO_2:B_2O_3 ratio successfully covered the C/SiC surface by sol-gel method combined with vacuum sintering (1000°C, 1 h)

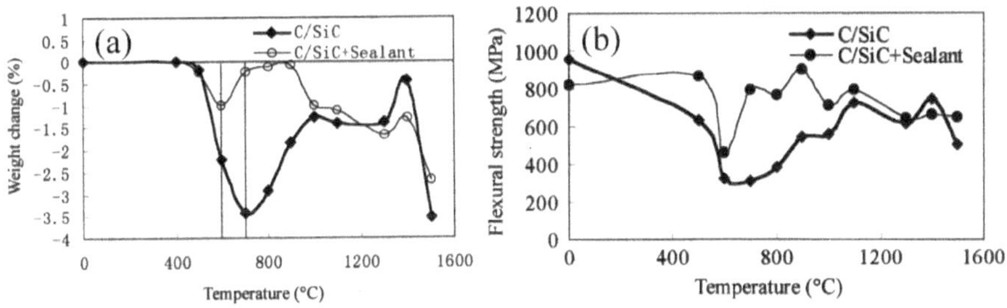

FIGURE 5.8 (a) The weight change and (b) flexural strength of the composites oxidated at different temperature [4].

process. Then the material was oxidized in air atmosphere at 400~1500°C for 5 h, and the curves of mass loss rate and bending strength respectively changing with oxidation temperature were obtained. As the thermal expansion coefficient of carbon fiber is much lower than that of SiC matrix, high residual tensile stress exists in the matrix during cooling from preparation temperature to room temperature, resulting in many microcracks in the matrix. When the temperature is lower than the preparation temperature, the cracks remain open, and oxygen diffuses rapidly into the material along the microcrack and oxidizes the interphase and fibers. In the range of 400~1200°C, borosilicate glass can seal cracks and effectively prevent the diffusion of oxygen into the material, and further prevent the oxidation of the carbon fiber. The mass loss rate of borosilicate glass modified C/SiC is much lower than that of C/SiC in Figure 5.8(a), and the flexural strength retention rate of the modified C/SiC is much higher than that of C/SiC in Figure 5.8(b). Above 1200°C, the volatilization of borosilicate glass is more intense, its sealing effect on the crack is weakened, and meanwhile, the cracks in C/SiC are closed, so both the mass loss rate and flexural strength retention rate of the two composites are similar. In addition, the melting point of borosilicate glass can be controlled effectively by adjusting the ratio of SiO_2 and B_2O_3, so as to adjust the effective operating temperature range of borosilicate glass. This also suggests that, for a specific service temperature, the content of B element in the self-healing component, that is, the content of B_2O_3 glass phase, can be properly adjusted to regulate the melting point of the glass phase for outstanding self-healing effect.

5.5.2 CMC-SiC Modified with Self-healing Component

According to the distribution form of self-healing components in the matrix, SHCMC can be divided into two categories. In one kind of SHCMC, the self-healing components are uniformly distributed in the continuous matrix phase. Because the distribution of glass phase is relatively dispersed, the effective self-healing temperature is higher. However, the formation rate of glass phase is slow and the resistance to environmental media is small, leading to poor self-healing effect. In another kind of SHCMC, self-healing components and matrix components are alternately distributed layer by layer. Because the glass phase with a fast formation rate can block

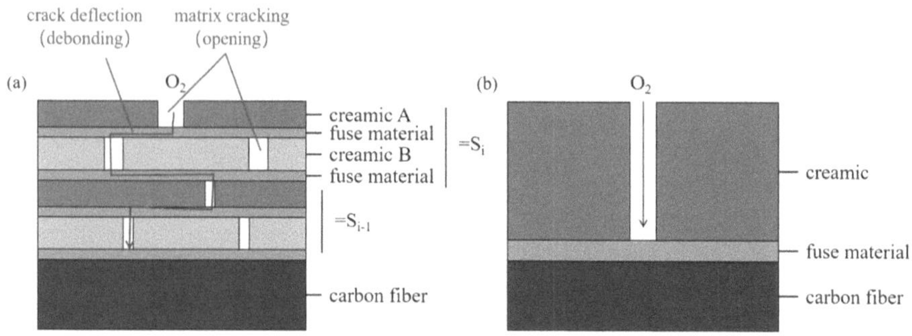

FIGURE 5.9 Schematic diagram of components, damage propagation and oxygen diffusion paths in (a) multilayered CMC and (b) monolayered CMC [5].

and consume the environmental media, the self-healing effect is good. However, the relatively concentrated glass phase is easy to volatilize at high temperature, so the effective self-healing temperature is low. At present, there are many researches on multilayered SHCMC, whose microstructure is shown in Figure 5.9 [5].

Multilayered matrix consists of alternating ceramic layers (usually is SiC), self-healing component layers, and thin layers of materials between them. The SiC matrix mainly plays the role of loading and transferring load, and the thin layer material is usually BC_x (boron-doped pyrolytic carbon) or BN phase, which take effect of mechanical fusing. Therefore, the cracks can be deflected when they propagate extending to thin layer material, the stress concentration at the crack tip can be alleviated, the crack propagation path can be effectively extended, and the crack propagation energy can be consumed. Compared with the single layer matrix, the multilayered matrix shows lower sensitivity of to crack opening. At the same time, with the extension of the crack propagation path, the oxygen diffusion path increases correspondingly. The glass phase generated by the rapid oxidation of the self-healing component can be used as a sealant to quickly seal the cracks, preventing the oxygen from continuing to spread inward, and delaying the oxidation of the interface and fibers to a certain extent when the matrix cracks initiate and the self-healing component is exposed to the service environment. The effective debonding, crack deflection and narrow crack width in multilayered matrix give SHCMC excellent self-healing effect.

5.5.2.1 CMC-SiC Modified by BCx

Liu et al. successfully prepared (SiC-BC_x) multilayered matrix modified C/SiC composite using CVI process. The pore inner fiber bundle is basically filled with SiC matrix, and the (SiC-BC_x) multilayered matrix is mainly located in the inter bundle region. The mechanical properties of multilayered matrix modified composite at room temperature are close to that of C/SiC. After oxidation at 700°C for 10 h in static air, the residual strength of the modified composite is close to that before oxidation. After respectively oxidation at 1000°C and 1300°C for 10 h, the residual strength of the modified composites even increased slightly compared with the original strength,

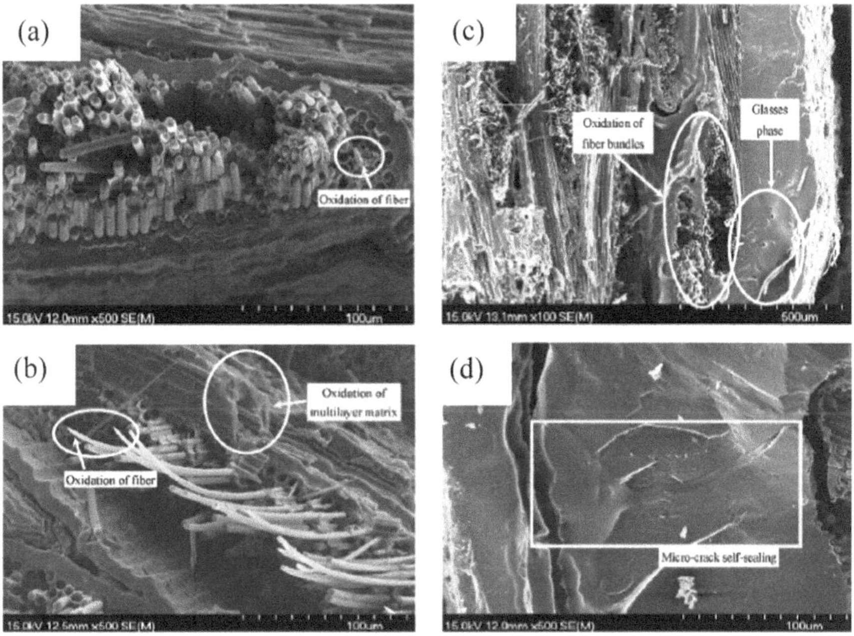

FIGURE 5.10 The tensile fracture morphology of the C/(SiC-BxC)m multilayered composite oxidized at 700°C for different time: (a) 10 h, (b) 25 h, (c) and (d) 60 h [6].

and the microcracks and holes on the surface of the composites were all sealed by the glass phase formed by the oxidation of multilayered matrix. After oxidizing the composites in 12% O_2/8% H_2O/80% Ar atmosphere, under 100 MPa load and at 700°C for 10 h, 25 h and 60 h respectively, the residual strength retention rates of the modified composites respectively were 95.44 percent, 93.55 percent and 83.25 percent. At the same time, the matrix cracks were sealed by the glass phase formed by oxidation of multilayered matrix (Figure 5.10). The strength attenuation rate of the (SiC-BC_x) multilayered matrix modified composite was only 0.95 MPa/h, while that of C/SiC reached 27.7 MPa/h. Through introducing (SiC-BC_x) multilayered matrix, the strength attenuation rate of the composite decreased by 96.57 percent and its service life can be extended significantly.

5.5.2.2 CMC-SiC Modified by SiBC

When the temperature is over 900°C, the self-healing effect of B_2O_3 attenuates and glass phase with higher melting point is required to heal the cracks. SiBC, which can form borosilicate glass by oxidization, is an excellent candidate for self-healing. Nowadays, there are two methods to prepare SiBC. The first method is CVI, which can obtain $(SiBC\text{-}SiC)_m$ multilayered matrix by alternating depositing SiBC and SiC; The second one is RMI process. The microstructures and compositions of SiBC prepared by the two processes are quite different, which will be discussed separately below.

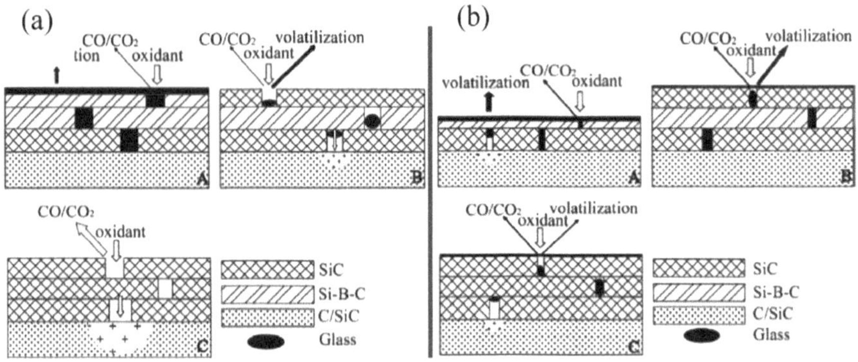

FIGURE 5.11 The oxidation models of C/SiC composites containing different coatings in air atmospheres when the temperature is (a) below 1000°C and (b) above 1000°C [7].

Zuo et al. used CVI process to make SiC/SiBC/SiBC (denoted as A), SiC/SiBC/SiC (denoted as B) and SiC/SiC/SiC (denoted as C) coatings on C/SiC surface, testing the oxidation resistance of the composites in static air atmosphere at different temperatures (700, 1000, 1200, 1300°C). For sample A, with the increase of oxidation temperature, the volatilization of borosilicate glass became severe, and the protective effect on the composite material was weakened. The internal carbon fiber was partly oxidized, leading to the increasing mass loss and the gradual decrease of bending strength. For sample B, the formed borosilicate glass with large viscosity was not enough when oxidized at 700°C, which made it difficult to seal cracks effectively. And the oxidation of carbon fiber was serious and a large mass loss showed. When oxidized at 1000~1300°C, sufficient borosilicate glass could be formed and cracks were quickly filled. At the same time, the external SiC layer effectively prevented the volatilization of glass phase, as shown in Figure 5.11, which helped to obtain well self-healing effect and excellent mass stability. Besides, sample B showed a good strength retention rate at 1000~1200°C and the highest bending strength after oxidation at 1200°C. As the oxidation rate of SiC was slow, it cannot generate enough glass phase to provide protection, so the C sample had the worst oxidation resistance and the lowest bending strength retention rate. Liu et al. further investigated the oxidation resistance of C/SiC composites containing SiC/SiBC/SiC coatings in a wet-oxygen environment for 50 h.

The oxidation temperatures were set at 700, 1000, 1200, 1300 and 1400°C, while the wet oxygen environment were set to 14 vol% H_2O/8 vol% O_2/78 vol% Ar and 21 vol% H_2O/8 vol% O_2/78 vol% Ar, respectively. The following chemical reactions may occur during the oxidation process:

600°C

$$C(s) + O_2(g) \rightarrow CO_2(g) \tag{5.14}$$

$$B_4C(s) + 4O_2(g) \rightarrow 2B_2O_3(g) + CO_2(g) \tag{5.15}$$

600–1000°C

$$B_4C(s) + 8H_2O \rightarrow B_2O_3(l) + CO_2(g) + 8H_2(g) \quad (5.16)$$

$$B_2O_3(l) \rightarrow B_2O_3(g) \quad (5.17)$$

$$B_2O_3(l) + H_2O(g) \rightarrow 2HBO_3(g) \quad (5.18)$$

$$B_2O_3(l) + 3H_2O(g) \rightarrow 2H_3BO_3(g) \quad (5.19)$$

$$2SiC(s) + 3O_2(g) \rightarrow 2SiO_2(l) + 2CO(g) \quad (5.20)$$

≥1000°C

$$SiC(s) + 3H_2O(g) \rightarrow SiO_2(s) + CO(g) + 3H_2(g) \quad (5.21)$$

1000°C

$$SiO_2(s) + 2H_2O(g) \rightarrow Si(OH)_4(g) \quad (5.22)$$

≥1000°C

$$B_2O_3(l) + nSiO_2(l) \rightarrow B_2O_3{\cdot}nSiO_2(g) \quad (5.23)$$

$$B_2O_3{\cdot}nSiO_2(g) \rightarrow B_2O_3(g) + nSiO_2(s) \quad (5.24)$$

At 700°C oxidation, the reaction (5.15)~(5.17) can occur and form B_2O_3 glass. Part of the matrix cracks are sealed, resulting in maintaining the bending strength of the material to a certain extent. When the oxidation temperature increases to 1000°C, all the reactions above can occur. The B_2O_3 and borosilicate glass generated by oxidation can effectively seal cracks, prevent the oxidation of interphase and fiber, and make the composite keep the highest strength retention rate which can go up to 106.7 percent. Above 1200°C, the speed of oxidation reaction is faster, the volatilization of glass phase is also more severe, so the self-healing effect is weakened. With the increase of water pressure, the reaction (5.18), (5.19) and (5.22) speed up. Meanwhile, the volatilization of the glass phase becomes more intense, the cracks are more difficult to be healed, and the strength of the composite goes down after oxidation.

Viricelle, J.P et al. studied the oxidation behavior of $SiC_f/C/SiBC_m$ composites containing multilayered ceramic matrix. The multilayered $SiBC_m$ matrix was an alternating structure consisting of SiC phase (matrix 3), B_4C phase (matrix 2) and SiBC phase (matrix 1). From the Figure 5.12(a), it can be seen that B_4C was closer to the fiber than the SiBC phase, which consisted of SiC grains and amorphous phase surrounding them, containing the B elements existing in amorphous boron carbide or the SiC lattice. When the temperature was 600°C, B_4C was slowly oxidated to produce glass phase, which was not enough to protect the PyC interphase from being oxidized.

So, the mass change curve in Figure 5.12(b) was approximately linear, showing the characteristics of continuous weight loss. When the oxidation temperature raised to 650°C and 700°C, B_4C speeded up to be oxidated, generating plenty of B_2O_3 glass which played a certain protective role to prevent the oxidation of PyC interphase and decrease the mass loss. At 800°C, the oxidation rate of B_4C was further increased and sufficient glass phase was formed to remain the PyC interphase intact and obtain the maximum increase in weight. When oxidized at 900°C and 1000°C, the B_2O_3 glass occurred decreasing viscosity and partial volatilization, which weakened the protection effect and leaded to the oxidation of PyC interphase and the decline of weight gain. In addition, a slight oxidation of SiBC at 1000°C was observed, supporting that the formation of borosilicate glass contributed to the increased mass and decreased porosity. When the oxidation temperature went up to 1200°C, due to the violent volatilization of B_2O_3 glass, there was about 0.07 percent weight loss in the first 50 min of oxidation, and then with the formation of borosilicate glass, the weight gain can reach 0.2 percent after 7h of oxidation. At 800°C, the PyC interphase mainly relied on the protection of B_2O_3 glass, while borosilicate glass played a major role at 1000~1200°C. The results showed that the combinational design of different self-healing components can obtain well self-healing effect in a wide temperature range.

SNECMA in France has carried out extensive research on multilayered self-healing composites and prepared a series of composite brands, such as A400, A410, A500 and so on [9][10][11][12]. At the stress level of 120 MPa When the temperature increased from room temperature to 800°C, the service life of A373 (SiC/SiC) rapidly decreases from more than ten hours to several hours; The service life of A400 (reinforced by Nicalon SiC fiber) and A500 (reinforced by T300 carbon fiber) reach 100 h at 1200°C; The A410 (reinforced by Hi-Nicalon SiC fiber) have the service life of 1000 h at 850°C and 500 h at 1200°C. Among them, A262 (C/SiC) has been served as the engine outer flaps of the M88-2 engine, and A410 has been successfully used in the flame stabilizer.

At present, SiBC ceramics prepared by RMI process have also attracted wide attention [13][14][15][16][17]. B4C is usually used as B source and C source, which reacts with liquid silicon in the RMI process to form the SiBC phase. Some

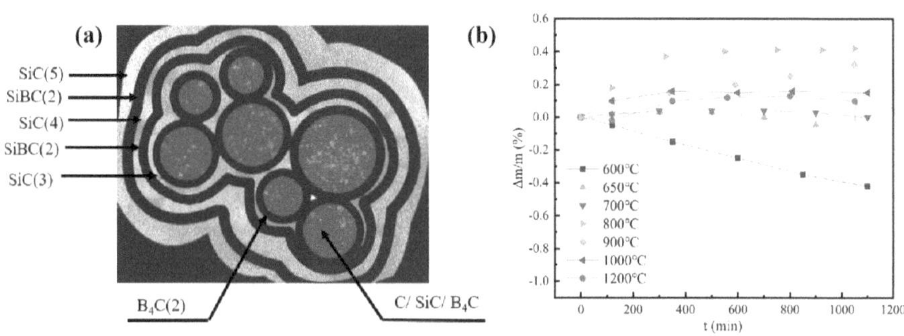

FIGURE 5.12 (a) The cross-section morphology of the SiCf/C/(SiBC)m multilayered composite; (b) the mass change ratio of the composite oxidated at different temperatures [8].

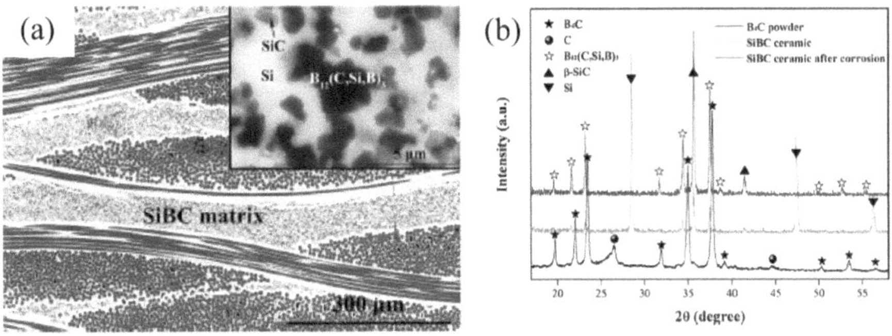

FIGURE 5.13 (a) Backscattered electron images at the interphase of C/SiC-SiBC composites prepared by RMI process; (b) XRD curves of B_4C powder and SiBC ceramics before and after pickling [18].

researchers have introduced RMI SiBC into C/SiC composites to study their mechanical properties and oxidation behavior. As shown in Figure 5.13(a), the intra-bundle holes of the carbon fiber bundles were basically filled by CVI SiC matrix, and the inter-bundle holes were mainly filled by RMI SiBC matrix. The microstructure of RMI SiBC ceramic matrix is shown in the upper right corner of Figure 5.13(a). Combined with XRD results in Figure 5.13(b), RMI SiBC consisted of SiC (gray phase), residual Si (white phase) and $B_{12}(C,Si,B)_3$ (black phase). Cao et al. studied the oxidation behavior of C/SiC-SiBC composites in static air at 800~1200°C. Because the thermal expansion coefficient of SiC and SiBC matrix are both higher than that of carbon fiber, residual tensile stress existing in the matrix caused the cracking of the matrix when the temperature reduced from the preparation temperature to room temperature. At 800°C, SiBC was oxidated slowly, and the oxidation behavior of the composite was mainly controlled by the diffusion rate of oxygen in the matrix cracks and coating defects. Meanwhile, the glass produced at this temperature had limited amount and high viscosity, which makes it difficult to seal the cracks.

Therefore, both the PyC interphase and the carbon fiber underwent oxidation (Figure 5.14(a)). The strength of composite decreased after it was oxidated at 800°C for 10 h. The SiBC oxidation rate increased at 1000°C, and the viscosity of the borosilicate glass decreased. This helped to seal some of the matrix cracks and prevent the fibers and interphase from being oxidized to some extent (Figure 5.14(c-d)), resulting in a significant increase of residual strength compared with that at 800°C. When the oxidation temperature went to 1200°C, the matrix cracks were basically sealed by borosilicate glass (Figure 5.14(e-f)), and the oxidation rate of the composite was mainly determined by the diffusion rate of oxygen in the glass. In this case, the interphase and fibers could be effectively protected, and the retention rate of bending strength could reach 110 percent.

5.5.2.3 CMC-SiC Modified by SiBCN

SiBCN ceramics are widely used in the preparation of ceramic fibers, coatings, and structural ceramic components due to the low density, low thermal expand coefficient,

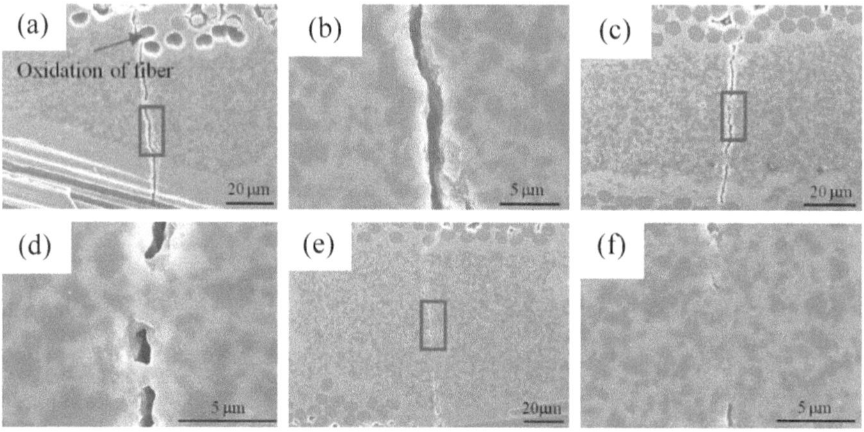

FIGURE 5.14 The cross-section morphology of C/SiC-SiBC composites after oxidation for 10 h at different temperatures: (a) and (b) 800°C, (c) and (d) 1000°C, (e) and (f) 1200°C [14].

and excellent thermal stability. In addition, SiBCN ceramics have gained a lot of attention for the oxidation resistance and self-healing behavior when it is introduced into CMC, being considered as a kind of promising material of performing well self-healing ability at 1200~1400°C.

At present, the key methods to prepare SiBCN are chemical vapor deposition (CVD), precursor derived ceramic (PDC), reactive magnetron sputtered (RMS) and mechanical alloying (MA) combined with hot pressing sintering (HPS). Studies on the preparation of SiBCN by CVD mainly focus on the deposition process, composition regulation, dielectric properties, electromagnetic wave absorption and shielding properties [19][20][21][22]. RMS is suitable for preparing not bulk ceramics or fiber-reinforced composites but coatings and films [23]. Studies on the oxidation resistance of SiBCN were focused on the SiBCN by precursor derived ceramic or mechanical alloying combined with hot pressing sintering. The results showed that the pre-cracks on the surface of the polymer derived SiBCN ceramic (PDC-SiBCN) were basically healed after being oxidized for 480 min at 1000°C. For SiBCN ceramics prepared by mechanical alloying combined with hot pressing sintering, the effects of B and C contents on their oxidation resistance were investigated [24]. It can be found that the oxidation of B-containing compounds is superior to that of Si-containing compounds, and the generated B_2O_3 can further promote the oxidation of SiC to form borosilicate glass and accelerate the oxidation process of SiBCN. Appropriately increasing the content of B elements can generate more glass phases in a short time and improve the self-healing rate of SiBCN. However, too much B elements may produce much B_2O_3 glass with low melting point, whose volatilization at high temperature would damage the integrity of the oxide layer, being adverse to the self-healing property of SiBCN. With the increase of free carbon content in SiBCN, the number of holes left increased after carbon oxidation, and the oxidation resistance decreased [25].

Mechanical alloying combined with hot pressing sintering is not suitable for the preparation of ceramic matrix composites, so the following studies are all about SiBCN modified CMC-SiC prepared by PIP process.

Luan et al. used the chemical vapor infiltration combined with polymer infiltration online pyrolysis (CVI+PIOP) to prepare the SiC/(SiC-SiBCN)$_x$ composites and studied the oxidation behavior in wet-oxygen environment [26]. Respectively using two kinds of fiber preforms of 2D plain weave and 2D satin weave, the SiC/(SiC-SiBCN)$_x$ composites were obtained by introducing SiBCN into porous SiC/SiC. SiBCN was mainly located in the inter bundle region and evenly distributed in the CVI SiC matrix. The SiC/(SiC-SiBCN)$_x$ prepared by using two-dimensional plain weaved porous SiC/SiC with higher density was denoted as DP, the SiC/(SiC-SiBCN)$_x$ prepared by using two-dimensional plain weaved porous SiC/SiC with lower density was denoted as P, and the SiC/(SiC-SiBCN)$_x$ with satin weave was denoted as S. The three composites were oxidized at 1300°C, 1400°C and 1500°C for 100 h with the oxidation atmosphere of 14 kPa H_2O/8 kPa O_2/78 kPa Ar, 14 kPa H_2O/20 kPa O_2/66 kPa Ar and 30 kPa H_2O/20 kPa O_2/50 kPa Ar, respectively. After oxidation, the linear segment of the flexural load-displacement curve (i.e., flexural modulus) and the fracture load (i.e., flexural strength) of several composites decreased, while the fracture displacement (i.e., fracture strain) increased, and all the composites exhibited pseudoplastic behavior. The degradation of fiber and interphase properties at high temperature leaded to the degradation of composite properties in some way. When the oxidation increased from 1300°C to 1500°C, the flexural strength retention ratios of S and P samples respectively decreased from 82 percent to 58 percent and from 69 percent to 48 percent, while DP sample had the highest strength retention ratio. After oxidizing at 1300°C for 100 h in an atmosphere of 14 kPa H_2O/8 kPa O_2/78 kPa Ar, the strength retention ratio of DP samples was closed to 100 percent. With the increase of O_2 partial pressure, the flexural strength of S and P samples basically remained, and their flexural modulus decreased slightly; With the increase of H_2O partial pressure, the flexural strength of S and P samples both increased slightly; With the synchronous lift of O_2 partial pressure and H_2O partial pressure, the flexural strength of DP samples decreased gradually, but the retention ratio of bending strength still kept at the highest value, which could reach 85 percent. The following reactions and reaction (5.17), (5.20), (5.25) mentioned above may occur during the oxidation process:

$$2C + O_2 \rightarrow 2CO(g) \tag{5.25}$$

$$4BN + 3O_2 \rightarrow 2B_2O_3(l) + N_2(g) \tag{5.26}$$

$$2B_2O_3 + O_2 \rightarrow 4BO_2(g) \tag{5.27}$$

$$Si_3N_4 + 3O_2 \rightarrow 3SiO_2(s) + 2N_2(g) \tag{5.28}$$

$$2SiC + 3O_2 \rightarrow 2SiO_2(s) + 2CO(g) \tag{5.29}$$

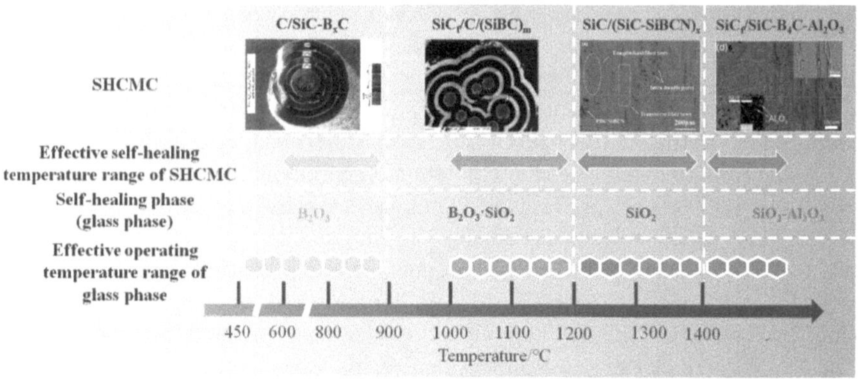

FIGURE 5.15 The effective self-healing temperature ranges of different self-healing components.

The glass phase generated by the oxidation of SiBCN matrix can protect the interphase and fibers from oxidation, so that the composite can maintain good mechanical properties.

According to the above results, the environmental performance of the composite is poor when SiBCN is used as the matrix alone without SiC matrix. The SiBCN introduced into the composite by the PIP method is usually amorphous. When the SiBCN is exposed to high temperature and oxygen-containing environment, one side of the SiBCN is oxidized. When the service temperature exceeds the preparation temperature of SiBCN, the other side of the SiBCN that are not exposed to oxygen will undergo atomic rearrangement and phase separation due to heat effect [27]. Complex changes occur in ceramic matrix composites, forming the stress field. In order to achieve good self-healing effect in the target temperature range (Figure 5.15), a systematic study of the interaction between the oxidation and crystallization processes of SiBCN is required to provide guidance for the wider application of SiBCN in SHCMC.

5.5.3 Glass Phases with High Melting Points

The violent volatilization of the glass phase will seriously damage the integrity of the oxide layer, promoting the oxygen to further diffuse into the material and oxidize the interphase and fiber. At high temperature, the viscosity of the glass phase decreases, the diffusion rate of oxygen in the oxide layer increases, which is not conducive to the oxidation resistance of the material. By introducing Al_2O_3, the stability of the glass phase can be significantly improved, and the diffusion rate of oxygen in the glass phase can be reduced, thus the oxidation resistance of the material is improved.

Some researchers introduced Al into SiBCN and studied its oxidation behavior [28][29]. Two shapes of SiBCN and SiBCNAl were prepared, one of which was foam-like ceramic particles obtained by direct pyrolysis of precursor, and the SiBCN and SiBCNAl obtained by this method were denoted as T2(1)c and Al3c, respectively.

The other one is the disc-shaped sample obtained by hot pressing, and the SiBCN and SiBCNAl obtained by this method were denoted as T2(1)c_{WP} and Al3c_{WP}, respectively. After 100 h oxidation at 1500°C, a large number of bubbles were generated at the edges of sample T2(1)c, while the oxidation layer at the rest parts was still dense and amorphous, but being full of cracks. In contrast, the oxide layer of Al3c was relatively uniform and smooth without cracks, and some foam-like pores were covered by glass phases. The oxide layer of T2(1)c_{WP} was full of bubbles and cracks with partially shed, while the oxide layer of Al3c_{WP} remained dense without bubbles and cracks. The results showed that with the formation of mullite, boroaluminate and $Al_2O_3/B_2O_3/SiO_2$ glass, the stability of the glass phase was significantly improved, while the fluidity of the glass phase was still maintained, which was conducive to improving the self-healing effect. Shan et al. studied the effect of Al_2O_3 on the oxidation behavior of Si-B-C ceramics. The ceramics were oxidized at 1200°C in different atmospheres (dry oxygen and wet oxygen) for 98 h. The introduction of Al_2O_3 could prevent the transition of amorphous SiO_2 glass to crystalline cristobalite, and the thermal expansion coefficient between amorphous SiO_2 glass and crystalline cristobalite was significantly different, and inhibiting this transition could effectively prevent the cracking of the oxide layer [30]. At the same time, the addition of Al_2O_3 can inhibit the conversion of bridging oxygen atoms (Si-O-Si, Si-O-Al) into non-bridging oxygen atoms (Si-O-H). And the protonated bridging oxygen (Al-O(H)-Si) can play a connecting role in the glass network, improving the viscosity of the aluminosilicate melt to a certain extent and reducing the reactivity of the glass phase with water vapor. Thus, the oxidation resistance of Si-B-C ceramics can be improved.

5.6 APPLICATIONS AND RESEARCH TRENDS

Long-term CMC, which can serve at thermo-mechanical-oxygenic coupling environment for hundreds and even thousands of hours, usually show several key characteristics, including high degree of densification, high σ_{mc}, and containing self-healing components or two of them. The pores in CMC act as crack source and promote the initiation of matrix cracks at low load. In addition, the pores provide diffusion pathway for oxygen, and it is harmful to the self-healing performance. Thus, the high degree of densification is the basic characteristic for CMC to obtain high σ_{mc} and excellent self-healing property.

Structure design, component selection and technological control should be taken into account for long duration CMC. The structural design mainly includes four aspects, namely fiber preform structure, interface shear strength, CTE match between fiber and matrix and matrix strengthening. The fiber preform architecture, which has the characteristic of high fiber volume fraction in loading direction and low fiber volume fraction perpendicular to loading direction, is chosen to ensure the excellent mechanical property. The higher interface shear strength is designed to ensure good load transfer capacity from matrix to fiber. Meanwhile, the interface debonding and crack deflection should be ensured to occur. The residual compressive stress can be applied to matrix in axial direction by using SiC fiber with higher CTE than that of SiC matrix, thus, the applied tensile stress is partially offset. The strong matrix can be obtained by introduction the nano-scale reinforcement into matrix. On one hand, the σ_{mc} can be improved by structure design, and the crack initiation is delayed. On the other hand, the matrix cracks density and cracks width are designed to bc decreased to decline the oxygen diffusion pathway.

According to the effective self-healing temperature range of different self-healing components, the matrix composition is highly designable for specific service environment. The self-healing component, which possesses suitable initial oxidation temperature, high oxidation rate and generates stable glass phase at service environment, is preferred for specific service temperature. Furthermore, the matrix composition can be designed with multiple self-healing components to meet the requirement for wide self-healing temperature range. Distance to fiber from far to near, the initial oxidation temperature of self-healing components should be more and more low, and the oxidation rate should be more and more fast. When the CMC is loaded, the microcracks initiate and propagate in matrix. Oxygen diffuses along microcracks, and it can be partly consumed by outer self-healing component. The microcracks can be partially healed by produced glass phase. With the propagation of matrix microcracks, the inner self-healing components with high oxidation rate are quickly oxidized and produce sufficient glass phase to seal the microcracks. The interphase and fiber can be protected from oxidation resulting from the gradual consumption of oxygen by multiple self-healing components when the cracks propagate from external to interior. The location of glass phase with low melting point is far from surface, and the evaporation of which is further hindered by outer high melting point glass phase. The excellent self-healing property can be obtained and the service life is obviously prolonged.

5.7 SUMMARY AND PERSPECTIVE

With the development of aeronautical and space engines, the service condition of thermal-structure components would be more and more severe. The higher service stress and temperature put higher demands on the mechanical property and self-healing performance of CMC. The SHCMC would be a kind of key thermal-structure material to meet future demand of high thrust-to-weight ratio engine, which also possesses high degree of densification or high σ_{mc} characteristic. In addition, the investigation about the wet-oxygen resistance of interphase and matrix are being done to reduce the dependence on EBC during the long service periods. Currently, the performance database of CMC is urgently needed to be built to provide basic data for structure and composition design. In addition, it is meaningful to develop the modeling tool to monitor the damage evolution and predict the service life of CMC.

REFERENCES

[1] Naslain RR. Design, preparation and properties of non-oxide CMCs for application in engines and nuclear reactors: an overview. *Composites Science & Technology*, 2004. 64(2): 155–170.

[2] Nakano K, Hiroyuki A, Ogawa K. Carbon fiber reinforced silicon carbide composites. In *Developments in the Science and Technology of Composite Materials: Fourth*

European Conference on Composite Materials September 25–28, 1990 Stuttgart-Germany, Füller, J., Grüninger, G., Schulte, K., Bunsell, A. R., Massiah, A. Eds.; Springer Netherlands, 1990. 419–424.

[3] Medina MC, Molina-Aldareguia JM, Gonzalez C, et al. Comparison of push-in and push-out tests for measuring interfacial shear strength in nano-reinforced composite materials. *Journal of Composite Materials*, 2016. 50(12): 1651–1659.

[4] Cheng L, Xu Y, Zhang L, et al. Effect of glass sealing on the oxidation behavior of three dimensional C/SiC composites in air. *Carbon*, 2001. 39(8): 1127–1133.

[5] Lamouroux F, Bertrand S, Pailler R, et al. Oxidation-resistant carbon-fiber-reinforced ceramic-matrix composites. *Composites Science & Technology*, 1999. 59(7): 1073–1085.

[6] Liu Y, Zhang L, Cheng L, et al. Preparation and oxidation resistance of 2D C/SiC composites modified by partial boron carbide self-sealing matrix. *Materials Science & Engineering A*, 2008. 498(1–2): 430–436.

[7] Zuo X, Zhang L, Liu Y, et al. Oxidation behaviour of two-dimensional C/SiC modified with self-healing Si–B–C coating in static air. *Corrosion Science*, 2012. 65: 87–93.

[8] Viricelle JP, Goursat P, Bahloul-Hourlier D. Oxidation behaviour of a multi-layered ceramic-matrix composite $(SiC)_f/C/(SiBC)_m$. *Composites Science & Technology*, 2001. 61(4): 607–614.

[9] Bouillon EP, Spriet PC, Habarou G, et al. Engine test and post engine test characterization of self-sealing ceramic matrix composites for nozzle applications in gas turbine engines. In *ASME Turbo Expo 2004: Power for Land, Sea, and Air, Vol. 2*. 2004, pp 409–416. DOI: 10.1115/gt2004-53976.

[10] Christin FA. A Global approach to fiber nD architectures and self-sealing matrices: from research to production. *International Journal of Applied Ceramic Technology*, 2010. 2(2): 97–104.

[11] Lamouroux F, Bouillon E, Cavalier JC, et al. An improved long life duration CMC for jet aircraft engine applications. In High Temperature Ceramic Matrix Composites, 2001, pp. 783–788.

[12] Bouillon E, Lamouroux F, Baroumes L, et al. An improved long life duration CMC for jet aircraft engine applications. In *ASME Turbo Expo 2002: Power for Land, Sea, and Air*, Vol. 4: Turbo Expo 2002, Parts A and B, 2002, pp 119–125. DOI: 10.1115/gt2002-30625.

[13] Cao X, Yin X, Fan X, et al. High-temperature flexural properties of SiBC modified C/SiC composites. *Ceramics International*, 2014. 40(4): 6185–6190.

[14] Cao X, Yin X, Ma X, et al. Oxidation behavior of SiBC matrix modified C/SiC composites with different PyC interphase thicknesses. *Ceramics International*, 2015. 41(1): 1695–1700.

[15] Cao X, Yin X, Fan X, et al. Effect of PyC interphase thickness on mechanical behaviors of SiBC matrix modified C/SiC composites fabricated by reactive melt infiltration. *Carbon*, 2014. 77: 886–895.

[16] Zhao D, Fan X, Yin X, et al. Oxidation behavior of Tyranno ZMI-SiC fiber/SiC-SiBC matrix composite from 800 to 1200°C. *Materials*, 2018. 11(8): 1367.

[17] Sun X, Yin X, Fan X, et al. Oxidation resistance of SiC/SiC composites containing SiBC matrix fabricated by liquid silicon infiltration. *Journal of the European Ceramic Society*, 2018. 38(2): 479–485.

[18] Ma X, Yin X, Fan X, et al. Improved tensile strength and toughness of dense C/SiC-SiBC with tailored PyC interphase. *Journal of the European Ceramic Society*, 2019. 39: 1766–1774.

[19] Liu, YS, Chai N, Liu XF, et al. The microstructure and dielectric properties of SiBCN ceramics fabricated via LPCVD/CVI. *Journal of the American Ceramic Society*, 2015. 98(9): 2703–2706.
[20] Zhao M, Liu Y, Chai N, et al. Effect of SiBCN content on the dielectric and EMW absorbing properties of SiBCN-Si3N4 composite ceramics. *Journal of the European Ceramic Society*, 2017. 38(4): 1334–1340.
[21] Qin H, Liu Y, Cheng Z, et al. Dielectric and microwave absorption properties of SiCnw-SiBCN composite ceramics deposited via chemical vapor infiltration. *Journal of Alloys and Compounds*, 2019. 771: 747–754.
[22] Wang C, Liu Y, Zhao M, et al. Three-dimensional graphene/SiBCN composites for high-performance electromagnetic interference shielding. *Ceramics International*, 2018. 44(18): 22830–22839.
[23] Vishnyakov VM, Ehiasarian AP, Vishnyakov VV, et al. Amorphous boron containing silicon carbo-nitrides created by ion sputtering. *Surface & Coatings Technology*, 2011. 206(1): 149–154.
[24] Li D, Yang Z, Jia D, et al. Effects of boron addition on the high temperature oxidation resistance of dense sSiBCN monoliths at 1500°C. *Corrosion Science*, 2017. 126: 10–25.
[25] Li DX, Yang ZH, Jia DC, et al. High-temperature oxidation behavior of dense SiBCN monoliths: Carbon-content dependent oxidation structure, kinetics and mechanisms. *Corrosion Science*, 2017. 124(1): 103–120.
[26] Luan X, Xu X, Zou Y, et al. Wet oxidation behavior of SiC/(SiC-SiBCN)x composites prepared by CVI combined with PIOP process. *Journal of the American Ceramic Society*, 2019. 102: 6239–6255.
[27] Ding Q, Ni D, Wang Z, et al., Mechanical properties and microstructure evolution of 3D Cf/SiBCN composites at elevated temperatures. *Journal of the American Ceramic Society*, 2018. 101: 4699–4707.
[28] Butchereit E and Nickel KG, Beneficial effect of aluminum on the oxidation behavior of precursor-derived ceramics. *The Electrochemical Society*, 2001. 16: 325–338, in *High Temperature Corrosion and Materials Chemistry IV*, edited by E. Opila, et al. The Electrochemical Society, Washington, DC, 2003.
[29] Müller A, Gerstel P, Butchereit E, et al. Si/B/C/N/Al precursor-derived ceramics: Synthesis, high temperature behaviour and oxidation resistance. *Journal of the European Ceramic Society*, 2004. 24(12): 3409–3417.
[30] Feng Q, Hu J, Zhang X, et al. Oxidation behavior in wet oxygen environment of Al2O3 added reaction-sintered Si-B-C ceramics. *Ceramics International*, 2018. 44: 4009–4015.

6 Functional Films and Coating Technology for Infrared Windows

Gang Gao and Jiaqi Zhu

6.1 INTRODUCTION

Infrared windows are an important component of infrared detection systems. These windows are widely used in many fields, such as industry, medical treatment, environment and military. However, the infrared detection system used outdoors urgently needs high transmission of infrared signals, and needs to be able to work stably in complex electromagnetic environment. Therefore, it is necessary to prepare infrared anti-reflection protective films and anti-electromagnetic shielding films on the infrared window of infrared detection system.

In this chapter, the scientific and technical problems of infrared anti-reflection protective films will be reviewed. In addition, the optical and mechanical properties of some typical infrared anti-reflection protective films are introduced. Then, the requirements of infrared windows for infrared transparent conductive films (ITCFs) will be described. Some ITCFs with potential applications in infrared windows will be highlighted. Finally, we will tell the reader the difficulties of coating technology for large-size planar/sharp arch infrared windows. Moreover, some new coating techniques developed by us will be shown.

6.2 INFRARED ANTIREFLECTIVE AND PROTECTIVE FILMS

6.2.1 Problems of Infrared Antireflective and Protective Films

As the speed of aircraft develop towards hypersonic, many materials that are used for infrared guidance and imaging system windows and fairings have been developed both domestically and abroad. However, up to now, there is still no shield and window material that can fully meet the comprehensive requirements of hypersonic flight conditions. Instead, composite material/structural shields are formed by coordinating the respective advantages and disadvantages of different materials. To address the issues facing infrared windows and shield materials, there are currently two main solutions. One is to modify existing infrared optical materials through various means to meet the practical requirements of infrared windows and fairings. The other is to coat infrared windows and shield materials with anti-reflective protective films. Infrared anti-reflective protective film materials need to have high infrared transmittance and a

DOI: 10.1201/9781003386223-6

certain strength to meet the practical requirements of infrared guidance and detection systems, in order to avoid a series of problems caused by harsh operating environments, such as film layer rupture, shedding, and thermal radiation interference.

Coating anti-reflective protective films on infrared windows have been shown in numerous studies to provide protection and increase transmittance of the windows. High-hardness films that are firmly attached to the infrared windows and fairing materials can protect the windows from raindrops or sand particles impact, minimizing the stress. This is in contrast to direct impact of stress on the window material itself. Anti-reflective protective films with contraction stress can increase the mechanical strength of infrared windows by balancing the expansion stress caused by the impact of atmospheric particles on the windows and shield surface. Currently, materials such as fluorides (PbF_2, ThF_4, BaF_2), sulfides (Sb_2S_3, ZnS, ZnSe), oxides (TiO_2, ZrO_2, Y_2O_3, HfO_2), and diamond-like materials are commonly used for coating.

There are several requirements for infrared anti-reflective protective films, including but not limited to:

1. The film should have sufficient mechanical strength to withstand the mechanical and thermal shocks caused by the rapid acceleration and deceleration of the aircraft. It should also possess properties that can resist sand and rain erosion, and be able to adapt to the specific working environment conditions.
2. The film should possess good optical properties, especially high transmittance and low emissivity, within the working band. It should also maintain good optical performance in harsh operating environments such as high-temperature and high-speed flights.
3. The film should have good adhesion to the window and shield materials. Under harsh operating conditions, the film should not detach or rupture from the surface, which could cause a series of issues.

In general, the selection of materials for infrared windows revolves around the three atmospheric windows. These windows are used to transmit infrared radiation through the atmosphere and into space. Commonly used infrared windows and shield materials include sapphire, magnesium fluoride, Y_2O_3, spinel, germanium (Ge), silicon (Si), aluminium oxynitride (AlON), zinc sulfide (ZnS), zinc selenide (ZnSe), among others [1–3]. However, currently, no single material possesses high transmittance in multiple spectral ranges. Additionally, most infrared windows materials are soft, easy to scratch, and prone to corrosion in harsh natural environments such as dust storms and acid rain [4]. Compounds with higher atomic mass have weaker bond strengths, leading to a larger infrared cut-off limit (λ_p) and a broader range of infrared transmission for window and shield materials. Therefore, materials with higher atomic mass are typically chosen as infrared windows materials for long-wave infrared transmittance. However, when the atomic mass is high, the bond strength between chemical bonds decreases, resulting in poorer mechanical properties. Hence, a comprehensive consideration of the infrared transmittance performance and mechanical properties of a material is essential.

Currently, the choice of material for infrared windows substrate in spacecraft primarily involves multi-spectral zinc sulfide (ZnS) (standard zinc sulfide treated using hot isostatic pressing). ZnS is a wide-bandgap material with stable chemical properties, capable of optical transmittance from near Infrared to long-wave infrared. ZnS comes in three categories based on the preparation process – hot-pressed ZnS, standard ZnS, and multi-spectral ZnS. The vacuum hot pressing method is used to produce hot-pressed ZnS, while standard ZnS is synthesized using chemical vapor deposition. Standard CVDZnS goes through hot isostatic pressing, transforming it into multi-spectral CVDZnS. The multi-spectral CVDZnS provides better transmission performance within the 3–5 μm and 8–12 μm spectral ranges. However, its mechanical properties are reduced. Although the hardness of multi-spectral CVDZnS is lower than that of hot-pressed ZnS and standard ZnS, it remains the material most commonly used for Infrared windows, given its excellent transmittance performance in the mid to long infrared regions [5].

Multi-spectral ZnS has shown excellent transmission performance in the long-wave infrared region, but its high refractive index results in an actual transmittance of about 70 percent. Moreover, during the actual manufacturing process, limitations in processing conditions result in a relatively soft texture of zinc sulfide, with mechanical strength falling short of theoretical values. Consequently, it becomes brittle, making it vulnerable to environmental erosion by external natural factors. Specifically, during high-speed flight, spacecraft infrared windows can be subjected to the harsh aerodynamic thermal/force environment, leading to thermal/force coupling effects, along with erosion by solid particles such as raindrops, sand, and ice particles in the atmosphere [6]. Such factors can easily cause surface wear, dents, and even lead to infrared windows structural and functional failure, ultimately leading to the entire infrared detection system becoming paralyzed. Furthermore, in addition to the requirements of increasing transparency both in mid-wave 3~5 μm and long-wave 8~12 μm, Infrared windows of anti-near spaceflight vehicles are even required to maintain high transmittance in the near infrared range so they can simultaneously integrate 1064nm solid-state lasers for ranging detection.

To meet these new technological demands, it is necessary to develop a hard, wide-band, high-temperature stable, anti-reflection coating material capable of transmitting light from near infrared to long-wave infrared. Anti-reflection coatings can maximize transmittance in the corresponding spectral region, improve infrared imaging quality, and reduce attenuation of the optical transmission performance in harsh environments.

6.2.2 Application Verification of Y_2O_3 Infrared Anti-reflective Protective Film

(1) Design and Implementation of anti-reflective film system

Zinc sulfide is a typical long-wave infrared material, which has been widely used as a component material for infrared windows and fairings. However, the refractive index of zinc sulfide is about 2.3, and it reflects 15 percent of the light waves when they pass through, resulting in a loss of about 30 percent in transmittance after passing through the zinc sulfide infrared windows due to the reflection from the surfaces. The

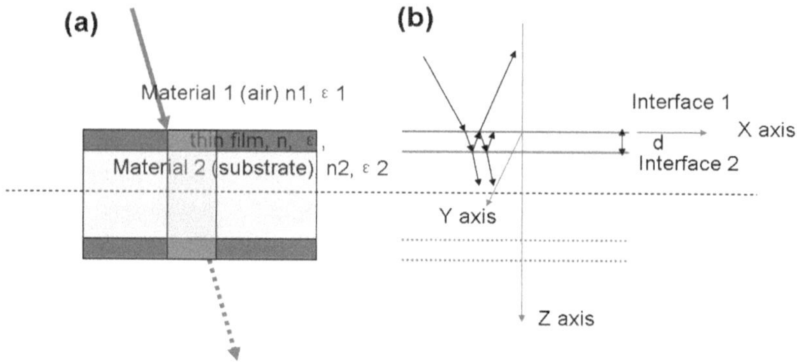

FIGURE 6.1 The schematic of transmittance of wave trough the film and substrate.

theoretically calculated transmittance of the zinc sulfide infrared windows is 70 percent, which is consistent with the measured value. Therefore, in order to improve the transmittance of the zinc sulfide infrared windows, it is necessary to deposit an anti-reflective film on the surface of the windows or fairings to increase the infrared transmittance. All the above-mentioned cases are for the vertically incident light. When the incident light is obliquely incident to the surface normal, the reduced transmittance due to enhanced reflection and polarization effects becomes more prominent. Therefore, the most effective and convenient method to achieve high transmittance is to deposit an anti-reflective film on the surface.

Based on the theory of thin-film optics, assuming that the incident light wave is a linear, monochromatic TEM wave propagating along the Z-axis in a homogeneous medium, and is normally incident to the interface (Z-Y plane) 1, the electric field E_1 and the magnetic field H_1 of the incident wave are respectively along the X-axis and Y-axis directions, as shown in Figure 6.1. The field distribution of the incident wave is:

$$E_1 = E_{10} e^{-jki_{1z}} x$$

$$H_1 = \sqrt{\frac{\varepsilon_1}{\mu_0}} E_{10} e^{-jki_{1z}} y \tag{6.1}$$

In the equation, $K^i = \omega\sqrt{\varepsilon_1 \mu_0}$, μ_0 is the magnetic permeability, which is the same for the film and substrate in this chapter since it is a non-ferromagnetic medium. ε is the capacitance of the material.

When the electromagnetic wave enters the film and reaches interface 2, it also generates a reflected wave. When the scattered wave reaches interface 1, secondary reflection occurs. This is repeated many times, resulting in countless electromagnetic waves propagating up and down in the film. Furthermore, in TEM waves, $k_{nx} = k_{ny} = 0$, $k_{nz} = k_n$(n = 0, 1, 2), so multiple downward-propagating waves in the film can be combined due to the mode function remaining unchanged during wave propagation, resulting in:

$$E_+ = E_{+0}e^{-jkt}{}_z x$$
$$H_+ = \sqrt{\frac{\varepsilon}{\mu_0}}E_{+0}e^{-jkt}{}_z y \tag{6.2}$$

Where $K^t{}_z = \omega\sqrt{\varepsilon\mu_0}$, z is between 0 and d.

In thin films, for waves proceeding downward, this can be written as:

$$E_- = E_{-0}e^{jk_z^r x}$$
$$H_- = -\sqrt{\frac{\varepsilon}{\mu_0}}E_{-0}e^{jk_z^r}y \tag{6.3}$$

Where $k_z^r = \omega\sqrt{\varepsilon\mu_0}$, z is between 0 and d. Also, $k_z{}^r = k_z{}^t = k$.

Similarly, the field distribution of the wave quantity synthesized by adding all the downward-propagating transmission waves in Material 2 can be expressed as:

$$E_2 = E_{20}e^{-jk_{2z}^t x}$$
$$H_2 = \sqrt{\frac{\varepsilon_2}{\mu_0}}E_{20}e^{-jk_{2z}^t y} \tag{6.4}$$

Where $k_{2z}^t = \omega\sqrt{\varepsilon_2\mu_0}$, z is larger than d.

At the dielectric intersection, the tangential electric field is continuous and a boundary relationship exists at the intersection:

$$n\times\left(E_{film} - E_1\right) = 0$$
$$n\times\left(H_{film} - H_1\right) = 0 \tag{6.5}$$

and

$$n\times\left(E_2 - E_{film}\right) = 0$$
$$n\times\left(H_2 - H_{film}\right) = 0 \tag{6.6}$$

By equation 6.5, at z = 0, we have:

$$E_{+0} = \frac{1}{2}\left(1+\sqrt{\frac{\varepsilon_1}{\varepsilon}}\right)E_{10}$$
$$E_{-0} = \frac{1}{2}\left(1-\sqrt{\frac{\varepsilon_1}{\varepsilon}}\right)E_{10} \tag{6.7}$$

By equation 6.6, at z=d, we have:

$$\begin{aligned} &E_{+0}e^{-jkd} - E_{-0}e^{jkd} - E_{20}e^{-jk_{2z}^{t}} = 0 \\ &\sqrt{\frac{\varepsilon}{\mu_0}}E_{+0}e^{-jkd} + \sqrt{\frac{\varepsilon}{\mu_0}}E_{-0}e^{jkd} - \sqrt{\frac{\varepsilon_2}{\mu_0}}E_{20}e^{-jk_{2z}^{t}} = 0 \end{aligned} \tag{6.8}$$

From equations 6.7 and 6.8, we can obtain:

$$\left(1+\sqrt{\frac{\varepsilon_1}{\varepsilon}}\right)\times\left(1-\sqrt{\frac{\varepsilon}{\varepsilon_2}}\right)e^{-j2kd} + \left(1-\sqrt{\frac{\varepsilon_1}{\varepsilon}}\right)\times\left(1+\sqrt{\frac{\varepsilon}{\varepsilon_2}}\right) = 0 \tag{6.9}$$

Since $n_1 = \sqrt{\varepsilon_{r1}} = \sqrt{\frac{\varepsilon_1}{\varepsilon_0}}, n_2 = \sqrt{\varepsilon_{r2}} = \sqrt{\frac{\varepsilon_2}{\varepsilon_0}},, n = \sqrt{\varepsilon_r} = \sqrt{\frac{\varepsilon}{\varepsilon_0}}$, equation 6.9 can be transferred:

$$\left(1+\frac{n_1}{n}\right)\times\left(1-\frac{n}{n_2}\right)e^{-j2kd} + \left(1+\frac{n_1}{n}\right)\times\left(1+\frac{n}{n_2}\right) = 0 \tag{6.10}$$

To make equation 6.10 hold true, both the real and imaginary parts in the equation should be equal to zero.

Therefore, we have:

$$\left(1+\frac{n_1}{n}\right)\times\left(1-\frac{n}{n_2}\right)\sin 2kd = 0 \tag{6.11}$$

As n is not equal to n2, the condition is only satisfied when sin2kd equals zero.

Solving this equation yields: d = mλ/4, m = 0, 1, 2, 3…

Moreover, since the real part is zero and 2kd = mπ, equation 6.10 can be transformed into:

$$\mp\left(1+\frac{n_1}{n}\right)\times\left(1-\frac{n}{n_2}\right) + \left(1-\frac{n_1}{n}\right)\times\left(1+\frac{n}{n_2}\right) = 0 \tag{6.12}$$

When m is an even number, it is found that $n_1 = n_2$, which does not satisfy the initial condition of this chapter.

When m is an odd number, $n = \sqrt{n_1 n_2}$ is obtained, indicating that when the refractive index of the film satisfies this relationship and the thickness of the film is d = mλ/4, m = 1, 3, 5... the effect of increasing the permeability is achieved.

In this chapter, the refractive index of zinc sulfide is 2.4. Therefore, the refractive index of the thin film needed to achieve the antireflection effect is around 1.5. Since antireflection is required in the infrared band, we can infer from the Cauchy formula

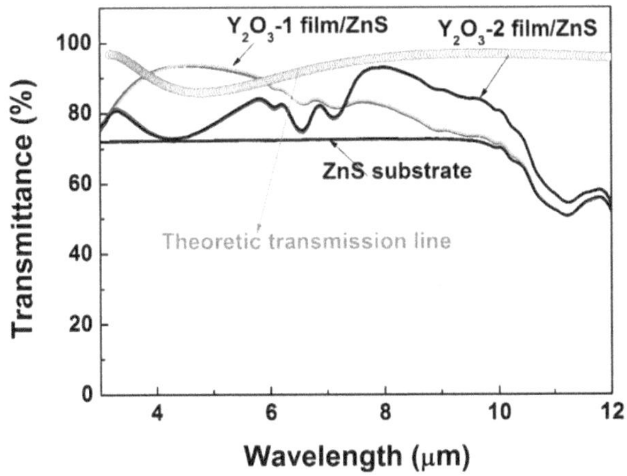

FIGURE 6.2 The transmittance of the Y_2O_3/ZnS/Y_2O_3.

fitted in chapter four that the refractive index of the thin film does not vary significantly between the infrared and visible light bands. Therefore, to achieve the best antireflection effect, the refractive index of the thin film should be as close to 1.5 as possible. However, in consideration of the mechanical properties of the thin film, it is necessary to select an appropriate process parameter for the thin film. The zinc sulfide antireflection protective films deposited at 600°C show good antireflection effects, as shown in Figure 6.2. As periodic thickness can achieve antireflection at different wavelengths, the thickness determines the central wavelength shift of the antireflection effect, as shown in Y_2O_3-1 film in Figure 6.1. It can be seen that the double-sided film prepared has a high transmittance (~93%), and the deviation from theoretical 100 percent antireflection arises from the refractive index deviation.

(2) Y_2O_3 Infrared Hard Protective Coating

The quality of zinc sulfide window is soft and has a low hardness, which cannot meet the demanding requirements of the working environment like rain and sand erosion without breaking. Therefore, it is necessary to deposit a hard film on the zinc sulfide window to improve its hardness. As we learned from the mechanical properties of the thin film in chapter five, both temperature and bias voltage can increase the hardness of the thin film. Therefore, films were deposited on the zinc sulfide substrate under different temperatures and bias voltages, and their mechanical properties were measured. Furthermore, in order to further determine the composition of the prepared films on the zinc sulfide substrate, EDS measurements were conducted on the film samples. As shown in Figure 6.3, the O/Y ratio of the film is close to 2, higher than the stoichiometric ratio of the film. This is because the surface of the film contains adsorbed oxygen and the Zn/S ratio of the substrate was 0.75, which is close to the stoichiometric ratio.

Thick Y_2O_3 films with a thickness of about 500 nm were prepared on zinc sulfide substrates using two different processing schemes, that is, substrate temperature and electric bias voltage, and nanoscale indentation tests were conducted, as shown in Figure 6.4. The film prepared at 400°C exhibited higher hardness of about 8 GPa,

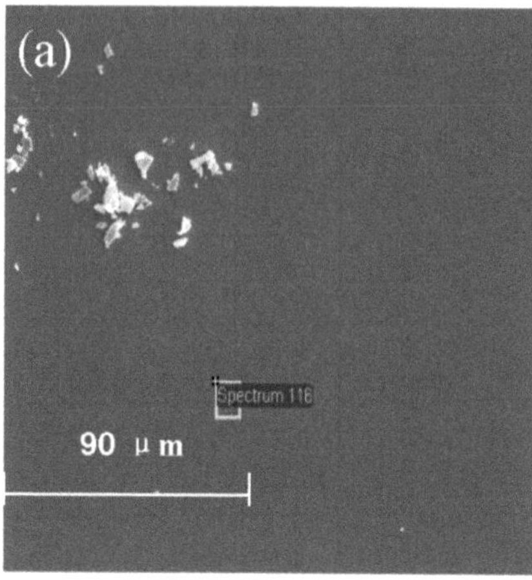

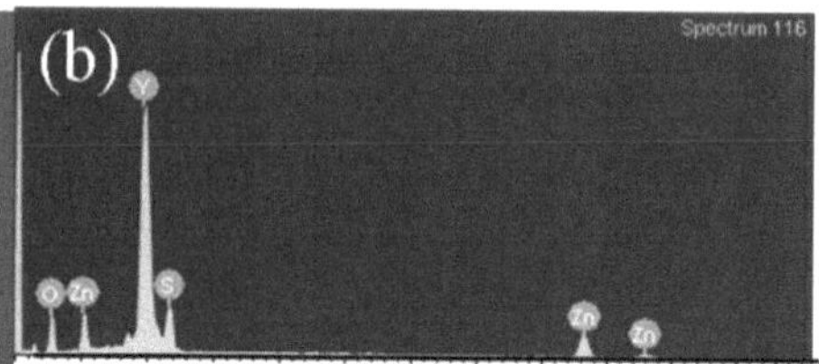

Element	Weight%	Atomic%
O K	19.32	52.08
S K	6.45	8.68
Zn K	18.43	12.16
Y L	55.80	27.07
Totals	100.00	

FIGURE 6.3 EDS results of yttrium oxide films on ZnS at bias voltage of −80 V.

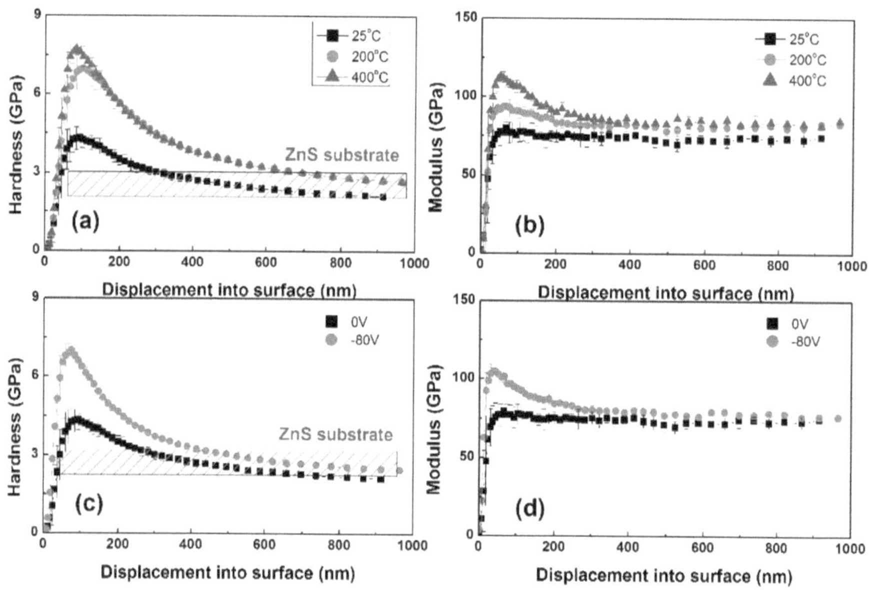

FIGURE 6.4 The hardness of ~500 nm yttrium oxide coated on ZnS substrate at different conditions: (a) the hardness at different substrate temperatures; (b) the modulus at different substrate temperatures; (c) the hardness at different substrate bias voltages; (d) the modulus at different bias voltages.

which was higher than that of the films prepared at room temperature and 200°C. When the depth of the indenter exceeded the thickness of the film, the hardness-displacement curve reflected the substrate's intrinsic hardness, which was low at about 3 GPa, as shown in Figure 6.4 (a). Similar to the hardness, at 400°C, the

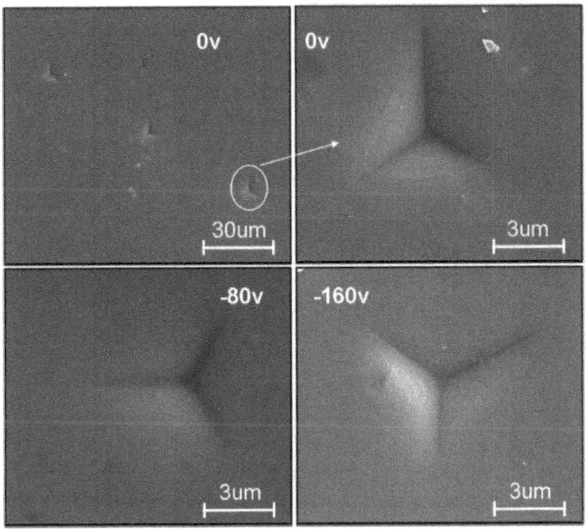

FIGURE 6.5 The nanoindentation SEM of yttrium oxide films on ZnS.

film demonstrated a higher elastic modulus, which was significantly higher than the intrinsic elastic modulus of zinc sulfide. Bias voltage can also improve the mechanical properties of the film, as shown in Figure 6.4 (c) and (d). The hardness and modulus of the film under −80 V bias were both close to 7.5 GPa and 110 GPa, respectively, and higher than the intrinsic hardness and modulus (75 GPa) of the zinc sulfide substrate. It can be seen that the films prepared at both temperature and bias voltage exhibit high mechanical properties. However, in comparison, the films prepared at higher temperatures demonstrate better mechanical and stability properties.

The indentation morphology of the film surface under different bias voltages is presented in Figure 6.5, and it is evident that the film left a relatively clear pyramidal groove after being pressed by the indenter. From the indentation photo of the film, it is noticeable that there are no obvious crack marks around the indentation, which indicates that the film has certain toughness during the plastic deformation process caused by the compression of the indenter, and therefore did not break. However, for films prepared under the condition of being grounded to the substrate, faint cracks appeared at the edge of the indentation groove, which signifies that the film has poor toughness under this condition and is prone to cracking. This can be attributed to the presence of numerous pores in the film under this condition, causing the collapse of the pores under compression, and leading to the appearance of cracks. As the bias voltage increased and the film gradually became denser, the film had a larger plastic deformation ratio, and no micro-cracks appeared around the indentation groove.

6.2.3 Application Validation of Nitrogen Y_2O_3 Enhanced Permeability Protective Film

A probe profilometer was used to measure the film thickness, and the ratio of film thickness to deposition time is the deposition rate of the film. Taking the target

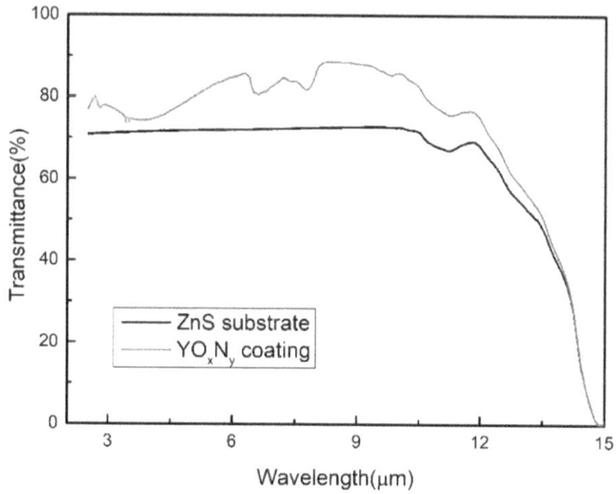

FIGURE 6.6 Transmission rates of Yttrium Nitride Oxide films in the infrared range.

wavelength λ = 4μm at the time of the penetration enhancement, the thickness of the film deposited on the ZnS substrate was calculated and the deposition time was obtained from the deposition rate. Since the deposition rate of nitrogen Y_2O_3 films is slow, the gas flow ratio of N_2:O_2:Ar = 10:3:50 was selected for the deposition of nitrogen Y_2O_3 films on ZnS substrate considering the deposition rate and the properties of the films.

The transmittance of the films in the infrared range was measured by FTIR spectroscopy, and the results are shown in Figure 6.6. As can be observed from the Figure, after the ZnS substrate was coated with Yttrium Nitride Oxide films on one side, the transmittance of the substrate and the films in the infrared range was significantly improved, and the maximum transmittance in the infrared range was up to 88%, and the films had a good effect of increasing the transmittance in the far infrared (8–12 μm) band. Compared with the 70% transmittance of ZnS substrate, the plated Yttrium Nitride Oxide films have achieved the purpose of increasing the transmittance of the substrate.

6.3 INFRARED TRANSPARENT CONDUCTIVE FILMS

6.3.1 The Bottleneck of Infrared Transparent Conduction Theory

Due to its good optical transparency and high electrical conductivity, Transparent Conductive Films (TCF) have been widely used in modern industries, such as computer displays, transistors, gas sensors, and aerospace fields. In military applications, TCF also has high value, especially in the application of aircraft cockpits or optical pods. The cockpit or optical pod of an aircraft is a core component for achieving functions such as searching, detecting, and tracking multiple targets. However, due to its special cavity structure, the cockpit or optical pod will produce a very large radar cross-sectional area, which greatly increases the chances of being detected. Therefore, in complex electromagnetic interference environments on modern battlefields, in order to improve the survival ability of new generation weapons and equipment, the optoelectronic detection system urgently needs to have strong radar stealth and anti-external electromagnetic wave interference capabilities.

With the development of modern weapons and aerospace fields, TCF not only needs to have good transparency in the visible light band, but also requires excellent optical transparency in the infrared light band. Therefore, higher performance requirements need to be proposed for the existing mature TCF. However, the electrical conductivity and optical transparency of TCF are mutually constrained. If good infrared light band high transmission is to be achieved, the concentration of free electrons in the film cannot exceed 10^{-19} cm^{-3}, which means poor electrical conductivity. Although many thin film materials currently have good near-infrared band transparency, such as oxide (CdO), superstructure films, metal grid films, and new carbon material films (Mg-C-O-H), the limitations of optoelectronic performance in the infrared light band make researchers skeptical about the possibility of achieving high electrical conductivity while achieving excellent transparency in the mid and far infrared light band for TCF. In order to distinguish from visible light transparent conductive film materials, the film that simultaneously has good visible-light-infrared light band transparency and electrical conductivity is referred to as the Infrared Transparent Conductive Film (ITCF). According to the Drude free electron theory, high concentration of free electrons will cause absorption and reflection of photons in the visible light band and the infrared light band, respectively, thereby reducing the transparency range of the visible light band and the transmission range of the infrared light band (plasma wavelength blue shift). If the carrier concentration is reduced to improve the optical transparency, this will have a negative effect on the electrical conductivity of the film. Thus, it can be seen that it is difficult to coordinate the high transmission of wide-band infrared light ranges of ITCF and the electrical conductivity that meets application requirements.

To more comprehensively explain the complex relationship between TCF's optoelectronic properties, it is necessary to combine the Maxwell equations with the Lorentz oscillator model. Using TCF's dielectric constant ε and plasma wavelength λ_p, the relationship between the film's electrical conductivity and optical transparency can be expressed in terms of the real and imaginary parts of the complex dielectric constant, denoted by formula $\varepsilon_r = \varepsilon_\infty \left(1 - \frac{1}{\omega^2 \lambda_p^2}\right)$ and $\varepsilon_i = \frac{\varepsilon_\infty}{\omega^3 \lambda_p^2 \tau}$. Here, τ represents the relaxation time of free electrons, ε_∞ represents the high-frequency dielectric constant, ε_0 represents the free space dielectric constant (8.8542×10^{-12} F/m), ε_r represents the real part of the complex dielectric constant, εi represents the imaginary part of the complex dielectric constant, and ω represents the vibration frequency of light. From these formulas, it can be inferred that for the plasma wavelength λ_p, when the vibration frequency of photons is lower than the vibration frequency of photons at the plasma wavelength ($\omega < \omega_p$), the real part of the complex dielectric constant ε_r is negative while the imaginary part of the complex dielectric constant ε_i is positive, indicating that TCF will exhibit strong reflective characteristics. Conversely, when the vibration frequency of photons is higher than that of the plasma wavelength

($\omega > \omega_p$), the real part of the dielectric constant ε_r approaches ε_∞ while the imaginary part of the dielectric constant ε_i approaches 0. At this point, only a small fraction of free electrons absorb photons, causing the optoelectronic film to exhibit high transparency. However, it is still challenging to reconcile the wide-band high transmission of mid-to-far infrared light and the electrical conductivity that meets the application requirements of ITCF due to the mutual constraint between TCF's electrical conductivity and optical transparency. When the real part of the dielectric constant is zero ($\varepsilon_r = 0$) at $\omega = \omega_p$, there is a significant discontinuity in the optical transmittance of the film. The corresponding wavelength to this point is defined as the plasma wavelength of the film, which is shown as point $\lambda_p = 2\pi c_0\sqrt{\frac{m^*\varepsilon_0\varepsilon_\infty}{ne^2}}$. Here, σ_0 is the electric conductivity in free space, c_0 is the speed of light, n is the carrier density, and m^* is the effective mass of the carrier. This formula clearly shows that the plasma wavelength is inversely proportional to the free electron concentration. Therefore, at high concentrations of free electrons, the mid-infrared transmission range of the film would become smaller. In order to achieve a wider range of transmission performance, it is necessary to reduce the carrier concentration of the film. However, this will negatively impact the electrical conductivity of the film, resulting in a challenging trade-off between the contradictory optoelectronic performances.

Irreconcilable contradictions between the photoelectric properties of TCFs are the main scientific problems faced by this field. In this section, we introduce the accumulation materials based on the idea of effective mass regulation of material carriers and coupled regulation of carrier concentration and mobility to achieve infrared transparent conducting, and analyze their structure, composition and properties.

6.3.2 LaSe$_2$ Films

6.3.2.1 Optical Properties of LaSe$_2$ Films

Figure 6.7 illustrates the mid-infrared transmittance spectra of LaSe$_2$ films after sulfidation treatment at various temperatures, namely, 600°C, 700°C, 800°C, 900°C, and 1000°C. The observed transmittance rate of all samples was above 60%, indicating excellent mid-infrared transmittance capability of LaSe$_2$ films. As the sulfidation temperature increased, a declining trend in transmittance became apparent, which is likely related to the grain size, surface roughness, and electrical characteristics of the films. Our study indicates that a higher sulfidation temperature led to a tendency of smaller grain sizes and increased crystallization of the films, contributing to more grain boundaries and photon scattering. Furthermore, the trend of surface roughness, as measured using AFM, was found to increase with sulfidation temperature. This change in roughness amplified the diffuse scattering of photons by the films, thereby becoming a significant factor influencing the transmittance of the films. Additionally, as per the Drude-Lorentz theory, photons undergo some scattering and absorption through the carrier of materials in semiconductors. High carrier concentration will lead to a shortened long-wavelength cut-off of the material. Because LaSe$_2$ films possess good P-type transparent conductive properties, and the carrier concentration in the films reaches 10^{19} cm^{-3}, the scattering and absorption of photons by the carriers are also a contributing factor to the decreased transmittance of the films in our study.

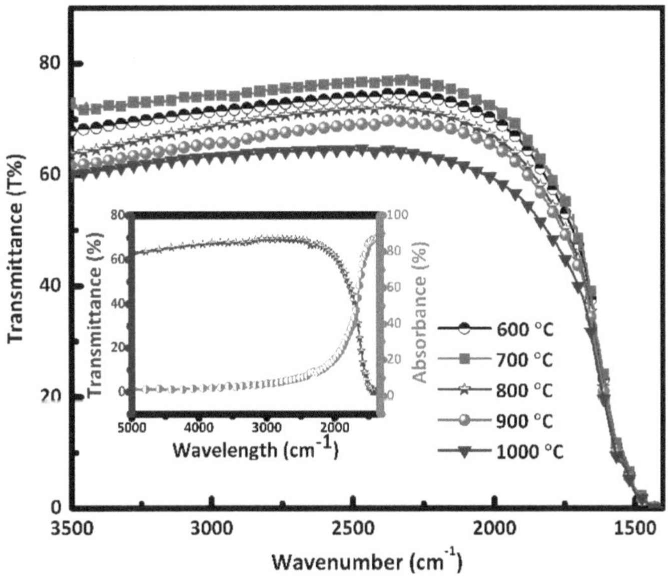

FIGURE 6.7 Mid-infrared transmittance of $LaSe_2$ films grown on sapphire substrates at different temperatures [7].

TABLE 6.1
Mid-infrared Optical and Electrical Properties of $LaSe_2$ Films Prepared at Different Selenization Temperatures

Temperature (°C)	Carrier concentration (cm^{-3})	Mobility ($cm^2 \cdot V^{-1} \cdot s^{-1}$)	Resistivity (Ω·cm)	Conductivity ($S \cdot cm^{-1}$)	Transmittance (%, 3–5μm)	Hall Coefficient
600	3.79×10^{17}	0.638	25.9	0.0386	73.85	16.49
700	7.26×10^{18}	0.337	2.56	0.391	76.57	0.86
800	1.18×10^{19}	1.95	0.271	3.69	71.83	0.53
900	1.14×10^{19}	1.09	0.504	1.99	68.94	0.55
1000	8.60×10^{17}	15.1	0.468	2.14	64.46	7.26

6.3.2.2 Electrical Properties of $LaSe_2$ Films

The electrical conductivity of thin films is inherently linked to the concentration and mobility of charge carriers within them. Table 6.1 provides a detailed summary of the specific parameters of the mid-infrared photoelectricity performance of $LaSe_2$ thin films. $LaSe_2$ thin films exhibit carrier concentrations of up to 1.18×10^{19} cm^{-3}, which is significantly higher than traditional *P*-type transparent conductive materials (10^{18} cm^{-3}). X-ray photoelectron spectroscopy (XPS) analysis suggests this is due to the presence of intergranular Se atoms and a low number of defects introduced during the Se sulfidation process in the prepared $LaSe_2$ thin films. With an increase in selenium sulfide temperature, the carrier concentration follows an upward trend, rising from 600°C to 800°C. This is due to the higher diffusion reaction of Se vapor with increasing temperature leading to a transient increase in Se vacancies within the film. However, as the sulfidation temperature further increases to 1000°C, the reaction's driving force increases, making the Se sulfidation process more complete and reducing the interstitial Se defects in $LaSe_2$ thin films. Additionally, the crystallinity of the thin film improves with the rising temperature, leading to reduced defect states within the grain interior. Therefore, the carrier concentration in this case gradually decreases and falls to as low as 8.6×10^{17} cm^{-3} when the temperature reaches 1000°C.

The carrier mobility of thin films is affected by multiple factors, including carrier, grain boundary, and defect scattering. Under various selenium sulfide processing conditions, the carrier mobility maintains a relatively constant state with an initial gradual rise, followed by a sudden increase as the sulfide temperature increases. This is likely because, even though internal defects within the thin film's grain size decrease at a sulfide temperature lower than 900°C, interstitial Se still exists between grains. Additionally, an increase in carrier concentration, small grain size, and an increase in grain boundaries would all contribute to scattering of the carriers' migration. Therefore, the carrier mobility in thin films should remain relatively stable. When the sulfide temperature reaches 1000°C, the interstitial Se in $LaSe_2$ thin films decreases, leading to a sharp decline in carrier concentration, weakens scattering effects on the carriers themselves and interstitial Se scattering. As a result, the carrier mobility in $LaSe_2$ thin films shows a sudden increase at this point.

From equation (1–13), we know that thin film's conductivity is proportionately related to both its carrier concentration (N) and mobility (μ). Hence, when the sulfide temperature is lower than 800°C, the thin film's conductivity considerably increases with an increase in carrier concentration. Though at a temperature above 800°C, the carrier mobility also increases sharply, while the carrier concentration decreases significantly, the effect of the decline in carrier concentration outweighs that of the rise in carrier mobility, leading to a drop in mobility. Thus, when the sulfide temperature exceeds 800°C, the film's mobility is subject to a certain decrease.

6.3.2.3 Comparative Analysis of the Optoelectronic Properties of $LaSe_2$ Films and Conventional Mid-infrared *P*-Type Transparent Conductive Materialss

Comparatively little recent research has studied *P*-type transparent conductive films in the mid-infrared band, and reported data on their photoelectric performance are not very promising. Table 6.2 summarizes and compares the photoelectric properties of $LaSe_2$ thin films with existing copper-iron-based transparent conductive materials in the mid-infrared band. It shows that traditional *P*-type materials either have good mid-infrared transmittance but poor conductivity, or poor mid-infrared transmittance and conductivity. Fortunately, $LaSe_2$ thin films prepared in this study have a high average transmittance (> 70%) and a conductivity of 3.7 S/cm, indicating significant potential for various applications.

TABLE 6.2
Comparison of the Photoelectric Properties of $LaSe_2$ Films with Traditional *P*-type Mid-infrared Transparent Conductive Films

Materials	Preparation methods	Transmittance (%, 3–5um)	Conductivity ($S{\cdot}cm^{-1}$)	Carrier concentration (cm^{-3})	Mobility ($cm^2{\cdot}V^{-1}{\cdot}s^{-1}$)
$LaSe_2$	RF sputtering and selenized annealing	71.83	3.69	1.18×10^{19}	1.95
$CuAlO_2$	sputtering	80	0.34	3.00×10^{18}	0.4
$CuAlO_2$	SOL-GEL	88	1.23	1.63×10^{17}	15.57
$CuAl_{0.95}Ca_{0.05}O_2$	SOL-GEL	65	0.83	6.12×10^{16}	7.23
$CuAl_{0.95}Fe_{0.05}O_2$	SOL-GEL	63	1.71	9.43×10^{16}	12.2
$CuScO_2$	Polymer assisted deposition (PAD)	90	0.955	1.12×10^{17}	53.2
$CuCrO_2$	SOL-GEL	82	1	4.07×10^{17}	15.4
$CuGaO_2$	SOL-GEL	80	0.02	1.70×10^{18}	0.23
$CuBO_2$	SOL-GEL	90	1.65	1.00×10^{17}	100

To further analyze the reasons for the exceptional mid-infrared photoelectric properties of the $LaSe_2$ thin films prepared in this study, we conducted a deeper analysis and investigation. Figure 6.8 presents the carrier mobility and concentration of different *P*-type transparent conductive films in the mid-infrared band. The inset shows a magnified image of the relevant data. Comparative analysis shows that traditional *P*-type transparent conductive films not only have low carrier mobility but also have low carrier concentrations, resulting in low conductivity. Even the rare materials with high mobility, such as $CuBO_2$, with a carrier mobility close to 100 $cm^2/(V{\cdot}s)$, still have low carrier concentrations, thereby leading to underwhelming conductivity. The advantage of $LaSe_2$ thin films is that while maintaining the same carrier mobility as conventional materials, their carrier concentration is higher by an order of magnitude. As indicated through equation (1–13) analysis, this is one of the reasons for their superior conductivity.

6.3.3 LaCuOS Films

6.3.3.1 Mid-infrared Transmission Properties of LaCuOS Films

The infrared transmission spectra of LaCuOS thin films prepared on sapphire substrates under different sulfurization times are illustrated in Figure 6.9. The films exhibit a mid-infrared transmittance of around 40 percent, which is much lower than that of LaCuOS thin films prepared under varying sulfurization temperatures. This effect can

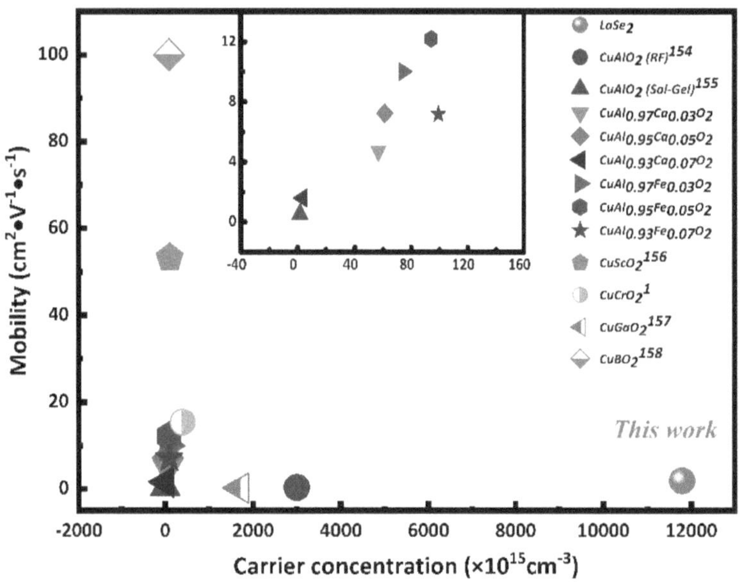

FIGURE 6.8 Comparison of carrier mobility and concentration between $LaSe_2$ films and traditional *P*-type mid-infrared transparent conductive films [7].

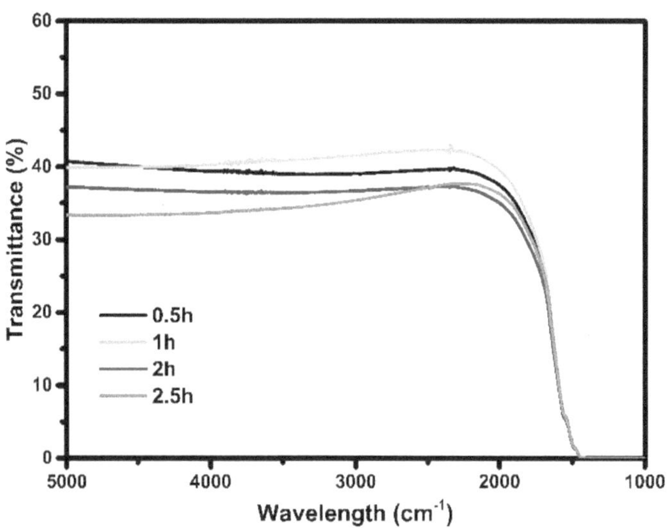

FIGURE 6.9 Mid-infrared transmittance of LaCuOS films prepared at 860°C for different sulfurization times.

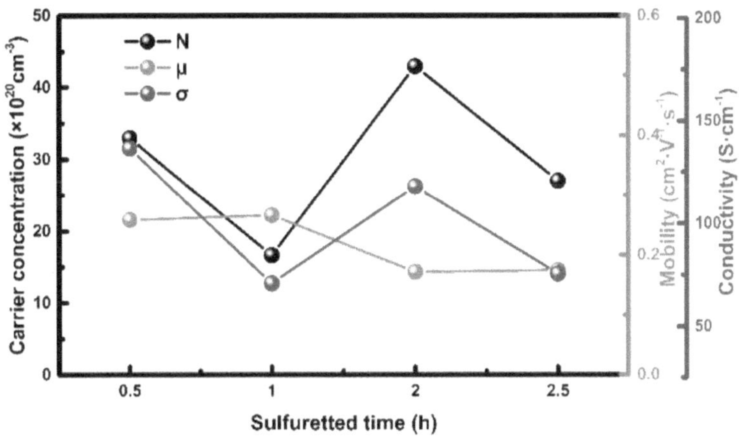

FIGURE 6.10 Electrical properties of LaCuOS films prepared at 860°C for different sulfurization times.

be attributed to the higher carrier concentration of LaCuOS thin films prepared under different sulfurization times, which increases the absorption and scattering of photons by carriers, resulting in a rapid decline in the transmittance of the films. Notably, the mid-infrared transmittance of LaCuOS thin films demonstrates a decreasing trend with prolonged sulfurization time. However, it should be emphasized that the variation in carrier concentration of LaCuOS thin films under different sulfurization times is not significant, as revealed by Figure 6.10. Therefore, the decrease in transmittance with prolonged sulfurization time may be ascribed to reduced film density resulting from the extended sulfurization time, leading to a porous structure in the film that scatters and reflects incident infrared light.

6.3.3.2 Electrical Properties of LaCuOS Films

Figure 6.10 illustrates the trend of electrical properties of LaCuOS thin film prepared under different sulfidation times. The test results indicate that the carrier concentration of LaCuOS thin films is over 1.5×10^{21} cm^{-3} at different sulfidation times. Additionally, by examining Figure 6.10, the electrical conductivity of LaCuOS thin film can reach about 140 S/cm, which is superior to the electrical properties of LaCuOS thin film prepared under different sulfidation temperatures. This result has fundamentally overcome the technological bottleneck of poor electrical performance in *P*-type transparent semiconductors, while also achieving exceptional optical performance, with a transmittance of over 70 percent in visible light. The only regret is the relatively low mobility of the thin film, which may be attributed to carrier scattering and impurities caused by the self-scattering of carriers, as well as the porous and defective structure of the film with increasing sulfidation time.

The photoelectric quality factors of LaCuOS thin films at various sulfuration times were calculated. The FOM values of the films remained stable in both visible and mid-infrared ranges, with slightly higher values for the sulfuration time of 0.5h than for other sulfuration times. This suggests that the optimal sulfuration time for LaCuOS thin films is approximately 0.5h.

6.3.4 *P*-type In_2O_3: N Films

6.3.4.1 Electrical Properties of *P*-type In_2O_3-N Films

After analyzing the surface morphology, crystal structure, and chemical bonding state of *P*-type In_2O_3 film, it can be concluded that N-type doping has a certain influence on its structure, which will further affect the optoelectronic properties of the film. The electrical performance of In_2O_3 film is a crucial component of its semiconductor performance, which determines its practical application value. The electrical properties of *P*-type In_2O_3 film were examined under varying nitrogen flow rates and are presented in Figure 6.11. Figure 6.11 (a) shows that the resistivity of the film decreases gradually with an increase in nitrogen flow rate, and it reached the minimum value at a nitrogen flow rate of 54 sccm. After the nitrogen flow rate exceeded 54 sccm, the resistivity started to increase again. This trend can be attributed to the dependence of the electrical properties of *P*-type In_2O_3 film on the carrier concentration and mobility, which are inversely proportional to the resistivity of the film. The hole mobility of *p*-type In_2O_3 film, as shown in Figure 6.11 (b), sharply decreases with a rise in nitrogen flow rate and starts to decline slowly after the nitrogen flow rate reaches 22 sccm. The high mobility observed at nitrogen flow rates below 22

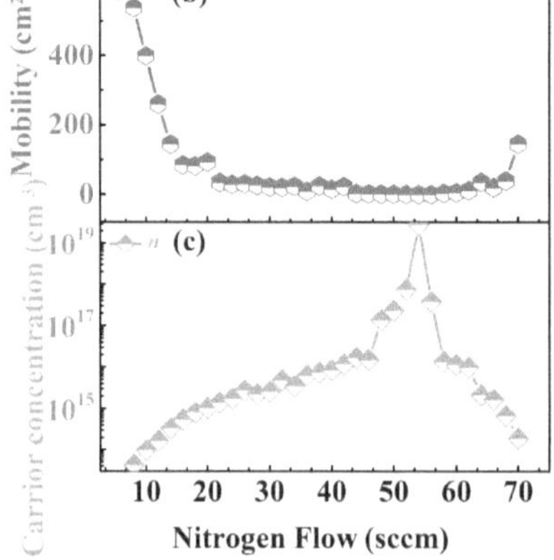

FIGURE 6.11 Electrical properties of *p* type In_2O_3-N films (a) Resistivity; (b) Mobility; (c) Carrier concentration.

sccm can be attributed to the lower concentration of defects and impurities in the film as well as its relatively larger grain size. However, as the nitrogen flow rate increases gradually, the concentration of defects and impurities in the film increases, leading to a decrease in the mobility. After the nitrogen flow rate exceeds 20 sccm, the mobility decreases slightly due to the consistent reflection of holes caused by comparable grain sizes. Nonetheless, with N-type doping, the impurity concentration gradually rises, resulting in a slow decrease in mobility.

Figure 6.11 (c) illustrates the gradual increase in carrier (hole) concentration in the film with an increase in nitrogen flow rate, reaching a maximum value of 2.803×10^{19} cm^{-3} at a nitrogen flow rate of 54 sccm, while the resistivity of the *P*-type In_2O_3 film reaches its minimum value of 0.62 Ω•cm at this point. Afterward, as the nitrogen flow rate continues to increase, the carrier concentration in the film starts to decrease gradually. The carriers in *P*-type film mainly originate from the holes generated by N-type doping. XPS analysis of the film reveals that the N-type doping concentration in the film increases gradually with an increase in nitrogen flow rate, leading to an increase in the hole concentration in the film. However, after the nitrogen flow rate exceeds 54 sccm, the doping concentration starts to decrease, consequently leading to a decline in the carrier concentration in the film. Under the poisoning mode, even though the film contains a certain concentration of oxygen vacancies, the low content of oxygen vacancies results in a low number of free electrons, which leads to the characteristics of a *P*-type semiconductor film.

6.3.4.2 Optical Properties of *P*-type In_2O_3-N Films

The optical transmittance of In_2O_3 film under different nitrogen flow rates is demonstrated in Figure 6.12 (a). This Figure shows that the film has a transmittance of over 80% in the visible to near-infrared wavelength range. The high-quality crystallinity, well-ordered columnar structure, smooth and flat surface with fewer defects that reflect light, contribute to the good transmittance performance of the film. Significant variations in the short wave absorption band of the In_2O_3 film can be observed under different nitrogen flow rates. The short wave absorption band is related to the optical bandgap of the film, which, in turn, depends on the carrier concentration in the film.

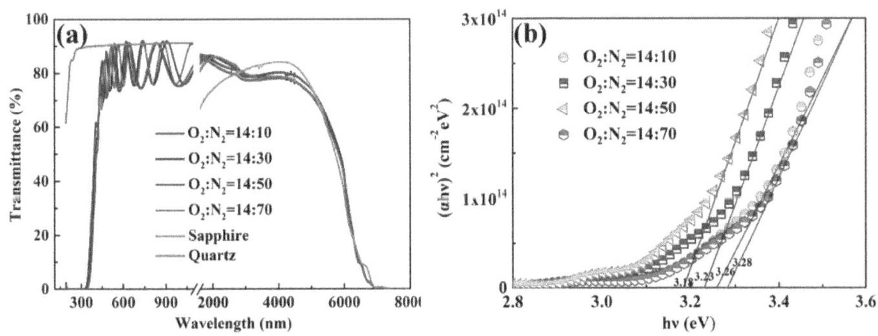

FIGURE 6.12 Optical properties of *p* type In_2O_3-N films (a) Transmittance; (b) Optical band gap.

The optical bandgap of the film was calculated by using the Tauc law based on the short edge absorption band from the transmittance spectrum, as shown in Figure 6.12 (b). The calculated optical bandgap of the In_2O_3 film decreases from 3.26 eV to 3.18 eV as the nitrogen flow rate increases from 10 sccm to 50 sccm, and then increases to 3.28 eV as the nitrogen flow rate further increases. The decrease in the optical bandgap of the film at high hole concentrations indicates that films with a larger concentration of holes have a smaller optical bandgap, while films with lower hole concentrations have a larger optical bandgap. The infrared transmittance of *P*-type In_2O_3 film is also included in Figure 6.12 (a), with an average transmittance of 80% in the visible to mid-infrared wavelength range, achieving a redshift of the plasma wavelength. In the near-infrared wavelength range, *P*-type In_2O_3 film has a certain enhancement effect on sapphire. Although the film is a *P*-type semiconductor film, it still contains a considerable concentration of free electrons, leading to a decrease in transmittance after the wavelength exceeds 3 μm, as the free electrons generate a certain degree of reflection of infrared photons.

6.3.5 Mg-C-O-H Films

6.3.5.1 Electrical Properties of Mg-C-O-H Films

Based on the analysis of the influence of substrate temperature on the film structure, it is known that the surface morphology and crystal structure of the film will change at different substrate temperatures. Therefore, the change in substrate temperature will also affect the electrical properties of Mg-C-O-H films. The electrical performance of Mg-C-O-H films is a key indicator for evaluating the performance of such films, and factors such as the elemental content and crystallinity of the film can affect its electrical performance. In terms of electrical properties, we mainly discuss the effects of different elemental ratios, reaction temperature, reaction time, and annealing temperature of the film on its electrical conductivity.

Figure 6.13 shows the relationship between the electrical properties of Mg-C-O-H films and substrate temperature. From the graph, it can be seen that the electrical performance of the film reaches its optimal value at a temperature of 400°C. As the substrate temperature increases, the carrier concentration of Mg-C-O-H films gradually increases, with a maximum value of 3.11×10^{16} cm^{-3} at a substrate temperature of 400°C. The carrier mobility is proportional to the substrate temperature, and as the substrate temperature slowly increases, the electron mobility of Mg-C-O-H films also gradually increases. The highest carrier mobility obtained at a substrate temperature of 400°C is 49.18 $cm^2V^{-1}s^{-1}$. Due to the increase in the concentration and mobility of free electrons in Mg-C-O-H films, the electrical resistivity of the film decreases with increasing substrate temperature. At a substrate temperature of 400°C, the lowest electrical resistivity obtained for Mg-C-O-H films is 4.086 Ω·cm.

As the substrate temperature increases, the crystalline quality of the Mgx-Cy film gradually improves, and the growth of grains reduces the concentration of grain boundaries, resulting in a decrease in the defect density in the Mg-C-O-H film generated after the reaction of the Mgx-Cy film with water vapor. The reduction of

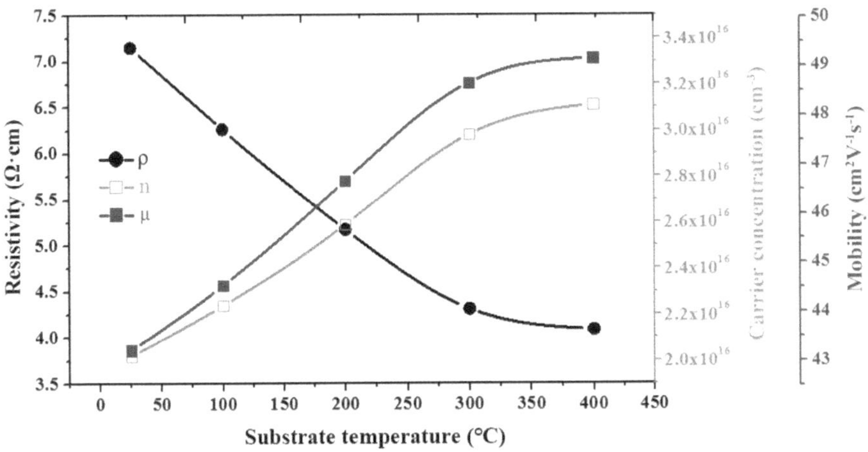

FIGURE 6.13 Electrical properties of Mg-C-O-H films at different substrate temperatures.

grain boundary concentration and defects decreases the scattering centers for free electrons in the Mg-C-O-H film, thereby improving the ability of charge carriers to move, resulting in an increase in its carrier mobility. As the substrate temperature increases, the number of C atoms entering the crystal lattice structure of the Mgx-Cy film increases. When the C in the Mg-C-O-H film generated from the reaction between the Mgx-Cy film and water vapor gains energy, it is more likely to replace the position of the H atom in the film, ionizing more free electrons, resulting in an increase in carrier concentration.

6.3.5.2 Visible Optical Properties of Mg-C-O-H Films

Based on the above analysis, it is known that the change in substrate temperature has a significant impact on the electrical properties of the film. Therefore, to obtain the optimal substrate temperature corresponding to the best performance, we also need to explore the impact of substrate temperature on the optical properties of Mg-C-O-H films.

The visible light transmittance of Mg-C-O-H films at different substrate temperatures was tested, as shown in Figure 6.14. From the graph, it can be seen that the transmittance of quartz substrates coated with Mg-C-O-H films is slightly lower than that of uncoated quartz substrates. However, the variation in the transmittance of Mg-C-O-H films in the ultraviolet-visible-near-infrared range due to the change in substrate temperature is not significant, and the visible light transmittance of all samples at different substrate temperatures is greater than 75%. Particularly, the transmittance in the near-infrared range of 800~1100 nm is greater than 80%, indicating a high transmittance. This is because the increase in substrate temperature has little effect on the surface roughness of the Mg-C-O-H film, resulting in a minimal impact on light scattering, resulting in little change in visible light transmittance with increasing substrate temperature. At the same time, it can be seen from Figure 6.14 that as the substrate temperature increases, the short-wavelength cut-off

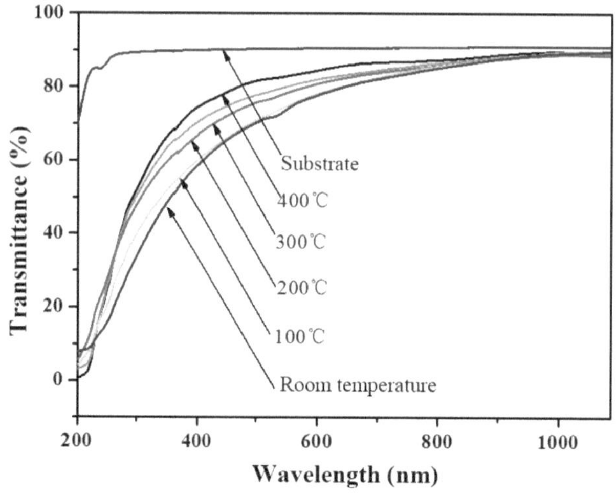

FIGURE 6.14 Visible light transmittance of Mg-C-O-H films with different substrate temperatures.

gradually shifts toward shorter wavelengths. This is because as the substrate temperature increases, the concentration of free electrons in the film gradually increases, enhancing the fill effect of the Mg-C-O-H film's optical bandgap of Moss-Burstein energy bands, increasing its optical bandgap, and causing the short-wavelength cutoff to blue-shift.

6.3.5.3 Infrared Optical Properties of Mg-C-O-H Films

Figure 6.15 illustrates the infrared transmittance of Mg-C-O-H films at various substrate temperatures. The graph indicates that the transmittance of Mg-C-O-H films decreases progressively with increasing substrate temperature, indicating that substrate temperature has a substantial influence on the transmittance of the Mg-C-O-H films. This decrease in transparency can be attributed to the increasing carrier concentration within Mg-C-O-H films due to the increasing substrate temperature of the Mgx-Cy film, resulting in a greater level of interaction between photons and electrons, which decreases the film's transmittance. It was found that when metallic Mg was directly deposited onto the substrate and reacted with water vapor, and the resulting $Mg(OH)_2$ film was formed on a CVD ZnS substrate, the transmittance in the infrared range of the Mg(OH)2 film was 3 percent higher than that of the Mg-C-O-H film, suggesting that the presence of free electrons decreases the transmittance of the film. Moreover, the transmittance of Mg-C-O-H films in the mid-infrared range of 3-5 μm and 8-10 μm is greater than 70%, indicating that the Mg-C-O-H film's refractive index is smaller than that of ZnS, making it effective in enhancing the substrate's transmittance. Furthermore, due to the relatively low carrier concentration of Mg-C-O-H films at varying substrate temperatures, the interaction between electrons and photons is weak, allowing for long-wavelength transmission.

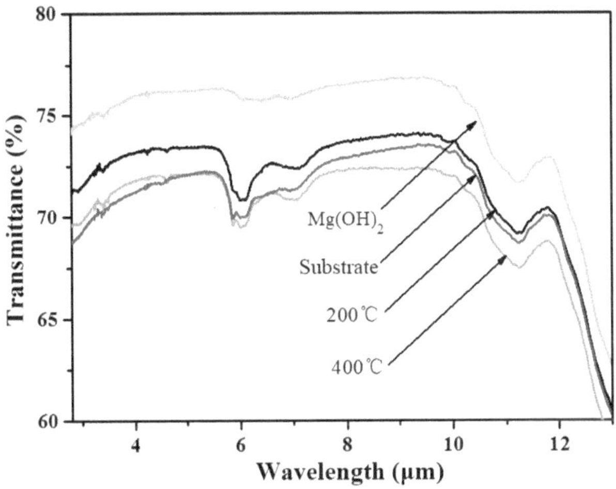

FIGURE 6.15 IR transmittance of Mg-C-O-H films at different substrate temperatures.

6.4 COATING EQUIPMENT AND TECHNOLOGY FOR INFRARED WINDOWS

As the trend of developing aircrafts with diversified service environments continues, the demand for functional films such as infrared enhancing protective films and infrared transparent conductive films for optical windows has increased. However, there exist significant technological bottlenecks in the preparation of functional films for optical windows, especially with regards to the fabrication technology for optical window functional films for complex surfaces. Various environmental conditions encountered during flight necessitate the use of optical windows with large-size planar, or pointed arch surfaces. Therefore, achieving the uniform preparation of functional films on such diverse and complicated curved surfaces presents a crucial challenge to ensure the normal operation of optical windows for aircrafts. In this section, we will discuss coating technologies for large-size planar, and pointed arch optical windows, respectively.

6.4.1 Coating Technology for Large Size Planar Infrared Windows

Currently, magnetron sputtering technology is widely used in industrial production as a simple, fast-coating-rate, and high-quality film preparation technique. It can be used to prepare various types of films, including conductors, semiconductors, and insulators. However, due to the non uniform distribution of the magnetic field and the resulting uneven plasma region, the deposited films often exhibit thinner areas in the middle and rear edges, which has been a key problem restricting its application. To improve the uniformity of the films, larger target materials are commonly used, resulting in a significant waste of target material and greatly increased production costs. Thus, there is an urgent need to develop a cost-effective method for producing large-area uniform films.

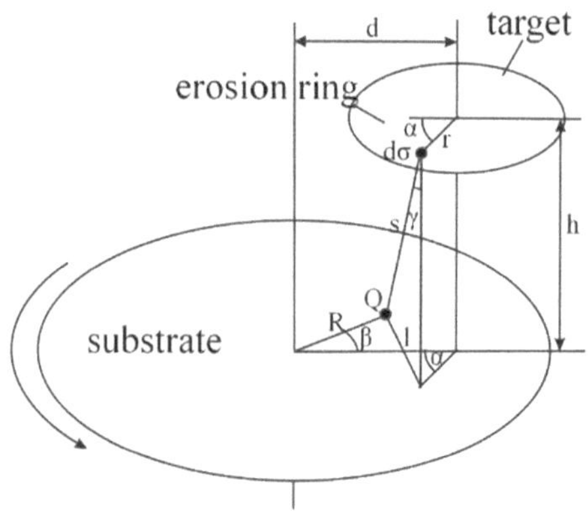

FIGURE 6.16 Schematic diagram of the magnetron sputtering system [8].

To address this issue, Professor Jiaqi Zhu's team at Harbin Institute of Technology has proposed a novel composite motion method of substrate self-rotation and magnetron sputtering target stepwise movement for depositing large-area uniform films on planar optical windows. In this method, a magnetron sputtering coating machine is designed with a self-rotating sample holder and two magnetron-sputtering targets with a diameter of 49mm, which can undergo stepwise motion. The team established a relationship between the thickness of the deposited film and the mechanical control parameters, including the dwell time ratio of the target material at different positions and the target material's step length. A schematic diagram of this magnetron sputtering method is shown in Figure 6.16.

Based on Figure 6.16, intricate geometric relationships among various parameters can be inferred. Specifically, h denotes the vertical distance between the target material and substrate, while d denotes the horizontal distance between them. Moreover, r represents the distance from the center of the sputtering unit dσ on the target material, and R represents the distance between a specific point Q on the substrate and the substrate center. In addition, s refers to the distance between the sputtering unit dσ on the target material and the point Q on the substrate, while γ is the sputtering angle:

$$\cos\gamma = h/s \tag{6.13}$$

$$s^2 = h^2 + l^2 \tag{6.14}$$

$$l^2 = (d - R\cos\beta - r\cos\alpha)^2 + (R\sin\beta + r\sin\alpha)^2 \tag{6.15}$$

For most planar magnetron sputtering, the incident particles possess low energy, while the operating pressure is high (around several pascals). Under this circumstance, four hypotheses were proposed as follows:

(1) As the magnetic field lines on the target material surface in magnetron sputtering are parallel to it, the ionized gas sputtering particles are accelerated

along the field lines to strike the target material surface. Consequently, the incident angle of the particles on the sputtering unit dσ is zero.

(2) All target material particles sputtered from the target material are originated from the non-uniform etching ring on the target material surface from r1 to r2, with r1 and r2 denoting the inner and outer radii of the magnetron cathode, respectively. In this region, where the magnetic field strength is extremely high, most of the secondary electrons are confined around the ring. Additionally, the gas near the ring is ionized by sputtering, resulting in the maximum sputtering of target material particles in this region.
(3) Typically, a simple cosine distribution is believed to describe the angular distribution of the sputtered target material particles in the preparation of sputtered films.
(4) During the sputtering deposition process in magnetron sputtering, the pressure in the vacuum chamber is usually several pascals. Therefore, sputtered particles are scattered by gas molecules in the vacuum chamber while flying toward the substrate, and the probability of depositing on the substrate is inversely proportional to the distance they fly in the vacuum chamber.

The moveable target surface is parallel to the rotatable substrate. let m be the mass of material ejected from the dσ region inside the etched ring on the target surface and ρ be the density of the deposited film. According to the above four assumptions, the deposition rate dv of the dσ region on the target to the Q point at a distance R from the center of the substrate can be expressed as:

$$dv = \frac{m(n+1)\cos^{n+1}\gamma}{2\pi\rho s^2} \tag{6.16}$$

Among them, n is the angular distribution coefficient that is related to multiple factors, but its value can be determined experimentally under other conditions. When the substrate undergoes rotational motion around the central axis, assuming that the etching region of the target material is from r_1 to r_2, the total film deposition rate for the point in question by all regions of the etching ring can be expressed as:

$$v = \int_0^{2\pi}\int_0^{2\pi}\int_{r_1}^{r_2} \frac{m(n+1)h^{n+1}r}{2\pi\rho s^{n+3}} d\alpha d\beta dr \tag{6.17}$$

6.4.2 Coating Technology for Sharp Arch Infrared Windows

Conformal shields provide significant advantages over traditional planar or hemispherical optical windows and fairings, such as reducing air resistance and expanding the scanning field. Due to their simple target configuration and step models, achieving uniform coating on large-sized planar surfaces is comparatively straightforward, resulting in a higher level of industrial production maturity. Similarly, uniform coating techniques for conformal surfaces with regular shapes, such as hemispherical surfaces, have been extensively studied, and the corresponding processes are well-established. However, with increasing optical performance demands, traditional planar or hemispherical optical windows are no longer sufficient to meet modern warfare needs. This has created an urgent need for uniform deposition techniques for conformal surfaces. However, research on coating conformal surfaces with changing radii of curvature is limited for two reasons: the equations for conformal surface curves are more complex, making it challenging to establish coating models, and experimental trials can be costly and ineffective. Therefore, developing a theoretical model for constructing uniform coating techniques for large-sized freeform surfaces is the future development trend and has significant implications.

Based on the surface properties of conformal samples, the team led by Professor Zhu Jiaqi has confirmed the sputtering coating scheme of small targets in magnetron sputtering systems. To achieve uniform film deposition on both flat and convex surfaces using smaller targets, some upgrades and modifications are required for the magnetron sputtering system. The main improvement lies in the change of motion mode, which is designed as a stepping and swinging mode for the target material, combined with the rotation and up-down movement of the sample. A schematic diagram of different positions of the target material and sample structure is shown in Figure 6.17(a), which demonstrates three typical positions: top, middle, and bottom. The sample can move up and down through the motion motor, and the target source adjusts the swinging angle according to the position of the coating sample, while stepping forward and backward to control the distance between the sample and the target source. To implement the above scheme, a special guide rail is designed to hold the target source, and there are no fasteners between the target source and the connecting rod. The curve of the guide rail can be changed according to different samples and is easy to replace. This enables the target source to move forward and backward on the track, adjust the tilt angle according to the curvature of the track, and control the distance between the target source and the sample, thus completing the deposition of thin films on relatively small and deep convex surfaces. Figure 6.17(b)

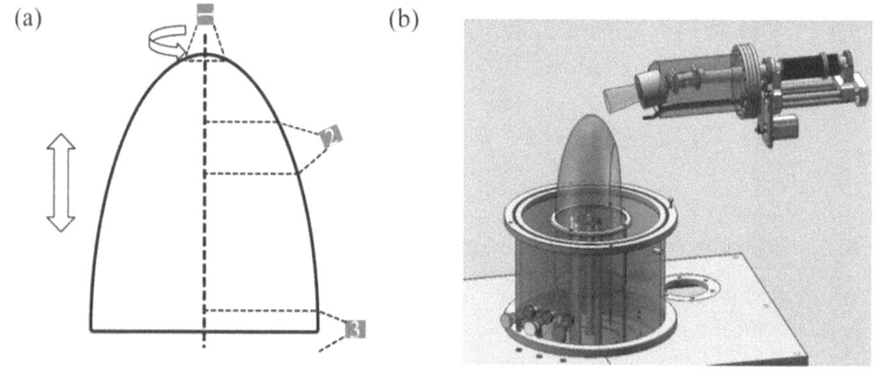

FIGURE 6.17 The schematic diagram of (a) relative position between target and substrate and (b) 3-D schematic of magnetron sputtering system [9].

shows a three-dimensional (3D) schematic diagram of the target source and substrate inside this magnetron sputtering system, from which the relative positional relationship between the target source, sample, and track can be clearly observed.

To achieve uniform film deposition on deep convex surfaces using smaller targets, it is first necessary to establish a suitable model for calculating the film distribution. Based on the basic theories and assumptions of spatial ion distribution in the deposition model, the motion parameters for depositing uniform films on conformal surfaces are obtained. The interrelationship between the relevant parameters and the deposition of uniform film layers are then studied. Subsequently, process parameters for depositing uniform films on conformal surfaces are determined, and the parameters are verified and adjusted. Finally, experiments are conducted in the designed magnetron sputtering system to confirm the accuracy of the model and calculation results.

In the process of vacuum deposition, the distribution of film deposition rate in different angular regions is mainly based on Knudsen's cosine law. Holland further summarized the distribution of film deposition rate under different types of target sources and proposed some assumptions to meet the deposition model. Pulker focused on magnetron sputtering technology and summarized the relationship between the uniformity of thickness distribution and the quality efficiency in the sputtering process. Both Holland and Pulker's works are based on equation 6.18.

$$v = \frac{K \cos^n \varphi \cos \theta}{S^2} \tag{6.18}$$

In Equation (6.18), v is the deposition rate, S is the distance between the sputtering point on the target surface and the deposition point on the surface, n is a real number determined by experimental conditions, φ is the angle between the plane perpendicular to the target surface and S, and θ is the angle between the plane perpendicular to the substrate surface and S. K is a constant.

In the uniform deposition process of magnetron sputtering, Meng et al. proposed four assumptions regarding plane magnetron sputtering based on the premise that the energy of the sputtering particles is low. Building upon the research of Pulker, Holland, and Meng et al., and based on experimental conditions presented in this study, four basic assumptions were pre-set according to Knudsen's cosine law to correspond to the specific deposition process:

(1) Due to the strong magnetic field of magnetron sputtering, sputtering occurs mainly in the annular magnetic field zone, so it is assumed that the sputtering particles come only from this region.
(2) In this experiment, collisions between sputtering particles from the target source and gas molecules in the chamber were neglected.
(3) It is assumed that there is no collision between the sputtering particles from different regions of the annular magnetic field.
(4) The number of sputtering particles from the target source is proportional to the distance between the target source and the substrate.

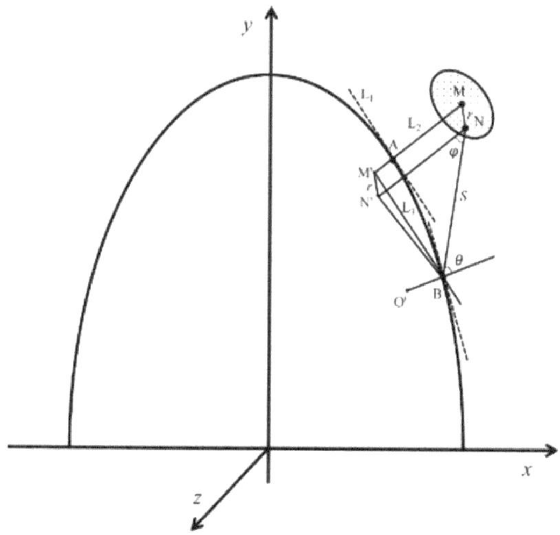

FIGURE 6.18 Geometrical relationship between convex surface substrate and target [9].

To provide theoretical guidance for the relevant parameters of uniform film deposition, the geometric relationship between the target base structure and the stepping method is abstracted based on the shape characteristics of the conformal sample and the movement mode of the stepped coating. A three-dimensional coordinate system is established for modeling and calculation. The abstracted model is illustrated in Figure 6.18, in which the curve of the conformal surfaces satisfies the elliptical curve relationship. The equation for the curve of the upper part is:

$$\frac{x^2}{a^2} + \frac{y^2}{b^2} = 1 \quad (y \geq 0) \tag{6.19}$$

The key to obtaining a uniform thin film in this study is to solve the corresponding relationship between the deposition rate of the point on the conformal surface and the corresponding point of the target material center on the conformal surface. By combining the relative deposition rates, stepping, and dwelling time parameters, a functional thin film satisfying the uniformity requirements is obtained.

The specific process of analyzing and deriving the model is as follows: In the process of evaporation deposition, such as magnetron sputtering or electron beam, the spatial distribution of vacuum particles follows the Knudsen cosine distribution law. The relationship between the film quality in a certain direction in space and the total mass of sputtering particles from the target source is given by:

(I) When the target source is a point source

$$\mathrm{d}m = \frac{m}{4\pi}\mathrm{d}\omega \tag{6.20}$$

where m is the mass and ω is the spatial direction.

(II) When the target source is a surface source

$$dm = \frac{m}{\pi}\cos^n \phi d\omega \tag{6.21}$$

When the base has a certain inclination angle with the space direction

$$v = K\frac{\cos^n \phi \cos\theta}{s^2} \tag{6.22}$$

where K is a constant, n is determined according to specific conditions, S is the distance from the source to somewhere in the substrate, φ is the spatial angle, and θ is the angle between the substrate and the spatial direction.

In the model developed in this study, the deposition rate of the center point A corresponding to the conformal substrate sample is easier to obtain. First, the mathematical relationship shows that the calculation of each part of equation 6.22, where:

$$\cos\varphi = \frac{h}{s} \tag{6.23}$$

$$s^2 = h^2 + r^2 \tag{6.24}$$

Bringing equations 6.23 and 6.24 into 6.22, the deposition rate at point A can be found as:

$$v_A = \int_0^{2\pi}\int_0^{r_2} K\frac{\cos^n \varphi\cos\theta}{r^2}\mathrm{d}r\mathrm{d}\alpha = K\int_0^{2\pi}\int_{r_1}^{r_2}\frac{\left(\frac{h}{\sqrt{h^2+r^2}}\right)^{n+1}}{h^2+r^2}\mathrm{d}r\mathrm{d}\alpha \tag{6.25}$$

Therefore, the difficulty and key point of this chapter is how to use the relationship to find the rate of any point B in the conformal surface. Introducing the tangent line L1 at point A, it satisfies the curve equation:

$$y - y_0 = -\frac{b^2 x_0}{a^2 y_0}(x - x_0) \tag{6.26}$$

At the same time, the curve L2 between the target center point M and point A, it is easy to get L1 and L2 are perpendicular to each other, then it satisfies the equation:

$$y - y_0 = \frac{a^2 y_0}{b^2 x_0}(x - x_0) \tag{6.27}$$

At any deposition point B on the conformal surface make a tangent line L_3 which satisfies the equation:

$$y - y_1 = -\frac{b^2 x_0}{a^2 y_0}(x - x_1) \tag{6.28}$$

L1 parallel line is made through the deposition point B and intersects with the line L2 at the point M'(x_2, y_2, z_2), then the coordinates of the point M' can be found by using the equations L2 and L3, so the distance between AM' can be further found and the result is:

$$AM' = \sqrt{(x_2 - x_0)^2 + (y_2 - y_0)^2} \tag{6.29}$$

Therefore, let the distance between the MM' points be h', the distance between the target base and AM be h, and the distance between BM' be d_1, then on the basis of equation 6.29, h' and d_1 can be found as follows, respectively:

$$h' = MM' = NN' = AM' + h = \sqrt{(x_2 - x_0)^2 + (y_2 - y_0)^2} + h \tag{6.30}$$

$$d_1 = BM' = \sqrt{(x_1 - x_2)^2 + (y_1 - y_2)^2} \tag{6.31}$$

At the same time, take any deposition point N on the target source, the distance from the center of the target M is radius r, and then cross the line BM' over for the target parallel plane, the corresponding point of N in the plane is N'. Then in ΔBM'N', it satisfies:

$$\begin{aligned} BN' &= M'N' + BM' - 2M'N' \times BM' \cos\alpha \\ &= r^2 + d_1^2 - 2rd_1 \cos\alpha \end{aligned} \tag{6.32}$$

Hence, on the basis of equations 6.30 and 6.32, the distance S^2 in equation 6.22, which characterizes the deposition velocity, can be calculated

$$\begin{aligned} s^2 &= NN'^2 + BN'^2 \\ &= h_1^2 + r^2 + d_1^2 - 2rd_1^2 \cos\alpha \end{aligned} \tag{6.33}$$

At the same time, cosφ can also be obtained.

$$\cos\varphi = \frac{h_1}{s} = \frac{h_1}{h_1^2 + r^2 + d_1^2 - 2rd_1^2 \cos\alpha} \tag{6.34}$$

Next, the deposition rate at point B can be found by simply obtaining cosθ. Let the center of curvature $O'(x_3, y_3, z_3)$ at point B, since it is in the x-y plane and therefore z_3 is 0, then x_3 and y_3 can be expressed as:

$$x_3 = x_1 - y'\frac{1+y'^2}{y''} \tag{6.35}$$

$$y_3 = y_1 + y'\frac{1+y'^2}{y''} \tag{6.36}$$

To facilitate the calculation, the coordinates at the center of the target source are introduced and set to AA, then using the relationship between AM, it is known that they satisfy equations 6.37 and 6.38, respectively:

$$\frac{y_5 - y_0}{x_5 - x_0} = \frac{a^2 y_0}{b^2 x_0} \tag{6.37}$$

$$(x_5 + x_0)^2 + (y_5 - y_0)^2 = h^2 \tag{6.38}$$

Taking the case of $x_5 \geq x_0$, then the values of x_5 and $y_5$can be solved, and let the angle between the target and the centerline be γ, then:

$$\tan\gamma = -\frac{b^2 x_0}{a^2 y_0} \tag{6.39}$$

Then, the coordinates N(x_4, y_4, z_4) of any point on the target can be expressed respectively as:

$$x_4 = x_5 + r\cos\alpha\cos(\pi - \gamma) \tag{6.40}$$

$$y_4 = y_5 - r\cos\alpha\sin(\pi - \gamma) \tag{6.41}$$

$$z_4 = r\sin\alpha \tag{6.42}$$

Therefore, the vector $O'B$ and the vector BN can be expressed in coordinates as follows, respectively:

$$\overrightarrow{O'B} = (x_1 - x_3, y_1 - y_3, 0) \tag{6.43}$$

$$\overrightarrow{BN} = (x_1 - x_3, y_1 - y_3, z_4) \tag{6.44}$$

Therefore, cosθ can be found using the relationship between coordinates:

$$\cos\theta = \frac{\overrightarrow{O'B} \cdot \overrightarrow{BN}}{\left|\overrightarrow{O'B}\right|\left|\overrightarrow{BN}\right|} \tag{6.45}$$

In summary, the deposition rate of any point B on the surface of a conformal surface can be expressed as:

$$v_B = \int_0^{2\pi}\int_{r_1}^{r_2} K\frac{\cos^n\varphi\cos\theta}{r^2}\,dr\,d\alpha \tag{6.46}$$

The parameters cos φ, cos θ, and r can be expressed in terms of coordinates and the distance between the target substrate and the substrate. Consequently, the deposition rate ratio between points A and B can be calculated, and the velocity distribution of the deposited particles on the outer surface of the conformal surface can be obtained during the deposition process. The problem is that the computational complexity of the formula is relatively high. Therefore, MATLAB software was used to program and analyze the model, resulting in the velocity distribution of all points on the conformal surface at different positions of the target substrate. This is beneficial for optimizing the deposition parameters.

6.5 SUMMARY AND PERSPECTIVE

Infrared windows serve as the forefront of infrared detection systems, with the primary challenge and advancement focusing on enhancing the performance of infrared systems in harsh electromagnetic environments. This chapter summarizes the main scientific issues currently faced by infrared windows, provides approaches and methods to address these issues in various scenarios, and points the way for future development. To resolve the issue that existing infrared windows cannot meet the comprehensive requirements of hypersonic flight conditions, the main strategy involves the fabrication of infrared anti-reflective protective films on the windows. This study, based on the design of the anti-reflective film system, has developed Y_2O_3 and N-Y_2O_3 coatings deposited on ZnS substrates, effectively achieving the protective effect on the infrared window substrate materials and increasing transmission. In response to the demand for electromagnetic interference resistance, there is an urgent need for infrared windows to integrate high transparency with high electrical conductivity. This chapter explores the complex relationship between electrical conductivity and optical transparency, thoroughly explains the core contradictions between optoelectronic properties, and realizes a series of material innovations based on the concept of balancing regulation. Factors controlling the optoelectronic properties of thin films are revealed, which can be utilized to develop materials with tailored performance for specific applications. Regarding coating technology, this research discusses the development of new methods for depositing uniform coatings on large-scale and complex surfaces. Through processes of modeling, simulation, and verification, the uniformity of the film is validated, and the design and manufacturing of the deposition system are achieved. This lays the foundation for the practical application of infrared anti-reflective protective films and transparent conductive films, which is of great significance for improving the performance and durability of infrared systems in practical applications.

Future work should focus on improving deposition processes, discovering new materials with enhanced performance, and further exploring the trade-offs between electrical conductivity and optical transparency. Additionally, the development of advanced coating technologies for complex surfaces could significantly impact the design and capabilities of the next generation of infrared systems, particularly in military and aerospace applications.

REFERENCES

[1] Dalchiele EA, Giorgi P, Marotti, RE, Martín F, Ramos-Barrado JR, Ayouci R, Leinen D. Electrodeposition of ZnO thin films on n-Si(100). *Solar Energy Materials & Solar Cells* 2001, 70:245–54.

[2] Yuankun Z, Rueben JM, Jiaqi Z, Jiecai H, André A. Transparent and conductive indium doped cadmium oxide thin films prepared by pulsed filtered cathodic arc deposition. *Applied Surface Science* 2013, 265:738–44.

[3] Saha B, Das S, Chattopadhyay KK. Electrical and optical properties of Al doped cadmium oxide thin films deposited by radio frequency magnetron sputtering. *Solar Energy Materials & Solar Cells* 2007, 91:1692–7.

[4] Canulescu S, Rechendorff K, Borca CN, Jones NC, Bordo K, Schou J, Nielsen LP, Hoffmann SV, Ambat R. Band gap structure modification of amorphous anodic Al oxide film by Ti-alloying. *Applied Physics Letters* 2014, 104:121910.

[5] Lakshminarasappa BN, Shivaramu NJ, Nagabhushana KR, Fouran S. Synthesis characterization and luminescence studies of 100 MeV Si^{8+} ion irradiated sol gel derived nanocrystalline Y_2O_3. *Nuclear Instruments and Methods in Physics Research B* 2014, 329:40–7.

[6] Lakshminarasappa BN, Jayaramaiah JR, Nagabhushana BM. Thermoluminescence of combustion synthesized yttrium oxide. *Powder Technology*, 2012, 217:7–10.

[7] Gang G, Lijia T, Lei Y, Chunqiang S, Liangge X, Fei X, Fangjuan G, Jingjing X, Hao G, Jiaqi Z, A P-type mid-infrared transparent semiconductor LaSe2 film with small hole effective mass and high carrier concentration. *Applied Physics Letters* 2021, 118:26.

[8] Chunzhu J, Jiaqi Z, Jiecai H, Pei L, Xunbo Y, Uniform film in large areas deposited by magnetron sputtering with a small target. *Surface and Coatings Technology* 2013, 229:222–5.

[9] Zhenhuai Y, Lei Y, Bing D, Xiuwei H, Qiang W, Yumin Z, Jiecai H, Jiaqi Z, Uniform films deposited on convex surfaces by magnetron sputtering with a small target. *Thin Solid Films* 2018, 665:1–5.

7 Laser Micro-processing of Materials for Transportation Industry Application

Quanjie Wang, Xinxin Li, Kuai Yang, Zhen Zhang, Yingchun Guan, and Xilun Ding

7.1 INTRODUCTION

Laser micro-processing techniques offer advantages such as non-contact properties, high adaptability, precision, and efficiency. In the transportation industry, where scenarios necessitate complex-surface micro-processing on materials, laser micro-processing techniques emerge as an essential approach.

A range of technologies exists to steer optical beams, encompassing galvanometers, polygonal scanners, micromirrors, piezo mirrors, and acousto-optic deflectors, among others. Each technology exhibits proficiency in specific applications. The primary method employed in transportation applications is galvanometers, which excel in tasks requiring vector scanning, high throughputs, and precision. It is noted that galvanometers suffer from a limited field of view, resulting in stitching errors and efficiency losses in large-area micro-processing. The first objective of this chapter is to present the advancement of the galvanometer-based system towards meeting the demands of large-area, stitchless, and highly uniform processing. Then, we introduce our recent findings of laser precision polishing, multifunctional surfaces fabrication and dissimilar materials joining. The microstructure, morphology, wear resistance, thermal stability, and wettability of induced surfaces are summarized. Several examples are provided to facilitate the design of the laser scanner-stage synchronized processing systems, and to introduce the applications of the aforementioned laser micro-processing techniques in transportation industry.

 DOI: 10.1201/9781003386223-7

7.2 A SCANNER-STAGE SYNCHRONIZED SYSTEM FOR LASER MICRO-PROCESSING

7.2.1 Advantages of the Scanner-stage Synchronized System in Laser Micro-processing

The future development trend of laser manufacturing equipment is high precision, high efficiency, and large-area processing [1, 2]. A crucial component facilitating this is the galvanometer scanner, swiftly steering the laser beam across the work surface. However, the scanner's field of view imposes limitations on processing area.

To overcome this, two methods are commonly employed. The first one involves improving the field of view of the scanner by designing the F-theta lens [3, 4] or by increasing the focusing distance [5]. Yet, wide-angle F-theta lenses offer limited improvement, while increased focusing distances reduce the motion resolution of galvanometers. The second method is adding new motion stages that can expand the processing area by enabling the movement of either the workpiece or the galvanometer scanner [6]. In this method, the motion stage moves to a predetermined position and then stops; subsequently, the galvanometer scanner initiates the processing and then stops. Next, the stage moves to the next position, and the cycle repeats. This repeated start-stop motion leads to stitching errors at image field boundaries, negatively impacting the surface quality of the processed material. Additionally, frequent acceleration and deceleration significantly reduce the throughput of the system during large-area processing.

Extensive research has targeted stitching errors in laser processing. Common practices involve compensating for F-theta lens distortion through calibration-based techniques [7, 8], such as using correction charts [9] or visual aid systems [10]. It is noted that these methods cannot fundamentally prevent stitching errors. A synchronized motion-control architecture is necessary to essentially avoid stitching errors. This architecture involves the synchronization of the stage and galvanometer scanner instead of repeated start-stop cycles. The "on-the-fly" concept, which utilizes a one-axis stage and galvanometer scanner, has been adopted for large-area laser processing [11] and laser-hardening processing [12] but has limitations in terms of processing flexibility. Another approach integrates motion stage and scanner over standard industrial Ethernet networks [13, 14]. This system is not an open architecture platform, limiting its convenience for customized applications.

Here, we present an open architecture scanner-stage synchronized system that integrates three important features. Firstly, the system enables large-scale laser micromachining without stitching errors. Secondly, it significantly improves system efficiency. Finally, based on the open architecture platform, it allows real-time compensation of servo errors.

7.2.2 The Setup of the Scanner-stage Synchronized System

Figure 7.1 illustrates a scanner-stage synchronized system that integrates the motion of the galvanometer scanner and the linear motion stage. A continuous wave neodymium-doped yttrium aluminum garnet (Nd: YAG) laser source is used to

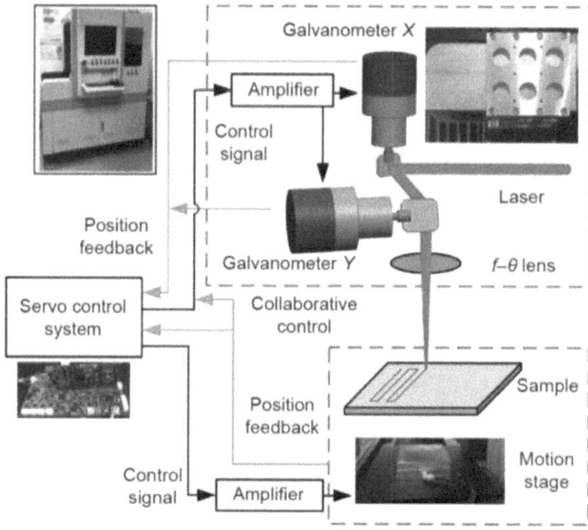

FIGURE 7.1 Schematic diagram of a scanner–stage synchronized system [15].

generate a nanosecond laser beam with a central wavelength of 1,064 nm and a maximum power of 100 W for the scanner-stage synchronized processing. The proposed system uses an XY galvanometer scanner with a 10 mm aperture and a focal length of 160 mm, allowing for high-speed scanning within a range of 100 mm × 100 mm. In order to realize large area laser processing, the proposed system integrates a custom-designed linear servo motion stage, with a range of 600 mm × 600 mm, a maximum speed of 800 mm/s, and a precision of 3 μm.

In contrast to commercialized products where separate servo systems control the galvanometer and motion stage, which makes it challenging to synchronize them for the fabrication process, the proposed system implements a single servo control system that integrates the controllers of the stage and the galvanometer scanner. The proposed servo control system generates control signals that are simultaneously sent to both the motion stage and the galvanometer scanner, avoiding stitching errors and maximizing the efficiency of large-area laser processing.

7.2.3 The Control Structure of the Scanner-stage Synchronized System

The control structure of the proposed system is shown in Figure 7.2, where r, $r_{scanner}$ and r_{stage} represent the reference signal, the reference pattern of the scanner, and the reference pattern of the stage, respectively; C_s, $C_{scanner}$, and C_{stage} represent the controllers of the synchronized system, the scanner, and the stage, respectively; $P_{scanner}$ and P_{stage} represent the models of the scanner and stage, respectively; and θ, x, and e represent the angular displacement of the scanner, the displacement of the stage, and the feedback error of the synchronized system, respectively.

The proposed synchronization strategy for the scanner and motion stage employs a master-slave control structure [16]. This approach considers the trade-off between the high speed but small stroke of the scanner and the low speed but large stroke of the motion stage. The commander obtains the reference patterns for both systems by decomposing the given reference vector. The laser controller is triggered at a predetermined time by the commander's instructions, and the input signals for both sub-motion systems are sent simultaneously to achieve synchronized motion between the scanner and stage. Each sub-motion system tracks their corresponding decomposed reference trajectories separately. The error synthesis module calculates the error of the synchronized system and feeds it back to the scanner system in real-time for adjustment to ensure optimal performance. It is worth noting that the motion stage serves as the master subsystem, while the galvanometer scanner functions as the slave subsystem due to its high dynamics. Detailed discussions on the master-slave control structure utilized in a scanner-stage synchronized system are referred to [17].

The motion decomposition principle aims to optimize the benefits of the galvanometer's high-speed scanning and the stage's large motion stroke by considering their distinct characteristics. To achieve this, various methods can be utilized, as long as they meet reasonable constraints. For instance, the galvanometer's scanning range should be within the field of view and not too large to prevent field distortion. The motion stage's acceleration should also be within an acceptable range to ensure its performance. Furthermore, the vector composition of the motion must satisfy the requirement that the vector sum of the stage motion and the scanner motion equals the reference target motion.

The aim of the error synthesis module in the Figure 7.2 is to improve the performance of the synchronized system. The angular displacement of the galvanometer scanner is converted into the linear displacement with respect to the workpiece using the optical system's parameters. And the synchronized system's error is calculated by utilizing the angular displacement of the galvanometer scanner (θ), the linear displacement of the motion stage (x), and the reference signal (r). Then the calculated error is fed back to the galvanometer scanner to compensate the servo error.

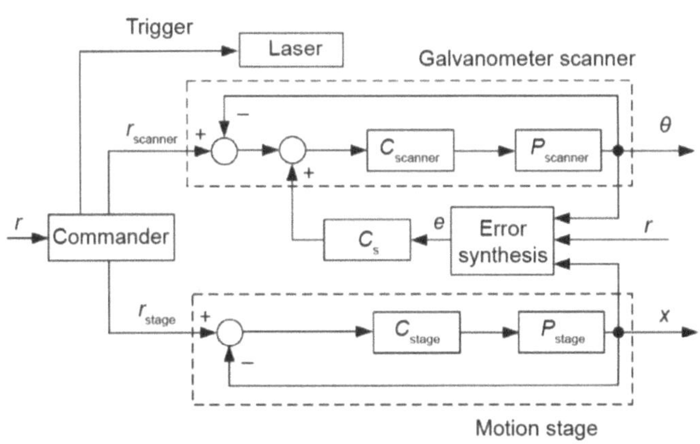

FIGURE 7.2 Block diagram of the scanner-stage synchronized control structure [15].

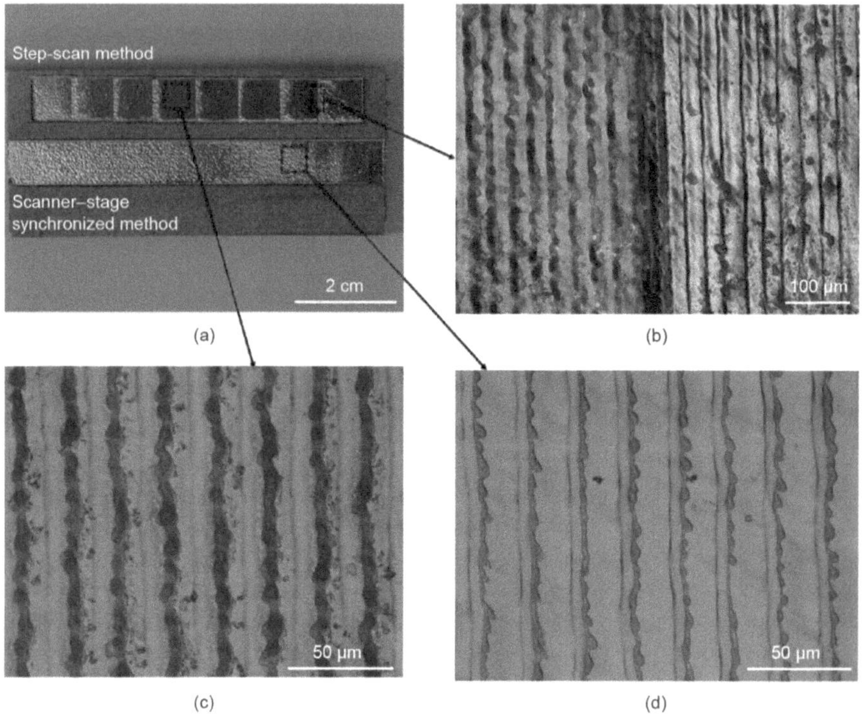

FIGURE 7.3 (a) Laser polishing of an IN718 workpiece fabricated by selective laser melting using different methods; (b) the stitching area of the step-scan method; (c) the laser polishing surface using the step-scan method; (d) the laser-polishing surface using the scanner–stage synchronized method [15].

7.2.4 Performances of the Scanner-stage Synchronized System

To assess the efficacy of the proposed scanner-stage synchronized system, a large area laser polishing experiment is conducted on an Inconel 718 (IN718) workpiece fabricated through selective laser melting. The experiment involved comparing the step-scan method with the scanner-stage synchronized method, as shown in Figure 7.3. The laser power used in this experiment is 100 W, and the scanning speed is set at 100 mm/s. The surface roughness (Ra) of the IN718 workpiece is measured as 10 μm, and laser polishing by both methods effectively improve the surface finishing. The step-scan method results in errors at the stitching boundaries as shown in Figure 7.3a and b. The smoothness of the polished surface by the two methods is similar, as shown in Figure 7.3c and d.

7.3 LASER PRECISION POLISHING OF METALLIC SURFACES

Laser precision polishing is a promising surface treatment technique to generate high-quality surfaces. Compared with traditional surface post-treatment technologies, such as mechanical polishing [18–20], chemical mechanical polishing [21], and electrochemical polishing [22, 23], laser polishing is promising for reducing surface roughness of complex and freeform surface components with high efficient, excellent adaptability, high precision, and high automation in an environmentally friendly fashion.

During laser polishing processes, smooth surfaces can be obtained by melting and evaporation rather than removal [24–26]. A controllable laser beam is focused onto the metal surface and laser energy converts into heat energy through the interaction between laser and metal lattice. The surface material will be melted rapidly and an extremely shallow molten pool is produced, in which the liquid flowing into the valley from convex peaks rapidly driven by surface tension and gravity [27–29].

Additive manufacturing (AM) is an important technique to produce complex or near-net-shaped components for various applications in aerospace, marine and offshore, nuclear and fossil fuel power plants, and biomedical industries. However, poor surface integrity and geometric tolerances usually occur at the surface of AM components due to the staircase effect [22], balling effect [30], and molten pool splashing during the AM process [31], which requires additional post-treatment to meet surface smoothness and in-service tolerance requirements [32]. Bhaduri et al. [33] used a fiber laser with a maximum average power of 50 W and wavelength of 1064 nm to polish 3D-printed stainless steel (SS316L) components, in which the surface roughness reduced by 94 percent. Obeidi et al. [34] using a CO_2 laser working under a continuous-wave (CW) mode polished AM 316L stainless steel cylindrical components, and the surface roughness was decreased from 10.4 to 2.7 μm.

For other metallic components manufacturing technique, a high-finish surface can also be obtained by laser polishing. Rosa and Mognol et al. [35, 36] concluded that laser polishing could reduce the surface roughness S_a from 21 to 0.79 μm for 316L parts manufactured via direct metal deposition. Bourell et al. [37] reported that laser polishing could produce a smooth finish on selective laser sintered (SLS) iron-copper parts by fast melting and resolidification with the percent reduction in surface roughness R_a up to 31 percent. Lamikiz et al. [38] proposed that laser energy density was the most important parameter during polishing process, and the surface roughness (R_a) of SLS parts (60 percent sintered AISI 420 stainless steel and 40 percent infiltrated bronze) could be reduced by 80 percent at optimized conditions.

Note that Ti alloys have been widely used to various components of transportation industry, such as aircraft gas turbines, turbine discs of aircraft engines, and fossil fuel fastenings. In this section, we report our recent work in laser precision polishing of Ti alloys (LMD TC11, SLM TC4). The alteration of micro-structure and roughness of the metallic surface before and after laser polishing are carefully studied. Finite element simulation is involved to analyze the thermodynamics and rapid solidification during the laser polishing process. Mechanical properties evolution, such as micro-hardness, wear resistance, and corrosion resistance of laser-polished layer and as-received surfaces are experimentally investigated. Finally, the application of laser precision polishing for transportation industry is introduced.

TABLE 7.1
The Chemical Compositions of TC11 and TC4 (wt.%)

Material	Al	V	Mo	Si	Zr	Fe	Ti
TC11	5.8-7.0	-	2.8-3.8	0.2-0.35	0.8-2.0	<0.25	Balance
TC4	5.5-6.8	3.5-4.5	-	-	-	<0.3	Balance

7.3.1 Laser Precision Polishing of Metallic Surfaces

It is reported that, high-finish Ti alloy surface can be obtained by laser polishing on the macro-level [39–41]. In the current work, LMD TC11 and SLM TC4 alloy specimens were polished using a nanosecond-pulsed fiber laser.

Table 7.1 exhibits the chemical compositions of TC11. Specimens with the length of 50 mm, width of 20 mm, and thickness of 10 mm were polished under the laser power density of 1.2×10^7 W/cm^2, scanning speed of 200 mm/s, and overlapping ratio of 50 percent respectively. Figure 7.4a shows the photographs of laser polished and as-received LMD TC11 surfaces. It can be observed that a reflective surface was successfully acquired after laser polishing due to the re-scanning and remelting of solidified layers, in which an extremely shallow molten pool is produced in a very short time interval. Figure 7.4b shows the SEM morphology of the as-received and laser-polished surfaces. Liquid spheroidizing and powder bonding exhibit on the initial surface and leading to the undesirable surface quality. As shown in Figure 7.4c, surface profiles before and after laser polishing were measured and only slight laser scanning track can be observed after laser polishing. Furthermore, the peak-to-valley height of TC11 surface reduced from 80 to 4.5 μm, and the average roughness decreased from over R_a 7.21 μm to less than 0.73 μm as exhibited in Figure 7.4d and e.

The chemical compositions of TC4 are also shown in Table 7.1. SLM TC4 alloy was annealed at 1003 K for 2 h and slow cooled in a vacuum furnace to 803 K, then a nanosecond fiber laser was used to polish the specimens. Figure 7.5 shows the photograph and surface topography of the laser-polished SLM TC4. Because of the staircase effect, bonding powder, and molten pool splashing in SLM process [42], the as-received surface exhibit a high roughness as shown in Figure 7.5b. A smoother surface was obtained with a significant surface roughness reducing (from R_a 6.53 to 0.32 μm) after laser polishing, as shown in Figure 7.5c and d.

Surface roughness alteration during the laser polishing process is mainly caused by the remelting of material surface under high energy absorption. The molten materials flow into the valleys driven by gravity and surface tension, and much smoother surface is obtained after a rapid cooling process [40, 43, 44].

7.3.2 Microstructure and Properties of Laser-polished Metallic Surfaces

The microstructure evolution before and after laser polishing LMD TC11 was carefully studied in the laser irradiated zone. To calculate the temperature effect in the polishing process, a commercial finite element solver ABAQUS was used to solve

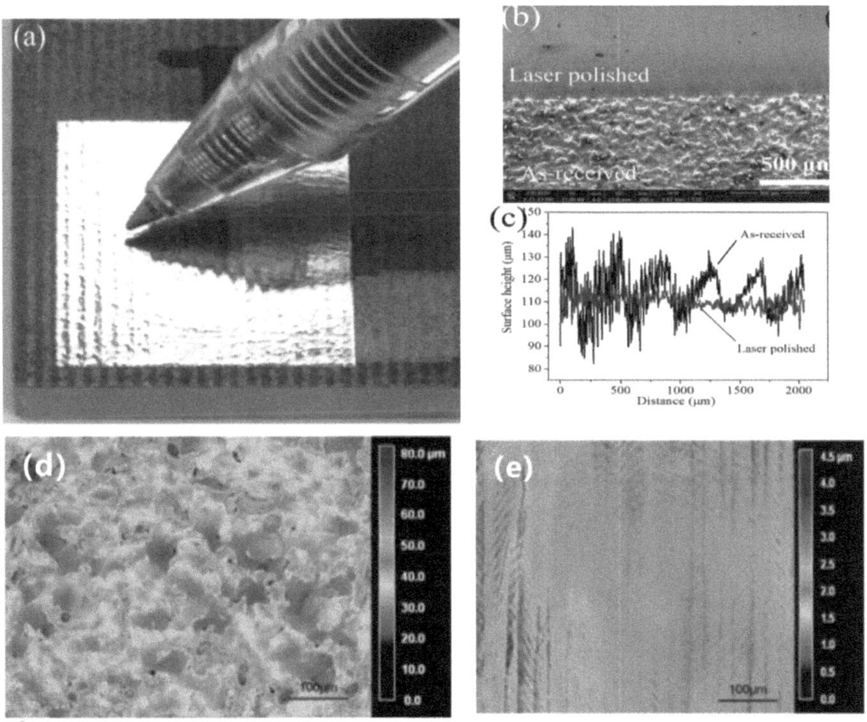

FIGURE 7.4 Effects of laser polishing on LMD TC11: (a) photographs of the laser irradiation zone and the as-received surface; (b) SEM morphology of micro-structure alteration; (c) the profiles of the as-received and optimized laser-polished regions; (d) LSCM topographic image of as-received surface; and (e) LSCM topographic image of laser polished surface [28, 29].

the thermal field and thermal cycle in TC11 plate model (4 × 2 × 1.5 mm). The upper surface, with the minor gridding size of 12 × 12 × 12 μm, was defined as laser polishing zone. The physical parameters of the materials were temperature-dependent. Heat transport through the bottom walls was neglected and convective was dominant in energy transmission during the laser polishing process. Linear elements were involved for the heat equation solving. The energy distribution was defined as Gaussian heat source [45]:

$$q = \frac{2P}{\pi r^2} \times eff \times e^{-2\times\frac{\left[(x1-x0)^2+(y1-y0)^2\right]}{r^2}} \tag{7.1}$$

where q is the heat flux density; P is the laser power (set as 85 W); r is the spot radius (25 μm); *eff* is the laser absorption rate (46.5%); x_0 and y_0 are the instantaneous positions of the spot center; and x_1, y_1 are the variables of the position.

The temperature distribution of the laser irradiated zone is shown in Figure 7.6a. Base on the fact that liquid and β-transus temperature of TC11 were approximately

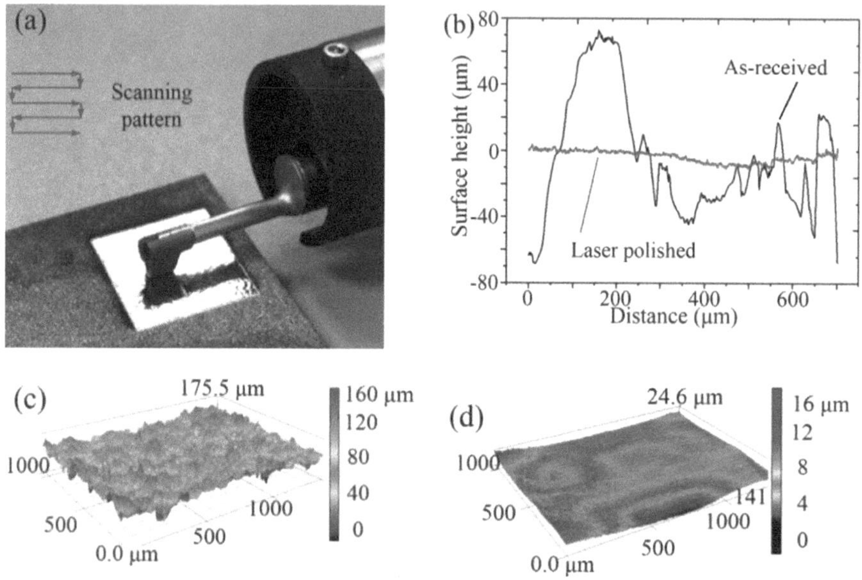

FIGURE 7.5 Effects of laser polishing on the SLM TC4: (a) the laser irradiation zone and the as-received surface; (b) the profiles of the as-received and optimized laser-polished regions; (c) LSCM topographic image of the as-received surface; and (d) LSCM topographic image of the laser-polished surface [46].

1660°C and 1000°C, respectively. Numeric simulation results indicate that the laser polishing pool forms with a diameter of about 800 μm and 250 μm in depth, and the dimensions of phase transition area were about 100 μm in diameter, 36 μm in depth, which match with the experimental results.

Figure 7.6b shows the thermal cycle of midpoint in laser polishing trace. It is clear that this position experienced three times of phase transition during the polishing process. In the first cycle, material was heated above the melting temperature, then molten metal generated β columnar crystal during solidification. Because of the much higher cooling rate (3.7×10^{5}°C/s) than the martensite transformation cooling rate (410°C/s), the primary phase (β phase) cannot be converted into the secondary phase (α phase). Therefore, bcc β phase transforms completely into metastable hpc martensite α' phase by diffusionless and shear-type transformation process.

Cross section of the laser polished LMD TC11 was examined using optical microscope and scanning electron microscope (SEM) to investigate the influence of laser irradiation on the subsurface material. It can be seen in Figure 7.6c that polished layer with the thickness of 90 μm is observed and the refined grains exhibit due to the rapid melting and solidification process. After the laser polishing process, the deposited basket-weave microstructure of LMD TC11 altered from ultrafine needle-shaped α phase (with an average width of 0.4 μm) to the thick lamellae α phase (with an average width of 2–3 μm) because of grain growth coarsening. It is seen in

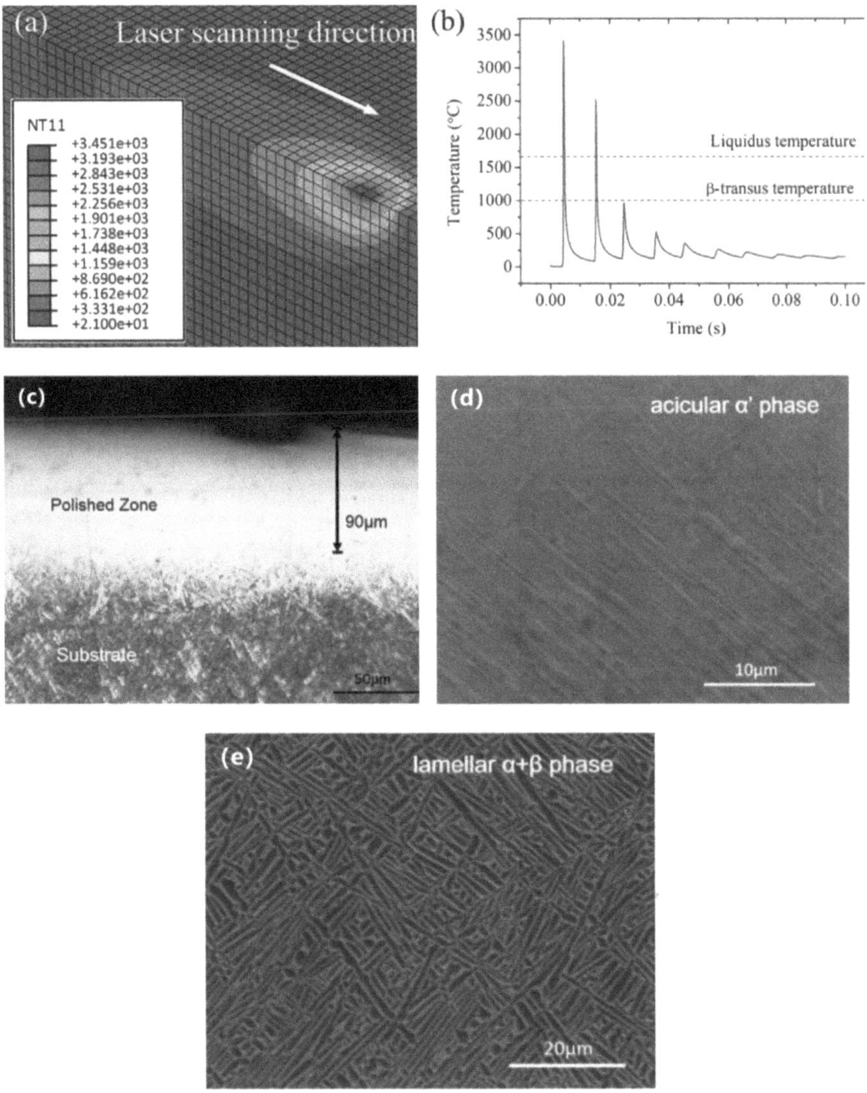

FIGURE 7.6 Numerical simulated (a) temperature distribution and (b) thermal cycle of surface temperature at a certain point; and the microstructure analysis of laser polished TC11 surface: (c) overview in cross section; (d) microstructure of laser polished zone; and (e) microstructure of substrate layer [28, 29].

Figure 7.6d and e that the LMD TC11 consists of coarsened α + β phase, while the laser polished surface consists of acicular α' martensitic phase without β phase.

The surface properties are significantly affected by the microstructure alteration due to the complex melting, vaporization, solidification, and other non-equilibrium phase transformation as well as thermal stress distribution on the surface during

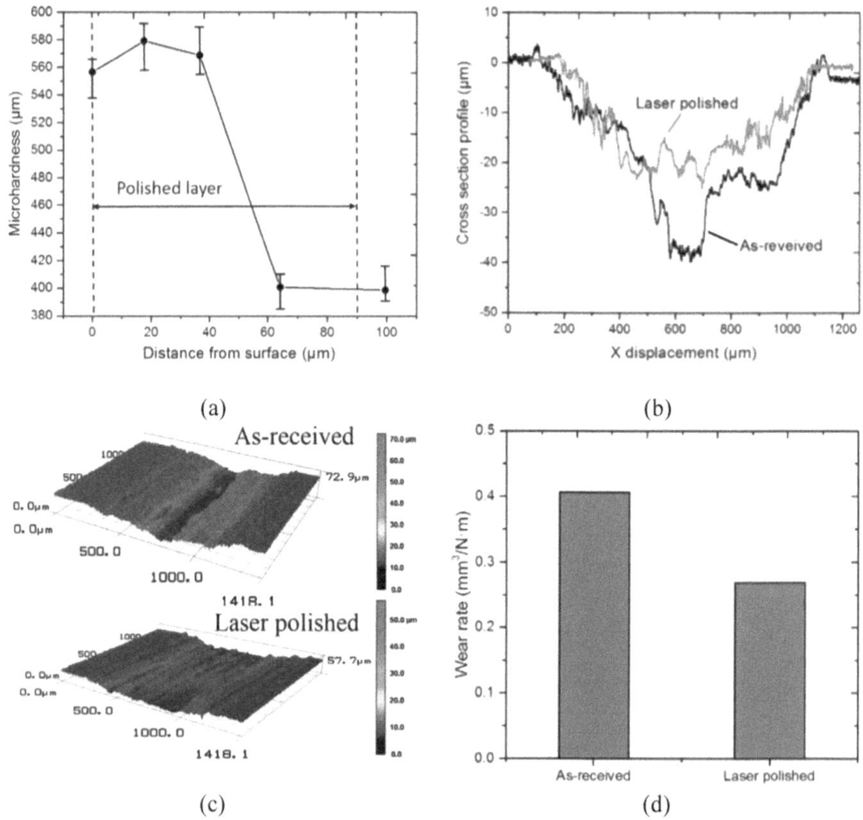

FIGURE 7.7 (a) Microhardness distributions of polished TC11 surface; (b) cross section profile of wear tracks; (c) LSCM morphology of wear tracks; and (d) wear rate of as-received and laser polished surfaces [29].

laser polishing process [29, 46]. In this work, microhardness test and dry sliding friction test were conducted to examine the improvement of mechanical properties. As shown in Figure 7.7a, the microhardness of the laser polished layer is 560 HV, increasing by 42 percent comparing with that of the initial TC11 (about 400 HV). The microhardness increasing is mainly because of the microstructure alteration, where α' martensitic has higher bulk modulus value than that of the β phase.

After dry sliding friction test, the cross-section profile of wear track before and after laser polishing was shown in Figure 7.7b. Better wear resistance property was attained after laser polishing due to the presence of α' martensitic, which matches the results of the LSCM morphology exhibited in Figure 7.7c. As exhibited in Figure 7.7d, the wear rate decreased from 0.41 mm^3/N·m to 0.27 mm^3/N·m after laser polishing, which indicate that the laser polishing is a promising surface strengthening technology.

To investigate the microstructure and properties of laser polished SLM TC4, the temperature field and thermal cycle were investigated by finite element simulation method. ABAQUS was used to calculate the temperature field within the TC4 surface during laser polishing process on a rectangle substrate with dimension of 4 × 2 × 0.5 mm, and with the minimum element size of 30 μm. The specific heat and thermal conductivity of TC4 were temperature-dependent. The heat input of CW laser spot was modeled as a Gaussian distribution with same intensity and scan strategy as aforementioned LMD TC11.

Figure 7.8a shows the numerical simulated temperature field and experimental microstructure in cross section of SLM TC4 alloy by laser polishing. It is seen that the depth of area heated above β-transus temperature was about 160 μm and the melting depth was about 60 μm. Figure 7.8a-d exhibit the microstructure evolution of laser polished TC4 surface. The melted layer with a thickness of 60 μm was observed, which fully consisted of fine acicular martensite α'. A heat-affected zone (HAZ) with thickness of 90 μm can be observed under melted layer, which is mainly composed of a coarser martensite α' than the melted layer caused by reheating and cooling by lower peak temperature than melted layer [28, 47–49]. Figure 7.8e exhibits the XRD patterns of the initial and laser polished surface. The initial surface contained α phase and tiny amounts of β phase, while the laser polished zone mostly consists of α' martensitic phase. Because of the rapid melting and cooling process in the laser polishing, TC4 alloy would undergo nonequilibrium phase transition β to α'.

Figure 7.8f and g exhibit the temperature cycle on the laser polished zone and HAZ, respectively. It can be seen that, material was heated above liquidus temperature rapidly and melted several times during the laser scanning process. When laser left, it will be quickly cooled to below 250°C in less than 0.01 s. Because of the much higher cooling rate (about 3×10^6 K/s) than the critical cooling rate (about 410 K/s), martensitic α' would formed in the laser irradiation zone.

The mechanical property of initial and laser polished SLM TC4 surfaces were investigated, and its relationship among initial roughness, laser parameters, and microstructure was experimentally established. Figure 7.8h shows the microhardness distribution on cross section of laser-polished TC4. It is seen that average microhardness in the laser polished zone is about 426 HV, which is 25 percent higher than that of as-received material (about 340HV). It is known that α' martensitic of TC4 has a hexagonal closed-packed (hcp) structure and higher bulk modulus value than α and β phase [50]. The microhardness improvement could be attributed to the formation of α' martensitic phase in laser polished zone. With the microhardness enhancement, wear resistance of polished surface was also increased. Wear rate could be calculated for quantitative comparison of wear resistance of the samples, and it is defined as:

$$W = \frac{\Delta V}{L \times D} \tag{7.2}$$

where ΔV (mm^3) is the volume of the wear trace, D (μm) is the sliding distance and L (N) is the applied load in the experiment [51]. As shown in Figure 7.8j, wear rate of TC4 surface decreased from 0.47 mm^3/N·m to 0.29 mm^3/N·m after laser polishing.

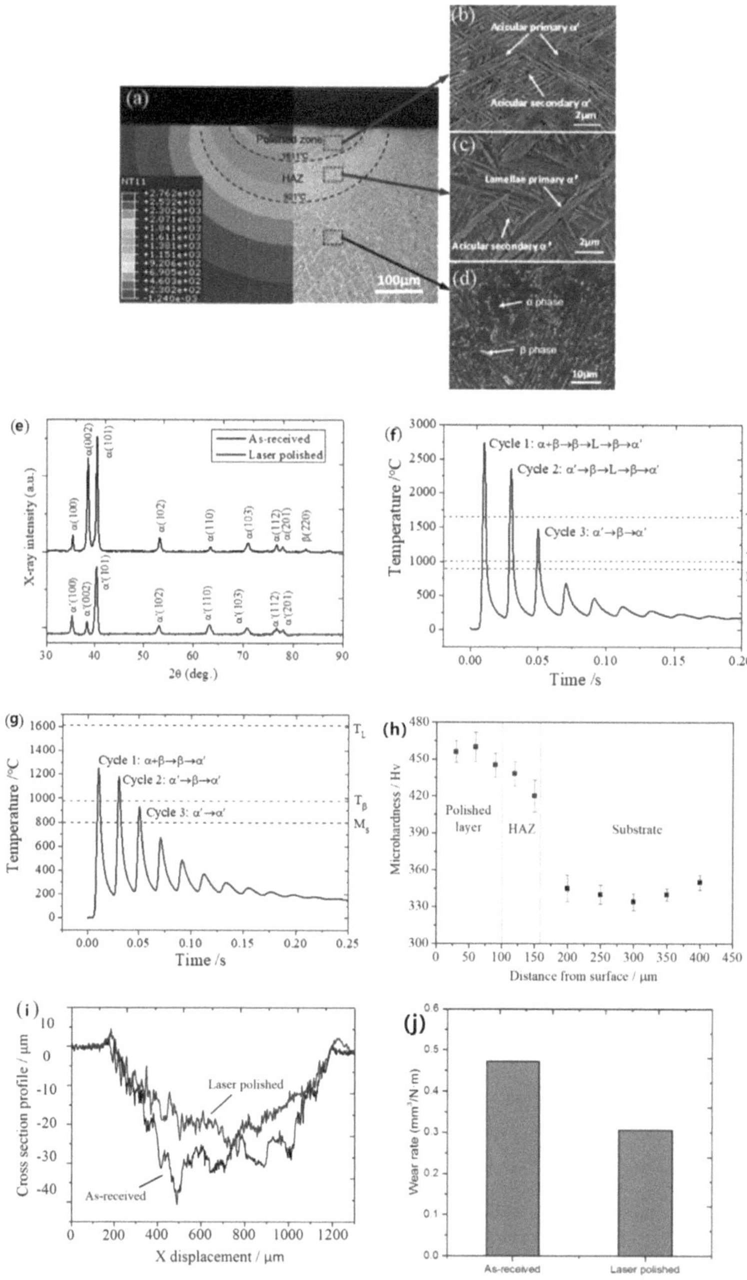

FIGURE 7.8 (a) Temperature field and experimental microstructure in cross section; and microstructure of the (b) polished zone; (c) heat-affected zone (HAZ); (d) as-received layer; phase diagram of SLM TC4 alloy by laser polishing: (e) XRD patterns for the initial and laser-polished surface; (f) thermal cycle in the polished layer; and (g) thermal cycle in HAZ; (h) microhardness distributions of the laser polished and as-received surface; (i) the cross-section profile of wear tracks and (j) wear rate of SLM TC4 surfaces before and after laser polishing [29, 46].

7.3.3 Application of Laser Precision Polishing for Transportation Industry

Infrared detector is one of the core devices of the Trace Hotbox Detection System (THDS), which is directly related to the safety of train. The laser welding line roughness of infrared detector dewar package significantly affects the life and performance. Fang et al. [52] used a commercial laser to polish the laser welding line of infrared detector. The surface roughness Ra of as-received laser welding line reduced from 0.25 to 0.03 mm after laser polishing.

7.4 ANTI-ICING AND SELF-CLEANING SURFACE FABRICATED BY LASER MICRO-PROCESSING

Surface treatments are important to enhance the surface-related properties without altering the bulk properties of the materials. Several surface modification treatments such as physicochemical, mechanical treatments, biomimetic process, vacuum processes, and high energy processes are widely used to fabricate the surface properties. Laser micro-processing is one of the precise techniques to achieve functional surfaces with micro/nanostructures on both metallic and polymer materials, owing to its high precision and non-contact process, high repeatability, pollution-free, small heat-affected zone, and controlled surface morphologies. The controlled laser-induced structures significantly alter the mechanical properties, chemical stability, optical properties and wettability of the surface [53–57].

In this section, multifunctional surfaces fabricated by laser micro-processing on both metallic and polymer materials were introduced. In particular, anti-icing and self-cleaning surfaces induced by laser micro-processing is capable of reducing the risk of traffic accidents for automobiles. The surface morphology, wettability, wear resistance, thermal stability, and potential application of induced surfaces are summarized.

7.4.1 Multifunctional Surfaces Fabricated by Laser Micro-processing

Nature provides many excellent examples of biological surface with superhydrophilic and superhydrophobic properties. Both micron and nanometer structure can alter the contact angle and therefore change the wettability of materials [58]. Xue et al. [59] used fs laser to directly prepare a variety of micro-nano structures on the surface of copper. It observed that the surface hydrophilic to hydrophobic transition was spontaneously achieved by putting them in the air without involving any chemical modification. A novel triple-scale superhydrophobic surface is reported by Pan et al. [60] using ultrafast laser fabrication. They indicated that the increased hierarchical level and structural complexity could effectively enhance the Cassie-state stability and the mechanical durability of the surface structures.

In the current work, we provide a facile one-step methodology to fabricate robust anti-icing surfaces on stainless steel using direct laser texturing technology. Commercial 302 stainless steel plates with the size of 25 mm × 25 mm × 1 mm and an average surface roughness of Ra 0.255 μm were used. A picosecond laser system (Time Bandwidth, Duetto) which produces 10.3 ps pulses at a repetition rate of 50 kHz with a central wavelength of 1064 nm was involved for the sample fabrication. The laser beam has a Gaussian profile with a TEM00 (M2 < 1.3) spatial mode and is directed to the sample surface using a galvanometric scanner with a telecentric f-theta lens. The beam was focused to a spot size of around 30 μm. Laser fluence was fixed at 5.13 J/mm^2. The optimized laser scanning speed of 10 mm/s and the scanning time of 15 were adopted.

Figure 7.9a and b exhibit the regular micro-grid structures with intersecting microgrooves of 60 μm in width fabricated on the stainless steel surface by laser scanning in two perpendicular directions. The laser ablation-induced plumes lead to the generation of numerous nano-sized particles [61, 62], and the average diameter of these particles is less than 500 nm. The initial stainless steel surface shows a contact angle of 87.2° as shown in Figure 7.9d. Immediately after laser texturing, the rough surfaces exhibit superhydrophilicity because of the intrinsic hydrophilicity of polar iron oxides and the amplification of rough solid surfaces [63, 64],

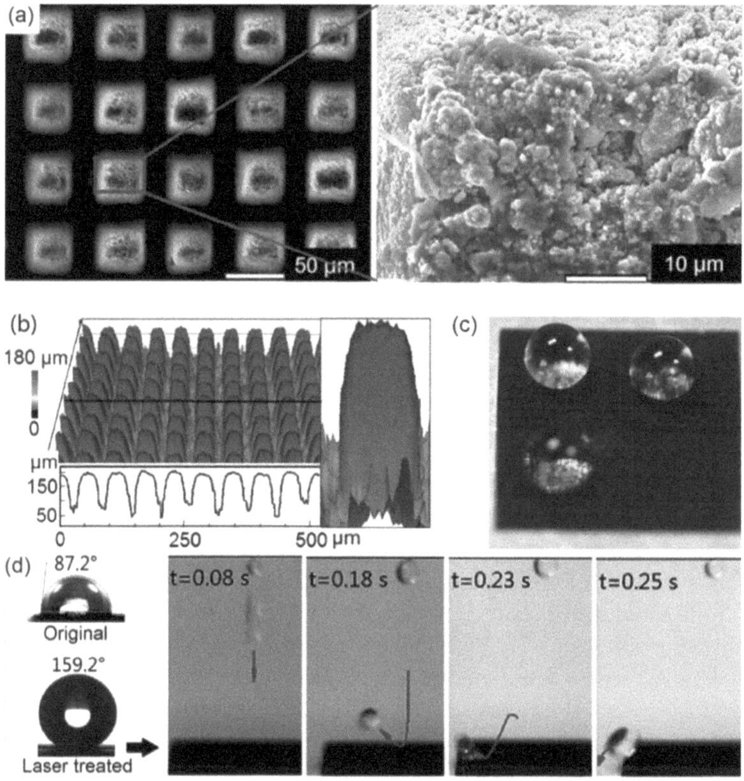

FIGURE 7.9 Obtained robust superhydrophobicity on laser-textured stainless steel surfaces: (a) Laser-fabricated regular micro-pillar structures on stainless steel with a groove pitch of 60 μm; (b) hierarchical surface morphology of the created micro-pillar structure with a groove pitch of 60 μm; (c) water droplets on the laser-textured surface; and (d) superhydrophobicity of the created hierarchical surface [61].

which has been confirmed by the rapid spreading of water droplets on surfaces. As shown in Figure 7.9c and d, after exposure to ambient air for 24 h, the textured surfaces become superhydrophobic with a contact angle of over 150° and then gradually stabilize over time. The surface wettability alteration is related to the decomposition of carbon dioxide in air [65]. In the laser texturing process, the generated active magnetite $Fe_3O_{4-\delta}$ ($0 < \delta < 1$) catalyzes the decomposition of CO_2 in air into monoxide and finally zero valence carbon, and the oxygen anions transfer into lattice vacancies of the substrate to form stoichiometric Fe_3O_4. After laser irradiation, the decomposition of CO_2 continues over time and the produced nonpolar carbons finally cover the rough surfaces. The produced nonpolar zero valence carbons accumulate on the rough surface and result in the decrease of surface energy. These hydrophobic nonpolar carbons confine a three-phase (solid-air-liquid) contact line (TCL) to prevent water from immersing the rough surfaces [66]. Together with the amplification effect of rough micro-/nanostructures, the laser-textured surfaces become superhydrophobic.

Superhydrophobic surfaces with different textures have been fabricated by controlling the groove pitch of the laser-created micro-grid structures, as shown in Figure 7.10. Figure 7.10a–c shows scanning electron microscopy (SEM) images of laser-fabricated typical micro-grid structures with groove pitches of 80, 140, and 240 μm, respectively. It can be observed that the amount of nanoparticles redeposited on the surface decreases with the increasing groove pitch. Three-dimensional (3D) surface morphologies and cross sections of the fabricated hierarchical structures were exhibited in Figure 7.10d–f. Wettability evolution with the increasing groove pitch of the fabricated hierarchical surfaces is characterized by contact angle and adhesive force measurements (Figure 7.10g). When the pitch increases to more than 160 μm, the contact angle on the surfaces shows a slight decrease from 156.3 ± 2.1° to 152.9 ± 3.0°. Figure 7.10h shows the recorded force-distance curves as a water droplet gradually approaches and retreats from superhydrophobic surfaces. The fabricated superhydrophobic surfaces show strong adhesion to a water droplet with adhesive force ranging from 41.7 ± 1.0 to 72.3 ± 8.4 μN, as shown in Figure 7.10i. The variation of surface adhesive force is mainly attributed to the changes of area fraction of the liquid-solid interface. The strong adhesion of superhydrophobic surfaces to the water droplet shows potential applications in liquid transportation and analysis of liquid samples.

7.4.2 Anti-icing and Self-cleaning Surfaces Performance

7.4.2.1 Anti-icing Surfaces Performance

Icephobicity/anti-icing property is of great significance to cold environment devices, such as aerofoils, power towers, and radars [67]. Once ice forms, these devices may fail to work or they may even be damaged. The fabrication of icephobic surfaces has aroused much interest in researchers since the 1950s [68]. Guo et al. [69] fabricate a biomimetic micro-/nanostructured surface composed of microratchets and ZnO nanohairs achieved superhydrophobic and icephobic properties with a 7 μL droplet on the surface taking about 120 min to freeze at −10°C. Zhan et al. [70] prepared superhydrophobic surfaces on polytetra-fluoroethylene by laser micro-processing,

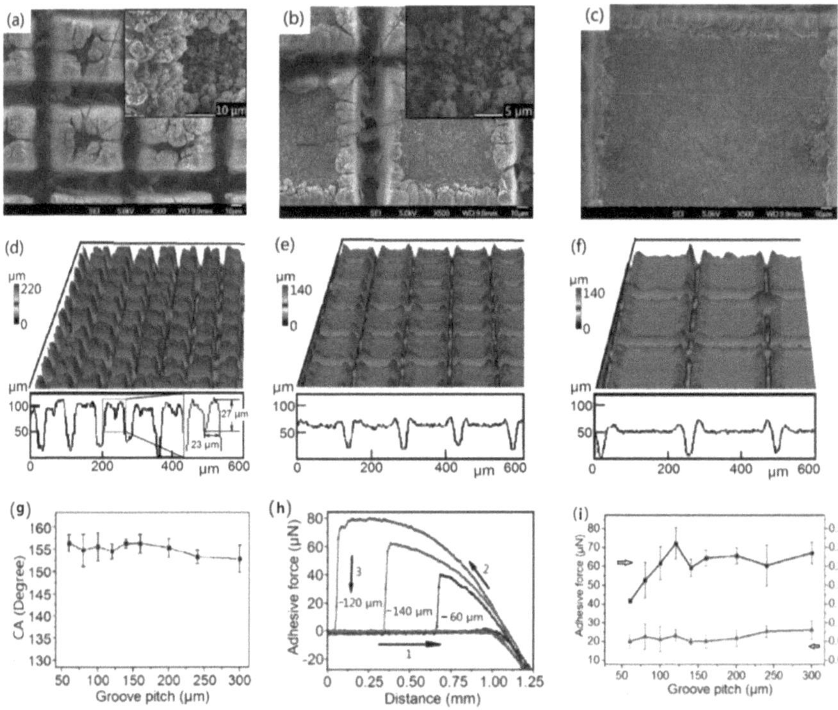

FIGURE 7.10 Surface morphologies of laser-created hierarchical micro-/nanostructures on the stainless steel surface: (a–c) SEM images of hierarchical micro-/nanostructures with groove pitches of 80, 140, and 240 µm, respectively; (d–f) confocal microscopy images of hierarchical micro-/nanostructures with a groove pitch of 80, 140, and 240 µm, respectively; and the wettability performance of laser-created hierarchical micro-/nanostructures on the stainless steel surface: (g) contact angles on superhydrophobic surfaces as a function of structural period; (h) force-distance curves recorded before and after water droplet contacting the superhydrophobic surface with the pitch of 60, 120, and 140 µm, respectively; (i) evolution of adhesive force and area fraction of the solid-liquid interface with increasing groove pitch [61].

which exhibit icephobicity to a 5 µL droplet even when the surface temperature was lower than −25°C. Wang et al. [71] carried out anti-icing tests on the fabricated superhydrophobic steel surfaces via a water-dripping process at −20°C. When a 50 µL droplet fell onto the superhydrophobic surface, the droplet rolled off easily without freezing within such a short time.

Anti-icing performance of the aforementioned superhydrophobic surfaces (section 7.4.1) has been explored in natural environment with a water droplet of 15 µL under ambient temperature of −8.5 ± 1°C and a humidity of around 30 percent. For comparison, a sample with initial surface was used as control. All the samples were completely cooled for 1 h prior to anti-icing tests. Figure 7.11a shows the anti-icing performance of the fabricated superhydrophobic surface on stainless steel in

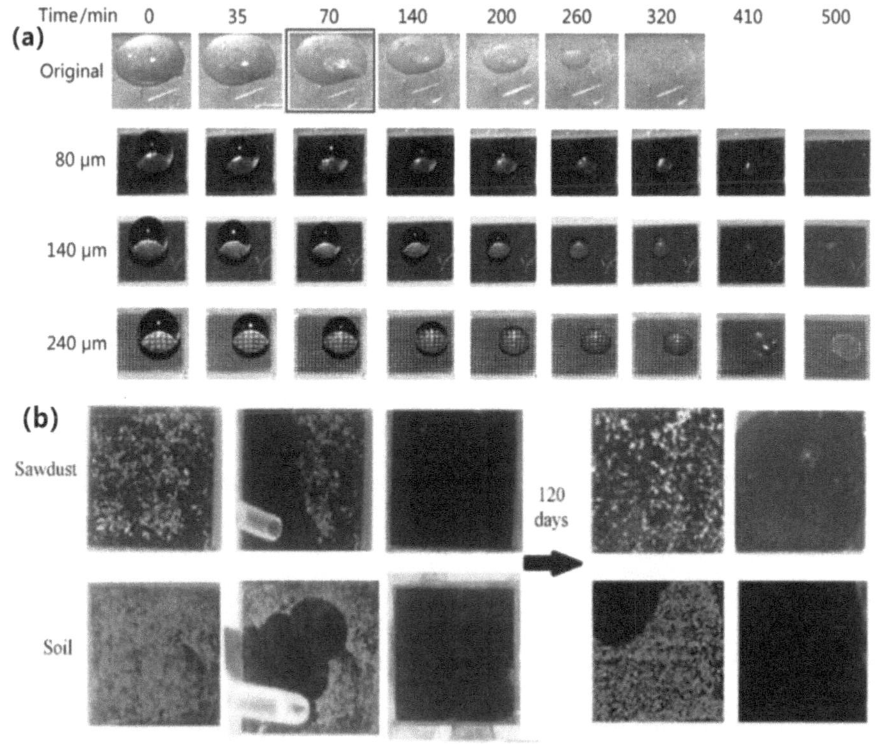

FIGURE 7.11 (a) Anti-icing performance of the fabricated superhydrophobic surface on stainless steel in natural environment at −8.5 ± 10°C; and (b) Self-clearing performance of fabricated superhydrophobic surfaces on stainless steel [61].

natural environment at −8.5 ± 10°C. We monitored evolution of water droplets on surfaces for the entire duration of experiments. At cooling state, the initial surface exhibits smaller contact angle compared to that at room temperature (25°C), whereas superhydrophobic surfaces present superhydrophobicity with the water droplet showing a transparent ellipsoid shape. After cooling for 70 min, the transparent droplet on the initial surface rapidly crystallized lost its transparency and transformed into a frozen state. After 320 min, the ice bulk on the initial surface completely disappeared, while the droplets on superhydrophobic surfaces retained transparent liquid state with ellipsoid shapes until the droplets were completely evaporated after cooling for 500 min. The fabricated superhydrophobic surfaces on stainless steel exhibit a better anti-icing property than those previously reported icephobic surfaces [69–73].

7.4.2.2 Self-cleaning Surfaces Performance

Bioinspired self-cleaning surface have drawn lots of interest in both fundamental and applied study, due to the extensive range of applications such as glasses, ceramic tiles, and solar panels. For transportation industry, self-cleaning surfaces can reduce or eliminate the need for cleaning, which lead to a great reduction in maintenance cost and the use of detergents. It is verified that superhydrophobic surface is good for self-cleaning of contaminations [74].

Zheng et al. [75] provide a hybrid method to prepare self-cleaning surface using picosecond laser pulses followed by chemical fluorinations on titanium alloy. The produced micro/nanostructured surfaces show better self-cleaning properties (with the cleanliness ratio of over 98 percent) compared with that of the as-received surface (with the cleanliness ratio of about 70%), whereas the fluorinated surfaces show slightly increasing of self-cleaning properties. Gong et al. [76] applied a picosecond laser to fabricate regular array of micro-holes on high strength steel substrate to form a lotus-leaf-like template, and superhydrophobic and self-cleaning polytetrafluoroethylene (PTFE) film surfaces were fabricated using a hot embossing process. Vanithakumari et al. [77] obtained a superhydrophobic (with average contact angle of over 160°) and self-cleaning titanium surfaces using nanosecond laser patterning.

The aforementioned superhydrophobic surfaces (section 7.4.1) also exhibit durable self-cleaning performance, as shown in Figure 7.11b. When dripping water droplets onto superhydrophobic surfaces, the soils or pencil shavings covered on the surfaces are absorbed by the droplets, and then the droplets roll away from the surfaces because of the instability of droplets on the surfaces. The contaminants on the surfaces are carried away by water droplets along the rolling trace. Even after exposure to air for 120 days, the fabricated superhydrophobic surfaces still exhibit good self-cleaning performance for both soils and pencil shavings.

7.4.2.3 Mechanical and Thermal Stability of the Anti-icing and Self-cleaning Surfaces

Robustness of the obtained anti-icing and self-cleaning surfaces is a major concern in many outdoor applications. In the current work, scratch tests were conducted to evaluate the mechanical stability and durability of the aforementioned superhydrophobic, anti-icing, and self-cleaning surfaces on stainless steel. Figure 7.12a shows the evolution of contact angle with increasing abrasion length at the applied pressure of 5.5 kPa. After abrasion for more than 2.3 m, superhydrophobic surfaces gradually become hydrophobic with contact angle less than 150°. Even after abrasion for 4.7 m, surfaces still exhibit high hydrophobicity with contact angle above 145°. These results demonstrate the excellent mechanical abrasion resistance of the fabricated anti-icing and self-cleaning surfaces. Adhesive force of superhydrophobic surfaces was also measured after abrasion tests for 4.7 m, as shown in Figure 7.12b. After abrasion tests, the rough surfaces show much higher adhesive force than those before abrasion.

As shown in Figure 7.12c–e, arrays of micropillars are retained after abrasion test, which is important for hydrophobic surfaces with high contact angles. For the fabricated hierarchical structures with small groove pitches (80 and 140 μm), the microaggregates stacked on micropillars are severely destructed and removed to adjacent grooves during abrasion tests (Figure 7.12c and d). In the case of the groove pitch of 240 μm, both microaggregates and nanoparticles covered on micropillars have been almost completely removed (Figure 7.12e), leading to the exposure of the initial steel surface.

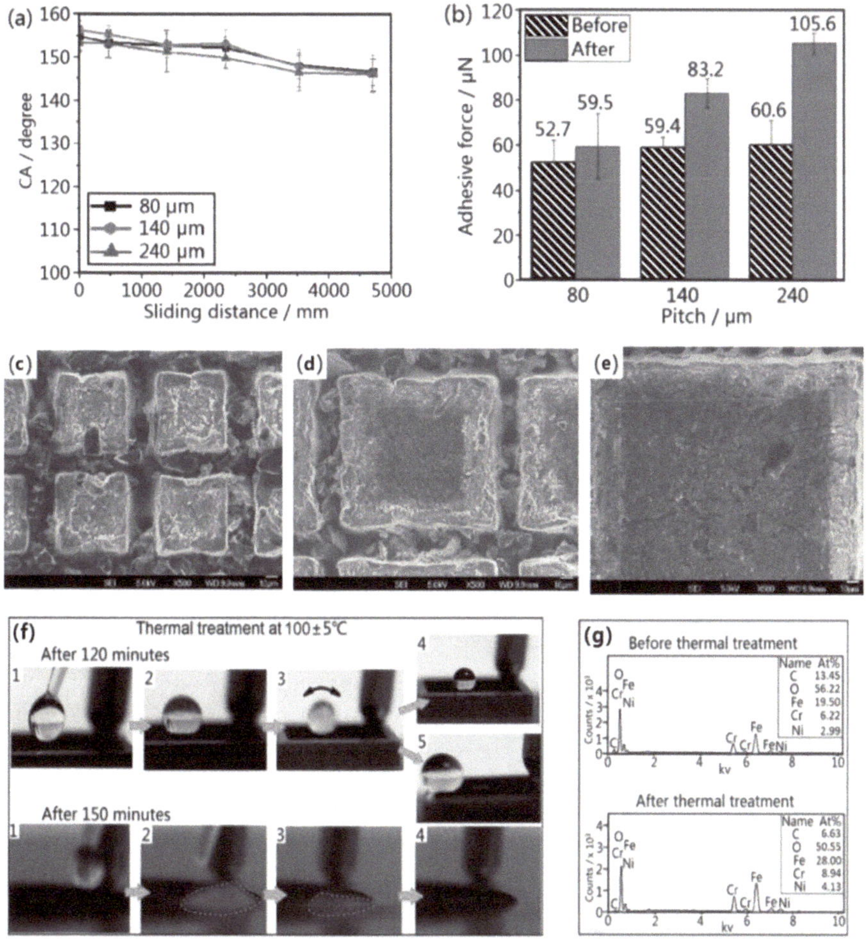

FIGURE 7.12 (a) Evolution of contact angle on superhydrophobic surfaces with increasing abrasion length at the applied pressure of 5.5 kPa; (b) adhesive force of superhydrophobic surfaces before and after abrasion for 4.7 m; SEM images of superhydrophobic surfaces after abrasion for 4.7 m: (c) pitch of 80 μm, (d) pitch of 140 μm, and (e) pitch of 240 μm; (f) thermal stability of the fabricated superhydrophobic surfaces on stainless steel; and (g) EDS results before and after thermal treatment for 150 min [61].

To elucidate the effect of thermal treatment on surface superhydrophobicity, we heated superhydrophobic surfaces at 100 ± 5°C and recorded dynamic behaviors of water droplet on the heated surfaces, as shown in Figure 7.12f. After being treated for 120 min, the superhydrophobic surfaces still retain their superhydrophobicity, showing good thermal stability. When the water droplet is set on the heated surface, it is instable and rolls away from the surface or shrinks in volume due to the high-temperature-induced bubbles at the liquid–solid interface. After being treated for more than 150 min, the fabricated surfaces lose their superhydrophobicity and become superhydrophilic. As soon as the droplet comes in contact with the heated surface, the droplet immediately spreads, evaporates, and rapidly disappears on the surface. Figure 7.12g shows energy-dispersive X-ray spectroscopy (EDS) spectra and elemental concentrations (atom %) on the fabricated surfaces before and after thermal treatment for 150 min. After treatment, carbon concentration on the surface decreases from 13.45 to 6.63 percent, and oxygen concentration decreases from 56.22 to 50.55 percent. The intrinsic hydrophilicity of the polar iron oxides is amplified by rough solid surfaces. Thus, the prepared superhydrophobic surfaces become superhydrophilic.

7.4.3 Application of Functional Surfaces for Transportation Industry

The superhydrophobic surfaces have drawn broad interest because of promising applications in the transportation industry, especially in the automobile industry. Lin et al. [78] prepared a superhydrophobic glass surface using ultrafast laser direct-writing technology. After chemical surface modification, the glass surfaces showed superhydrophobicity with a static contact angle of $161.2 \pm 0.4°$, a sliding angle of $2 \pm 1°$, and a transparency of over 92 percent in the visible light and near-infrared spectrum. The as-prepared glass shows good potential of application in the windshield and rearview mirror of automobile, which can significantly reduce the fogging behavior and therefore reduce the risk of traffic accidents.

Additionally, we fabricate anti-drag surfaces on the airfoil (NACA0015) using ultrafast laser processing. The "V" type micro-grooves were formed with the depth and width of 50 μm, and with a groove pitch of 100 μm. The result of wind tunnel test shows that, the laser-processed surfaces (with the resistance of about 1.97 N) show a drag reduction rate of 11.73 percent compared with that of the as-received surfaces (with the resistance of about 2.24 N), under the attack angle of 5°.

7.5 LASER MICRO-PROCESSING METHOD FOR JOINING DISSIMILAR MATERIALS

Weight reduction of parts is of great importance in the automotive and aerospace industries. In order to achieve lightweight components in these industries, the production of hybrid components consisting of dissimilar materials is necessary [79, 80]. This is also important in biomedical applications [81, 82]. The most common hybrid components consist of metal and plastic, as plastic offers low weight and excellent plasticity while metal maintains superior mechanical properties and can withstand high local mechanical loads. Various connection methods have been developed and applied to obtain these hybrid components, including adhesive bonding [83], mechanical fastening [84, 85], and injection molding [86]. However, conventional methods have been restricted due to low bonding strength, increased weight, or stress concentration. In comparison, the laser joining process offers more flexibility without adding weight or causing contamination to the environment [87]. A variety of laser-assisted joining methods have been developed for bonding plastic to metals.

The modification of interface properties is necessary due to the different physical and chemical properties of the joined materials. Sandblasting pretreatment has been shown to be an effective method for increasing joint strength between transparent plastic and structural steel, due to the higher laser absorption of the treated surface [88]. In recent years, the formation of micromechanical interlocking structures between dissimilar materials has been investigated by combining other technologies with the laser joining process to improve bonding strength. Significant enhancement of bonding quality has been achieved through these efforts [89–92]. For example, Klotzbach et al. created a macroscopic structure of mild steel that allowed for mechanical fastening of polymer at the metal surface, resulting in improved shear strength of the metal-polymer joint to more than 20 MPa [93]. Interlocking structures have also been generated between glass-fiber-reinforced polyamide and metallic bulk (steel and aluminum) using grid blasting, laser structuring, and thermal spray coating on metallic surfaces. These methods increased the joint strength from around 2 MPa to over 10 MPa [94]. Byskov-Nielsen et al. produced a strong interlocking structure between plastic and stainless steel using laser structuring of the metallic surface followed by injection molding of plastic into the surface, resulting in significantly increased bonding strength and high mechanical stability [95]. Therefore, controlling the surface structural morphology is essential for achieving the desired bonding strength in joints.

7.5.1 Direct Laser Joining Process of Metal and Polymer

Direct laser joining of dissimilar materials has been investigated previously to obtain lightweight characteristics while retaining excellent local plasticity and mechanical properties. Guan et al. adopted the direct laser joining method to bond Ti6Al4V and polyamide (PA66) [96]. Oblique microgrooves were designed and fabricated on the metal surface, and the laser-assisted metal and plastic (LAMP) joining process was performed using a nanosecond-pulsed laser system. The laser-textured area had dimensions of approximately 10 mm in length and 20 mm in width. The designed texture consisted of ten microgrooves with a pitch of around 1.05 μm. To fabricate oblique microgrooves, the Ti6Al4V sheet was fixed on a 3D mobile platform set at an angle of θ with the horizontal plane. Oblique microgroove arrays were fabricated by controlling the precise movement of the 3D platform. After laser surface texturing, the LAMP process was performed. The Ti6Al4V and PA66 sheets were held together by two artificial clamps with Ti6Al4V on top of PA66 and the textured Ti6Al4V surface in contact with PA66. The overlapped area was tightly clamped by tightening the bolts to provide clamping pressure through press plates. After assembly, laser irradiation was performed on the opposite side of the textured Ti6Al4V surface.

The depth of the microgrooves greatly affects the bonding strength of joints. Different microgrooves with a width of 500 μm and depths of 0 μm, 100 μm, 350 μm, and 550 μm were fabricated on the Ti6Al4V surface, as indicated in Figure 7.13. Then, the laser irradiation joining process was performed to produce Ti6Al4V-PA66 joints. All the microgrooves with different depths on the Ti6Al4V surface were fully filled by the melts during the joining process, and strong interlocking structures formed between PA66 and Ti6Al4V alloy. Compared to the joint with the as-received

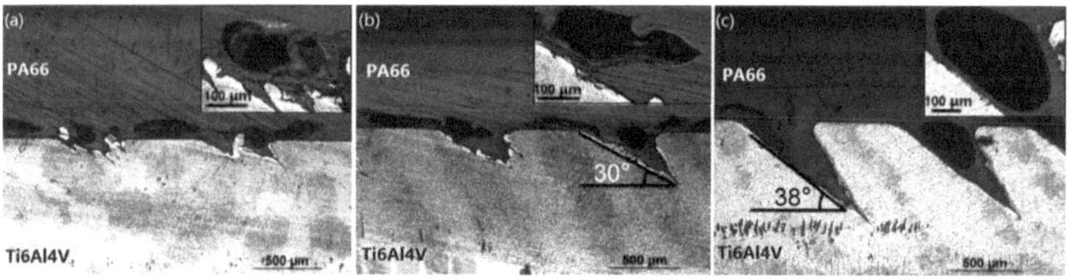

FIGURE 7.13 LAMP produced PA66-Ti6Al4V joints with a microgroove depth of 100, 350, and 550 µm [96].

Ti6Al4V surface, all the joints with laser-textured Ti6Al4V surfaces showed a significant enhancement of breaking force. The joint with the as-received Ti6Al4V surface exhibited poor breaking force of only 194 N. On the other hand, the breaking force of the joints with the laser-textured Ti6Al4V surface increased from 683 to 942 N as the microgroove depth increased from 100 to 550 µm, which is around 3.5–4.9 times the breaking force of the joint with the as-received Ti6Al4V surface. The enhancement of the breaking force of PA66-Ti6Al4V joints is mainly due to the mechanical anchor effect of the formed interlocking structure between PA66 and Ti6Al4V alloy [96].

During the joining process, laser energy is absorbed by the Ti6Al4V surface and conducts through its thickness. The PA66 material contacting the textured Ti6Al4V surface is heated by the accumulated energy near the interface and quickly melts. In this process, pyrolysis of PA66 occurs inside the melted PA66 due to the high temperature, producing a large number of gaseous products such as ammonia and carbon dioxide [97]. The gaseous products inside the melts lead to the generation of high-pressure microbubbles. Together with the clamping effect of the clamps, these high-pressure microbubbles promote the flow of melts. Thus, the created microgrooves on the Ti6Al4V surface are filled by the melted plastic, leading to the formation of the interlocking structure between PA66 and Ti6Al4V alloy after the solidification of the melts. Therefore, the bonding strength of the joints is significantly improved.

7.5.2 Direct Laser Joining Process of Metal and Fiber Reinforced Polymer Composites

Fiber reinforced polymer composites are a highly promising option for lightweight construction applications, given their lightweight nature and expected thermomechanical properties. In fact, carbon fiber- and glass fiber-reinforced thermoplastics (CFRTP/GFRTP) have increasingly been used to replace metals [98–101]. However, when it comes to practical applications of CFRTP/GFRTP, there is often the challenge of achieving high-strength heterojunctions between metals and thermoplastic composites. To address this issue, direct laser joining has been identified as an effective connection mode. The formation of micro-pores inside plastic during the laser joining process is a major concern, as it can have a negative impact on joint strength [102, 103].

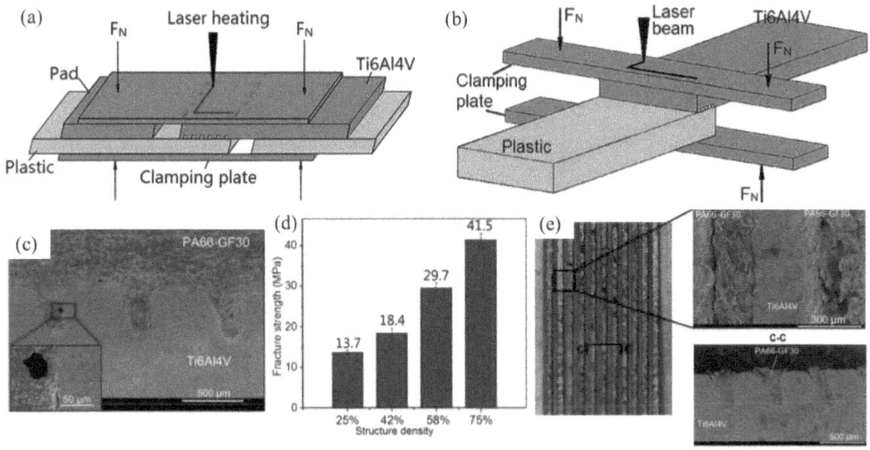

FIGURE 7.14 Schematic diagrams of laser joining process and Cross-sections and fracture strength of Ti6Al4V-PA66-GF30 joints produced by modified laser joining process: (a) conventional direct laser joining process; (b) the modified laser joining process; (c) Cross-section of the produced Ti6Al4V-PA66-GF30 joints; (d) fracture strength of Ti6Al4V-PA66-GF30 joints as a function of structure density; and (e) surface and cross morphologies of Ti6Al4V-PA66-GF30 joint after tensile shear test [82].

To eliminate micro-pores generated in LAMP-produced joints, a modified laser joining process was proposed (see Figure 7.14b) [82]. Two preset Ti6Al4V clamping plates, perpendicular to the overlapping direction, were used to provide suitable compressive pressure for the overlapping area. During the laser joining process, laser irradiates the Ti6Al4V plate surface and heat conducts from the plate surface to the Ti6Al4V-PA66-GF30 interface. The plastic near the interface melts and flows into the grooves created on the Ti6Al4V surface under the clamping pressure. Since the preset clamping force is always applied to the overlapping area, the solidification shrinkage-induced pores near the boundary of the melted zone (see Figure 7.14a) can be significantly reduced. As seen from Figure 7.14c, no large size pores were generated in the joint and only a few small pores with sizes around 30 μm were observed. Figure 7.14d shows that the fracture strength of the produced joints monotonically increased with the increasing structure density, reaching up to around 41.5 MPa as the structure density increased to 75 percent. The fracture morphology of the joints after a tensile shear test is shown in Figure 7.14e, which suggests a smooth and neat fracture interface. Therefore, as the structure density on Ti6Al4V increases, the fracture strength of the joints improves due to the increasing effective bearing area.

After undergoing the laser joining process, analysis of the cross-sectional microstructures of the Ti6Al4V-CF30/PEEK joints produced revealed that the melted plastic composites had fully infiltrated the microgrooves, as well as the gaps between micropillars (see Figure 7.15d). It is worth noting that the structure density of the Ti6Al4V surface had no significant bearing on this outcome. Therefore, despite the low flowability and high viscosity of the melted PEEK [104], there are superior

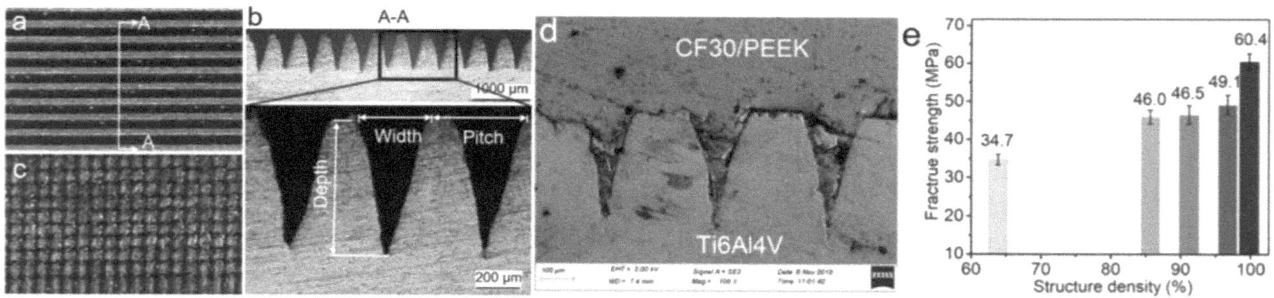

FIGURE 7.15 (a–c) Surface morphology and cross-sectional microstructure of laser-fabricated microgrooves (Sdf 63.8%), (a,b) and micropillars (Sdp 85.6%, (c) on Ti6Al4V surface. (d) Typical cross-sectional microstructures of LAJ-produced Ti6Al4V-CF30/PEEK joints (Sdf 63.8%). (e) Fracture strength of LAJ-produced Ti6Al4V-CF30/PEEK joints with increasing structure density [105].

mechanical interlocks formed between Ti6Al4V and CF30/PEEK. This is primarily due to the pre-applied clamping pressure on the joining area (see Figure 7.15b). During the laser irradiating process, when the temperature of the PEEK matrix near the interface reached its melt point, the PEEK matrix melted, causing the melted polymer matrix, along with the short carbon fibers blended in it, to be forced to flow and fill the microgrooves or gaps between micropillars under the clamping pressure. This process was accompanied by the expulsion of air among the microstructures. The robust mechanical interlocks formed near the interface significantly contribute to the joint's strength. As depicted in Figure 7.15e, the fracture strength of the joints increases from 34.7 MPa to 46.5 MPa as the structure density increases from 63.8 percent to 91.2 percent, with the highest fracture strength reaching 60.4 MPa. This is notably higher than previously reported results (<45 MPa) [104, 106–109].

7.5.3 Application of Direct Laser Joining Method in Automobile Industry

The generation of high-performance heterojunctions between high-strength composites and metals is of great significance for lightweight applications in automobile engineering field. Laser joining technology is used in the connection of composites and metals with the advantages of non-contact, high efficiency and good processing flexibility [110–112]. For example, in the FlexHyJoin project funded by the European Union's Horizon 2020 research and innovation program, the direct laser joining method was successfully employed to join the metal bracket and composite rib on the Fiat city car [113]. Simply melting the plastic onto the steel brackets would not produce a join with enough strength to meet structural requirements. Thus, a laser surface treatment was required prior to joining. The treated brackets are placed onto the composite part. Heat and pressure are then applied during joining, causing the plastic matrix from the composite to flow into the laser-induced undercuts. Thus, in order to join this thermoplastic fibre reinforced polymer composites (TP-FRPC) roof stiffener into the Panda's steel body-in-white, a set of side brackets (left and right) made from 0.7-millimeter-thick DC04 steel were laser joined onto the ends [114].

Moreover, the direct laser joining technology has been used in the bulbholder of the rearlamp of the Renault Twingo by VALEO [115]. In this field, the laser joining process is expected to be tested as alternative to the stamping + riveting process currently in use. The metal circuit is made of steel DX53 and the plastic of grey PP with 20 percent of talc. In order to achieve this, the geometry of each individual groove was studied as a function of different microstructure parameters. Only the number of iterations (Ntracks) and alignment angle (θ) affect the shape of the grooves, while the $d_{c\text{-}c}$ and their orientation (transversal or longitudinal) play no role. Based on the finding concerning the dependence of the internal cavity angle on the number of iterations, the groove geometry of the microstructures are produced at $\theta = 45°$ and Ntracks = 4 [116]. The highest shear load was achieved with this structure. Based on this microstructure, the metal circuit of the bulbholder was joined to the plastic body by means of laser joining technology.

In addition, this process can be used for many automotive parts such as stabilizer links, bumpers and doors.

7.6 SUMMARY AND PERSPECTIVE

This chapter has presented the components, performance characteristics, and design principles for scanner-stage synchronized laser processing systems. Such systems provide an alternative for achieving high-efficiency, large-area, stitchless, and highly uniform micro-processing. Notably, ensuring the synchronization of the galvanometers and stages is essential for the accuracy of this system. Moreover, laser polishing is promising for reducing surface roughness and enhancing the properties of metallic surfaces without altering or affecting bulk properties. The anti-icing and self-cleaning surface induced by laser micro-processing is provided to reduce the risk of traffic accidents for automobiles. Finally, a direct laser joining method for metal and polymer/fiber reinforced polymer composites is proposed for the increasing demands of tailor product properties and enhance the flexibility in automobile design.

It can be anticipated that in the future, the laser micro-processing techniques will continue to evolve and thrive among increasing competition from other technologies. An in-depth understanding of this scanner-stage synchronized system will aid in reducing development time for new systems, and extremely broad application prospects of the laser micro-processing techniques can be expected in the transportation industry.

REFERENCES

[1] J. Yeo et al., Next generation non-vacuum, maskless, low temperature nanoparticle ink laser digital direct metal patterning for a large area flexible electronics. *PLoS ONE* 7 (2012) e42315.

[2] Y. Li, M. Hong, Parallel laser micro/nano-processing for functional device fabrication. *Laser Photonics Rev.* 14 (2020) 1900062.

[3] S. Thibault, F-sin(theta) lens system and applications. *Curr. Dev. Lens Design Optical Eng. II* 4441 (SPIE, 2001) 98–105.

[4] W.-S. Sun, C.-L. Tien, Y.-H. Chen, P.-Y. Chu, Ultra-wide angle lens design with relative illumination analysis. *J. Eur. Opt. Soc. Rapid Publ.* 11 (2016) 16001.

[5] L. Yu, K. Yang, Z. Zhang, Galvanometer driven optical system for laser dynamic focusing. *Opt. Express* 31 (2023) 673–683.

[6] P. Penchev, S. Dimov, D. Bhaduri, S. Soo, B. Crickboom, Generic software tool for counteracting the dynamics effects of optical beam delivery systems. *Proc. Inst. Mech. Eng. Pt. B J. Eng. Manufact.* 231 (2015) 48–64.

[7] Y. Li et al., A study of correction method to the pincushion distortion based on dual galvanometer LiDAR scanning system. *Optik* 181 (2019) 555–561.

[8] P. E. Verboven, Distortion correction formulas for pre-objective dual galvanometer laser scanning. *Appl. Opt.* 27 (1988) 4172–4173.

[9] J. Ai, M. Lv, M. Jiang, J. Liu, X. Zeng, Focused laser lithographic system for efficient and cross-scale fabrication of large-area and 3D micro-patterns. *Opt. Lasers Eng.* 107 (2018) 335–341.

[10] M. A. Ortega Delgado, A. F. Lasagni, Reducing field distortion for galvanometer scanning system using a vision system. *Opt. Lasers Eng.* 86 (2016) 106–114.

[11] K.-H. Yoon, K.-H. Kim, J.-H. Lee, One-axis on-the-fly laser system development for wide-area fabrication using cell decomposition. *Int. J. Adv. Manuf. Technol.* 75 (2014) 1681–1690.
[12] S. Martínez, A. Lamikiz, I. Tabernero, E. Ukar, Laser hardening process with 2D scanning optics. *Phys. Procedia.* 39 (2012) 309–317.
[13] K. Kim, K. Yoon, J. Suh, J. Lee, Laser scanner-stage synchronization method for high-speed and wide-area fabrication. *J. Laser Micro Nanoeng.* 7 (2012) 231–235.
[14] C. Jolliffe, H. Schlüter, Z. Kirshenboim, Combined galvanometer scanners and motion platforms over standard industrial ethernet networks. *Proceedings of the 19th International Symposium on Laser Precision Microfabrication, Edinburgh, UK.* (2018) 1–6.
[15] M. Cui, L. Lu, Z. Zhang, Y. Guan, A laser scanner–stage synchronized system supporting the large-area precision polishing of additive-manufactured metallic surfaces. *Engineering* 7 (2021) 1732–1740.
[16] M. F. Heertjes, B. Temizer, M. Schneiders, Self-tuning in master–slave synchronization of high-precision stage systems. *Control Eng. Pract.* 21 (2013) 1706–1715.
[17] Y. Cao, Z. Zhang, Enhanced contour tracking: A time-varying internal model principle-based approach. Preprint at https://doi.org/10.48550/arXiv.2203.12232 (2022).
[18] L. E. Murr, E. Martinez, K. N. Amato, S. M. Gaytan, J. Hernandez, D. A. Ramirez, P. W. Shindo, F. Medina, R. B. Wicker, Fabrication of metal and alloy components by additive manufacturing: Examples of 3D materials science. *J. Mater. Res. Technol.* 1 (2012) 42–54.
[19] T. M. Mower, M. J. Long, Mechanical behavior of additive manufactured, powder-bed laser-fused materials, *Mater. Sci. Eng. A.* 651 (2016) 198–213.
[20] S. Bagehorn, J. Wehr, H. J. Maier, Application of mechanical surface finishing processes for roughness reduction and fatigue improvement of additively manufactured TC4 parts. *Int. J. Fatig.* 102 (2017) 135–142.
[21] E. Łyczkowska, P. Szymczyk, B. Dybała, E. Chlebus, Chemical polishing of scaffolds made of Tie6Ale7Nb alloy by additive manufacturing. *Archiv. Civil Mech. Eng.* 14 (2014) 586–594.
[22] Z. Baicheng, L. Xiaohua, B. Jiaming, G. Junfeng, W. Pan, S. Chen-nan, N. Muiling, Q. Guojun, W. Jun, Study of selective laser melting (SLM) Inconel 718 part surface improvement by electrochemical polishing. *Mater. Des.* 116 (2017) 531–537.
[23] V. Urlea, V. Brailovski, Electropolishing and electropolishing-related allowances for powder bed selectively laser-melted TC4 alloy components. *J. Mater. Process. Technol.* 242 (2017) 1–11.
[24] L. B. Boinovich, E. B. Modin, A. R. Sayfutdinova, K. A. Emelyanenko, A. L. Vasiliev, A. M. Emelyanenko, Combination of functional nanoengineering and nanosecond laser texturing for design of superhydrophobic aluminum alloy with exceptional mechanical and chemical properties. *ACS Nano.* 11 (2017) 10113–10123.
[25] L. Jiang, A. D. Wang, B. Li, T. H. Cui, Y. F. Lu, Electrons dynamics control by shaping femtosecond laser pulses in micro/nanofabrication: Modeling, method, measurement and application. *Light Sci. Appl.* 7 (2018) 17134.
[26] T. Deng, J. Li, Z. Zheng, Fundamental aspects and recent developments in metal surface polishing with energy beam irradiation. *Int. J. Mach. Tool Manufact.* 148 (2020) 103472.
[27] M. V. Chao Ma, N. A. Duffie, F. E. Pfefferkorn, X. Li, Melt pool flow and surface evolution during pulsed laser micro polishing of Ti6Al4V. *J. Manuf. Sci. Eng.* 135 (2013) 061023–061021–061028.

[28] Y. C. Guan, Y. H. Li, H. M. Wang, Laser polishing of additive manufactured Ti alloys and Ni alloys. Additive Manufacturing. *ELSEVIER Inc*. Chap. 10 (2019) 343–368.
[29] C. P. Ma, Y. C. Guan, W. Zhou, Laser polishing of additive manufactured Ti alloys. *Optic Laser. Eng*. 93 (2017) 171–177.
[30] S. Marimuthu, A. Triantaphyllou, M. Antar, D. Wimpenny, H. Morton, M. Beard, Laser polishing of selective laser melted components. *Int. J. Mach. Tool Manufact*. 95 (2015) 97–104.
[31] L. L. Fang Zhihao, L. Chen, Y. Guan, Laser polishing of additive manufactured superalloy. *Proc. CIRP*. 71 (2018) 150–154.
[32] S. Bagehorn, J. Wehr, H. J. Maier, Application of mechanical surface finishing processes for roughness reduction and fatigue improvement of additively manufactured TC4 parts. *Int. J. Fatig*. 102 (2017) 135–142.
[33] D. Bhaduri, P. Penchev, A. Batal, S. Dimov, S. L. Soo, S. Sten, U. Harrysson, Z. Zhang, H. Dong, Laser polishing of 3D printed mesoscale components. *Appl. Surf. Sci*. 405 (2017) 29–46.
[34] M. A. Obeidi, E. McCarthy, B. O'Connell, I. Ul Ahad, D. Brabazon, Laser polishing of additive manufactured 316L stainless steel synthesized by selective laser melting. *Materials* (2019) 12.
[35] P. M. Benoit Rosa, J. yves Hascoët, Laser polishing of additive laser manufacturing surfaces. *J. Laser Appl*. 27 (2015) S29102.
[36] B. Rosa, P. Mognol, J. Y. Hascoët, Modelling and optimization of laser polishing of additive laser manufacturing surfaces. *Rapid Prototyp. J*. 22 (2016) 956–964.
[37] D. L. B. J. A. Ramos-Grez, Reducing surface roughness of metallic freeform-fabricated parts using non-tactile finishing methods. *Int. J. Mater. Prod. Technol*. 21 (2004) 297–315.
[38] A. Lamikiz, J.A. Sánchez, L. N. López de Lacalle, J. L. Arana, Laser polishing of parts built up by selective laser sintering. *Int. J. Mach. Tool Manufact*. 47 (2007) 2040–2050.
[39] F. E. Pfefferkorn, N. A. Duffie, X. Li, M. Vadali, C. Ma, Improving surface finish in pulsed laser micro polishing using thermocapillary flow. *CIRP Annal*. 62 (2013) 203–206.
[40] L. Giorleo, E. Ceretti, C. Giardini, Ti surface laser polishing: Effect of laser path and assist gas. *Proc. CIRP*. 33 (2015) 446–451.
[41] H. L. T. Chen, Fundamental study of the bulge structure generated in laser polishing process. *Optic Laser. Eng*. 107 (2018) 54–61.
[42] Y. Wang, R. Chen, X. Cheng, Y. Zhu, J. Zhang, H. Wang, Effects of microstructure on fatigue crack propagation behavior in a bi-modal TC11 titanium alloy fabricated via laser additive manufacturing. J. *Mater. Sci. Technol*. 35 (2019) 403–408.
[43] Q. Wang, J. D. Morrow, C. Ma, N. A. Duffie, F. E. Pfefferkorn. Surface prediction model for thermocapillary regime pulsed laser micro polishing of metals. *J. Manuf. Process*. 20 (2015) 340–348.
[44] C. S. Chang, T. H. Chen, T. C. Li, S. L. Lin, S. H. Liu, J. F. Lin, Influence of laser beam fluence on surface quality, microstructure, mechanical properties, and tribological results for laser polishing of SKD61 tool steel. *J. Mater. Process. Technol*. 229 (2016) 22–35.
[45] L. Chen, Y. Yang, F. Jiang, C. Li, Experimental investigation and FEM analysis of laser cladding assisted by coupled field of electric and magnetic. *Mater. Res. Exp*. 6 (2018) 016516.

[46] Y. H. Li, B. Wang, C. P. Ma, Z. H. Fang, L. F. Chen, Y. C. Guan, S. F. Yang, Material characterization, thermal analysis, and mechanical performance of a laser-polished Ti alloy prepared by selective laser melting. *Metals*. 9 (2019) 112.
[47] J. Yang, H. Yu, J. Yin, M. Gao, Z. Wang, X. Zeng, Formation and control of martensite in TC4 alloy produced by selective laser melting. *Mater. Des*. 108 (2016) 308–318.
[48] E. Brandl, D. Greitemeier, Microstructure of additive layer manufactured Tie6Ale4V after exceptional post heat treatments. *Mater. Lett*. 81 (2012) 84–87.
[49] A. R. Nassar, E. W. Reutzel, Additive manufacturing of TC4 using a pulsed laser beam. *Metall. Mater. Trans*. 46 (2015) 2781–2789.
[50] C. N. L. K. Iyakutti, S. Anuratha, S. Mahalakshmi, Pressure-induced electronic phase transitions and superconductivity in titanium. *Int. J. Mod. Phys. B*. 23 (2009) 723–741.
[51] G. Guo, G. Tang, X. Ma, M. Sun, G. E. Ozur, Effect of high current pulsed electron beam irradiation on wear and corrosion resistance of Ti6Al4V. *Surf. Coating. Technol*. 229 (2013) 140–145.
[52] Z. H. Fang, Z. K. Fu, L. Zhang, S. J. Li, Y. J. Sun, Application exploration of laser polishing technology in package of infrared detector assembly. *Infrared*. 41 (2020) 28–32.
[53] A. H. A. Lutey, L. Gemini, L. Romoli, G. Lazzini, F. Fuso, M. Faucon, R. Kling, Towards laser-textured antibacterial surfaces. *Sci. Rep*. 8 (2018) 10112.
[54] F. Mangano, L. Chambrone, R. van Noort, C. Miller, P. Hatton, C. Mangano, Direct metal laser sintering titanium dental implants: A review of the current literature. *Int. J. Biomater*. (2014) 1–11.
[55] Q. Pan, Y. Cao, W. Xue, D. Zhu, W. Liu, Picosecond laser-textured stainless steel superhydrophobic surface with an antibacterial adhesion property. *Langmuir*. 35 (2019) 11414–11421.
[56] F. H. Rajab, Z. Liu, L. Li, Long term superhydrophobic and hybrid superhydrophobic/superhydrophilic surfaces produced by laser surface micro/nano surface structuring. *Appl. Surf. Sci*. 466 (2019) 808–821.
[57] L. A. Dobrzański, A. Drygała, Laser application in photovoltaics for surface texturization of silicon and front electrode deposition. *Mater. Perform. Charac*. 8 (2019) 20190061.
[58] Z. J. Chen, J. Yang, H. B. Liu, Y. X. Zhao, R. Pan, A short review on functionalized metallic surfaces by ultrafast laser micromachining. *Int. J. Adv. Manuf. Tech*. 119 (2022) 6919–6948.
[59] L. Xue, J. Yu, X. Ma, Z. Liu, H. Tao, J. Lin, Femtosecond laser fabricated wetting copper surfaces and their anti-icing properties. *Aeronautical Manufactur. Technol*. 61 (2018) 74–79.
[60] R. Pan, H. Zhang, M. Zhong, Triple-scale superhydrophobic surface with excellent anti-icing and icephobic performance via ultrafast laser hybrid fabrication. *ACS Appl. Mater. Interfaces*. 13 (2020) 1743–1753.
[61] H. P. Wang, M. J. He, H. Liu, Y. C. Guan, One-step fabrication of robust superhydrophobic steel surfaces with mechanical durability, thermal stability, and anti-icing function. *ACS Appl. Mater. Interfaces*. 11 (2019) 25586–25594.
[62] J. Lopez, M. Faucon, R. Devillard, Y. Zaouter, C. Honninger, E. Mottay, R. Kling, Parameters of influence in surface ablation and texturing of metals using high-power. J. *Laser Micro/Nanoeng*. 10 (2015) 1–10.
[63] Y. Tamaura, M. Tahata, Complete reduction of carbon dioxide to carbon using cation-excess magnetite. Nature. 346 (1990) 255–256.
[64] A. M. Kietzig, S. G. Hatzikiriakos, P. Englezos, Patterned superhydrophobic metallic surfaces. *Langmuir*. 25 (2009) 4821–4827.

[65] X. Wang, B. Wang, H. Xie, H. Zheng, Y. Lam, Picosecond laser micro/nano surface texturing of nickel for superhydrophobicity. *J. Phys. D: Appl. Phys.* 51 (2018) 115305.
[66] K. K. S. Lau, J. Bico, K. B. K. Teo, M. Chhowalla, G. A. J. Amaratunga, W. L. Milne, G. H. McKinley, K. K. Gleason, Superhydrophobic carbon nanotube forests. *Nano Lett.* 3 (2003) 1701–1705.
[67] M. Martínez-Calderon, M. Manso-Silván, A. Rodríguez, M. Gómez-Aranzadi, J. P. García-Ruiz, S. M. Olaizola, R. J. Martín-Palma, Surface micro- and nano-texturing of stainless steel by femtosecond laser for the control of cell migration. *Sci. Rep.* 6 (2016) 36296.
[68] C. C. Ryerson, Ice protection of offshore platforms. *Cold Reg. Sci. Technol.* 65 (2011) 97–110.
[69] L. E. Raraty, D. Tabor, The adhesion and strength properties of ice. *Proc. R. Soc. A.* 245 (1958) 184–201.
[70] P. Guo, Y. Zheng, M. Wen, C. Song, Y. Lin, L. Jiang, Icephobic/anti-icing properties of micro/nanostructured surfaces. *Adv. Mater.* 24 (2012) 2642–2648.
[71] Y. L. Zhan, M. Ruan, W. Li, H. Li, L. Y. Hu, F. M. Ma, Z. L. Yu, W. Feng, Fabrication of anisotropic PTFE superhydrophobic surfaces using laser microprocessing and their self-cleaning and antiicing behavior. *Colloids Surf. A.* 535 (2017) 8–15.
[72] N. Wang, D. Xiong, Y. Deng, Y. Shi, K. Wang, Mechanically robust superhydrophobic steel surface with anti-icing, UV-durability, and corrosion resistance properties. *ACS Appl. Mater. Interfaces.* 7 (2015) 6260–6272.
[73] M. He, J. Wang, H. Li, Y. Song, Super-hydrophobic surfaces to condensed micro-droplets at temperatures below the freezing point retard ice/frost formation. *Soft Matter.* 7 (2011) 3993–4000.
[74] L. Wang, Q. Gong, S. Zhan, L. Jiang, Y. Zheng, Robust antiicing performance of a flexible superhydrophobic surface. *Adv. Mater.* 28 (2016) 7729–7735.
[75] E. Chibowski, L. Hołysz, K. Terpilowski, M. Jurak, Investigation of super-hydrophobic effect of PMMA layers with different fillers deposited on glass support. *Colloids Surf. A. Physicochem. Eng. Aspects.* 291 (2006) 181–190.
[76] B. Zheng, G. Jiang, W. Wang, X. Mei, Fabrication of superhydrophilic or superhydrophobic self-cleaning metal surfaces using picosecond laser pulses and chemical fluorination. *Radiat. Eff. Defects Solids.* 171 (2016) 461–473.
[77] D. W. Gong, J. Y. Long, P. X. Fan, D. F. Jiang, H. J. Zhang, M. L. Zhong, Thermal stability of micro–nano structures and superhydrophobicity of polytetrafluoroethylene films formed by hot embossing via a picosecond laser ablated template. *Appl. Surf. Sci.* 331 (2015) 437–443.
[78] Y. Lin, J. Han, M. Cai, W. Liu, X. Luo, H. Zhang, M. Zhong. Durable and robust transparent superhydrophobic glass surfaces fabricated by a femtosecond laser with exceptional water repellency and thermostability. J. *Mater. Chem. A.* 175 (2018) 9049–9056.
[79] D. W. Zhang, Q. Zhang, X. G. Fan, S. D. Zhao, Review on joining process of carbon fiber-reinforced polymer and metal: Methods and joining process. *Rare. Metal. Mat. Eng.* 47 (2018) 3686–3696.
[80] J. Ahamed, M. Joosten, P. Callus, S. John, C. H. Wang, Ply-interleaving technique for joining hybrid carbon/glass fibre composite materials. *Compos. Part A Appl. Sci. Manuf.* 84 (2016) 134–146.
[81] F. Lambiase, P. Alfonso, Mechanical behavior of AA5053/polyetheretherketone (PEEK) made by friction assisted joining. *Compos. Struct.* 189 (2018) 70–78.
[82] H. P. Wang, Y. Chen, Z. Y. Guo, and Y. C. Guan, Porosity elimination in modified direct laser joining of Ti6Al4V and thermoplastics composites. *Appl. Sci.* 9 (2019) 411.

[83] C. Alía, J. M. Arenas, J. C. Suárez, R. Ocaña, J. J. Narbón, Mode II fracture energy in the adhesive bonding of dissimilar substrates: carbon fibre composite to aluminium joints. *J. Adhes. Sci. Technol.* 27 (2013) 2480–2494.
[84] X. Liu, X. J, Wang, B. S. Wang, L. L. Zhang, C. Yang, T. X. Chai, The role of mechanical connection during friction stir keyholeless spot welding joints of dissimilar materials. *Metals.* 7 (2017) 217.
[85] F. Lambiase, D.-C. Ko, Two-steps clinching of aluminum and carbon fiber reinforced polymer sheets. *Compos. Struct.* 164 (2017) 180–188.
[86] J. Gebauer, M. Fischer, A. F. Lasagni, I. Kühnert, A. Klotzbach, Laser structured surfaces for metal-plastic hybrid joined by injection molding. J. *Laser Appl.* 30 (2018) 032021.
[87] J. Kwang-Woon, Y. Kawahito, S. Katayama, Laser direct joining of CFRP to metal or engineering plastic. *Trans. JWRI.* 42 (2013) 5–8.
[88] A. Bauernhuber, T. Markovits, and J. Takács, Investigating the pulse mode laser joining of overlapped plastic and metal sheets. *Phys. Procedia.* 83 (2016) 1094–1101.
[89] A. Roesner, S. Scheik, A. Olowinsky, A. Gillner, U. Reisgen, and M. Schleser, Laser assisted joining of plastic metal hybrids. *Phys. Procedia.* 12 (2011) 373–377.
[90] S. Chanthapan, P. Wattanapornphan, C. Phongphisutthinan, Y. Kawahito, T. Suga, Effects of oxide layer on adhesion and durability of titanium and transparent polyamide joint by laser joining. *J. Laser Appl.* 30 (2018) 042005.
[91] M. A. Aripin, Z. Sajuri, S. Abdullah, M. Z. Omar, W. F. H. Zamri, W. N. M. Jamil, M.F. Abdullah, Microscale groove effect on shear strength of epoxy-bonded dissimilar metal plate. *J. Adhes. Sci. Technol.* 30 (2016) 2001–2012.
[92] Z. Zhang, J. G. Shan, and X. H Tan, Evaluation of the CFRP grafting and its influence on the laser joining CFRP to aluminum alloy. *J. Adhes. Sci. Technol.* 32 (2018) 390–406.
[93] A. Klotzbach, M. Langer, R. Pautzsch, J. Standfuß, E. Beyer, Thermal direct joining of metal to fiber reinforced thermoplastic components. *J. Laser Appl.* 29 (2017), 022421.
[94] A. Schuberth, M. Göring, T. Lindner, G. Töberling, M. Puschmann, F. Riedel, I. Scharf, K. Schreiter, S. Spange, T. Lampke, Effect of new adhesion promoter and mechanical interlocking on bonding strength in metal-polymer composites. *IOP Conf. Ser. Mater. Sci. Eng.* 118 (2016), 012041.
[95] J. Byskov-Nielsen, J. V. Boll, A. H. Holm, R. Højsholt, P. Balling, Ultra-high-strength micro-mechanical interlocking by injection molding into laser-structured surfaces. *Int. J. Adhes. Adhes.* 30 (2010), 485–488.
[96] H. P. Wang, W. H. Feng, Z. Zhang, Y. C. Guan, H. Y. Zheng, Hybrid laser technique for joining of polymer and titanium alloy. *J. Laser Appl.* 31 (2019) 042017.
[97] J.-N. Leichtnam, D. Schwartz, and R. Gadiou, The behaviour of fuel-nitrogen during fast pyrolysis of polyamide at high temperature. *J. Anal. Appl. Pyrolysis.* 55 (2000) 255–268.
[98] J. Cheon, M. Kim, Impact resistance and interlaminar shear strength enhancement of carbon fiber reinforced thermoplastic composites by introducing MWCNT-anchored carbon fiber. *Compos. Part B Eng.* 217 (2021) 108872.
[99] B. Fox, A. Subic, An industry 4.0 approach to the 3D printing of composite materials. *Engineering.* 5 (2019) 621–623.
[100] T. Hirahara, H. Hidai, Fiber implantation for interfacial joining of polymer to metal. *ACS Appl. Polym. Mater.* 2 (2020) 3049–3053.

[101] A.-S. Adham, L. Prashant, J. Bardon, P. Hirchenhahn, L. Houssiau, P. Plapper, Laser joining of titanium alloy to polyamide: Influence of process parameters on the joint strength and quality. *Int. J. Adv. Manuf. Technol.* 107 (2020) 2917–2925.

[102] S. Katayama, Y. Kawahito, Laser direct joining of metal and plastic. *Scr. Mater.* 59 (2008) 1247–1250.

[103] J. Cheon, S.-J. Na, Relation of joint strength and polymer molecular structure in laser assisted metal and polymer joining. *Sci. Technol. Weld. Join.* 19 (2014) 631–637.

[104] X. P. Gao, Z. G. Huang, H. M. Zhou, D. Q. Li, Y. Li, Y. M. Wang, Higher mechanical performances of CF/PEEK composite laminates via reducing interlayer porosity based on the affinity of functional s-PEEK. *Polym. Compos.* 40 (2019) 3749–3757X.

[105] X. H. Tian, J. Zhang, J. G. Shan, S. L. Yang, J. L. Ren, Characteristics and formation mechanism of porosities in CFRP during laser joining of CFRP and steel. *Compos. Part B Eng.* 70 (2015) 35–43.

[106] Y. J. Chen, T. M. Yue, Z. N. Guo, Laser joining of metals to plastics with ultrasonic vibration. *J. Mater. Process. Technol.* 249 (2017) 441–451.

[107] Z. J. Wu, A. Romeijn, J. Wardenier, Stress expressions of single-lap adhesive joints of dissimilar adherends. *Compos. Struct.* 38 (1997) 273–280.

[108] M. A. Elahi, M. Koch, J. Bardon, F. Addiego, P. Plapper, Failure mechanism analysis based on laser-based surface treatments for aluminum-polyamide laser joining. *J. Mater. Process. Technol.* 298 (2021) 117318.

[109] H. P. Wang, P. Yan, Y. C. Guan. Robust heterojunctions of metallic alloy and carbon fiber-reinforced composite induced by laser processing. *Materials.*14 (2021) 7469.

[110] K. Schricker, A. Drebing, M. Seibold, J. P. Bergmann, Laser-assisted joining of AISI 304 thin sheets with polymers. *Proc. CIRP.* 94 (2020) 531–536.

[111] F. Lambiase, S. Genna, Laser-assisted direct joining of AISI304 stainless steel with polycarbonate sheets: Thermal analysis, mechanical characterization, and bonds morphology. *Opt. Laser Technol.* 88 (2017) 205–214.

[112] D. J. Jung, J. Cheon, S. J. Na, Effect of surface pre-oxidation on laser assisted joining of acrylonitrile butadiene styrene (ABS) and zinc-coated steel. *Mater. Des.* 99 (2016) 1–9.

[113] Automated Joining of Hybrid Metal-Thermoplastic Composite Structures. Available online: automated-joining-of-hybrid-metal-thermoplastic-composite-structures (accessed on 24 November 2021).

[114] Flexible Production Cell for Hybrid Joining. Available online: https://cordis.europa.eu/project/id/677625

[115] Development of a Direct Laser Joining of hybrid Plastic-Metal components for industrial applications. Available online: https://cordis.europa.eu/project/id/677625

[116] E. Rodriguez-Vidal, C. Sanz, C. Soriano, J. Leunda, G. Verhaeghe, Effect of metal micro-structuring on the mechanical behavior of polymer–metal laser T-joints. *J. Mater. Process Tech.* 229 (2016) 668–677.

8 Applications of PIIID Technology on Surface Strengthening of Basic Components in Industry

Yonghao Yu and Langping Wang

8.1 INTRODUCTION

Stainless steel bearings are often used in the Chinese aerospace industry as it possesses good corrosion-resistant properties. Bearings are subjected to high-cycle contact usages, wear and fatigue damages often take place on the surfaces of these components. A failed bearing has a significant impact on a machine and the reliability of the whole system. In aerospace applications, the main mechanisms of bearing failure are surface wear and corrosion [1]. Since wear and fatigue damages often take place at the surfaces of bearings, surface coating techniques are often utilized to increase its wear and fatigue resistance [2–4]. Hence, improving the surface wear and corrosion properties by surface treatment is a good way to prolong the lifetime of bearings in the field.

Surface modification techniques, such as nitriding and carbonization, are usually utilized to increase bearings' wear and fatigue resistance [5–7]. However, such treatments must be achieved at high temperatures, the properties of the heat-treated bearing steel, such as GCr15 and AISI 440C steels [8, 9], will be softened when the temperature exceeds half of their melting points. And an extra machining process has to be carried out to reduce the dimension variation derived from the high temperature. In addition, surface modifications of industrial components, such as the balls, inner and outer rings of bearings, are very difficult to perform because of their sophisticated shapes.

Plasma immersion ion implantation and deposition (PIIID) circumvents the line-of-sight restriction and the retained dose problem characteristic of conventional ion beam implantation [10–14]. PIIID offers an effective way for surface treatment of such components. It can fabricate a uniform surface modification layer in a sophisticated-shape component, and it has undergone rapid development in recent years [14, 15]. In a PIIID process, because ion implantation and film deposition techniques, including vacuum arc deposition, magnetron sputtering, et al., can be combined together, a high adhesion layer is easy to obtain. PIIID also retains the advantages of beam-line ion implantation, that is, the ability to introduce multiple elements at concentrations

DOI: 10.1201/9781003386223-8

exceeding the solubility limits of conventional alloys [16]. It produces high adhesion strength coatings and combines excellent bonding strength between the deposited film and substrate with a complex-shape treatment. It is therefore an excellent technique to treat large and irregular components such as aerospace bearings.

8.2 APPLICATIONS OF PIIID IN PRECISION BEARINGS

We have conducted in-depth research on the surface modification of various bearing steel materials, including 2Cr13, GCr15, 9Cr18 martensitic stainless steel and M2 tool steel, which are widely used in the bearing industry, especially in the aerospace industry.

8.2.1 Adhesion Strength of TiN Films Synthesized on GCr15 Bearing Steel

To figure out how to obtain high adhesion strength between a film and its substrate in the PIIID process, TiN films were synthesized on the GCr15 bearing steel using PIIID, and the effects of the bias voltage, the ratio of implantation to deposition and the surface pretreatment on the adhesion strength were investigated.

Test coupons of GCr15 bearing steel (composition listed in Table 8.1) in the quenched-and-tempered state often used in bearing applications were studied. The samples to be implanted received a final polish to a surface roughness, Ra, of 0.04 μm, and were cleaned ultrasonically in acetone. PIIID was carried out in a multi-purpose plasma immersion ion implanter [17]. In the research, nitrogen and titanium plasmas were generated in the implanter simultaneously. The nitrogen plasma was generated by RF glow discharge, and the titanium plasma was produced by a cathodic arc plasma source. The bias pulse applied to the samples and the arc discharge pulse are synchronous. In the PIIID process, since the duration of the bias pulse is often shorter than that of the arc discharge pulse, we can obtain two treatments: (i) When the bias voltage pulse exists, ions are implanted into samples, which can be defined as the implantation process. (ii) When only the arc discharge pulse exists, ions are deposited on the samples, which can be characterized as the deposition process.

In our experiments, the base pressure in the vacuum chamber was 5×10^{-4} Pa. Parameters of the cathodic arc metal plasma source and the RF glow discharge are shown as follows: arc discharge current - 120A, duct bias - 12V, working pressure - 4.0×10^{-2} Pa, RF power - 500W, pulse repetition frequency - 30Hz. In order to study the effects of the surface pretreatment, the bias voltage, the ratio of implantation to

TABLE 8.1
Chemical Composition of the GCr15 Bearing Steel

C	Si	Mn	Cr	S	P
0.9~1.05	0.1~0.35	0.25~0.45	1.40~1.65	≤0.025	≤0.025

TABLE 8.2
Experimental Details of the Samples

Sample No.	Sputtering time (min)	N+ implantation voltage (kV)	Bias voltage (kV)	Implantation pulse duration (μs)	Cathodic arc pulse duration (μs)	Ratio of implantation to deposition	PIIID Time (hour)
A1	0	--	-20	30	330	1:10	2
A2	15	--	-20	30	330	1:10	2
A3	30	--	-20	30	330	1:10	2
B1	30	--	-10	30	330	1:10	3
B2	30	--	-20	30	330	1:10	3
B3	30	--	-30	30	330	1:10	3
C1	30	--	-26	30	330	1:10	3
C2	30	--	-26	30	630	1:20	3
C3	30	--	-26	30	930	1:30	3
C4	30	--	-26	30	1230	1:40	3
D1	30	--	-22	30	1230	1:40	3
D2	30	-8	-22	30	1230	1:40	3
D3	30	-20	-22	30	1230	1:40	3

deposition and the interface layer on the adhesion strength, experiments were divided into four groups. These variable parameters are listed in Table 8.2. In group A, the effect of sputtering time is studied. In group B, the bias voltage is studied. In group C, the cathodic arc pulse duration is studied. In group D, implantation voltage is studied.

Scratch tests were used to determine the adhesion strength between film and substrate, and the testing parameters are shown as follows: scratching speed -2 mm/min, loading rate - 50N/min.

In scratch test, critical load, Lc, can be used to calculate the adhesion strength. In order to obtain the magnitude of critical load, acoustic signal, friction curve and microscope observation were utilized. Acoustic signal produced by the delamination of the film can be used to characterize the critical load, *Lc*. However, the fracture of the film can also generate acoustic signals, but this mode is not very accurate. Therefore, the friction curve and microscope observation were used in the research. When film is delaminated, the friction coefficient between the indenter and the sample will be changed; hence the inflexion of friction curve can be used to determine the critical load. Through observing the scratch trace, the critical load can be defined as the normal load in the start point of delamination.

The tribological test results reveal that no inflexion can be found in friction curves of sample C1, C2 and all samples in group A and B, and samples C1 and C2, which means that the films can not be delaminated in the scratch process. For, sample C3, C4 and all samples in group D, inflexion can be found in their friction curves (samples C3, C4 and all samples in group D). In addition, for the samples with the inflexion in their friction curves, delamination can be found in their scratch traces.

According to above test results, scratch traces can be classified into two types: crack and delamination.

Scratch track of the crack type (sample A1) at the beginning and end is shown in Reference [18]. The scratch process of this mode can be described as follows: When the normal load applied on the indenter increases, plough-like channels appear in the scratch trace. Cracks originate from the plough-like channel and spread with the increasing load. As cracks expand gradually, cracks between two plough-like channels converge, thereby leading to chipping. No delamination can be found in their scratch traces, and none inflexion is shown in their friction curves. For the crack mode, rapid increasing of acoustic signal intensity represents the emergence of cracks. Consequently, the critical load is defined as the normal load when the crack emerges. To some extent, the critical load can be used to estimate the adhesion strength between film and substrate for the crack type.

The scratch track of the delamination type (sample C4) at the beginning and end is shown in Reference [18]. The scratch process of this mode is described as follows: When loading is applied on the indenter, the film fractures at first; then big delamination is found in the scratch traces. In addition, as shown in Figure 8.8 (a), some obvious transversal cracks and some worm-like crack spreading into the film around the track can be observed. When the load increases gradually, the film delaminates from the substrate at the margins of the scratch trace. Furthermore, the entire film delaminate the substrate when the load is large enough. At the same time, the rapid increase of the acoustic signal is also in agreement with the fracture of the film. In order to compare the critical load with the test results in crack mode, the critical load is also characterized according to the acoustic signal.

The critical loads tested by the acoustic signal are showed in Figure 8.1. As shown for samples A1–A3, the critical load of the samples treated by Ar+ sputtering is higher than that of the untreated sample, and when sputtering time increases, the critical load increases also. In addition, when the bias voltage is increased from −10kV to −30kV(samples B1-B3), the adhesion strength rises also. Especially when the voltage raises from −20kV to −30kV, the critical load increases from 2.74N to 14.6N. Furthermore, when the ratio of implantation to deposition (samples C1–C3) decreases, the critical load increases at first and then decreases. The optimism ratio should be

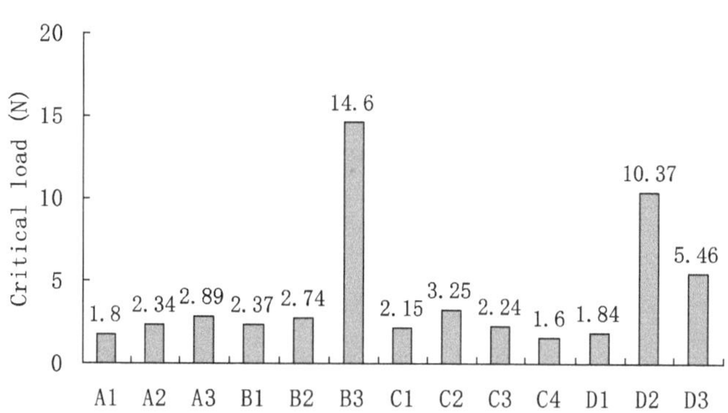

FIGURE 8.1 Critical load characterized by acoustic signal.

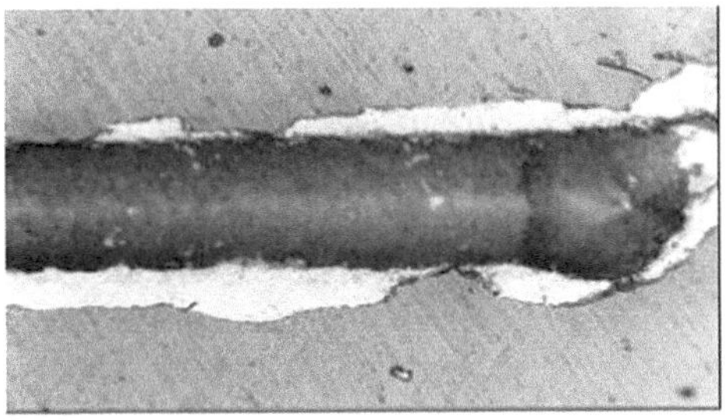

(a) treated

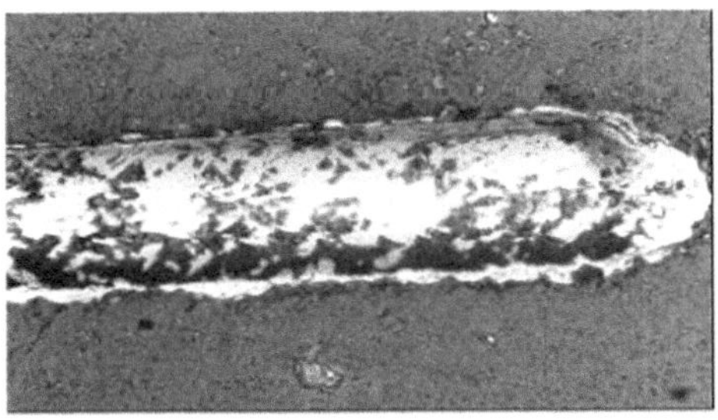

(b) untreated

FIGURE 8.2 Comparison of nitrogen pre-implantation on scratch trace.

adjusted between 1:20 and 1:30. Finally, nitrogen pre-implantation has a great influence on critical load (samples D1–D3). After −8kV nitrogen pre-implantation, the critical load increases from 1.84 to 10.37N. However, when the implantation voltage increases to −20 kV, the critical load decreases to 5.46N. This may be explained as follows. Because high implantation voltage results in a high implantation depth, the nitrogen concentration in the interface will decrease when the implantation voltage is increased, and the adhesion strength decreases also. In addition, when nitrogen pre-implantation is applied the type of scratch trace is changed. Figure 8.2 compares the scratch traces of sample D2 and D1. The scratch track of sample D2 shows a crack mode, where sample D1 exhibits a delamination mode.

Scratch trace of TiN films synthesized on GCr15 bearing steel using PIII can be divided into two types – crack and delamination mode – and the delamination is better than crack. The critical load of samples treated by Ar+ sputtering is higher than that of untreated samples and the optimism ratio of implantation to deposition should be adjusted between 1:20 and 1:30 [18]. After nitrogen pre-implantation is employed, the critical load will increase.

8.2.2 Structure and Mechanical Properties of TiAlSiN Coating on M2 Tool Steel

Hard coatings, such as TiN deposited on machining tools, have been widely used to increase the lifetime and performance of cutting [19, 20]. However, TiN begins to oxidize in air at 500°C, which limits industrial applications of high speed cutting. Some researchers incorporated high temperature element Al, Si, Cr and B into TiN to prepare nanocomposite coatings [21–24], which can remarkably improve the oxidation resistance and hardness. Si is the most frequently added element so far because of its hardness and thermal stability [25–27]. The content of Si has significant influence on the structure and hardness of composite coatings. Several theories have been developed to explain the mechanism of hardness enhancement, such as solid solution hardening [28] and grain boundary hardening by spindle phase segregation [29].

Up to now, various approaches have been used to prepare TiAlSiN coatings, compared with other methods, PIIID has the following merits: low temperature processing (T<200°C), high adhesion coating by ion implantation [18]. In our research, a PIIID facility equipped with a multiple-cathodes vacuum arc plasma source was used to fabricate TiAlSiN coatings. In addition, the structure and mechanical properties of these coatings were studied.

Si (100) wafer and polished M2 tool steel were used as substrates. Fabrication of the TiAlSiN coating was carried out in a multi-purpose plasma immersion ion implantation and deposition facility [30]. The multiple-cathodes plasma source in the facility, as shown in Reference [31], equipped with four individual cathodes, can synchronously produce composite plasma with different ions. The composite plasma was guided into the processing chamber through a straight duct. Pure Ti (99.9%), Al (99.9%) and SiAl alloy cathodes (70:30 wt.%) were used for this study. During the experiment, the vacuum chamber was evaluated to a base pressure of 5.0×10^{-3} Pa. Prior to deposition, the substrates were cleaned by Ar+ sputtering at a bias voltage of 6 kV for 30 min so as to remove the residual pollution and native oxide. To form TiAlSiN coating, nitrogen (N_2) was introduced into the chamber when the composite plasma was formed. In addition, through turning the discharge duration time of different cathodes, the composition of the composite plasma can be controlled. The TiAlSiN coatings were synthesized with the following parameters: working pressure –0.3 Pa, pulse bias voltage –20 kV, pulse repetition frequency –50 Hz, pulse bias voltage duration time –60μs, and the experimental details of the multiple-cathodes plasma source were listed in Table 8.3.

TABLE 8.3
Experimental Detail of TiAlSiN Coatings

Sample	Ti pulse duration time (ms)	Si/Al pulse duration time (ms)	Al pulse duration (ms)	Time (h)
S0	3	-	-	3
S1	3	0.5		3
S2	3	1		3
S3	3	1.5		3
S4	3	2		2.5
S5	2	2		2.5
S6	1	3		3.5
S7	3	1.5	1	2.5
S8	3	1.5	2	2.5

TABLE 8.4
Compositions of TiAlSiN Coatings Determined by EDX

Sample	Ti at. %	Si at. %	Al at. %	N at. %
S0	45.03	-	-	54.97
S1	46.07	0.64	0.46	52.83
S2	45.17	1.46	0.84	52.53
S3	45.80	2.33	0.68	51.19
S4	46.83	3.99	0.73	48.45
S5	48.25	5.33	0.89	45.53
S6	35.55	15.8	3.56	45.09
S7	47.56	2.74	2.26	47.44
S8	43.66	2.51	4.74	49.09

The composition of the as-deposited TiAlSiN coatings was determined by energy dispersive X-ray (EDX). A high-resolution field emission SEM (FEI-Quanta-200) was used to study the cross sectional microstructure of the coatings. The crystalline structure of the films was determined by X-ray diffraction (XRD, Philips-X'pert) with Cu K radiation, and the diffraction angles scanned from 20° to 90°. In addition, an X-ray photoelectron spectroscopy (XPS, Thermo, K-Alpha) was employed to detect the chemical composition of the elements in the coating. A MTS XP Nano-indentor equipped with a Berkovich diamond tip was used to measure the hardness of the coatings in a mode of continuous stiffness measurement, and six indentations were conducted in each sample.

Table 8.4 shows the composition of the TiAlSiN films analyzed by EDX. Through changing the main arc duration time of different cathodes, the Si contents can be varied from about 0.64 to 15.8 percent, the Al contents can be varied from 0.46 to 4.74 percent.

Figure 8.3 displays the XRD patterns of different TiAlSiN coatings. TiN (200) and TiN (311) peaks can be found in all patterns. In addition, with the increasing of Si and Al contents, the intensity of TiN(200) and TiN(311) peaks obviously reduce, which hints that the TiAlSiN coating tends to become a nanocrystal or amorphous structure when the content of Al or Si is increased.

The chemical shifts of Ti,Al,Si and N in TiAlSiN film determined by XPS are displayed in Reference [31]. Ti element in the coating exists as the TiN (455.2eV), Ti_2O_3 (457.4eV) and TiO_2 (458.7eV) and most of the Si element exists as Si_3N_4 (101.9eV), and some behaves as SiO_2 (103eV). TiN, AlN and Si_3N_4 can be found according to the N1s binding energy.

Figure 8.4 presents the SEM cross-section of TiN and TiAlSiN coatings on Si wafer, it can be seen that TiN coating (S0) has a columnar morphology. With the increase of Si, the coating also shows a columnar structure, but the columnar morphology becomes slightly blurry. This microstructure change may be attributed to the Si element incorporated into coatings as the bonding state of Si_3N_4, which stopped the grain growth and grain boundary mobility [29].

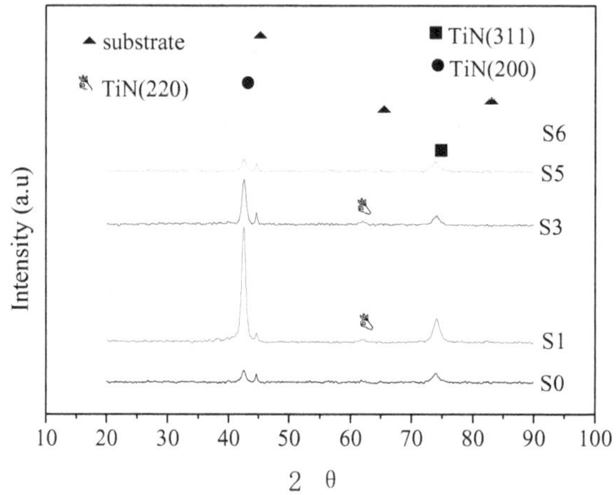

FIGURE 8.3 X-ray diffraction patterns of TiAlSiN coatings.

Hardness of TiAlSiN coatings on M2 tool steel is shown in Reference [31]. The hardness of the TiN film is approximately 15GPa. As the Si content is 0.64%, the hardness increases to 32GPa. When the Si content increases from 0.64 to 2.33%, the hardness reduces gradually to 23GPa. Furthermore, increasing the Si content from 2.33 to 5.33 percent, there is a rapid increase in hardness from 23 to 30GPa. When the Si content is 15.8 percent, the hardness decreases to 10GPa. For sample S3, S7 and S8 with the approximate Si content, when the Al content increased from 0.68 percent (S3) to 4.74 percent (S8), the hardness reduced from 24GPa to 15GPa. The hardness difference of the TiAlSiN coatings with different Si content may result from different hardening mechanisms. For sample S1 (Si-0.64%), the Si and Al atoms dissolved in the TiN bring the lattice distortion duo to different atom radius. The hardness enhancement could be resulted from a solid solution hardening effect [28]. For sample S5 (Si-5.33%), the hardness enhancement may base on a stable nanostructure by self-organization duo to the spindle phase segregation [29]. The amorphous Si_3N_4 is thin and envelops TiN grain, the interfaces between different phases can effectively hinder dislocation formation or movement and inhibit of the crack propagation. For the hardness decrease of sample S6 (Si-15.8%), one probable explanation is that the ratio of Si_3N_4 phase to the specific interface area of the nanocrystalline TiN phase exceeds a certain optimum value, which limits the phase grain separation and blocking effect of grain boundaries [26, 29].

The optical micrographs of the Rockwell indentation of sample S1 and S5 are displayed in Reference [31]. Visible peelings and spallings of the coating (S1) can be found around the indentation margin, which indicates that fracture toughness and adhesion strength of the coating on sample S1 is small. However, for sample S5, except some slight cracks around the indentation margin, few peeling and spalling can be found, which shows that the coating on Sample S5 has larger fracture toughness and adhesion strength.

TiAlSiN composite coatings with different Si and Al contents were prepared by multi-plasma immersion ion implantation and deposition technology using Ti, Al and

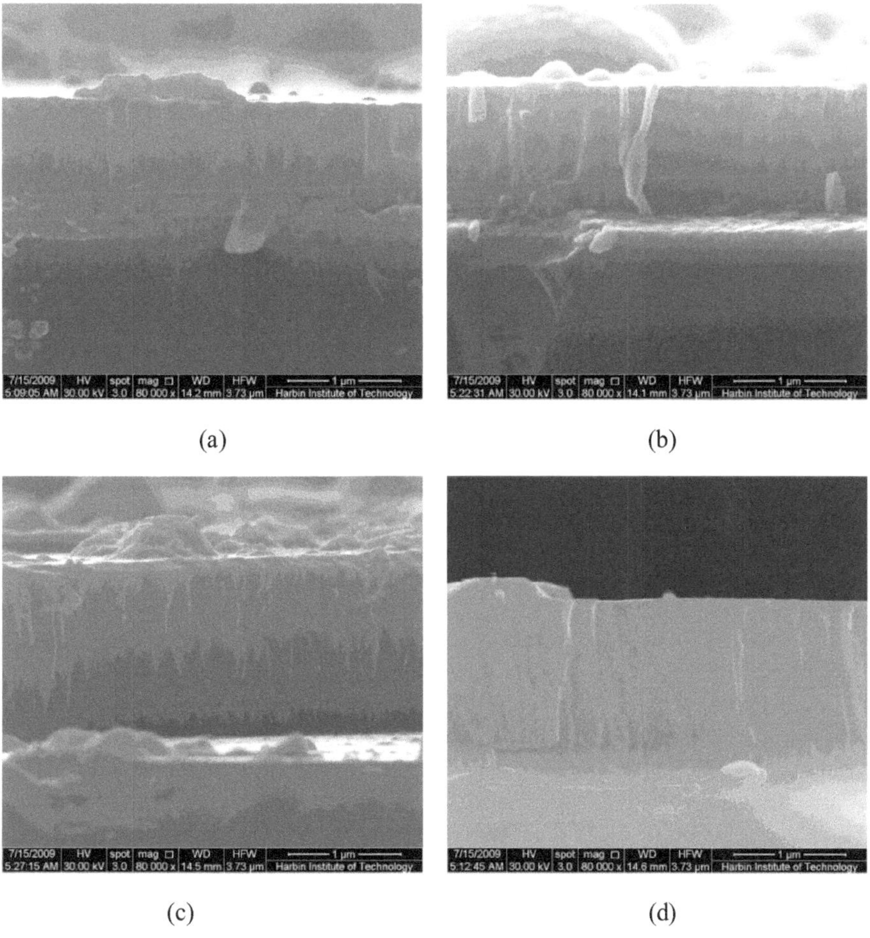

(a) S0 Si-0 at.% (b) S7 Si-2.74at.% (c) S5 Si-5.33at.% (d) S6 Si-15.8at.%

FIGURE 8.4 SEM cross-sections of TiAlSiN coatings deposited on Si substrate (a) S0 Si-0 at.%; (b) S7 Si-2.74 at.% (c) S5 Si-5.33 at.% (d) S6 Si-15.8 at.%.

SiAl (70:30 wt.%) cathodes. Through adjusting the main are duration of different working cathodes, the TiAlSiN coatings with different Si and Al compositions can be effectively controlled. The results revealed that TiAlSiN coatings have a columnar morphology and show the strong orientations of TiN (200) and TiN (311). The content of Si and Al elements has remarkable influence on mechanical properties of TiAlSiN coatings [31]. The film has the largest hardness at the Si content of 0.64%, but the fracture toughness and adhesion strength are small. When the Si content increases to 5.33 percent. the film shows a good comprehensive property with high hardness and excellent fracture toughness and adhesion strength.

8.2.3 Diamond-like Carbon and TiAlSiCN Nanocomposite Coatings

Diamond-like carbon (DLC) is another type of super-hard coating, and it has high wear resistance and an ultra low friction coefficient of about 0.1 [32–35]. Compared with graphite, DLC and TiN or TiCN nanocomposite coatings should possess better tribological properties.

We synthesized DLC-TiAlSiCN nanocomposite coating using multi-plasma immersion ion implantation and deposition method. The structural, mechanical and tribological properties of the DLC-TiAlSiCN nanocomposite coating at room and high temperature were studied.

In our experiments, Al foil and M2 tool steel (W18Cr4V) were used as substrates. The hardness and modulus of the M2 steel were about 9.41 and 210 GPa, respectively. The samples received a final polish to a surface roughness, Ra, of about 0.03 µm, and they were cleaned ultrasonically in acetone. The fabrication of the DLC-TiAlSiN coating was carried out in a multi-purpose plasma immersion ion implantation and deposition facility [30]. The multiple-cathodes plasma source in the facility, equipped with four individual cathodes, can synchronously produce composite plasma with different ions. The composite plasma was guided into the processing chamber through a duct. Pure Ti (99.9%) and SiAl alloy cathodes (70:30 wt.%) were used for this study. In the experiment, the vacuum chamber was evacuated to a base pressure of 5.0×10^{-3} Pa and before deposition the substrates were cleaned by Ar+ sputtering at a bias voltage of 6 kV for 30 min to remove residual pollution and the native oxide. To form the DLC-TiAlSiCN coating, nitrogen (N_2) and acetylene (C_2H_2) were introduced into the chamber simultaneously for composite plasma formation. The DLC-TiAlSiCN coatings were synthesized using the following parameters: a working pressure of 0.3 Pa, a pulse bias voltage of 20 kV, a pulse repetition frequency of 50 Hz, a pulse bias voltage duration time of 60 µs, a Ti cathode pulse duration time of 2ms, a Si/Al cathode pulse duration time of 2 ms, a nitrogen (N_2) mass flow of 25 sccm, and an acetylene (C_2H_2) mass flow of 25 sccm. The total deposition time was 4 h, and the thickness and surface roughness of the coating were about 1.8 µm and 32.2 nm, respectively. After the deposition, some of the samples were annealed at different temperatures for 1 h in a vacuum chamber at a pressure of 1×10^{-3} Pa.

The microstructure of the coating was investigated by transmission electron microscopy (TEM FEI-Tecnai-30) with an acceleration voltage of 300 kV. The TiAlSiN coating for TEM observation was deposited on a 20-µm-thick Al foil. The Al foil substrate was dissolved using a sodium hydroxide solution after the deposition and the coating left was used for the TEM observation. The crystalline structure of the coating was determined by X-ray diffraction (XRD; Philips-X'pert) with Cu Kα radiation and an X-ray incident angle of about 5°, while the diffraction angles were scanned from 20° to 90°. X-ray photoelectron spectroscopy (XPS, Thermo, K-Alpha) was used to detect the chemical state of each component in the coating. The energy and the X-ray incident angle were about 1000 eV and 45°, respectively. A laser Raman spectrometer with a wavelength of λ = 532 nm was used to identify the bonding structure within the as-deposited coatings. The nanohardness of the composite coating was evaluated using a nano-indentation system (Nano Indenter® G200, MTS Corp., USA). To minimize the influence of the substrate on the hardness,

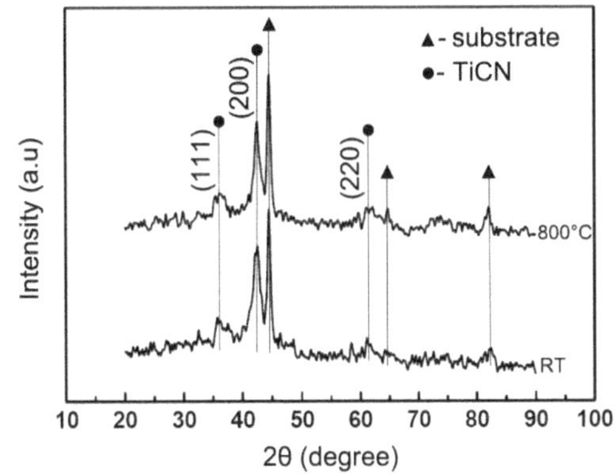

FIGURE 8.5 XRD patterns of the as-deposited and annealed DLC-TiAlSiCN coatings.

the indentation depth was kept to less than 10 percent of the coating thickness. During the ball-on-disc test, a 2 N load was applied to the sample using an Si_3N_4 ball with a diameter of 6.3 mm, and the line speed was 7.5 m/min.

Figure 8.5 shows the XRD patterns of the as-deposited and annealed samples.

For the as-deposited DLC-TiAlSiCN coating, the main reflection corresponded to a strong TiCN (200) preferred orientation and weak TiCN (111) and (220) orientations. The annealed DLC-TiAlSiCN coating at 800°C had similar peaks to the untreated coating, which indicates that the crystalline structure of the DLC-TiAlSiCN coating remained stable after the high temperature process.

Figure 8.6 shows depth profiles of Si, Ti, C, N, and Al in the DLC-TiAlSiCN coatings that were annealed at 800°C and compares the Ti_2p, Si_2p and Al_2p XPS spectra of the DLC-TiAlSiCN coatings before and after annealing. As shown in Figure 8.6 (a), the depth profiles of Ti, Si, Al, C and N were stable throughout the coating and the relative ratio of these elements did not change much, which indicates that no obvious outward diffusion occurred during the high-temperature annealing process. For the as-deposited DLC-TiAlSiCN coating, according to Figure 8.6 (b), Ti was present as Ti(CN) (454.9eV) and Ti_2O_3 (457.4eV). Figure 8.6 (c) shows that Si was present as SiC (100.5 eV), Si_3N_4 (101.9 eV) and SiO_2 (103 eV). Figure 8.6 (d) reveals that AlN (74.3 eV) and Al2O3 (75.2 eV) had formed. In addition, we found that Ti_2p, Al_2p and Si_2p, after annealing, mostly retained the chemical states of the untreated sample, which is similar to the XRD tests.

Figure 8.7 shows the C1s spectra of the as-deposited and annealed coatings. C in the as-deposited coating was present as TiC (281.9 eV), SiC (283.1 eV), C-sp2 (284.3 eV), C-sp3 (285.3 eV) and CN (286.5 eV). After annealing at 800°C, TiC, SiC, C-sp2, C-sp3 and CN were also found, which indicated that the DLC phase still remained in the coating, although the annealing temperature exceeded the critical value of graphitization. In addition, the contents of TiC, SiC, C-sp3 and C-sp2 before and after annealing were calculated and are shown in Figure 8.7. We found that the TiC and SiC content remained stable, which shows that C did not interact with Ti and Si atoms during the annealing process. However, the content of C-sp3 decreased and

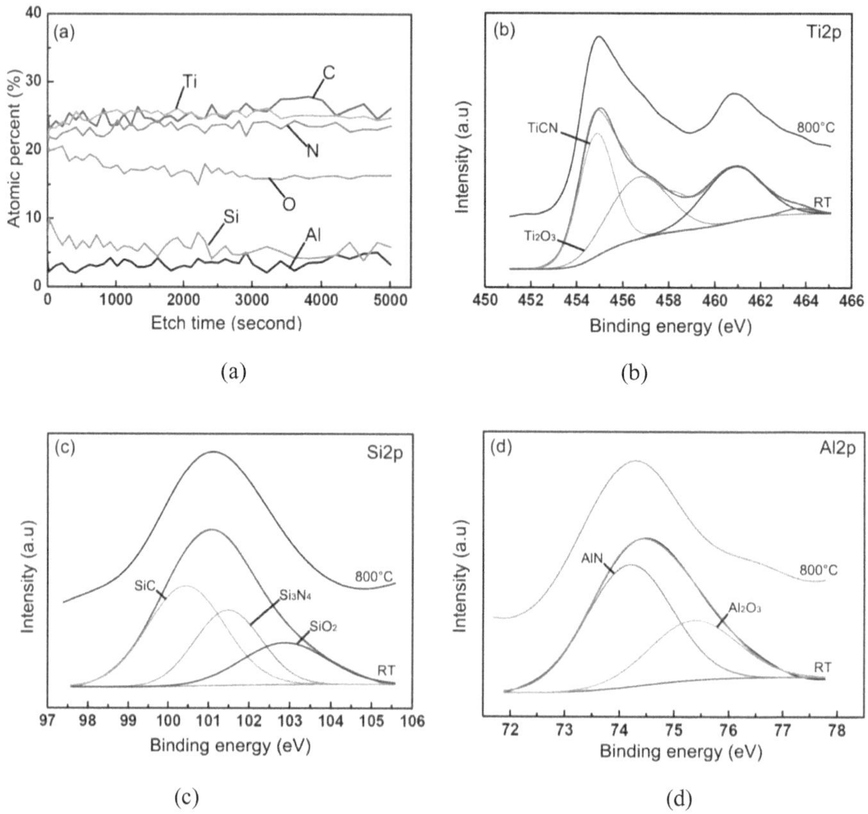

(a) depth profile; (b) Ti_2p; (c) Si_2p; (d) Al_2p

FIGURE 8.6 XPS results for the DLC-TiAlSiN coating (a) depth profile; (b) Ti_2p; (c) Si_2p; (d) Al_2p.

the content of C-sp2 increased after the annealing process, which shows that a graphitization conversion from C-sp3 to C-sp2 occurred.

Evolved Raman spectra of the samples that were annealed at different temperatures are shown in Reference [36]. All the samples gave a broad asymmetric Raman scattering band from 1100–1700 cm^{-1}, indicating the formation of a typical DLC structure. The Raman spectra were fitted by two Gaussian peaks, the G peak around 1550 cm^{-1} and the D peak around 1350 cm^{-1} [37]. According to Figure 8.8 (a)–(c), no obvious differences in the Raman spectra were present when the annealing temperature was increased from room temperature (RT) to 400°C. However, as shown in Figure 8.8 (d) and (e), the Raman spectra changed obviously as the annealing temperature was increased to 600 or 800°C. In addition, as shown in Figure 8.8 (f), the ID/IG ratio increased slowly from 0.89 to 1.2 when the annealing temperature

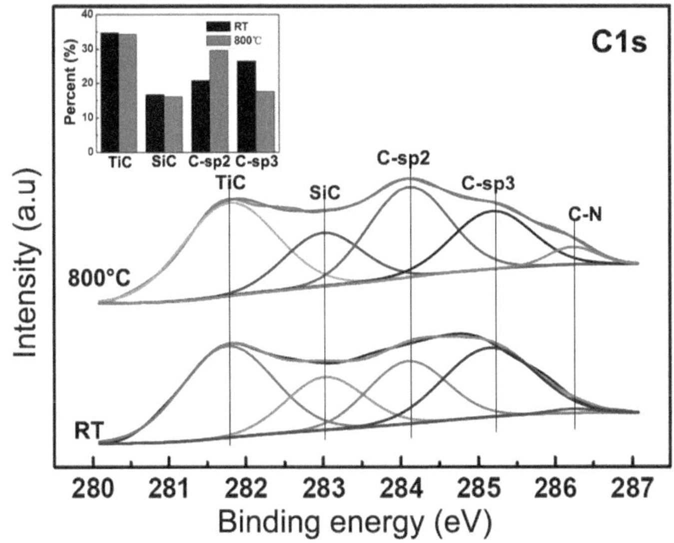

FIGURE 8.7 C1s spectra of the as-deposited and annealed DLC-TiAlSiCN coatings.

was increased from RT to 400°C and rapidly from 1.2 to 2.33 when the annealing temperature was increased from 400 to 800°C. The rapid increase in ID/IG further shows that serious graphitization occurred when the annealing temperature exceeded 400°C [37].

Raman shifts and the full width at half maximum (FWHM) of the G-peak for the coatings annealed at different temperatures is shown in Reference [36]. It is clear that the G-peak shifted from 1458 to 1578 cm^{-1} and the FWHM decreased from 190 to 98 cm^{-1} as the annealing temperature was increased from RT to 800°C, which also proves that the degree of graphitization increases with the annealing temperature [38, 39].

Figure 8.8 shows plain view TEM images of the as-deposited and annealed coatings. It is clear from Figure 8.8 (a) that the grains were randomly oriented. The corresponding selected area diffraction pattern (inserted in Figure 8.8 (a)) reveals that the DLC-TiAlSiCN coating had a polycrystalline cubic structure, and the TiCN (111), (200) and (220) reflections were identified. In addition, the high resolution transmission electron microscopy (HRTEM) images shown in Figure 8.8 (b) reveal that the as-deposited DLC-TiAlSiCN coating had a typical nanocomposite structure consisting of nanocrystals and amorphous phases. According to XRD, XPS and Raman results, the nanocrystal phase consists of TiCN grains, and the amorphous phase is composed of a-CN, a-Si_3N_4, a-SiC and DLC. The annealed DLC-TiAlSiCN coating (800°C) also had a polycrystalline cubic structure consisting of the TiCN (111), (200), (220), (311) and (400) phases according to the corresponding selected area diffraction pattern inserted in Figure 8.8 (c). According to XRD, XPS and Raman

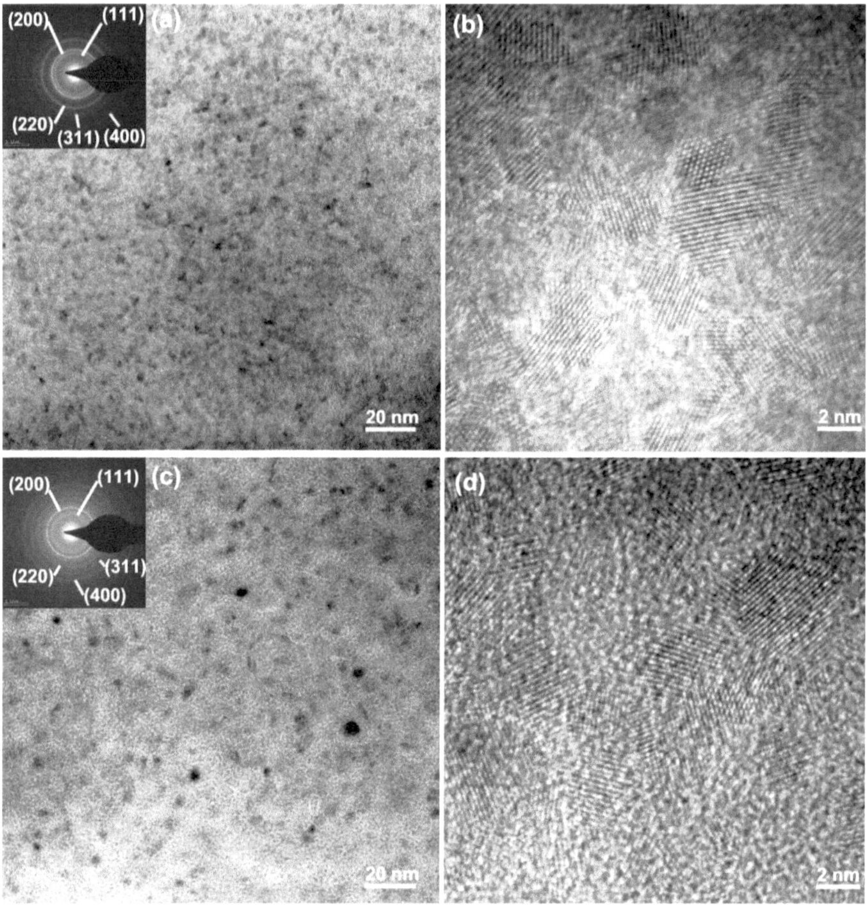

FIGURE 8.8 TEM micrographs of the DLC-TiAlSiCN coatings (a) plain view and SADP of the as-deposited coating; (b) HRTEM of the as-deposited coating; (c) plain view and SADP of the annealed coating at 800°C; (d) HRTEM of the annealed coating at 800°C.

results, only the graphitization of the DLC phase occurred after the annealing, and consequently the structure did not change obviously after the annealing process. The HRTEM image in Figure 8.8 (d) further shows that the annealed coating remained similar in the nanocomposite structure.

The nanohardness of the DLC-TiAlSiCN coatings at different annealing temperatures is shown in Reference [36]. The as-deposited DLC-TiAlSiCN coating shows a super hardness of 45 GPa. When the annealing temperature was increased, the hardness of the DLC-TiAlSiCN coating decreased gradually and was 36 GPa after annealing at 800°C. Since the hardness of the DLC is proportional to its sp3 content [37, 40], the change in hardness of the DLC-TiAlSiCN coating at different annealing temperatures should be attributed to the degree of graphitization.

The friction curves of the as-deposited and annealed coatings is shown in Reference [36]. The friction coefficient of the M2 tool steel substrate is about 0.8 and the as-deposited DLC-TiAlSiCN coating has a low and stable friction coefficient of about 0.21. In addition, the coating that was annealed at 400°C and then at 800°C had a similar friction coefficient to the as-deposited coating, but the friction coefficient of the annealed sample was not very stable.

The images and cross-section profiles of the wear tracks of the samples and the Si3N4 balls after the friction tests are shown in Reference [36]. As shown in Figure 8.29 (a) and Figure 8.29 (b), the as-deposited DLC-TiAlSiCN coating has a smooth and narrow wear track and the corresponding wear track on the sliding ball is difficult to find. However, Figure 8.29 (c) and Figure 8.29 (d) reveals that the degree of wear for the annealed coating and the corresponding sliding ball were more serious than those of the as-deposited coating. Figure 8.29 (e) and Figure 8.29 (f) compares the cross-section profiles of the wear tracks after the friction tests. As shown in Figure 8.29 (e), the as-deposited DLC-TiAlSiCN coating has a shallow wear track and the corresponding depth of the wear track is about 0.4μm. Figure 8.29 (f) shows that the depth of the wear scars is 2.3μm. It is clear that the as-deposited coating possesses better wear resistance than the annealed coating. Since only the sp3 content in the coating changed after the annealing, the DLC-TiAlSiCN coating with a higher sp3 content is beneficial for wear resistance.

From above the experiments, we can draw the conclusion that vacuum annealing at 800°C has a negligible effect on the chemical states of Ti, Si and Al in DLC-TiAlSiCN nanocomposite coatings. When the annealing temperature exceeds 400°C, vacuum annealing induces graphitization of the DLC in the DLC-TiAlSiCN nanocomposite coating, and the graphitization process evolves quickly. The hardness of the DLC-TiAlSiCN nanocomposite coating decreases gradually with an increase in the annealing temperature owing to this graphitization. The annealed DLC-TiAlSiCN coating retains a small friction coefficient after annealing at 800°C, but its wear resistance decreases after the annealing process.

8.3 FOURTH GENERATION PIIID FACILITY FOR HYBRID SURFACE MODIFICATION LAYER FABRICATION

Hybrid PIIID is a new-developed and useful technique for surface modification of soft substrates, such as aluminum and titanium alloys, because this method can synthesize a gradient layer on such substrate, which shows a larger load bearing capacity [41, 42]. In addition, current researches have proved that PIIID can improve the life time of many industrial components. However, since common PIIID facilities were designed for research, the efficiency of the PIIID batch treating process of industrial components is very low.

To take full advantage of this technique, we designed and built 3 multipurpose plasma immersion ion implanters several years ago. However, these facilities can not meet the requirement of the hybrid PIIID process and PIIID batch treatment of industrial components. In light of these requirements, we have designed and constructed a fourth-generation plasma immersion ion implantation and deposition facility. In this article, we describe the design principles, important features of the instrument, and application examples in the industrial components.

(a) (b)

1) sputtering target; 2) pulsed cathodic arc plasma source; 3) revolution-rotation sample-holder

(a) Schematic (b) Photograph

FIGURE 8.9 Schematic and photograph of the fourth-generation PIIID facility.

The fourth generation machine was built based on the knowledge gained from our design and construction of our first to third generation machines, and our research and development and commercial work in the past several years. In order to realize the hybrid PIIID treatment, the functions of the fourth-generation facility includes gaseous and metallic ion implantation, metal and its compound film deposition as well as the Diamond-like carbon film fabrication. In addition, In order to treat industrial components in batch, large area gas and metallic plasma and sample manipulation under high voltage are necessary characteristics.

To meet the requirements of the PIIID hybrid technique, the main parts of the fourth-generation PIIID facility includes: the vacuum chamber and pumps, one gaseous plasma source, two pulsed cathodic arc plasma sources, one magnetron sputtering system, one revolution-rotation sample-holder and the high voltage power source. As shown in Figure 8.9 (a), four ports locate uniformly on the wall of the vacuum chamber, and the cathodic arc plasma source as well as the magnetron sputtering system can be installed on each port. In addition, the revolution-rotation sample-holder is set in the center of the vacuum chamber, and the components on the holder can adjust its position to the cathodic arc plasma source or the magnetron sputtering system through the revolution of the sample-holder. The built machine is shown in Figure 8.9 (b).

Considering many components in the fourth generation facility are similar to those described previously [43, 44], we only listed those features that are not available or not as advanced in our previous machine.

(1) In order to overcome the short lifetime disadvantage of the pulsed cathodic arc plasma used in our previous PIIID facility, two sets of multi-cathodes pulsed cathodic arc plasma sources are symmetrically placed on the side of the vacuum chamber (shown in Figure 8.9 (b)). As shown in Figure 8.10, each plasma

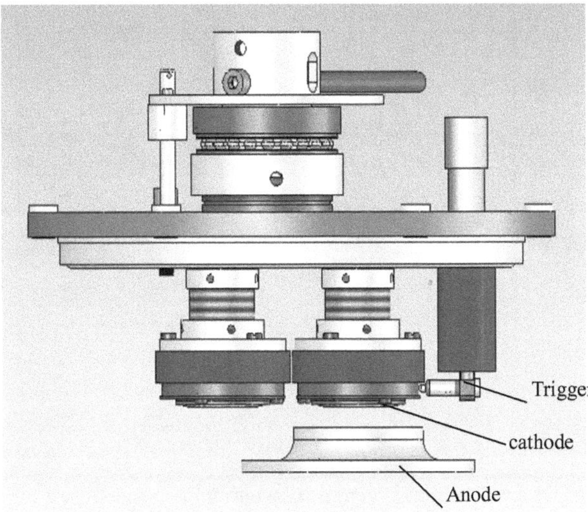

FIGURE 8.10 Drawing of the multi-cathodes plasma gun.

> source has three cathodes, and the working cathode can be changed through a handle in the outside without breaking vacuum. If same cathodes are installed in one source, the cathode working for 6 hours (its lifetime) can be substituted by another new cathode, thus a long steady working time (about 18 hours) can be realized, which is three times of that in our previous PIIID facility. If different cathodes are installed in one source, different ion implantation or deposition can be obtained through changing the working cathode. In addition, in order to acquire a large area vacuum arc plasma, the diameter of the duct is increased from previous 90 to 209 mm. Our experimental results exhibit that the diameter of the deposition area at the distance of 200mm from the exit exceeds 170 mm. The repetition frequency of the vacuum arc can be changed from 0 to 100 Hz, and the pulse duration time can be varied from 100 to 5000 μs.

In order to eliminate macro-particles, both 45 and 90°C duct are used. Figure 8.11 shows the measured titanium ion deposition current as a function of duct magnetic field current and bias voltage, for an arc current of 120 A, where the deposition current is measured on a sample 200 mm from the exit of the 45°C magnetic duct. It can be found that the optimum duct bias voltage is about 40 V, and this optimum is independent of duct magnetic field current over the range investigated. We measured a maximum titanium ion deposition current of 2.0 A for a duct magnetic field current of 3 A and duct bias voltage of 40 V. We point out that since the power dissipated in the magnetic field coil varies as the square of the coil current, and because of heat removal constraints, there is in practice always an upper limit to the duct field current that can be used. For our system a duct magnetic field current of 2.5A was chosen as a safe upper limit. However, when the 90°C duct is used and the arc discharge is 120A, the pulse ion deposition current is only 0.8A.

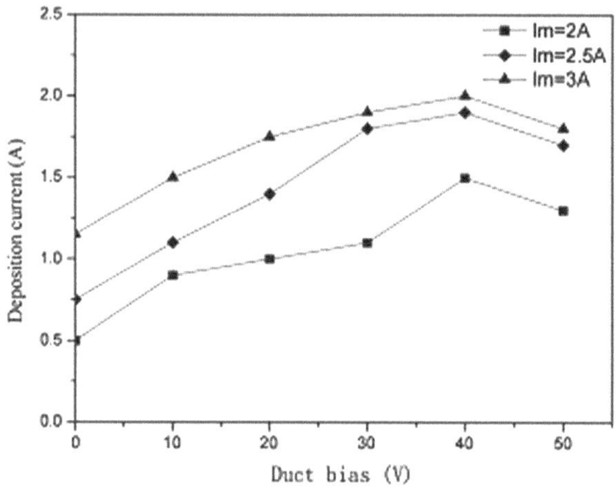

FIGURE 8.11 Measured titanium ion deposition current as a function of duct magnetic field current and duct bias voltage.

(2) In our previous PIIID facility, a rotating shaft using as the sample-holder for complicated-shape components was installed in front of one cathodic arc plasma source and driven by a motor in the outside. However, since the component on the sample-holder is fixed in the same position, a breaking vacuum process must be employed when a hybrid PIIID process is needed. In order to change the position of the sample without breaking vacuum, a complicated revolution-rotating sample-holder was designed for our latest facility and located in the center of the vacuum chamber. As shown in Reference [41], 24 shafts are set symmetrically on the sample-holder. Through an outside motor, a driven shaft and gears, the shaft can rotate at a speed of 1–5 r/min. The revolution of the sample-holder is driven by another outside motor through the center driven shaft, which can revolute at a speed of 0–1 r/min and locate at four fixed points. In addition, the sample-holder and driven shaft are insulated by ceramic tubes, thus a high voltage (up to 50 kV) can be applied on the component even if it is rotating. Through the revolution of the sample-holder, the components installed on the sample-holder can change its position conveniently. In addition, because the component with an irregular shape installed on the rotating shaft can gyrate during the PIIID process, the line-of-sight restriction of the metal ion implantation and deposition can be improved significantly.

(3) In our previous PIIID facility, a custom designed magnetron sputtering source is installed on the top cover, a breaking vacuum process must be carried out for a PIIID hybrid process. In this machine, the custom designed magnetron sputtering source is set on the side of the vacuum chamber, and it locates at the same height of the pulsed cathodic arc plasma source. When the component changes its position from the front of the pulsed cathodic arc plasma source to the magnetron sputtering target, a duplex PIIID process involving gaseous or

metal ion implantation, pulsed cathodic arc deposition as well as magnetron sputtering deposition can be realized without breaking vacuum. To avoid contaminating of the target from other sources, the magnetron sputtering target is completely shielded by a moveable stainless plate. In addition, when the magnetron sputtering system is not in use, it can be easily removed from the side flange.

(4) Two kind of pulse modulators are used in this machine. One based on the hard tube technique can generate a 5–40 kV and 1–400 Hz pulsed bias, another one based on the serial IGBTs technique can obtain a 1–4 kV and 1–10 kHz pulsed bias. The latter one is mainly used in the inner surface PIIID process [45].
(5) Similar to our third-generation machine, the main vacuum chamber is a stainless steel cylinder. However, because we found the metal plasma attenuate rapidly in the vacuum chamber, in order to decrease the travel distance of the metal plasma, the diameter and height of the vacuum chamber are changed from 1 to 0.7m and 1.2 to 1m, respectively. A 1200 L/s turbomolecular pump and a 15 L/s rotary vane pump are used to evacuate the vacuum chamber. The vacuum pumps can evacuated the vacuum chamber to a base pressure of 5.0×10^{-3} Pa in 40 minutes, and the measured ultimate vacuum is $<5.0\times10^{-4}$ Pa.
(6) A custom designed radio-frequency (RF) capacitive glow discharge system is installed in the vacuum chamber. The power of this system can be changed from 100 to 1500 W. In addition, a gas feeding system is employed to transport different gases into the RF glow discharge system. There are two mass flow meters in this system, and adjusting the mass flow meter can change the pressure of each individual gas in the chamber. Up to 4 different gases can be fed into the chamber and 2 gases can be premixed before bleeding into the chamber.

Comparing with the PIII facilities built by other research groups [46–48], the main advantages of this facility are also listed as follows:

(1) The combination of the multi-cathodes pulsed cathodic arc plasma sources and the magnetron sputtering target provides versatile methods for ion implantation and deposition.
(2) The batch treating of commercial components can be obtained in this machine.

8.3.1 Gradient Layer Fabrication

Using different vacuum arc plasma sources, we synthesized a gradient layer on TC4 substrate. In our experiments, TC4 samples were placed on the sample-holder and located initially in front of the pulsed cathodic arc plasma source with the Ti cathode. At first, a TiN layer of 0.75μm was synthesized on TC4 substrate using Ti plasma and N_2. Then a 0.75μm Ti(CN) layer was deposited on TiN layer using Ti plasma and N_2 and C_2H_2. At last, the sample was revolved to the front of the pulsed cathodic arc plasma source with the carbon cathode, and 1μm Diamond like carbon (DLC)

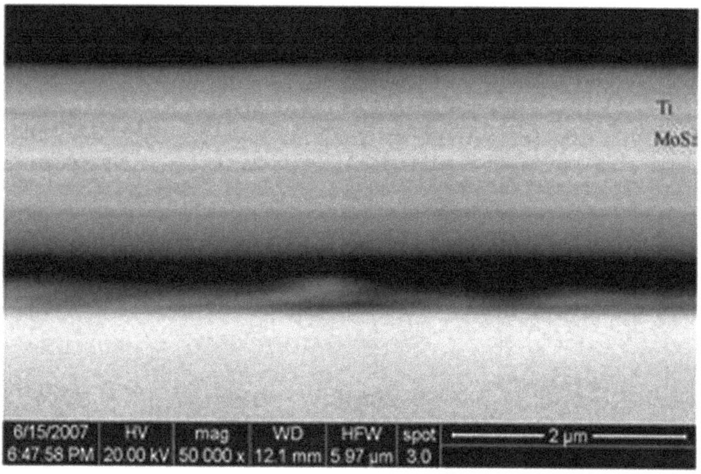

FIGURE 8.12 Cross section of the multi (MoS2/Ti) layers on 2Cr13 substrate.

film was synthesized on Ti(CN) layer. The wear-resistance of the gradient layer is improved significantly.

8.3.2 Multilayer Deposition

Using one vacuum arc plasma source and the magnetron sputtering system, we synthesized (MoS_2/Ti) multilayer on 2Cr13 substrate. In our experiment, a Ti transition layer was firstly deposited on 2Cr13 substrate using the vacuum arc plasma source, and then MoS_2 and Ti layers were deposited in turns using the magnetron sputtering and vacuum arc plasma source, respectively. Figure 8.12 shows the cross section of the MoS_2/Ti multilayer on 2Cr13 substrate, it can be found that a clear multilayer structure can be formed by this PIIID facility [31].

8.4 BATCH TREATMENT OF INDUSTRIAL COMPONENTS USING PIIID

PIIID offers many advantages for surface modification, such as low temperature treatment, none line-of-sight restriction and potential for sophisticated-shape components. However, the dose uniformity of these components is not reasonable in conventional PIIID conditions, especially in the batch treatment [14, 49–52]. In order to obtain commercial applications of the PIIID technique, we had successfully developed several novel methods to improve the dose uniformity.

The vacuum chamber used in our experiments has the radius of 0.5 m and the height of 1.2 m. Radio-frequency (RF) glow discharge was used to generate gaseous plasma, and its power was set to 600W. One pulsed cathodic arc plasma source was used to obtain metallic or carbon plasma, and the deposition rate can reach about 3 Å/s. In addition, two high-voltage power sources can be utilized in our experiments. One power source can generate a 5–40 kV pulsed bias, and its repetition frequency can be adjusted from 1 to 500 Hz. Another one can generate a 1–10 kV pulsed bias, and its frequency can reach 10 kHz.

In addition, we used a rotating target holder for the batch treatment of these industrial components. Revolution and rotating of many sophisticated-shape components can be acquired by this rotating target. Using the proper process window and rotating target holder, batch treatments of these components can be finished conveniently.

8.4.1 The Surface of a Ball-Shape Component

A particle-in-cell (PIC) model was used to simulate the PIIID process of a ball. During the simulation, the implantation voltage, pulse width, ball radius and plasma density are −20 kV, 6 μs, 4.8×109 /cm^3, and 16 mm, respectively. The simulation results reveal that the implantation dose of the ball is really un-uniform. As shown in Figure 8.13(a), the dose at the point near the holder is evidently smaller than the upper surface. In order to improve the uniformity, we also calculated the dose distribution along the ball surface when the ball is rotating. Figure 8.13 (b) shows the dose distribution along the circumference perpendicular to the rotating plane. The simulation result reveals that the rotating of the ball can improve the dose distribution significantly. In addition, along the circumference of rotation, the implantation dose is uniform obviously.

According to the simulation results, in order to improve the dose uniformity, the ball must be rotated during the treatment. However, since the PIIID process employs a high voltage pulse, in order to prevent arcing, an intimate contact between the ball and target holder must be promised. The revolving principle of a ball using the rotating target holder is shown in Reference [41], which is mainly composed of two plates: the rotated and fixed. In the fixed plate, many holes are drilled, whose radiuses are smaller than that of the balls to be treated.

In order to obtain the PIIID batch treatment of balls, the balls are put into the holes of the fixed plate. When the radius of the hole is proper, balls can drop to contact the

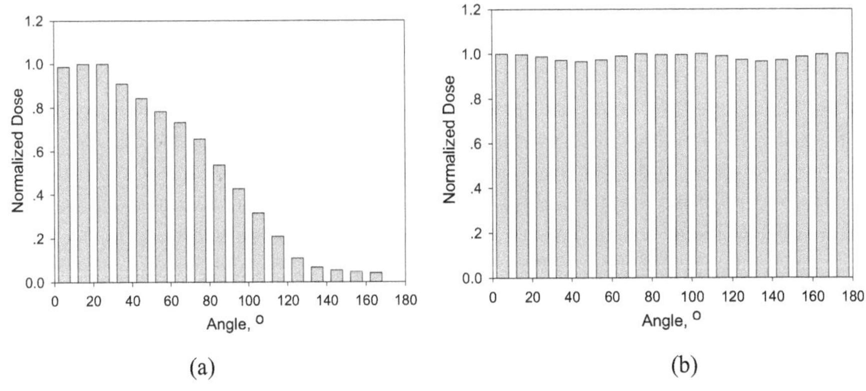

(a)Without rotation (b) with rotation

FIGURE 8.13 Dose distributions along the circumference perpendicular to the rotating plane.

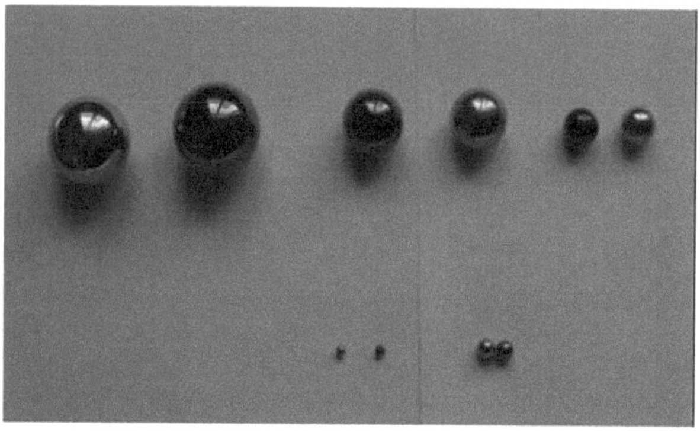

FIGURE 8.14 PIIID Batch treated balls with different diameters.

rotated plate intimately under natural gravity. When the rotated plate revolves, the ball will gyrate around its center by the friction force between the ball and the rotated plate. The gyrating speed can be controlled through adjusting the revolving speed of the rotated plate as well as the distance between the ball and the center of the rotated plate. Figure 8.14 shows the nitrogen-implanted balls with different diameters using the rotated target holder. It is found that the surface color due to nitrogen implantation in each ball is uniform.

8.4.2 The Outer Surfaces of a Cylinder-like Component

During the PIIID process of a cylinder-like component, when the sheath shape is similar to the outer surface of the component, a uniform surface modification layer can be achieved. In order to get high shape conformability, the sheath evolution during the PIIID process of the inner ring of a bearing was calculated. The simulation results reveal that the uniformity can be improved when the ion density is increased and the pulse width is reduced. Therefore, we can use a high plasma density and small pulse width for the PIIID process of the outer surface of a cylinder-like component.

In order to acquire the PIIID batch treatment of the inner rings of bearings, many inner rings are assembled intimately along the axis. Since the metallic plasma produced by a vacuum arc plasma source has a large dream velocity and results in line-of-sight restriction, the inner ring must be rotated during the treatment. Consequently, the uniformity of the track along the rotating circumstance will be improved through the revolving. Using this technique, we can treat 20 or more inner rings in one time, thus the efficiency is increased greatly. When the metallic and gaseous plasma are generated simultaneously in the vacuum41chamber, single or composite films will be synthesized on the outer surface of the cylinder-like component.

The surface modification layer acquired by PIIID should be varied according to the substrate. For instance, Diamond-like-carbon (DLC) film deposited on 9Cr18 steel can achieve high quality wear-resistance, while deposited on the titanium alloy (TC4) indicate un-reasonable characteristic. However, TiN+Ti(CN)+DLC film synthesized

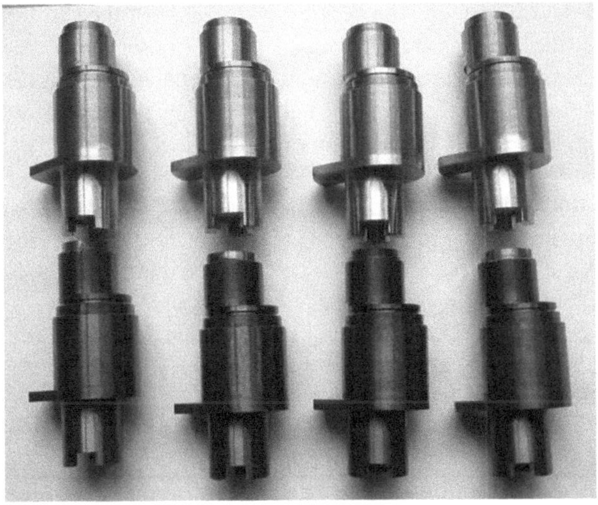

FIGURE 8.15 Comparison of special TC4 shafts treated by PIIID and untreated specimens.

on TC4 alloy exhibits a high wear resistance. Tribological properties of DLC and TiN+Ti(CN)+DLC films deposited on TC4 samples are displayed in Reference [41]. It is found that the cut-through number of the TiN+Ti(CN)+DLC film is evidently larger than that of the DLC film.

Because the diameter of the effective metal plasma is larger than 170mm and the distance between different rotating shafts is 39 mm, 4 rotating shaft can be used in a PIIID process. If components can be fixed on these shafts without any interference, at least 4 components can be treated in one time. Figure 8.15 depicts the PIIID batch treated outer surfaces of special TC4 shafts. The shaft was treated by a TiN+Ti(CN)+ DLC film deposition process by this machine. It is found the TiN+Ti(CN)+DLC film is very uniform and practical tests reveal that each one of the PIIID batch treated TC4 shafts can meet the operational requirements.

8.4.3 The Inner Surface of a Cylinder-shape Component

Because conventional PIIID processes of the inner surface of a cylinder-shape component exist sheath overlapping in a short time, the impact energy and retained dose will be affected greatly. Flow dynamic simulation results reveal that the dose uniformity of the inner surface of a bearing is only 40 percent in a conventional nitrogen PIII process. Furthermore, uniform film deposition on the inner surface is difficult because the metal plasma has a large dream velocity. Consequently, the PIIID batch treatment for the component with inner surface is very difficult to perform in conventional conditions.

Recently, we used grid-enhanced inner surface PIIID [53] to treat the outer rings of bearings. A stainless grid and titanium rod were put into the center of the cylinder. The grid was grounded, the cylinder was biased by a high negative voltage, and the titanium rod was connected to a RF power supply. Therefore, plasma will be generated between the grid and the center rod. Because of the self-bias effect of RF glow discharge, the rod was biased with −1000 V. Consequently, atoms in the rod will be sputtered off and deposited onto the inner surface. The test result shows that the film thickness on the inner surface is uniform. When we put many cylinder-shape components together, it will form a long cylinder. Therefore, the batch treatment of these components can be achieved using grid-enhanced inner surface PIIID.

In addition, middle-frequency and –voltage glow discharge PIIID was also used to treat the inner surface of a cylinder-shape component. During this process, a middle-voltage (4–8 kV), middle-frequency (6–10 kHz) bias is applied on the component, and glow discharge will occur in the cylinder-like component. When C_2H_2 is introduced to the vacuum chamber, C_2H_2 plasma will be generated. Consequently, the thickness of 3μm DLC film can be synthesized on the inner surface through this method. The cut-through number of the DLC film deposited on GCr15 steel is larger than 10000 cycles, which exhibits a high wear resistance [41]. The treated outer rings of bearings using grid-enhanced inner surface and middle-frequency and –voltage glow discharge PIIID processes are shown in Reference [41]. It is found that a uniform surface modification layer can be obtained by either of these two methods.

Using the rotated target holder and proper PIIID process parameters, the batch treatment of balls and the outer surfaces of cylinder-shape components can be obtained easily. In addition, with grid-enhanced inner surface or middle-frequency and voltage glow discharge PIIID techniques, the batch treatment of the inner surfaces of cylinder-shape components can be easily achieved.

8.5 SUMMARY AND PERSPECTIVE

In summary, this chapter presented a comprehensive overview on the applications of PIIID technology on surface modification of basic components in industry, especially in the aerospace industry. We synthesized films, multilayers and nanocomposite coatings on various bearing steel materials and studied the structural, mechanical and tribological properties of the coatings. The potentials of the fourth-generation PIIID facility are illustrated for treating industrial components in batch, large area gas and metallic plasma. With the continual research efforts being put into the development of PIIID, it should become an indispensable modification technology in the bearing industry.

REFERENCES

[1] Xu F.L., Ding N., Li N., et al. A review of bearing failure modes, mechanisms and causes. *Eng. Fail. Anal.* 2023,152:107518.

[2] Voevodin A.A., O'Neill J.P., t6 Zabinski J.S. Nanocomposite tribological coatings for aerospace applications. *Surf. Coat. Technol.* 1999, 116–119: 36–45.

[3] Wang D.Y., Weng K. W., Chang C. L., et al. Tribological performance of metal doped diamond-like carbon films deposited by cathodic arc evaporation. *Diam. Relat. Mater.* 2000, 9(3–6): 831–837. www.sciencedirect.com/science/article/pii/S0925963500002053?via%3Dihub

[4] Zeng Z.M., Zhang T., Tang B.Y., et al. Surface modification of steel by metal plasma immersion ion implantation using vacuum arc plasma source. *Surf. Coat. Technol.* 1999, 120–121: 659–662.

[5] Yoshida A., Fujii M. Influence of soft surface modification on rolling contact fatigue strength of machine element. *Tribol. Int.* 2002, 35(12): 837–847.

[6] Costa J.D., Ferreira J.M., Ramalho A.L. Fatigue and fretting fatigue of ion-nitrided 34CrNiMo6 steel. *Theor. Appl. Fract. Mec.* 2001, 35(1): 69–79.

[7] Rie K.T. Recent advances in plasma diffusion processes. *Surf. Coat. Technol.* 1999, 112(1–3): 56–62.

[8] Dommarco R.C., Bastias P.C., Dall'O H.A., et al. Rolling contact fatigue (RCF) resistance of Austempered ductile iron (ADI). *Wear.* 1998, 221(1): 69–74.

[9] Dommarco R.C., Bastias P.C., Hahn G.T., et al. The use of artificial defects in the 5-ball-rod rolling contact fatigue experiments. *Wear.* 2002, 252(5–6): 430–437.

[10] Conrad J. R., Baumann S., Fleming R., et al. Plasma source ion implantation dose uniformity of a 2×2 array of spherical targets. *J. Appl. Phys.* 1989, 65(4): 1707–1712.

[11] Chen A., Scheuer J. T., Ritter C., et al. Comparison between conventional and plasma source ion-implanted femoral knee components. *J. Appl. Phys.* 1991, 70(11): 6757–6760.

[12] Baoyin T. Plasma source ion implantation (I)—principles and technology. *Physics.* 1994, 23(1): 41–45.

[13] Wang S.Y., Chu P.K., Tang B.Y., et al. Radio-frequency plasma nitriding and nitrogen plasma immersion ion implantation of Ti-6A1-4V alloy. *Surf. Coat. Technol.* 1997, 93(2–3): 309–313.

[14] Conrad J.R., Radtke J.L., Dodd R.A., et al. Plasma source ion-implantation technique for surface modification of materials. *J. Appl. Phys.* 1987, 62(2): 4591–4596.

[15] Anders A. Metal plasma immersion ion implantation and deposition: A review. *Surf. Coat. Technol.* 1997, 93 (2–3): 158–167.

[16] Mayer J.W., Eriksson L., Davies J.A. *Ion implantation in semiconductors: Silicon and germanium.* Academic Press, New York, 1970.

[17] Chu P.K., Tang B.Y., Cheng Y.C., et al. Principles and characteristics of a new generation plasma immersion ion implanter. *Rev. Sci. Instrum.* 1997, 68(4): 1866–1874.

[18] Tang B.Y., Wang Y.H., Wang L.P., et al. Adhesion strength of TiN films synthesized on GCr15--bearing steel using plasma immersion ion implantation and deposition. *Surf. Coat. Technol.* 2004,186(1–2): 153–156.

[19] Sproul W.D. Reactively sputtered nitrides and carbides of titanium, zirconium, and hafnium. *J. Vac. Sci. Technol.* A. 1986, 4: 2874–2878.

[20] Veprek S., Veprek-Heijman, Maritza J.G. Industrial applications of superhard nanocomposite coatings. *Surf. Coat. Tech.* 2008, 202(21): 5063–5073.

[21] Oliveira J.C., Manaia A., Dias J. P., et al. The structure and hardness of magnetron sputtered Ti-Al-N thin films with low N contents (< 42 at.%). *Surf. Coat. Tech.* 2006, 200(22–23): 6583–6587.

[22] Bendavid A., Martin P.J., Preston E.W., et al. and Hoffman M. Deposition of nanocomposite thin films by a hybrid cathodic arc and chemical vapour technique. *Surf. Coat. Tech.* 2006, 201(7): 4139–4144.

[23] Karvankova P., Veprek-Heijman M.G.J., Azinovic D., et al. Properties of superhard nc-TiN/a-BN and nc-TiN/a-BN/a-TiB2 nanocomposite coatings prepared by plasma induced chemical vapor deposition. *Surf. Coat. Tech.* 2006, 200(9): 2978–2989.

[24] Tam P.L., Zhou Z.F., Shum P.W., et al. Nuclear instruments & methods in physics research section B-beam interactions with materials and atoms. *Thin Solid Films.* 2008, 516(16): 5725–5731.

[25] Veprek S., Niederhofer A., Moto K., et al. Composition, nanostructure and origin of the ultrahardness in nc-TiN/a-Si3N4/a- and nc-TiSi2 nanocomposites with HV=80 to ≥105 GPa. *Surf. Coat. Tech.* 2000, 133–134: 152–159.
[26] Yu D.H, Wang C.Y., Cheng X.L., et al. Microstructure and properties of TiAlSiN coatings prepared by hybrid PVD technology. *Thin Solid Films*. 2009, 517(17): 4950–4955.
[27] Sun Kyu Kim, Pham Van Vinh, Jae Wook Lee. Deposition of superhard nanolayered TiCrAlSiN thin films by cathodic arc plasma deposition. *Surf. Coat. Tech.* 2008, 202(22–23): 5395–5399.
[28] Hoerling A., Sjölén J., Willmann H., et al. Thermal stability, microstructure and mechanical properties of $Ti_{1-x}Zr_xN$ thin films. *Thin Solid Films*. 2008, 516(18): 6421–6431.
[29] Veprek S., Veprek-Heijman M.G. J., Karvankova P., et al. Different approaches to superhard coatings and nanocomposites. *Thin Solid Films*. 2005, 476(1): 1–29.
[30] Wang L.P., Huang L., Xie Z.W., et al. Fourth-generation plasma immersion ion implantation and deposition facility for hybrid surface modification layer fabrication. *Rev. Sci. Instrum.* 2008,79(2): 023306.
[31] Xie Z.W., Wang L.P., Wang X.F., et al. Influence of Si content on structure and mechanical properties of TiAlSiN coatings deposited by multi-plasma immersion ion implantation and deposition. *Trans. Nonferrous Met. Soc. China*. 2011, 21: 476–482.
[32] Lee K.R., Eun K.Y. Tribology of diamond-like carbon coated VCR head drums. *Mater. Sci. Eng. A*. 1996, 209(1–2): 264–269.
[33] Robertson J. Diamond-like amorphous carbon. *Mater. Sci. Eng. R*. 2002, 37(4–6): 129–281.
[34] Voevodin A. A., Zabinski J. S. Superhard, functionally gradient, nanolayered and nanocomposite diamond-like carbon coatings for wear protection. *Diamond Relat. Mater.* 1998, 7(2–5): 463–467.
[35] Hauert R., Patscheider J. From alloying to nanocomposites - Improved performance of hard coatings. *Adv. Eng. Mater.* 2000, 2(5): 247–259.
[36] Xie Z.W., Wang L.P., Wang X.F., et al. Influence of high temperature annealing on the structure, hardness and tribological properties of diamond-like carbon and TiAlSiCN nanocomposite coatings. *Appl. Surf. Sci.* 2011, 258(3): 1206–1211.
[37] Richter A., Scheibe H.J., Pompe W., et al. About structure and bonding of laser generated carbon-films by Raman and electron-energy loss spectroscopy. *J. Non-Cryst. Solids*. 1986, 88(1): 131–144.
[38] Beeman D., Silverman J., Lynds R., et al. Modeling studies of amorphous carbon. *Phys. Rev. B*. 1984, 30(2): 870–875.
[39] Lin Y.H., Lin H.D., Liu C.K., et al. Annealing effect on the structural, mechanical and electrical properties of titanium-doped diamond-like carbon films. *Thin Solid Films*. 2009, 518(5): 1503–1507.
[40] Lossy R., Pappas D.L., Roy R.A., et al. Filtered arc deposition of amorphous diamond. *Appl. Phys. Lett.* 1992, 61: 171–173.
[41] Wang L.P., Wang Y.H., Yu Y.H., et al. Batch treatment of industrial components using plasma immersion ion implantation and deposition. *Surf. Coat. Technol.* 2007, 201(15): 6585–6588.
[42] Takaki K., Mukaigawa S., Fujiwara T., et al. Shunting arc-produced hybrid plasma and its magnetic drive for PBII&D. *Surf. Coat. Technol.* 2007, 201(15): 6490–6494.
[43] Chu P. K., Tang B. Y., Cheng Y. C., et al. Principles and characteristics of a new generation plasma immersion ion implanter. *Rev. Sci. Instrum.* 1997, 68: 1866–1874.
[44] Chu P. K., Tang B. Y., Wang L. P., et al. Third-generation plasma immersion ion implanter for biomedical materials and research. *Rev. Sci. Instrum.* 2001, 72(3): 1660–1665.

[45] Wang Y.H., Wang L.P., Yu Y.H., et al. DLC film fabrication on the inner surface of a cylinder by carbon ion implantation. *IEEE Tran. Plasma. Sci.* 2006, 34(4): 1116–1120.
[46] Scheuer J. T., Walter K. C., Adler R. A., et al. Commercial plasma source ion implantation facility. *Surf. Coat. Technol.* 1997, 93(2–3): 192–196.
[47] Prozesky V. M., Meyer K., Alport M., et al. Plasma-based ion implantation facility and research at the National Accelerator Centre, Faure. J. *Vac. Sci. Technol. B.* 1999, 17: 808–812.
[48] Oztarhan A., Brown I., Bakkaloglu C., et al. Metal vapour vacuum arc ion implantation facility in Turkey. *Surf. Coat. Technol.* 2005, 196(1–3): 327–332.
[49] Chu P. K., Qin S., Chan C., et al. Plasma immersion ion implantation—a fledgling technique for semiconductor processing. *Mater. Sci. Eng.: R.* 1996, 17(6–7): 207–280.
[50] Wang L. P., Tang B. Y., Tian X. B., et al. Principle and process window of cerium dioxide thin film fabrication with dual plasma deposition. *J. Mater. Sci. Technol.* 2001, 17: 29–30.
[51] Wang L. P., Tang B. Y., Huang N., et al. Low temperature growth of CeO_2(111) layer on Si(100) using dual plasma deposition. *Mater. Sci. Eng.: A.* 2001, 308 (1–2): 176–179.
[52] Keidar M., Brown I. G. Sheath expansion in a drifting, nonuniform plasma. J. *Vac. Sci. Technol. B.* 1999, 17: 2648–2650.
[53] B. Liu, C. Liu, D. Cheng, et al. A new method for inner surface modification by plasma source ion implantation (PSII). *Nucl. Instrum. Meth., B.* 2001, 184(4): 644–648.

9 Materials Science in Space Technology

Qiang Wei, Tingting Gao, Yue Liu, Chaoqun Xia, and Haichao Hu

9.1 INTRODUCTION

Space travel is challenged not only by the difficulties of launching, but also by the potential threats presented by the space environment. Space environmental threats to materials vary greatly, based not only on the material but also its environment. A degradation of materials used on exterior spacecraft surfaces will happen in the space orbit environment. Each space environment has the potential to cause severe interactions with other spacecraft, which includes photon radiation, charged particle radiation, temperature effects and thermal cycling, impacts from micrometeoroids and debris, contamination, and low Earth orbit atomic oxygen. Furthermore, environmental variables include orbital parameters for the mission, mission duration, the solar cycle and solar events, view angle of spacecraft surfaces to the Sun, and orientation of spacecraft surfaces with respect to the spacecraft velocity vector in low Earth orbit. The spacecraft mission obviously has its own unique set of environmental exposure conditions that must be well understood for purposes of selecting durable spacecraft materials and interpreting observed degradation.

Within the space environment, the damage to and failure of spacecraft materials are the important content in investigations of space material science. Understanding the behavior of spacecraft materials can be determined through actual space orbital exposures experiments, ground laboratory simulation, and computer simulation. Each has advantages and disadvantages. The space environment is kind of extreme environment condition, and the space exposure test is a combined environmental factor test, which is true environment and can get the determinate effects. However, opportunities for spaceflight testing are rare. Dedicated space flight experiments are expensive and require long lead times from planning to flight. It is difficult to understand the failure mechanism through orbital exposure tests. Ground simulation tests are called laboratory accelerating simulation tests, which provide the chance to study the environmental effects and failure mechanism. Meanwhile, the ground simulation can save substantive resources and time, which are necessary for material behavior in the space environment. Laboratory tests can be conducted in a timely manner using accelerated levels for some environmental effects but, due to the difficulties in exactly simulating the space effects, complex calibrations and cautious interpretation of the results are required. In addition, with the computer technology development, the

 DOI: 10.1201/9781003386223-9

computer numerical simulation can be used for the conjecture of space environment effects on materials and mechanisms. Therefore, a combination of space exposures, ground laboratory studies and computational modeling is most useful for assuring durability of spacecraft materials.

This chapter is the related document collection in order to describe the development trend of space environment ground simulation technology, and explicate the relation between space environment and materials, including "Materials Behavior in Space Environment" and "Materials Sciences for Space Environment."

9.2 GROUND SIMULATION TECHNOLOGY OF SPACE ENVIRONMENT

The former Soviet Union and the United States were the first to develop thermal vacuum equipment, which is used to simulate the effects of vacuum and thermal environment on spacecraft. With the continuous growth of space flight time and in-depth understanding of space environment, it is found that various space environment factors have non-negligible impacts on spacecraft [1]. Therefore, a variety of different environmental simulation equipment has appeared successively to simulate proton, electron and other charged particle irradiation environment, space ultraviolet irradiation environment, atomic oxygen and micrometeorites, space debris environment and so on.

However, due to various restrictions on the conditions of ground simulation test equipment, a large number of test results on space material property evolution are mostly obtained under the condition of a single factor, which is greatly different from the influence of comprehensive factors of the actual space orbit environment [2–3]. This makes it difficult to reveal the behavior and failure mechanism of space materials at a deep level. Therefore, it is inevitable for space environment simulation technology to develop from single factor to multi-factor. In order to improve the effectiveness of environmental effect assessment tests, domestic and foreign efforts have been made to improve test methods and techniques, including comprehensive environmental test equipment and techniques [4–5].

Ground integrated simulation equipment of the space environment simulates several environmental factors at the same time and acts on sample materials together, which is one of the important ways to realize multi-factor ground simulation of space environment. NASA (National Aeronautics and Space Administration), ESA (European Space Agency) and Russia have built relatively complete supporting ground test equipment for integrated space environment. At present, the environmental factors simulated by the existing integrated environmental simulation equipment in the world are generally two to five factors. China, too, is keeping pace with the world in this field. At present, the China Academy of Space Technology, the Space Center of the Chinese Academy of Space Technology, the Harbin Institute of Technology and other units have several sets of integrated space environment simulation equipment. For example, in recent years, the national major science and technology infrastructure under construction, "Solar System Environment and Material Action Scientific Comprehensive Research Facility," can realize the comprehensive simulation of the solar system environment on the ground.

On the basis of the comprehensive simulation equipment, aiming at the systematic study of the spatial comprehensive environmental effect and the coupling effect of material damage mechanism, some countries in the world have carried out experiments on the synergistic effect of some environmental factors, and obtained some meaningful research results. For example, synergistic effects of proton and electron irradiation, atomic oxygen and ultraviolet, space debris and induced discharge and pollution, and atomic oxygen and vacuum environment synergistic effects [6–8].

At present, large space environment ground simulation equipment around the world is based on the principle of ground reproduction of a space environment. By reproducing the real space environment as much as possible, the action mechanism of materials in the simulated environment is the same as that in orbit [9]. The sameness or similarity between a simulated environment and the real space environment is emphasized. This research method has high engineering reliability, which helps to reveal the basic law of material structure evolution under the action of space environment, and provides important scientific support for the reliable, stable and long-life operation of spacecraft and human safety activities.

However, the operational cost of large-scale ground simulation equipment for space environment is generally high, which requires strong support from the state to carry out relevant research. For a few individuals or groups, it is difficult to conduct space science and technology research because of lack of funds. Especially in recent years, with the rapid development of commercial spaceflight, the miniaturization of spacecraft will be a development trend. Micro-satellite is a new generation of spacecraft with clear purpose in space science and technology, which has shown great application value. Many scholars have proposed, and are carrying out research on, small-scale ground simulation devices for a space environment [10–12]. At present, many small sample service behavior and commercial space materials service evaluation test requirements appear in the research and development of new spacecraft materials. The ground simulation method of space environment has gradually changed from the principle of reproducing space environment factors on the ground (oriented to environmental factors) to the principle of studying the consistency between ground and in-orbit (oriented to material factors) of the main properties and mechanism of materials.

The single-particle effect on spacecraft is mainly caused by heavy ions and protons, and protons also produce heavy ions through nuclear interaction with semiconductor materials and then induce single-particle effect by heavy ions. In the process of simulating single-particle effects with pulsed laser, although the laser is used as an environmental factor – one which is quite different from the heavy ion environment – the physical mechanism generated in semiconductor materials is similar to that of heavy ions [13]. At the same time, atomic oxygen does not exist in the form of neutral atomic oxygen in the ground atomic oxygen simulation methods of ion beam type and plasma type, but its response mechanism is similar to that of space flight tests and experimental phenomena and reaction laws can be obtained. The space atomic oxygen equipment developed by Beihang University belongs to plasma type [14]. Guimin Yang et al., Tianjin University, took ozone as the corrosive medium and realized the relative scouring of samples by high-speed relative movement. Similar phenomena and laws were also obtained [15].

9.3 MATERIALS BEHAVIOR IN SPACE ENVIRONMENT

Space is often incorrectly thought of as a vast, empty vacuum. In fact, space is not empty. The type of "space" that is encountered by an orbiting spacecraft may contain significant amounts of neutral molecules, charged particles, micro-particulates, and electromagnetic radiation. All of these components influence the design and operation of space systems [16].

Generally speaking, the space environment refers to the natural and man-made environment encountered by spacecraft during orbit. Low Earth orbit environment and geosynchronous orbit environment are the main places for human exploitation and utilization. Through continuous exploration and research, human beings have gained a considerable understanding of the space environment [17–18].

Temperatures on the sunny side of an object can be very high, yet extremely low on the shady side, just a short distance away. Charged particles continually bombard exposed surfaces. Some have so much energy that they pass completely through an object in space. Magnetic fields can be intense. Each of these environments has the potential to cause severe interactions with spacecraft surfaces or subsystems and, if not anticipated, may severely impact mission effectiveness.

9.3.1 Thermal and Thermal Cycling Effects

9.3.1.1 Thermal Environment

Space thermal environment primary includes the solar radiation environment, earth reflection and infrared radiation environment, as well as cold sink environment [19]. Spacecraft orbited in LEO (low earth orbit) will receive radiant thermal energy from three primary sources: the incoming solar radiation (described by the solar constant), reflected solar energy (Earth albedo energy), and outgoing long wave radiation emitted by the Earth and atmosphere. Portions of this energy will be reflected by the vehicle, and the vehicle radiates energy into the cold sink of space at 3K.

Earth orbital environments are capable of producing significant temperature variations as the spacecraft passes from sunlight to shadow. The number of thermal cycles expected for a mission depends upon the orbit. There is obvious difference on the surface temperature of spacecraft, according to the orbital height, season and heat shielding measure. Usually, spacecraft in LEO complete one orbit approximately every 90 minutes, and the surface temperature of the spacecraft undergoes a cycle change −101–93°C (or −122–122°C at limiting case) [20]. Spacecraft in geosynchronous orbit complete one orbit each Earth day (24 hours), and undergo the thermal cycle −150–150°C. Many factors decide the range of temperature which a material experiences during thermal cycling, such as its thermo-optical properties (solar absorptance and thermal emittance), its view of the Sun, its view of the Earth, its view of other surfaces of the spacecraft, durations of time in sun and in shadow, and the influence of equipment or components that produce heat [21].

9.3.1.2 Thermal Effects on Materials

On-orbit temperatures and thermal cycling produce threats to materials durability for various reasons. First, for inhomogeneous materials in intimate contact with one another, such as in composites or coated materials, the cracking, delamination or change in dimension appear for a mismatch in coefficients of thermal expansion and difference in elastic modulus when the material experiences significant temperature excursions.

Second, mechanical properties of polymer materials can be a strong function of temperature. Therefore, during the course of on-orbit thermal cycling, a polymer may experience temperatures at which it has decreased strength or ductility, making it more vulnerable to damage [22].

9.3.2 Atomic Oxygen (AO) Effects

9.3.2.1 Atomic Oxygen Environment

As a space vehicle orbits the Earth at orbital speed (7.8 km/sec at low altitudes), it undergoes energetic collisions with atoms and molecules in the orbital environment. AO is formed by solar ultraviolet (UV) radiation (wavelengths < 243 nm) dissociating oxygen molecules (O_2) into free oxygen atoms in the outer ionosphere from altitudes greater than 100 km [23].

Atomic oxygen (AO) is an important constituent. It is generally the dominant constituent at altitudes below about 500 km (becoming as much as 90 percent of the atmosphere at 500 km), and dominates to altitudes beyond 1000 km during periods of high solar activity. Other atoms are present but are far less reactive, such as H, He, N_2, O_2. The number density of AO at about 250-km altitude is 10^9 atoms·cm^{-3}, which corresponds to the density of residual gas in a vacuum of 10^{-7} mm Hg. In additional, the number density of AO also depends on the solar activity and location, orbital altitude, atmosphere motion following earth's rotation.

9.3.2.2 Interaction Mechanism of AO with Materials

AO is produced due to photodissociation of the diatomic oxygen molecules present in the upper atmosphere by absorption of solar UV radiation [24]. Photo-dissociated AO has a very high probability of long-term survival in the low earth altitudes because there is an appropriate O_2 density here to facilitate reasonable AO production and low probability of interaction with neighboring atoms or molecules for long mean free path enough (~10^8 m).

In the LEO environment, the density and energy of AO are very low, respectively (~10^8 atoms/cm^3) and (~0.1 eV), which may not initiate any changes in surfaces. However, the spacecraft orbiting in the LEOs ram into the AO environment at a velocity of 8 km/s and produce collision energies between the spacecraft surfaces and oxygen atoms of 5 eV, which is equivalent to high temperature effects (5×10^4 K). Meanwhile, the AO is a kind of strong oxidizer. Therefore, The AO collision with the spacecraft surfaces with this energy initiates numerous chemical and physical events on the surfaces. In general, the effects and erosion mechanism on spacecraft due to AO mostly include the following physical or chemical process [25].

[1] The elastic scattering accompanying by diffractive
[2] The elastic scattering directly produced by AO

[3] The indirection scattering, which is also called capture and desorption process.
[4] The impinging AO may be physically captured by a potential well at or below the surface where it chemically reacts to form an oxide, which migrates from the surface into the bulk of the material.
[5] The AO can be captured by chemical adsorption, and further dissociation or forming new type molecule directly or indirectly. For example, it may chemically react with nitrogen atoms on the spacecraft surfaces or while impinging upon the surface to form nitrous oxide in the excited state, which can de-excite by producing glow.
[6] The AO may simply scatter off the surface either in its original or altered charged state.

In fact, the interaction mechanism between AO and materials is complexities, which are concurrent results for multi-process above. Therefore, it is difficult to generalize the interaction mechanism.

Polymeric materials are obviously subject to atomic oxygen erosive [26]. Reactions of AO with different polymeric materials have been shown to occur by various mechanisms. The basic mechanisms are abstraction, addition, elimination, insertion and replacement. In company with continued AO exposure, the volatile oxidation products eventually appear through all mechanism of oxygen reaction, and further hydrocarbon-based materials are eroded gradually. Meanwhile, the oxygen content of polymer surface augments under AO environment. Atomic oxygen can also oxidize the surfaces of metals to produce nonvolatile metal oxides. However, for most metals, the oxides tend to shield the underlying metal from oxidation. Silver is one exception, because silver oxide tends to spall from the metal substrate.

The interaction of AO with spacecraft surfaces will result in mass loss or gain and changes in surface morphology and optical, mechanical and thermal properties [27]. Surfaces of materials with volatile oxidation products (such as hydrocarbon polymers) that are oriented in a fixed position with respect to the ram direction gradually develop left-standing cones, which point in the direction of arriving atomic oxygen, which causes the surface roughness to increase.

The sensitivity of hydrocarbon materials to reaction with atomic oxygen is quantified by the atomic oxygen erosion yield of the material [28]. The volume or mass loss is quantified and used to calculate an erosion or surface recession per incident atom. Depending on the functional requirement of the material surface, the necessary property changes are also measured and correlated with the total effect of AO exposure. The most well-characterized atomic oxygen erosion yield is that of polyimide Kapton® H, which has an erosion yield of 3.0×10^{-24} cm^3/atom for LEO 4.5 eV atomic oxygen.

9.3.3 Space Radiation Effects

9.3.3.1 Space Radiation Environment

The energy deposition in spacecraft materials, particularly electronic materials, can cause significant damage on materials and system. Generally, the natural space radiation environment consists of both electromagnetic radiation such as UV-light, X-rays and γ-rays, and charged particles radiation such as electrons (e-), protons (p), alpha particles (α) and other heavier ions.

Particulate radiation sources [29–34] mainly include Solar particles, geomagnetically trapped charged particles radiation near the Earth (Van Allen Belts), high-energy galactic cosmic rays, and a low-energy plasma consisting of electrons and a variety of positive ions. Solar particles are referred to high-energy electrons and protons ejected from the sun. Cosmic rays consist of energetic protons, electrons, alpha particles, and heavy ions of all the elements in the Periodic table. Particles trapped by the magnetosphere include protons, electrons, and heavy ions [35–38].

Spacecraft outside of the Earth's atmosphere are also subject to the solar electromagnetic radiation. The main source can be classified in to three sources: The largest source is the direct solar flux. The mean value of this solar flux at the mean Sun–Earth distance is called the "solar constant." The two others radiation sources are the fraction of the incident sunlight reflected off the planet, termed albedo, and the Earth infrared radiation. Earth's atmosphere absorbs all ultraviolet radiation from the Sun, which is less than 0.3 microns in wavelength. However, spacecraft outside of the Earth's atmosphere and with a view of the Sun are subject to the full solar spectrum. Additionally, spacecraft surfaces without a direct view of the Sun, but with a view of Earth, may still experience solar ultraviolet effects due to Earth albedo, which is the Sun's energy (~31%) reflected back to space by the Earth's atmosphere. Relevant to the study of space environment effects on materials, the ultraviolet range of wavelengths can be conveniently divided into two bands: Near UV (NUV) as the 200–400 nm range and vacuum UV (VUV) as the 100–200 nm range. During solar maximum conditions the output of the sun in the vacuum ultraviolet (VUV) region of the spectrum ($\lambda < 200$ nm) increases by 50–100 percent relative to solar minimum conditions [39]. Solar flares are releases of intense energy from the Sun occurring over a short duration (minutes to hours), observed as sudden brightening of the chromosphere of the Sun and producing energy throughout the electromagnetic spectrum from radio waves to gamma rays. During the approximately 11-year solar cycle, periods of high solar activity are correlated with enhanced X-ray emission.

9.3.3.2 Charged Particle Radiation Damage Mechanism on Materials

When the energetic charged particle penetrates the target materials, the part of particle's energy will be transferred to the nuclei or electron of target materials by the coulomb interaction between charged particles and the nuclei or electron of target materials [40–53]. Eventually, the injected particles stop in the target materials, and result in ionization phenomenon of target materials, which induced the physical or chemical change of target materials. Therefore, energetic particles must cause radiation damage, as is well known. The charged particle interactions can be subdivided into two cases based on mass: (1) heavy charged particles such as alphas and protons, and (2) light electrons (both positrons and negatrons). Protons and electrons are the primary particles source.

Protons and electrons of a given energy penetrate and do their damage to a definite depth or range in materials. Generally, the higher the particle energy, the greater its range will be. The proton range also varies approximately as the square root of the atomic number of the material. For electrons, the nature of the material has little effect on range, except that the range in hydrogen is about half that in other materials. Electron ranges are greater than those of protons of the same energy.

Bremsstrahlung is the creation of X-rays from electron deceleration. When fast electrons interact with matter, a small percentage produces X-ray photons, which may have energies up to that of the electrons. This process is important because the photons may also cause radiation damage, and their range in matter is much greater than the range of the electrons producing them. Photon range decreases with increasing atomic number.

The general types of radiation effects on materials can be primary categorized into two mechanisms: ionization and atomic displacement.

Ionization is the removal of electrons from atoms in the material and the formation of ion pairs in the path of the charged particles, which is the chief mechanism of damage to insulators: plastics, elastomers, oils and greases, glasses, and ceramics. Almost all the energy of fast protons and electrons and of bremsstrahlung photons ends up as ionization when these particles interact with matter. So, the total amount of ionization is determined by the amount of incident energy. The lower the particle energy is, the relatively shorter the penetration depth of particles into materials is. Therefore, the energy absorbed and deposition particles in materials mainly concentrate in their surface layer. Consequently, the effects of proton radiation with such low energies on optical materials are important, especially for long term spacecraft.

Radiation damage to metals and semiconductors is primarily due to the second mechanism: atomic displacement. Atomic displacement can occur ballistically through kinetic energy transfer, or radiolytically by the conversion of radiation-induced excitation into atom motion (i.e., recoil). As a charged particle passes through matter, the particle energy dissipates by exciting orbital electrons and by elastic collisions with the material nuclei. An elastic collision can eject an atom from its normal lattice position. Displacement damage is the result of nuclear interactions, typically scattering, which cause lattice defects. Displacement damage is due cumulative long-term non-ionizing damage from the ionizing radiations. Displacement is also important in inorganic insulators, such as glasses and other ceramics. Displacement consists of atoms knocked out of their positions in the crystal lattice by collisions with the incident particles, directly or indirectly. The intensity of displacement production falls with increasing proton energy but rises with increasing electron energy. Displacements are conveniently expressed as the fraction of atoms displaced.

9.3.3.3 Electromagnetic Radiation Damage Mechanism on Materials

Radiation damage to materials can also induced by energetic photons. The threshold for atomic displacement by photons is above 10^5 ev, corresponding to wavelengths below 10^{-1} Å. The threshold for ionization, 12 to 25 ev, corresponds to 1000 to 500 Å. when absorbed in a solid, photons of longer wavelengths cause electronic excitation. Sunlight above 3000 Å and the long wave-length emission from planets are not likely to damage generic engineering materials (except for photographic film). Damage through electronic excitation need then be considered for wavelengths between 500 and 3000 Å. (Excitation occurs together with ionization in materials exposed to wavelengths shorter than 500 Å, but need not be separately treated.) During irradiation, both excitation and ionization can cause photoconductivity, photovoltaic effects, and photoemission.

With its wavelength and material kind, the penetration depth (range) of electromagnetic radiation in solids varies greatly. As for the gamma-ray region, at wavelengths below 10^{-1} Å, the penetration ranges can be over 10 cm. For wavelengths between 100 and 1000 Å, penetration in solids is very small, 10^{-4} to 10^{-7} cm. Near a wavelength of 3000 Å, the range in some insulators may again be 10 cm or more. Contrary, the absorption coefficient is the reciprocal of the range.

At the longer wavelengths, at least, the amount of radiation entering a solid is decreased by reflection at the surface. This decrease is not more than a factor of 10 at normal incidence and is often only a few percent.

Incident electromagnetic radiation can affect the properties of a material only if it is absorbed in the material. Major changes to occur the radiation must be strongly absorbed, giving a high dose rate per gram. The radiation will probably penetrate the material only to a shallow depth for the strong absorption. Accordingly, changes in materials induced by electromagnetic radiation are most likely to occur in surface layers. Damage by electromagnetic radiation in space is thus primarily to spacecraft external surfaces and coatings and to other exposed optical surfaces.

9.3.4 Micrometeoroid and Orbital Debris Impact Effects

9.3.4.1 Micro-particle Environment

The micro-particle environment encountered by a spacecraft in low and medium Earth orbit is defined by two sources: man-made debris from space activity since October 1957, and naturally occurring micrometeoroids [54–56].

Meteoroids are solid particles moving in interplanetary space and originate from both cometary and asteroid sources. Micrometeoroids are of extraterrestrial origin and, as such, will have a flux which is reasonably constant with time. Their velocity is typically in the 4–51 km/sec range, with an average velocity near 20 km/sec. As shown in Figure 9.1, as micrometeoroid particle mass decreases, the flux of particles increases.

Space debris differs from natural meteoroids. Orbital debris refers to man-made particles orbiting the Earth, which is a continuous changing environment. It ranges in size from microscopic particles of rocket propellants, to fragments created during explosions and collisions in space, to large spent rocket motor cases or even derelict satellites. It is estimated that there are about 3,000,000 kg of man-made orbiting objects within about 2,000 km above the Earth's surface. Orbital debris has an average velocity of 8.7 km/sec. Debris population distribution is largely a function of launch frequencies and sites, with subsequent perturbations caused by accidental or deliberate explosions, collisions, fragmentations, surface erosion, and manned or unmanned mission-related debris. The orbital debris size distribution can also be seen in Figure 9.1.

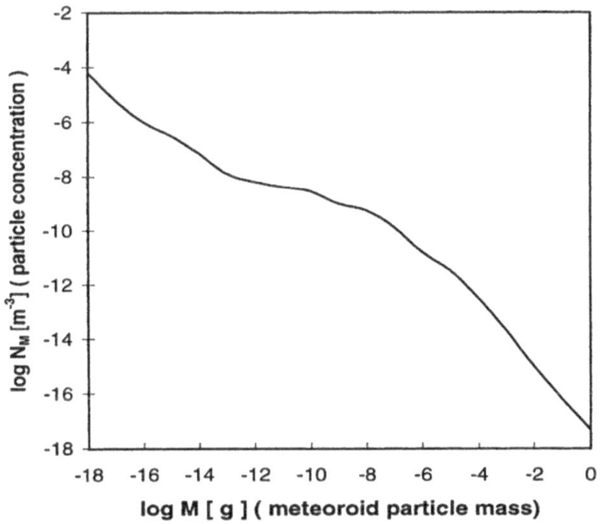

FIGURE 9.1 Micrometeoroid and debris particle flux, versus mass for low Earth orbit (LEO) [57].

9.3.4.2 Micrometeoroid and Debris Impact Damage Behavior

The impact of micrometeoroid or orbital debris with spacecraft materials is usually sufficiently energetic to cause vaporization of the impacting particle, as well as produce an impact crater of volume an order of magnitude greater than the impacting particle [58–63]. The different phenomena, such as Microware Emission and Light Flash, observed with hypervelocity impacts in materials depend on several factors: the impact velocity, the relative sizes of the impactor and target, and the material properties of the target. The physical response of any target to a micrometeoroid or debris impact depends on the material, induced stress level, material temperature, number of projectiles and the system configuration.

9.3.4.3 Penetration and Crater Formation

When a hypervelocity particle impacts spacecraft surface it can create a crater or penetration on the surface (also referred to as the target). If the targets are thick relative to impactor sizes, craters will be formed that generally have lips resulting from plastic flow to molten spatter. In general, impacts into ductile materials, like metal, will form craters, which have diameters averaging about five times the impact diameter. The exact size of the crater is a function of impact diameter, impact speed, and relative ratio of impact density to target density. However, as for very thin targets, such as foils, which are much smaller than the impactor diameter, perforations occur resulting in a hole only slightly larger than the impactor diameter.

Secondary or collateral damage can occur from the impactor remnants and the punched-out section. For high-velocity impacts, both the target foil and the impactor are vaporized. However, for lower velocities, the impactor and foil can remain molten or solid, and collateral damage is possible.

Impact damage can degrade the performance of exposed spacecraft materials and, in some cases, destroy a satellite's ability to perform or complete its mission. Large particles can penetrate through protective wall surfaces. With a relative impact velocity of 10 km/s, a piece of aluminum debris which is ~0.7 mm in diameter can penetrate through a typical 2.5 mm (1000 mil) thick aluminum satellite wall.

While particles greater than 1 mm can penetrate typical satellite skins and cause catastrophic damage, the more common smaller particles mostly cause a gradual degradation of a satellite surfaces.

9.3.4.4 Spallation

Spallation is another impact phenomenon. In addition to crater formation, surrounding areas can experience spallation, undercutting, cracking or delamination of an attached layer. That the structural strength decreasing, thermal and optical property degradation and erosion of underlying materials can result from these damage effects.

Brittle materials, such as glasses or ceramics, often show a central crater, usually filled with finely crushed material, and exhibit little or no crater lips. These craters are usually surrounded by conchoidal fracture areas, which act as spallation zones.

Layered targets, such as coated substrates, often exhibit delamination around or near the crater. Averaging over all impacts, the ratio of crater size to impactor size is about five. For local spall regions, the spall radius to impactor radius ratio is about 20. Star cracks, when formed, can extend outward over a huindred times the impactor diameter.

9.4 MATERIALS SCIENCES FOR SPACE ENVIRONMENT

9.4.1 Optical Materials

Optical materials and systems that serve primarily to transmit or to reflect electromagnetic radiation in the ultraviolet, visible, and/or infrared wavelength regions are described in this chapter in terms of their use in space. Included are lenses, windows, mirrors, and optical filters. Applications of optical systems in space include still and television camera lenses; windows; selective wavelength filters for detectors used in light-sensing devices such as star trackers and meteorological experiments; and reflective devices such as mirrors to direct light beams and to function as telescope reflectors.

9.4.1.1 Proton Radiation Damage Behavior on Quartz Glass

As the most appropriate optical material for high stable optical systems in spacecraft, quartz glass has a series of excellent properties such as optical, mechanical, thermal-physics and anti-radiation stability and is extensively used as optical windows and optical lens in spacecraft [64–67]. The optical elements or materials would be degraded in properties as the spacecraft flies in orbits subjected to the radiation of protons and electrons with a wide energy spectrum of the Van Allen radiation belts.

Wei Qiang and others study the quartz glass behavior under space radiation in detail. The change in transmittance spectrum of the JGS3 optical quartz glass irradiated with

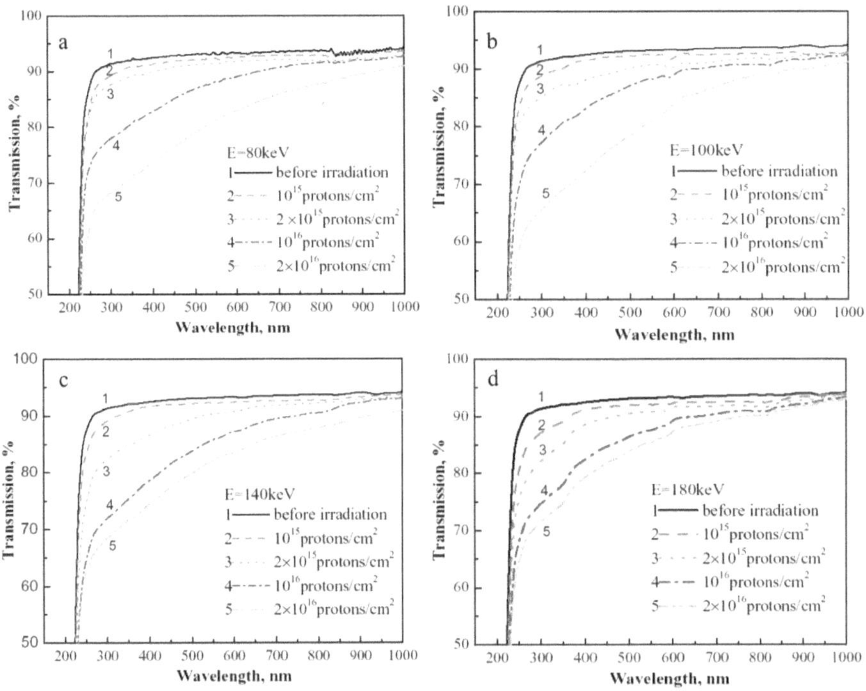

FIGURE 9.2 The spectral transmittance of JGS3 quartz glass before and after protons radiation with 80 keV, 100 keV, 140keV and 180 keV [66].

the protons is shown in Figure 9.2 curve 1 was the transmittance before radiation, while the curves 2–5 showed the transmittance after radiation with different fluences. With increasing fluence, the transmittance of the quartz glass begins to decrease, and such an effect gradually spread from the near-ultraviolet to the visible region, but little influence is observed in the infrared region, as shown in Figure 9.2.

The Electron Spin Resonance analysis (ESR) study shows that the color center E′ was responsible for the 209 nm absorption peak, while color center E′ was disturbed by hydrogen for the 220–240 nm absorption peak. During the radiation not only the breaking of Si-O bonds and the formation of para-magnetism E′ but also the transition from the [≡Si-] and [≡Si-O] radicals into the [≡Si-H] and [≡Si-OH] should occur. Thus, a varied color center E′ is formed.

The additional absorption of quartz glass in some spectra induced by proton radiation was determined by the absorption dose of quartz glass. The absorption dose and the radiation fluence are correlated with each other by following the formula

$$D = E\Phi / R\rho$$

here D is the absorption dose in Gy, E the particle energy in MeV, Φ the radiation fluence in protons/cm^2, R the incidence particle penetration depth in material in cm, and ρ the material density in g/cm^3.

Reference [66] further illustrates the relationship between optical density change and absorption dose or irradiation energy of JGS 3 optical quartz glass. With the increase of the absorbed dose of irradiation, the change of optical density increases when the trend of the time density increases gradually slows down, and finally tends to saturation. For 140 keV and 180 keV proton irradiation in the figure, the optical density change of quartz glass has gradually shown a saturation trend within the dose range used.

Based on analyzing experimental results, a kinetic model for the evolution of color centers in quartz glass irradiated with protons is proposed, and the change in optical density can be given. The mathematic fitted curve is similar to experimental ones. It is believed that the proposed kinetic model can be used in the quantitative description for the change in optical property of quartz glass with increasing absorption dose under proton radiation with low energy.

9.4.1.2 Radiation Damage Behavior on Al Film Reflector

Aluminum reflectors have been widely used in space optics field and serve as the optical payload on spacecraft [68–73]. Under space radiation environment, the reflectors could be deteriorated, resulting in the decreases of service lifetime of spacecraft. It is of great significance to investigate the damage effects and mechanisms of space radiation on Al film reflectors.

The experimental results show that the influence on the optical properties of Al film reflectors induced by proton or by electron is different. After proton radiation the specular reflectance of Al film reflectors decreases with increasing radiation fluence in the wavelength interval 200–800 nm (Figure 9.3). The reflectance decreases greatly with increasing proton fluence when the radiation damage primarily occurs in the multilayer films of Al film reflector. There is no change in optical properties can be seen after electron radiation, because the penetration depth is too long to result in

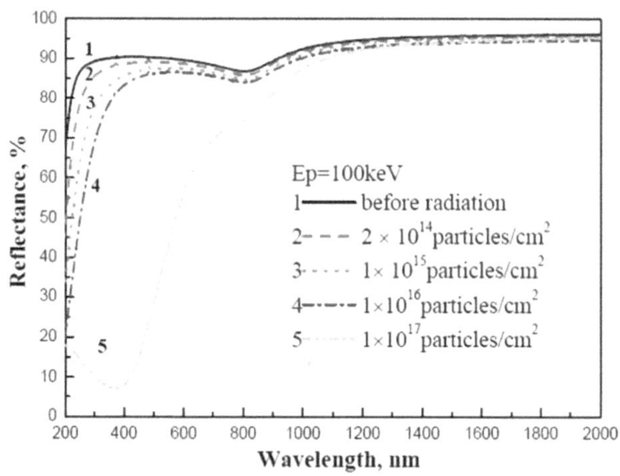

FIGURE 9.3 The concentration distribution of proton with different energy in Al film reflector [69].

obvious effects. However, the synergistic effects situated between the two after the combined radiation of protons and electrons. Therefore, proton radiation is the primary reason for the damage of Al film reflectors.

After proton radiation, the different changes appear in the multilayer films [70]. The augment of surface roughness was also responsible for the increase of light scattering and decrease of specular reflectance. After radiation the color centers form in the SiO_2 film. With larger fluence, the number of color centers increases, and the coloration of SiO_2 protective film is mainly attributed to the decrease of specular reflectance. The filling effect of implanted protons induces the distortion in the crystal structure of Al film, and changes electronic density. As a result, the specular reflectance of Al film reflectors eventually depredates in the range from UV to visible light. Therefore, the influence on reflectance induced by radiation attributes to the comprehensive effects of multilayer films in reflector.

Based on the analysis of properties evolution and microstructural mechanisms, the energy loss of incidence light in different multilayer films is calculated, and under proton radiation the phenomenon model of reflectance evolution is founded [71–74]. The phenomenon model is expressed in formation of index function, and reflects the influence on reflectance attributed to different defects in multilayer films. According to the mathematic analysis of radiation experimental results, the integral reflectance evolution model expressed as index function is founded, which indicates the relation between integral reflectance and radiation fluence. The simulation results of the integral reflectance evolution model are similar to experimental data, which suggests the integral reflectance evolution model can describe the property evolution of reflector under proton radiation. The internal damage mechanisms and external property evolution can be associated by the above two models.

9.4.1.3 Space Dust Impact on Optical Glass

With the development of space technology, the cumulative hypervelocity impact damage effects of micro-size space debris and micrometeoroid (space dust) are getting more attention.

Dong Shangli and others investigated [75] the effects of hypervelocity impact of space dust (micro-sized orbit debris and micrometeoroids) on surface-coated silica glass, optical lens and reflector by using an electrostatic accelerator. The research results indicate that the surface of each impacted sample is characterized by discontinuous damage defects, such as impacted craters, cracks or punctures of the surface layer, local melting, and attaching of micro-particles. Spectrophotometer measurements show that accumulated impacts from micro-particles would lead to degradation of optical properties of the investigated optical lens or reflector. The creation of ionization plasma or light flash during the hypervelocity impact of micro-particles on the optical components was also detected by in-situ measuring techniques.

Jiang Lixiang and others study [76] the effect of the high-speed space dust collision on transmittance of optical glass. The results of dust collision test on K8 quartz glass show that under the impacting of dust particles, the transmissivity decreases obviously and is by the flux, impact speed, species of dust particles and the incident wavelength. The transmittance of optical glass increases with increasing the flux of dust particles

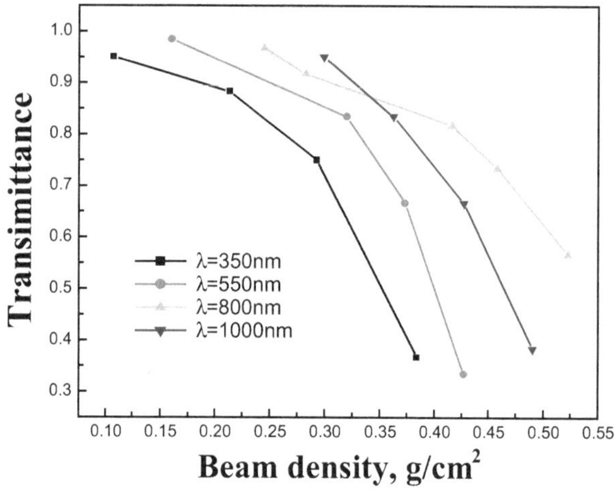

FIGURE 9.4 Transmittance of optical glass versus beam density of Al micro-particles [76].

and decreasing the incident wavelength (as shown in Figure 9.4). The analysis on the relationship between different impact speeds and transmittance showed that absorption or craters on the optical glass surface produced by dust particles would result in a degradation of transparence, for which the extent is proportional to the damaged area (as shown in reference [76]). A physic-mathematic model for the interaction between a single crater and the incident ray was set up based on the optical principle. The mechanism of the impact damage defects influencing the optical transparence was illustrated in terms of this model. Analysis indicates that transmittance decreases with increasing the depth of the hollow crater, while in the case of crater containing metallic particles the transmittance decreases with decreasing the crater depth.

9.4.2 Thermal Control Coatings

Thermal control coatings are important components of the passive thermal control system on spacecraft, which can adjust the surface balance temperature of spacecraft through their own thermo-physical properties such as solar absorptance as and semispherical emissivity. Under the exposure of space environment, the optical properties of thermal control coatings are degraded, influencing the lifetime and reliability of spacecraft.

9.4.2.1 Optical Degradation on ZnO/Silicone White Paint by Radiation

ZnO/Silicone white paint is an important type of thermal control coatings used in spacecraft. Under the exposure of space charged particles, the degradation in optical properties occurs for the ZnO/Silicone, which would affect the lifetime and reliability of thermal control systems or even spacecraft. It is of significance to characterize the damage effects and mechanisms caused by space radiation environment. In Xiao Haiying' study [77–81], the effects of <200 keV protons and electrons on degradation in optical properties of the ZnO/Silicone are investigated separately and combined,

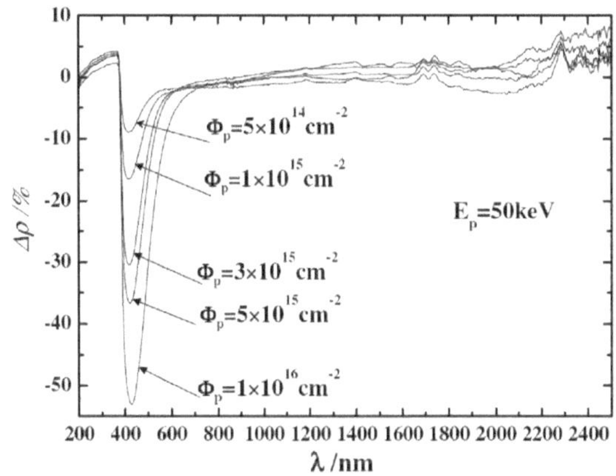

FIGURE 9.5 The change in spectral reflectance $\Delta\rho\lambda$ versus incident wavelength λ for the ZnO/silicone irradiated by 50 keV protons for various fluences [78].

using a ground-based simulation equipment for a large number of charged particles with lower energies existing in the Earth radiation belts. The formation mechanisms of the radiation-induced defects are revealed, and a kinetic model for the concentration of color centers is established through the analyses of absorption bands in the reflective spectra.

It is indicated that the responses of the ZnO/Silicone white paint are different to the protons, the electrons and the protons plus electrons, which have energies less than 200 keV. Under the proton exposure, the degradation in optical properties occurs mainly in the visible region, while in the near-infrared region primarily for the electron irradiation (Figure 9.5). Under the simultaneous irradiation of protons and electrons, the degradation in the visible region is mainly controlled by the protons and that in the near-infrared one primarily by the electrons. The effects caused by the sequential and simultaneous irradiation are similar, and the combined exposure of protons and electrons shows a synergistic effect on the degradation in optical properties of ZnO/Silicone white paint. Under lower fluences, the combined exposure makes the degradation stronger; otherwise, the situation is the opposite. Air exposure leads to an obvious recovery of the degradation in optical properties. The recovery magnitude is the largest after electron irradiation, the next after the simultaneous irradiation and the lowest after the proton one. In the case of the range less than the optical absorption layer for the ZnO/Silicone, the protons with higher energies lead to more obvious degradation, whereas the electrons with higher energies are not benefit to the formation of color centers in the visible region under larger fluences.

The X-ray photoelectron spectroscopy further shows that with increasing proton fluence, both the amount of the lattice oxygen in ZnO and the Zn3d peak area decrease, demonstrating that the oxygen vacancies increase during the exposure. The photoluminescence spectrum analyses indicate that the singly-ionized oxygen vacancies gradually become the dominant radiation-induced defects with increasing proton fluence. This result is also demonstrated by the analysis of slow positron annihilation spectra. The analyses of Fourier transform infrared spectroscopy show that under the exposure, the degraded product of the silicone binder would be changed from the cyclotrisiloxane (D3 structure) into Si-C=O groups, due to the existence of ZnO. The formation of Si-C=O groups causes a red-shift in the $\Delta\rho_\lambda$ spectrum of the ZnO/Silicone. SRIM analysis illustrates that the damage of the organic silicone binder enhances the degradation of ZnO/Silicone white paint.

Based on the analyses of absorption bands in reflective spectra, it is revealed the under the proton and electron exposures, 9-types of color centers can be generated in the ZnO/Silicone, including: the recombined emission bands, V_{Zn}, p^-, V^-, VO^{--}, Frenkel, F^+, F and F^-. Also, the proton irradiation could cause the degradation of t silicone binder to form the Si-C=O groups, resulting in the $n\rightarrow\pi^*$ jumps. Under the exposure of < 200keV protons, the kinetics of cumulation for color centers in the ZnO/Silicone show a general trend that rapidly increases first and then tends increase slowly. The cumulative rates of various color centers are different. The concentrations of the color centers related with oxygen vacancies increase faster and that for the $n\rightarrow\pi^*$ jump type of color centers rises slowly. Based on the evolution kinetics of color centers, a quantitative expression for the change concentration of color centers with proton fluence is established, which is in good agreement with the change in $\Delta\alpha_s$.

9.4.2.2 Radiation Damage on Second Surface Mirror of Kapton/Al

The second surface mirrors of metalized polymer film type are a very important kind of thermal control coatings used on spacecraft. Under space environment, the metalized polymer films would be damaged to influence the thermal balance of spacecraft surface, resulting in the decreases of service reliability and lifetime of spacecraft. It is of great significance to investigate the damage effects and mechanisms of space radiations on the metalized polymer surface mirrors.

Li Ruiqi's experimental results [82–86] show that under irradiation of space energetically charged particles, the degradation in reflective property of Kapton/Al mainly occurred in the visible light and near infrared light region, in which the spectral reflectance reduces with increasing fluence (in Figure 9.6). Superposition effects of single action are higher than synthetic irradiation effects. The optical property of Kapton/Al shows synergistic effect after the synthetic irradiation. When the irradiation energy is given, the change in solar absorptance under proton irradiation with fluence conforms to linear functional relation: $\Delta\alpha_s = k\Phi$. Under electron irradiation the change in solar absorptance with fluence conforms to power functional relation: $\Delta\alpha s = \alpha\Phi^\beta$.

The obvious change in color appeared on the surface of samples after sole proton irradiation and synthetic irradiation. The charged and discharged lacework is obvious by the observation of SEM when electronic fluence is up to $3 \times 10^{15}cm^{-2}$. It was found by AFM that some finger-like bulges (proton irradiation and synthetic irradiation) or hummock-like bulges (electron irradiation) appeared on the surface of Kapton/Al, which results the increase of surface roughness.

After irradiation of charged particles, the edge of optical absorption gradually shifts from UV to the visible region and a strong increase in absorbance in the visible region was observed with increasing fluence. The increase of the absorbance in the visible region is attributed to the formation of new chromophores and auxochromes.

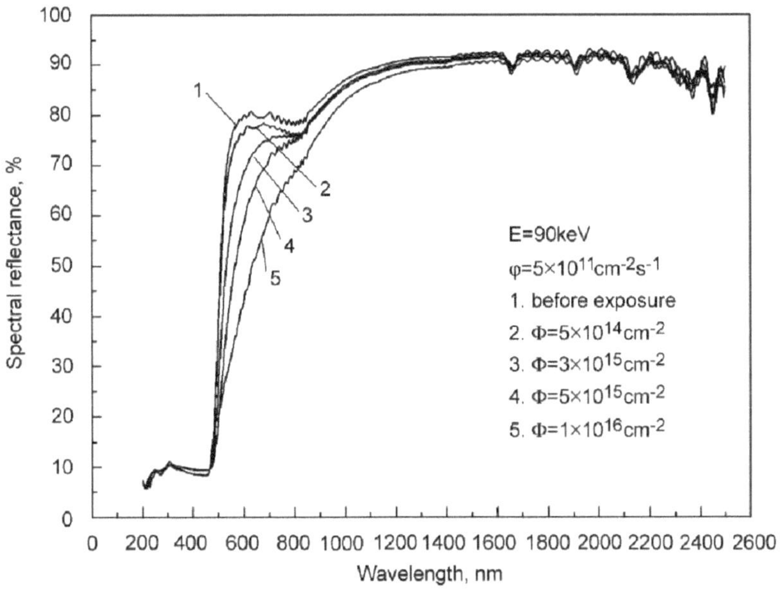

FIGURE 9.6 Spectral reflectance of aluminized polyimide film before and after proton irradiation [84].

The red shifting of the end absorption edge is related to the gradually reducing of the optical energy gap.

Due to the bond breaking of imide and ether during the irradiation, free radicals and ion fragments from ionization and excitation combined to volatilizable micromolecules that escaped from the material, which resulted in the decrease of N and O and a surplus of carbon atoms. Some new components formed such as isocyano (–N=C=O), ketimine (C=C=N–) and hydroxylamine (C–O–N). Benzene ring may be destroyed to some extent and change to carbonizing products similar to graphite structure.

According to saturated track model, some bands which represent typical function group were selected to analyze quantitatively. The results indicate that all function groups in Kapton participate in the degradation process with similar damage cross-section during the irradiation. The damage cross-section increased with the increase of electronic energy loss, indicating that irradiation damage strongly depends on electronic energy loss, this conclusion is coincident with TRIM calculation.

The electrons with the energy of 50 keV mainly destroyed the C–N bonds in the skin of Kapton, forming N^+ ion fragment, accompany the break of partly ether linkage. The fragments of N^+, O^+ and C^+ or radicals will combine to volatilizable micromolecules that escaped from the material, which resulted in the relative increase of carbon contents in benzene ring. Under the electron irradiation of 70 keV not only the above debonding reaction will occur, but also the volume concentration of ion fragments and radicals from debonding will increase, which may increase the possibility of crosslinking between molecules and forming new structure. With the increase of irradiation energy, the trend of carbon enrichment in the surface of Kapton film will increase.

On the basis of quantum chemistry theory, five kinds of bond breaking models on irradiation damage were built by simulation software. The thermodynamic calculation indicates that five kinds of possible debonding reaction mode are easy to occur. The probability of debonding is arranged descendingly: C–N (imide ring)>C–C (imide ring)>C–N (joined to the carbon atom of benzene ring)>C–O–C.

9.4.2.3 Space Dust Impact on Thermal Control Coatings

Bai Yu and others investigate [87–90] the space dust impact effects on thermal control coatings. The results of collision test of dust particles on thermal control coatings of the solar reflection type show that the absorptance α_S is significantly increased, while the emissivity ε changes a little. The solar absorptance α_S increases with increasing the flux and the impact speed, and under the perpendicular impacting by dust particles is larger than that under inclined impacting. It would be concluded that the impact of dust particles results in surface damage of thermal control coatings, and the variation in solar absorptance α_S of thermal control coatings could be directly related to the damaged area *Sn*. The $\Delta\alpha_S$ is linearly increased with the increase of *S*n, and the α_S could reach on extreme of α_S^{∞} until the surface damage of thermal control coatings is accumulated to a certain level. The effect of the solar absorptance variation $\Delta\alpha_S$ on the thermal balance of spacecraft was briefly analysis. The experimental results indicate that under the impact of the dust particles, the thermal control coatings would transform from the solar reflector-type to the solar absorber-type, which could provide a theoretic base for selecting thermal control coatings in spacecraft.

9.4.3 Electronic Materials

9.4.3.1 Radiation Damage on Solar Cell

Solar cell is a key element of the electric energy providing system in spacecraft. It is of theoretical and practical significance to deeply study the performance and damage mechanisms of the solar cells under space environment.

The degradation and damage mechanisms of GaAs/Ge and GaInP/GaAs/Ge solar cells caused by irradiation of space charged particles are investigated in Hu Jianmin's research [91–96]. Experimental results show that under exposure of protons with various fluxes, the degradation in electric characteristics of the GaAs/Ge single-junction solar cells can be related to the damage in the emitter, the base and the space charge regions, respectively, as shown in Figure 9.7. The 70keV protons primarily lead to the damage in the space charge region and an obvious degradation in the open voltage. The 40keV protons mainly result in the damage in the emitter and a considerable degradation in short circuit current. Under the >100keV protons exposure, the degradation of electric properties gradually decreases with increasing proton energy. Compared at a given fluence of electrons, the electric characteristics of the GaAs/Ge solar cells increase with increasing electron energy. The exposure sequence of 1MeV electrons and 170keV protons does not influence irradiation damage to the GaAs/Ge

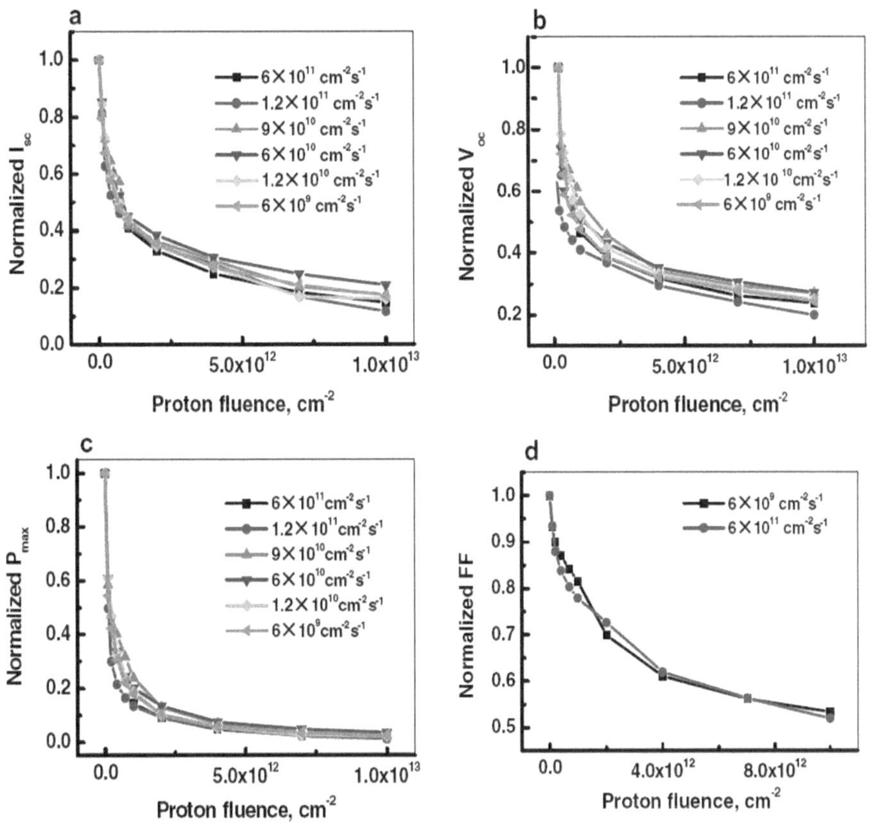

FIGURE 9.7 Effects of proton fluxes on the normalized short circuit current Isc (a), open-circuit voltage Voc (b), maxim output power Pmax (c) and filled factor FF (d) versus fluences of 100 keV protons [93].

cells. Deep level transient spectroscopy (DLTS) analysis shows that the irradiation-induced defects caused by <200keV protons in the GaAs/Ge solar cells are mainly the E_c-0.31eV and E_c-0.47eV, and their concentration decreases with increasing proton energy. Based on the irradiation-induced damage effects, two physical-mathematic models on the degradation in electric characteristics of the GaAs/Ge single-junction cells are set up, which are separately responsible for the short circuit current and the open circuit voltage.

Under <200keV proton exposure, the degradation in electric characteristics of the GaInP/GaAs/Ge triple-junction cells have a close relationship with proton energy. Compared at a given fluence, the 170keV proton exposure leads to the largest degradation magnitude of the open circuit voltage, short circuit current and maximum power. In contrast, the degradation magnitude in the electric characteristics caused by the 40,100 and 130keV protons is lower. Under the irradiation of protons with 4MeV and 10MeV, the degradation magnitude in electric characteristics decreases with increasing proton energy. At a given fluence, the degradation magnitude in electric characteristics increases with increasing electron energy. DLTS analysis shows that the <200keV protons can introduce the deep-energy levels of E_c-0.26, E_v+0.18eV and E_v+0.42eV in the GaInP sub-cell, and those of E_c-0.015eV, E_c-0.15eV and E_v+ 0.33eV in the GaAs sub-cell.

9.4.3.2 Radiation Damage on Bipolar Devices

Li Xingji and others examine [97–102] the degradation on bipolar devices caused by protons, electrons, Br ions with various energies and Co-60 γ-ray radiation. Experimental results show that with increasing irradiation fluence, the degradation trends of different bipolar transistor and bipolar ICs are similar, under a given condition for different particles, fluxes and energies (the example is shown in Figure 9.8). The susceptibility of various bipolar transistors to radiation damage can be assessed by using a given parameter of electric properties (e.g., the current gain). The effect of irradiation flux on the degradation in electric properties of bipolar devices is limited, in the chosen flux range. However, the 1MeV electron exposure with high fluxes aggravates the degradation of some electric property parameters.

Based on the analyses of degradation in electric properties due to radiation damage for bipolar transistors and the calculation of ionization and displacement doses per unit fluence of incident particles (designated as D_i and D_d, respectively), a parameter of $D_d/(D_d+D_i)$ is proposed to evaluate the displacement damage ability of an incident particles. The bigger the ratio of $D_d/(D_d+D_i)$, in the sensitive region caused by incident particles, the more severe the displacement damage, and thus showing an obvious feature of displacement damage for bipolar devices. The displacement damage caused by various particles can be normalized by using a displacement dose equivalent method based on the NIEL in the sensitive region.

It is shown that under the exposure of 70keV and 110keV electrons and 70keV protons, the change in the reciprocal of the current gain of bipolar transistors increases and is gradually saturated with increasing the fluence, the saturation is induced by ionization damage. Under a given fluence, due to their different abilities for ionization damage, the different type particles with various energies would lead to different ionization damage magnitudes. Based on the theoretical calculation and analyses on the electric property degradation, it is indicated that the larger the ionization dose per unit fluence of incident particle, the smaller the ionization damage magnitude for the bipolar transistors.

During the exposure of charged particles, it is observed that the influence of bias condition on the ionization and displacement damage is different. If voltage on the base-emitter junction is forward, the exhausted region becomes narrow, resulting in weaker effect of ionization damage on the bipolar transistors. On the contrary, the ionization damage gives stronger effect on the transistors. The difference in the bias condition gives little contribution to the bulk defects caused by displacement damage. However, when the base-emitter voltage is forward, the electric charges will be injected into the base region, inducing an annealing effect on the displacement damage.

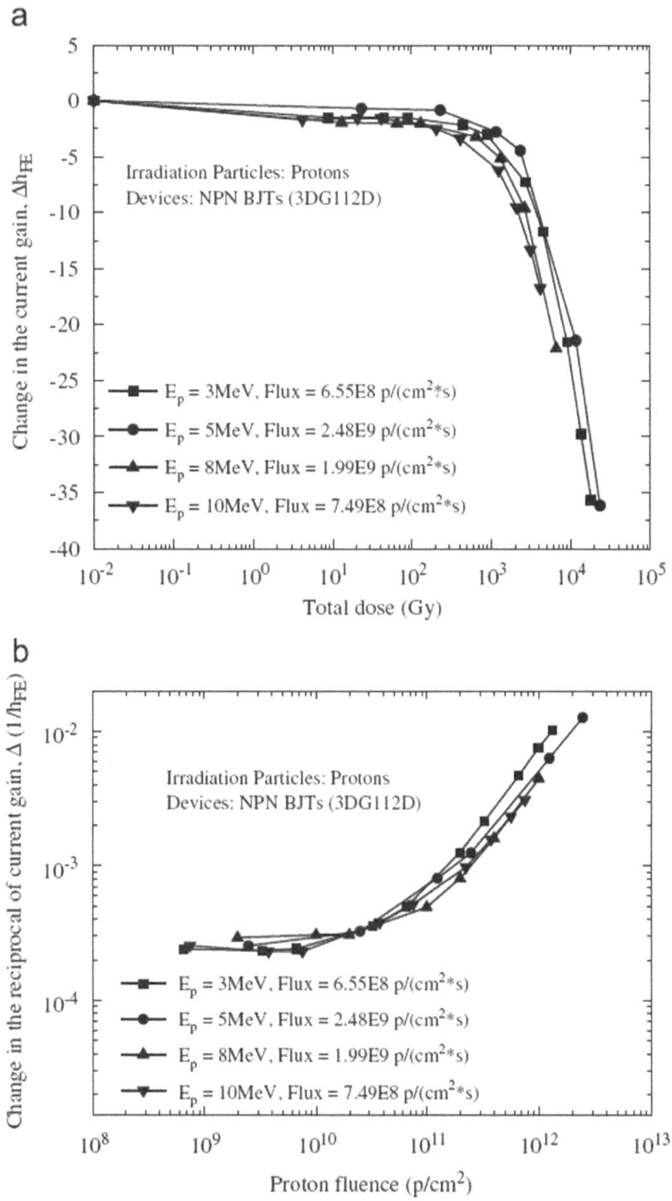

FIGURE 9.8 The change in current gain as a function of total dose (a) and the change in reciprocal of the current gain versus fluence (b) for NPN BJTs irradiated by protons with various energies [99].

Through the combined irradiation experiments of protons and electrons, it is revealed that the bipolar transistors show a synergistic effect of displacement damage with ionization damage. Under the combined exposure of 170 keV protons and 110 keV (or 70 keV) electrons, the displacement damage dominates the current gain degradation for the transistors, and the ionization gives both annealing and aggravation effect on the displacement damage. The photocurrent induced by ionization damage in Si bulk will passivate a portion of the bulk defects, inducing the annealing effects. Meanwhile, the interface state and oxide charge caused by ionization damage in the Si/SiO_2 interface and oxide layer will move the peak of the surface recombination current into Si bulk, leading to more server displacement damage. The trend of degradation of current gain, caused by the synergistic effect of displacement damage with ionization damage, for NPN and PNP transistors is the same.

9.4.3.3 Thermal Stress of Solar Panel under Thermal Cycling

Wang Xiaoyan and others construct [103–108] the unidimensional and the pseudo-two-dimensional models simulating the distribution of stress in the multilayer bonded structure related to solar panel, and the examples are shown in Figure 9.9. The effect of size of cellular construction and physical properties of constructional materials on the stress distribution of cellular construction for solar panel after one thermal cycle was simulated using analytical method.

The measuring and analytic results showed that the thermal stress of cellular construction at the holding stage of low temperature was far larger than that at the holding stage of high temperature. The normal stress maxima of cellular construction occurred in polyimide film. The shear stress maxima of cellular construction occurred in silicon rubber between polyimide film and carbon-fiber composite. That main reason was that the thermal expansion coefficient of the polyimide film was greater than that of other constructional materials. The direction of thermal stress in constructional material decided on the interval of thermal coefficient of expansion on adjacent materials, and displayed cyclic variation.

The maximum of the normal and shear stress in cellular construction increased with the increase of structural length, and stabilized when the structural length exceeded

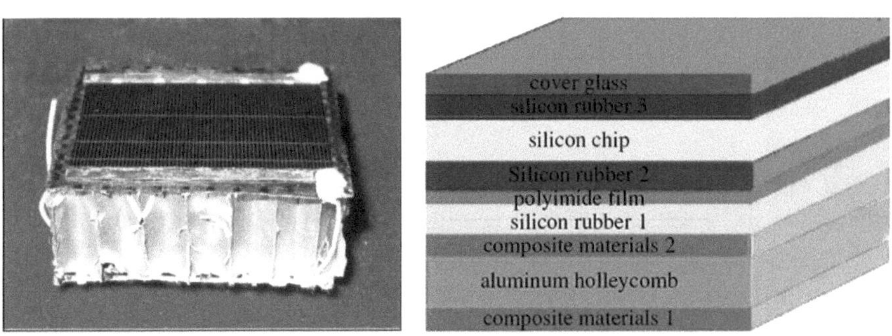

FIGURE 9.9 Sample pattern of solar battery and structural representation of the solar panel [105].

40mm. The maximum of the normal stress in cellular construction increased and that of the shear stress decreased when the thickness of the polyimide film decreased. The maximum of the normal and shear stress in cellular construction decreased remarkably by using the polyimide film with lower elastic modulus and thermal coefficient of expansion.

Both the maximum of the normal stress and the shear stress in cellular construction increased with the increase of thermal cycles. There exists a higher concentration of normal stress in the polyimide film and shear stress in silicone rubber between the polyimide film and carbon fiber composite. There exists the residual strain with different degree in the silicon chip, carbon fiber composite, polyimide film and the two layers of the silicone rubber adjacent to the polyimide film. The maxima thermal strain of polyimide film increased with the increase of thermal cycles. And the maxima thermal strain of cover glass and silicon chip changed little.

9.4.4 Organic Materials

9.4.4.1 Proton Radiation Damage on Methyl Silicon Rubber

Silicone rubber is applied widely in spacecraft as binders and sealant due to its excellent characteristics of electric isolation, good performance at both low and elevated temperatures, and resistance to ozone and radiation. However, long-term irradiation with protons and electrons in space environment may induce microstructure changes and property degradation of the silicone rubber, even affecting spacecraft reliability and lifetime.

The research results from Zhang Lixin [109–111] show that at lower fluence, the proton irradiation induced cross-linking, resulting in an increase in tensile strength and hardness of the methyl silicone rubber (Figure 9.10). However, at higher proton fluence, radiation-induced degradation, which decreased the tensile strength and hardness, became dominant. A macromolecular-network destruction model for silicone rubber irradiated with protons was proposed.

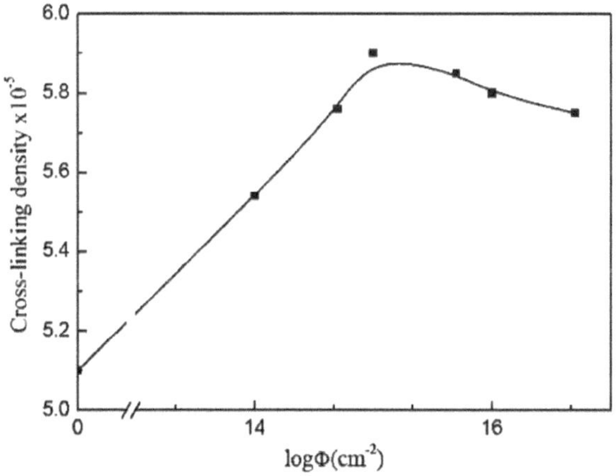

FIGURE 9.10 Cross-link density of silicone rubber versus proton irradiation fluence [109].

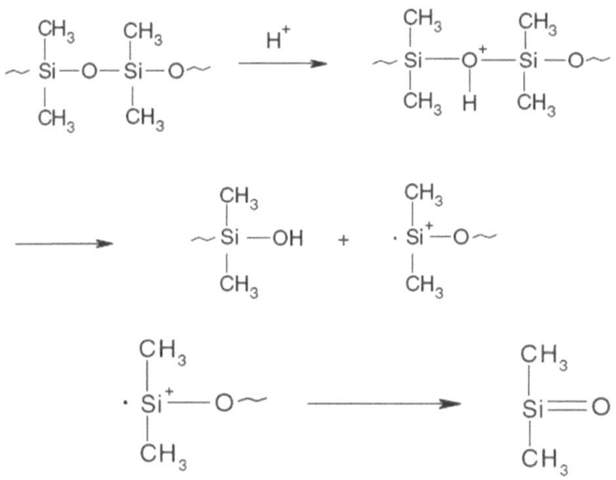

FIGURE 9.11 Degradation process of silicone rubber under proton radiation [109].

Under the proton fluence less than $10^{15}cm^{-2}$, the cross-linking was the main damage effect in the silicone rubber, showing the increase in cross-linking density and in glass transition temperature, the closer stacking of molecular chains, the decrease in free volumes between molecules and the increase in amount of surface Si-O_x structure. The change in microstructure led to an increase in the hardness, the tensile strength, the dynamic storage modulus at low temperatures, the thermal stability, and the alternating breakdown voltage, but the bulk resistivity, the dielectric constant and the dielectric loss angle tangent changed a little. After the fluence more than $10^{15}cm^{-2}$, degradation became a dominant effect for the damage of silicone rubber, showing the decrease in cross-linking density and in glass transition temperature, the increase in free volumes, and the decrease in the amount of surface Si-O_x structure. As a result, the changes in properties were opposite to those mentioned above for the case of the fluence less than $10^{15}cm^{-2}$. Through calculation of degradation activated energy and a comparison of damage effects, it was indicated that the addition silicone rubber had better resistance to proton radiation than that of the condensation rubber. The radiation resistance of the rubber could be related to the electron numbers of internal shells in the absorption atoms, the structure of molecular chains and cure processing.

Proton irradiation can produce the gas product CH_3SiOCH_3 which is different from those generated during pyrolysis. The proton-induced damage mechanism for the methyl silicone rubber could be related to the higher ionicity of the Si—O bonds. Under the irradiation, the protons prefer to attack the O atoms in the Si—O bonds, resulting in the rupture of the macromolecule chains (Figure 9.11).

9.4.4.2 Atomic Oxygen on Polymer

Polymers are extensively used as construction materials in space systems due to their high strength-to-weight ratio and a variety of mechanical, thermal, electrical and thermo-optical properties. Typical polymers are polysiloxanes (silicones), epoxies, polyurethanes, polyesters, acetals, acrylics, polyamides, fluorocarbons,

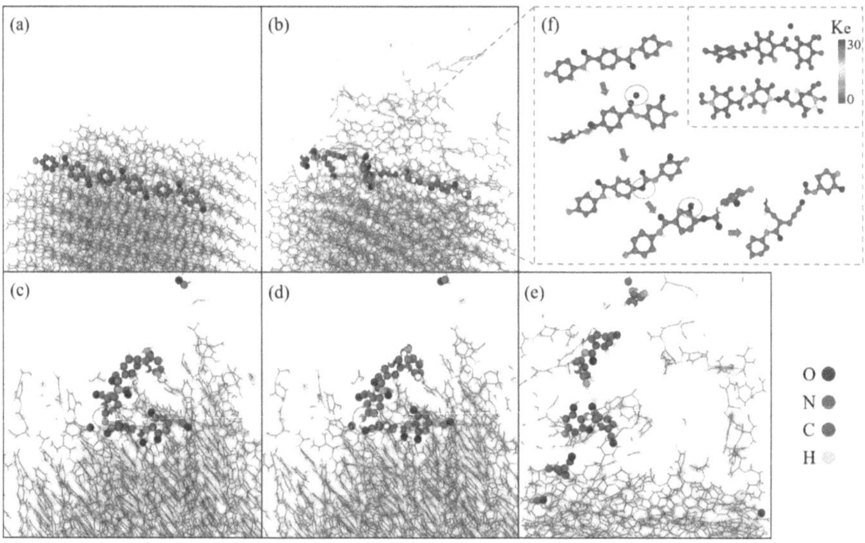

FIGURE 9.12 Atomic oxygen reaction process with polymers [113].

polyimides, etc. The space atomic oxygen environment is especially harsh towards polymers.

Atomic oxygen can react with polymers and carbon to form oxygen bonds with atoms on the surface being exposed [25]. For most polymers' hydrogen abstraction, oxygen addition or oxygen insertion can occur (Figure 9.12). The interaction of AO with polymers may result in surface erosion, changes in chemical composition and surface morphology, changes in optical properties, and formation of particulate and molecular contamination of the spacecraft surfaces. Surfaces of polymers exposed to atomic oxygen also develop an increase in oxygen content [112].

Polyimide is used extensively on spacecraft primarily as flexible substrates for light weight, high-power solar arrays because of its inherent strength, temperature stability, excellent insulation properties, ultraviolet stability and infrared transparency. However, polymeric materials undergo severe degradation after being exposed on atomic oxygen in LEO. AO interacts with PI materials is important in predicting performance characteristics such as in-space durability.

Wang Zhao [114] investigated the dependence of incidence angle and flux density in the damaging effect of atomic oxygen on Kapton film by ground-based experiments in combination with ReaxFF MD simulations. The results show that the AO effect in the lower incidence angle range (0°–60°) is independent of the incidence angle and depends only on the number of impacted atomic oxygen. AO in the higher incidence angle range (60°–90°) has a surface stripping effect, which causes more significant mass loss and temperature rise while stripping raised macromolecules from rough surfaces, and the erosion effect increases with increasing incidence angle and number of impacted atomic oxygen. There is a critical value for the influence of flux density on the AO effect. Above this critical value, AO has a reduced erosive capacity due to a lower chance of participating in the reaction. The amount of each main product from the AO effect varies with the incidence angle and flux density. Still, the total content of the main products is essentially constant, around 70 percent. This work will contribute to understanding the incidence angle and flux density dependence of the AO effect and provide valuable information for developing standards for ground simulation tests.

Donghua Jiang from Tianjin University carried out relevant research titled "Atomic Oxygen Adaptability of Flexible Kapton/Al_2O_3 Composite Thin Films Prepared by Ion Exchange Method" [115]. In contrast with the C–C bond rupture in the oxydianiline (ODA) benzene in Kapton films, the Kapton/Al_2O_3 nanocomposite film mainly destroyed the C=C bond in the pyromellitic dianhydride (PMDA) benzene ring. On exposure to an atomic oxygen environment for a short period, the Kapton/Al_2O_3 nanocomposite film exhibited improved atomic oxygen erosion resistance because the Al_2O_3 layer inhibited atomic oxygen diffusion.

Tong Peiyi et al., Hebei University of Technology, focused on the damage effect and mechanism of Kapton thin film under the synergistic environmental factors of atomic oxygen and micro debris [116]. The performance evolution of Kapton films under the conditions of MD (micro debris), AO (atomic oxygen) single factor load spectrum and MD + AO, AO + MD asynchronous synergistic load spectrum were studied by laser driven flyer and microwave atomic oxygen technology. The macro morphology, optical properties and quality changes of Kapton films before and after each load spectrum were compared, and the mechanism of micro morphology and structure changes was explored.

9.4.5 Composite Materials

9.4.5.1 Degradation Behavior of B/Al Composite Induced by TMC

Because of high specific stiffness and strength and good-dimensional stability over a relatively wide temperature range, unidirectional boron fiber-reinforced aluminum matrix composites (B/Al) are considered as a potential structural material for advanced aerospace applications. The characterization of mechanical properties, thermal expansion behavior and thermal residual stress of the material has drawn much attention. Thermal–mechanical cycling (TMC) is one of the most severe environmental conditions for the application of the composites. The interface plays a key role in the load transfer and stress relaxation. Substantial degradation of the interfacial strength such as sliding or debonding may result from the stress variation in the vicinity of interfaces during TMC. Therefore, it is important to study the interfacial degradation of B/Al composites during TMC.

Qin Yaochun and others investigate [117–120] the degradation behavior and microstructure evolution of B/Al composite subjected to thermal-mechanical cycling (in Figure 9.13).

The experimental results and analysis further show that the ultimate tensile strength (UTS) and fracture morphology of the B/Al composite are influenced by TMC noticeably. The UTS versus TMC cycle curves exhibit a UTS peak, which corresponds to different cycles under different TMC. It takes more cycles to exhibit the UTS peak under smaller applied stresses or elevated upper temperatures. After a given cycle, the UTS of the B/Al composite is higher under lower applied stresses

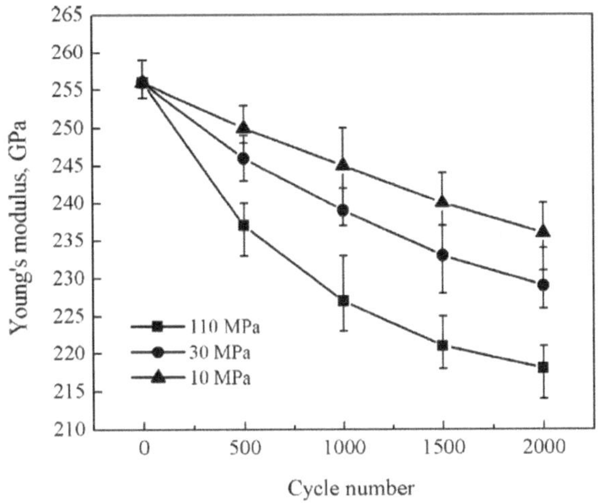

FIGURE 9.13 Young's modulus versus TMC cycles for the B/Al composites [120].

at the same temperature interval and cycles. The interfaces in the B/Al composite change from the strongly-bonded toward the appropriately-bonded, and then to the weakly–bonded ones with increasing the cycles. The combined effect of the changing in microstructure of the matrix and the interfacial degradation during TMC is found to play an important role in controlling the changes of the UTS and the fracture behavior of the composites. TMC would provide a new approach of improving the UTS of the B/Al composite under some TMC conditions. It is safe for some application situations in which the B/Al composite is subjected to TMC for a certain cycle under smaller applied stresses.

The damping capacity increases with increasing TMC cycles and rises more strongly at lower TMC cycles. The damping capacity of B/Al composite increases more greatly under elevated applied stresses at the same temperature interval and cycle and would be increased 10 times due to the TMC. The increase in damping capacity under the present test condition (100–200Hz) is primarily associated with the degradation of the interfacial strength, particularly with the interfacial slip occurring at the local weakly-bonded interfaces. Therefore, TMC can be a probe to analyze the interface degradation in CMMCs.

In the early stages of TMC, the dislocation density in the matrix of the B/Al composite is gradually increased and a highly-tangled dislocation configurations gradually develops. The dynamic recovery, recrystallization and even grain growth would occur in the aluminum matrix with increasing TMC cycles. Different TMC conditions make different extents of the microstructure evolution in the matrix. The thermal mismatch stress depending on the temperature interval and cycles and the applied stress are the stress factor to influence the microstructure evolution and the upper cyclic temperature provides the thermal activation condition. The mechanism of the dynamic recovery and recrystallization can be related to the decreases of the storage energy and the sum of the thermal mismatch stress and the applied stress.

The "knee" in the thermal expansion curves appears at lower temperatures for the B/Al composite subjected to TMC. With increasing the external tensile load and/or TMC cycles, the maximum thermal strain decreases during heating, and the compressive residual strain increases after one time of thermal cycling. The thermal expansion behavior of the B/Al composite is related to the thermal mismatch stress and the degree of the interfacial degradation after TMC, as well as the change in thermal stress during thermal expansion test. A model of calculating thermal mismatch stress in the fibers of CMMCs was established. The changes in generation, accumulation and relaxation of the thermal mismatch stress in the fibers and the extent of the interfacial degradation of the B/Al composite could be explained by the model and the CTE curves on heating and cooling.

9.4.5.2 Radiation Damage on M40J/Epoxy Composites

Carbon/epoxy composites are extensively used in structural components of spacecraft, such as in the truss structure, antennas and solar cell panels. Gao Yu etc. study [121–123] the damage effects caused by space radiation for a carbon fiber reinforced epoxy matrix composite.

The 160 keV proton irradiation leads to non-monotonous changes in mechanical properties of the M40J/AG-80 composite, including bend strength and modulus, as well as interlayer shear strength. Under the fluence lower than approximately 5.5×10^{15} cm^{-2}, the bend strength and modulus increase with the fluence, while decrease with further increasing the fluence (in Figures 9.14–9.15). The interlayer shear strength shows a similar trend to some extent, as shown in reference [121].

Dynamic Mechanical Thermal Analysis (DMA) analysis shows that a transit also takes place at the proton fluence of approximately 5.5×10^{15} cm^{-2} for the change in the cross-linking density in the AG-80 epoxy resin. Before the transit, the cross-linking density increases with the fluence, while descends with further increasing the fluence. The increase and decrease in the cross-linking density before and after the transit could account for the changes in the mechanical properties caused by the proton irradiation for the M40J/ AG-80 composite.

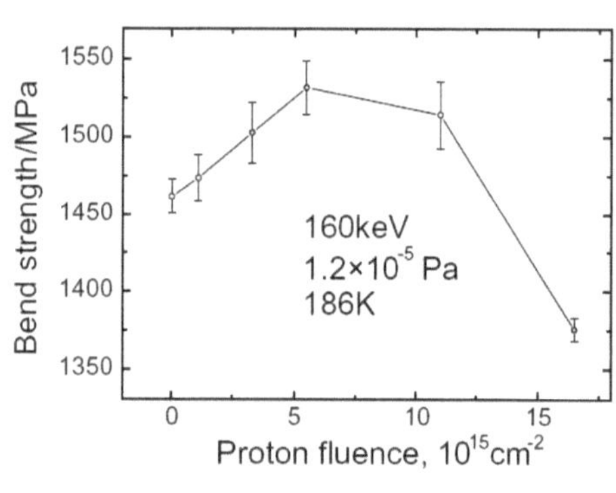

FIGURE 9.14 Bend strength versus fluence for M40J/AG-80 composite irradiated with 160 keV protons [121].

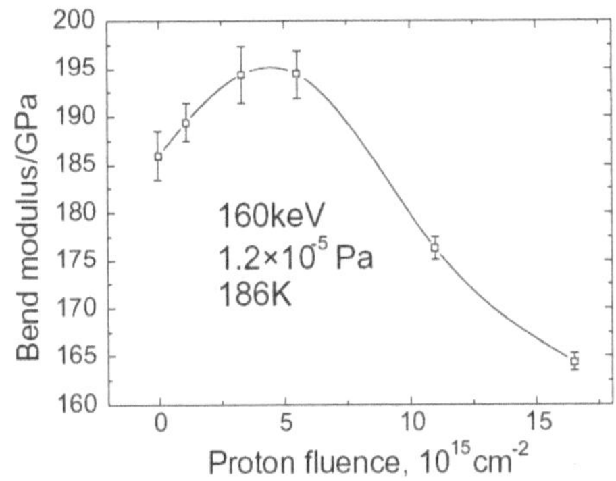

FIGURE 9.15 Bend modulus versus fluence for M40J/AG-80 composite irradiated with 160 keV protons [121].

9.5 CONCLUSION AND PROSPECTS

Spacecraft materials exposed to low Earth orbit environments have been found to undergo degradation or damage due to environmental threats including atomic oxygen, contamination, radiation, temperature effects and temperature cycling, and micrometeoroids and orbital debris. The degree to which the space environment degrades or damages materials depends upon the unique conditions of an individual spacecraft environment and the susceptibility of the material to being altered by these environmental exposures. As has been shown in this chapter, space environment interactions with materials are complex, often producing combined or synergistic effects. As space missions are becoming longer in duration, it is critical that advancements continue in understanding material degradation mechanisms and in the development of simulation laboratory durability testing and modeling capabilities. These advancements will ensure accuracy in predicting long-term material durability in the space environment for future missions.

In the future, complex equipment is a multi-scale system formed from condensates, material continuum and moving assembly. Therefore, the service behavior of space materials is not only dependent on space environment factors, but also closely related to working load conditions. With the increase of spacecraft moving parts and the complexity of electromechanical system service environment, the failure of moving parts has become a key factor affecting the life and possibility of spacecraft. The future development and evaluation of spacecraft new materials and electromechanical systems will be a multi-field coupling service environment where space environment factors and working load conditions act together. In the emerging space technology, the serviceability of materials will be represented by massive data. The traditional expert manual analysis method will inevitably make the decision-making process slow and possibly biased. Efficient data-driven alternative models for the service behavior of complex systems, such as machine learning, will become a key approach to quickly achieve data-driven and/or material performance prediction [124–125].

REFERENCES

[1] Huang B.C. *Space Environment Engineering*. Beijing: China Aerospace Press, 1993.
[2] Feng W.Q. ESA/ESTEC research on space environment experiment capability. *Spacecraft Environmental Engineering*, 2010, 27(5): 543–551.
[3] Song Y.G. Experiment on integrated effects of electron, proton, ultraviolet and atomic oxygen in space environment. *Spacecraft Environmental Engineering* , 1994, 39(2): 14–22.
[4] Fang B., Shen R.J., He W., et al. Research summary of the multiparameter combined environmental test of spacecraft component reliability, *China Mechanical Engineering*, 2009, 20(22): 2766–2771
[5] Huang B.C., Ma W.L. *Experimental Technology of Spacecraft Space Environment*. Beijing: National Defense Industry Press, 2002: 194–201.
[6] Han J.-H., Kim C.-G. Low earth orbit space environment simulation and its effects on graphite/epoxy composites, *Composite Structures*, 2006, 72: 218–226.
[7] Ghosh L., Fadhilah M.H., Kinoshita H., Ohmae N. Synergistic effect of hyperthermal atomic oxygen beam and vacuum ultraviolet radiation exposures on the mechanical degradation of high-modulus aramid fibers. *Polymer*, 2006, 47: 6836–6842.
[8] Peiyi T., Qiang W., Ning H., Xueguang C. Asynchronous synergistic damage effect of atomic oxygen and space micro debris on Kapton film. *Coatings*, 2022, 12(179): 1–14.
[9] He S., Dezhuang Y., Jiao Z. *Handbook of Space Materials (Vol. 1) – Space Environment and Effect Calculation and Ground Simulation Test*. Beijing: China Astronautics Press, 2021.
[10] Fugen X. Development of small space environment simulation equipment is a trend in the future. *Spacecraft Environmental Engineering*, 1999, 3: 38–43.
[11] Johnson R.H., Montierth L.D., Dennison J.R., Dyer J.S., Lindstrom E.R. Small-scale simulation chamber for space environment survivability testing. *IEEE Transactions on Plasma Science*, 2013, 41(12): 3453–3458.
[12] Li-hong S., Qiang W., Bai Y., Zhihao W. Review on application of laser technology in ground simulation of space environment. *Optoelectronics Technology*, 2013, 33(2): 96–102.
[13] Jianguo H., Jianwei H. Mechanism of pulsed laser induced single particle effect. *Science in China: Series G*, 2004, 34(2): 121–130.
[14] Zhigang S., Xiaohu Z., Xin W. *Atomic Oxygen Effect and Ground Simulation Experiment*. Beijing: National Defense Industry Press, 2006.
[15] Qiang W., Guimin Y., Gang L., Haifu J., Tingting Z. Effects and mechanism on Kapton film under ozone exposure in a ground near space simulator. *Applied Surface Science*, 2018, 440: 1083–1090.
[16] Pippin G. Space environments and induced damage mechanisms in materials[J]. *Progress in Organic Coatings*, 2003, 47(3–4): 424–431.
[17] Heng D., Zonghai Y. *Space Environment Manual for Low-Orbit Spacecraft*. Beijing: National Defense Industry Press, 1996:12–404.
[18] Richard D.L. Spacecraft system failures and anomalies attributed to the natural space environment. AIAA 1995 Space Programs and Technologies Conference, September 26–28, 1995. Huntsville, AL: AIAA 95-3564, 1995: 1–5.
[19] Jianyin M., Qi Z., Qiwei Z., Xin Z. *Spacecraft Thermal Control Technologies*. Beijing: Beijing Institute of Technology Press, 2020.
[20] Peters P.W.M., Andersen S.I. The influence of matrix fracture strain and interface strength on cross-ply cracking in CFRP in the temperature range of –100°C to +100°C. *Journal of Composite Materials*, 1989, 23(9): 944–960.

[21] Schwinghmaer, R.J. *Space Environmental Effects on Materials [R].* Huntsville: NASA Marshall Space Flight Center, 1980: 1–45.
[22] Han J.H., Kim C.G. Low earth orbit space environment simulation and its effects on graphite/epoxy composites. *Composite Structures*, 2006, 72(2): 218–226.
[23] Chambers A.R., Harris I.L., Roberts G.T. Reactions of spacecraft materials with fast atomic oxygen. *Materials Letters*, 1996, 26(3): 121–131.
[24] Jiang H.F., Jiang L.X., Li T., et al. Current situation and development trend of neutral atmosphere simulation test technology. *Spacecraft Environment Engineering*, 2022, 39(5): 460–467.
[25] Reddy M.R. Effect of low earth orbit atomic oxygen on spacecraft materials. *Journal of Materials Science*, 1995, 30: 281–307.
[26] Degroh K.K., Terlep J.A., Dever T.M. Atomic oxygen durability of solar concentrator materials for space station freedom [C]. Annual Air Force Workshop on Surface Reactions in the Space Environment. Evanston, USA: NASA, 1990: 1–12.
[27] Banks B.A., Miller S.K.R., deGroh K.K., et al. Atomic oxygen effects on spacecraft materials [C]. Ninth International Symposium on Materials in a Space Environment. Noordwijk, USA: NASA, 2003: 1–7.
[28] Grossman E., Gouzman I. Space environment effects on polymers in low earth orbit[J]. *Nuclear Instruments and Methods in Physics Research Section B: Beam Interactions with Materials and Atoms*, 2003, 208: 48–57.
[29] Garnett H., Hastings D. The space radiation environment [C]. 32nd Aerospace Sciences Meeting and Exhibit. Reno, Nevada, USA: AIAA Meeting Paper, 1998: 590.
[30] Kovalev E.E. Radiation protection during space flight. *Aviation, Space, and Environmental Medicine*, 1983, 54(12): S16–S23.
[31] Evans S.W.. Space environment hazards. 43rd Structures, Structural Dynamics, and Materials Con. AIAA/ASME/ASCE/AHS/ASC, AIAA, Reston, 2022: 1610–1616.
[32] Maurer R.H., Fraeman M.E., Martin M.N., Roth D.R. Harsh environments: space radiation environment, effects, and mitigation. *Johns Hopkins APL Technical Digest*, 28(1), 2008: 17–29.
[33] Newell H.E., Naugle J.E. Radiation environment in space. *Science*, 132(3438): 1465–1472.
[34] Wilson J.W. Space radiation interaction mechanisms in materials [C]. 21st Aerospace Sciences Meeting. Reno, Nevada, USA: AIAA Meeting Paper, 2013: 1–7.
[35] Blue M.D. Degradation of optical components in space. NASA, N94-31029, 1994: 217–225.
[36] Nicolettn C.A., Eubanks A.G. Effects of simulated space radiation on selected optical materials. NASA TN D-6758, N72-24836, 1972: 1–14.
[37] Becher J., Fowler W. The simulated space proton environmet for radiation effects on space telescope imaging spectrograph. NASA-CR-190618, N92-33745, 1992: 1–43.
[38] Durante M., Kronenberg A. Ground-based research with heavy ions for space radiation protection. *Advances in Space Research*, 2005, 35: 180–184.
[39] Johnson F.S. The solar constant. *Journal of the Atmospheric Sciences*, 1954, 11(6): 432–439.
[40] Cahn R.W. Irradiation effects in crystalline solids. *Journal of Nuclear Materials*, 1979, 84(1–2): 343–344.
[41] Khan K.A., Farooq R., Iqbal M.N. Chemical effects of UV radiation on aqueous solutions of heparin-Ca salt. *Polymer Photochemistry*, 1985, 6(6): 465–474.
[42] Oshima A., Tabata Y., Kudoh H., et al. Radiation induced crosslinking of polytetrafluoroethylene. *Radiation Physics and Chemistry*, 1995, 45(2): 269–273.

[43] Attix F.H., Roesch W.C. *Radiation Dosimetry* (2nd ed). New York: Academic Press, 1968: 462.
[44] Seitz F. Displacement of atoms during irradiation. *Solid State Physics*, 1956, 2: 307–442.
[45] Merkle K.L. Fission-fragment tracks in metal and oxide films. *Physical Review Letters*, 1962, 9: 150–153.
[46] Toulemonde M., Dufour C., Paumier E. Transient thermal process after a high-energy heavy-ion irradiation of amorphous metals and semiconductors. *Physical Review B*, 1992, 46(22): 14362.
[47] Wang Z.G., Dufour C., Paumier E., et al. The Se sensitivity of metals under swift-heavy-ion irradiation: a transient thermal process. *Journal of Physics: Condensed Matter*, 1994, 6(34): 6733.
[48] Wang Z.G., Dufour C., Jin Y.F., et al. Modification of defects induced by nuclear collisions in Fe and Ni in electronic stopping power regime. *Nuclear Instruments and Methods in Physics Research Section B: Beam Interactions with Materials and Atoms*, 1998, 146(1–4): 290–295.
[49] Fleischer R.L., Price P.B., Walker R.M. *Nuclear Tracks in Solids: Principles and Applications* (1st ed.). California: University of California Press, 1975: 23–41.
[50] Ritchie R.H., Claussen C. A core plasma model of charged particle track formation in insulators. *Nuclear Instruments and Methods in Physics Research*, 1982, 198(1): 133–138.
[51] Lesueur D., Dunlop A. Damage creation via electronic excitations in metallic targets part II: a theoretical model. *Radiation Effects and Defects in Solids*, 1993, 126(1–4): 163–172.
[52] Kaganov M.I., Lifshits I.M., Tanatarov L.V. Relaxation between electrons and crystalline lattice. *Soviet Physics JETP*, 1960, 12(1–2): 69–78.
[53] Lifshits I.M., Kaganov M.I., Tanatarov L.V. On the theory of radiation-induced changes in metals. *Journal of Nuclear Energy*, 1960, 12(1–2):69–78.
[54] Portnyagin Y.I., Klyuev O.F., Shidlovsky A.A., et al. Simulation of cosmic man-made dust effects on space vehicle elements in rocket and laboratory experiments. *Advances in Space Research*, 1991, 11(12): 89–92.
[55] Potter A.E. Measuring the orbital debris population. *Earth Space Review*, 1995, 4(3): 21–29.
[56] Breisacher P., Mahadevan P. Impact of liquid hydrazine on a heated surface in the low-pressure space environment. Shuttle Environment and Operations Meeting. Washington: AIAA Meeting Paper, 1983: 2630.
[57] Miao J., Stark J.P.W.. Direct simulation of meteoroids and space debris flux on LDEF spacecraft surfaces. *Planetary and Space Science*, 2001, 49(9): 927–935.
[58] Eichhorn G. Measurements of the light flash produced by high velocity particle impact. *Planetary and Space Science*, 1975, 23(11): 1519–1525.
[59] Takano T., Murotani Y., Toda T., et al. Microwave emission experiment with hypervelocity impacts and applications of its results. Proceedings of the Third European Conference on Space Debris. Netherlands: ESA Publications Division, 2001: 569–575.
[60] Auer S. Two high resolution velocity vector analyzers for cosmic dust particles. *Review of Scientific Instruments*, 1975, 46(2): 127–135.
[61] Neish M.J., Kibe S. Hypervelocity impact damage equations for Kapton multi-layered insulation and Teflon second-surface mirrors. Proceedings of the Third European Conference on Space Debris. Netherlands: ESA Publications Division, 2001: 577–582.

[62] Thoma K., Riedel W., Schfer F., et al. Hypervelocity impacts and protection [C]. Proceedings of the Third European Conference on Space Debris. Netherlands: ESA Publications Division, 2001: 223.
[63] Schaefer F., Schneider E., Lambert M. Impact damage on shielded gas-filled vessels. Proceedings of the Third European Conference on Space Debris. Netherlands: ESA Publications Division, 2001: 607–614.
[64] Wei Q., Liu H., He S.Y. Study about the coloration of quartz glass induced by proton radiation with 80 keV. *Radiation Effects and Defects in Solids*, 2004, 159(3): 195–201.
[65] Wei Q., He S.Y., Liu H., et al. Kinetics of radiation damage of quartz glass by low-energy protons. *Journal of Spacecraft and Rockets*, 2015, 43(3): 514–517.
[66] Wei Q., He S.Y., Liu H., et al. Damage kinetics of quartz glass under proton radiation with low energy. *Acta Optica Sinica*, 2005, 25(1): 83–87.
[67] Victor P.S., Ryskul B.T., Mark P.K., et al. The peculiar alterations of the optical and mechanical characteristics of the quartz glass plates with a reinforcing coating under the collision with a hyper-velocity flux of microparticles. *Acta Astronautica*, 2024, 216: 428–436.
[68] Liu H., Wei Q., He S.Y., et al. Proton radiation effects on optical constants of Al film reflector. *Chinese Physics*, 2006, 15(5): 1086–1089.
[69] Wei Q., He S.Y., Liu H., et al. Radiation effects of low-energy particles on optical properties of Al film reflector. *Opto-Electronic Engineering*, 2006, 33(5): 141–144.
[70] Wei Q., Liu H., He S.Y., et al. Slow positron annihilation study of Al film reflector after proton irradiation. *Acta Physica Sinica*, 2006, 55(10): 5525–5530.
[71] Wei Q., Liu H., He S.Y., et al. Characterization of surface morphology changes induced by proton irradiation of an aluminum film reflector. Proceedings of the 9th International Conference: Protection of Materials and Structures from Space Environment. Toronto, Canada: American Institute of Physics, 2009: 657–664.
[72] Wei Q., Liu H., Wang D., et al. Degradation in optical reflectance of Al film mirror induced by proton irradiation. *Thin Solid Films*, 2011, 519(15): 5046–5049.
[73] Zhou Z.X., Wang H.L., Shen Y.Q., et al. Study on the optical property of quartz glass and Al film reflector under charged particles irradiation. *Journal of Physics*, 2008, 57(1): 592–599.
[74] Bitetti G., Marchetti M., Mileti S., et al. Degradation of the surfaces exposed to the space environment. *Acta Astronautica*, 2007, 60(3): 166–174.
[75] Dong S.L., Liu H., Lu G., et al. Investigation of effects of hypervelocity impact of space dust on optical components. *Spacecraft Environment Engineering*, 2011, 28(2): 115–120.
[76] Jiang L.X., Bai Y. A study on effect of high-speed space dust collision on transmittance of optical glass. *Spacecraft Environment Engineering*, 2005, 22(4): 215–218.
[77] Xiao H.Y., Li C.D., Yang D.Z., et al. A study on damage effects of < 200 keV protons on ZnO/silicone white paint. *Journal of Materials Research*, 2006, 21(12): 3022–3028.
[78] Xiao H.Y., Li C.D., Sun M.R., et al. An analysis on optical degradation of ZnO/silicone white paint under proton exposure. *Nuclear Instruments and Methods in Physics Research Section B: Beam Interactions with Materials and Atoms*, 2008, 266(1): 86–92.
[79] Xiao H.Y., Li C.D., Yang D.Z., et al. Optical degradation of silicone in ZnO/silicone white paint irradiated by <200 keV protons. *Nuclear Instruments and Methods in Physics Research Section B: Beam Interactions with Materials and Atoms*, 2008, 266(15): 3375–3380.
[80] Xiao H.Y., Li C.D., Yang D.Z., et al. Optical degradation of polydimethylsiloxane under 150 keV proton exposure. *Journal of Applied Polymer Science*, 2008, 109(6): 4060–4064.

[81] Xiao H.Y., Sun M.R., Li C.D., et al. Formation and evolution of oxygen vacancies in ZnO white paint during proton exposure. *Nuclear Instruments and Methods in Physics Research Section B: Beam Interactions with Materials and Atoms*, 2008, 266(14): 3275–3280.
[82] Li R.Q., Li C.D., He S.Y., et al. Radiation effect of keV protons on optical properties of aluminized Kapton film. *Radiation Physics and Chemistry*, 2007, 76(7): 1200–1204.
[83] Li R.Q., Li C.D., He S.Y., et al. Effects of keV protons irradiation on polyimide films. *Rare Metals*, 2007, 26: 124–128.
[84] Li R.Q., Li C.D., He S.Y., et al. Damage effect of keV proton irradiation on aluminized Kapton film. *Radiation Physics and Chemistry*, 2008, 77(4): 482–489.
[85] Di M.W., He S.Y., Li R.Q., et al. Radiation effect of 150 keV protons on methyl silicone rubber reinforced with MQ silicone resin. *Nuclear Instruments and Methods in Physics Research*, 2006, 248(1): 31–36.
[86] Li R.Q., Li C.D., He S.Y., et al. Energy transfer of keV proton in Kapton/Al. *Chinese Physics C*, 2007, 31(1): 114–118.
[87] Somkin N.D., Bai Y. Introduction of high-speed particles with thermo-regulating coatings. *Physics and Chemistry of Materials Processing*, 2002, 6: 42–48.
[88] He S.Y., Yang D.Z., Bai Y., Semkin N.D. Interplay of flows of high-velocity fragments with members of designs of a space vehicle. The MSTC. Physics and technical applications of wave processes (in Russian). Moscow, Russia. Samara, 2003: 333–334.
[89] Bai Y., Yang D.Z., Liu Y., et al. Damage effect of micrometeoroid on ZrO_2 coating. *Chinese Journal of Materials Research*, 2005, 19(6): 669–672.
[90] Semkin N.D., Bai Y., Piyakov A.V. Injector for the particle accelerator. *SGAU Bulletin, Series: Current Problems of Radio Electronics*, 2002, 7: 39–43.
[91] Hu J.M., Wu Y.Y., Xiao J.D., et al. Degradation behaviors of electrical properties of GaInP/GaAs/Ge solar cells under < 200 keV proton irradiation. *Solar Energy Materials and Solar Cells*, 2008, 92(12): 1652–1656.
[92] Hu J.M., Wu Y.Y., Zhang Z.W., et al. A study on the degradation of GaAs/Ge solar cells irradiated by < 200 keV protons. *Nuclear Instruments and Methods in Physics Research Section B: Beam Interactions with Materials and Atoms*, 2008, 266(2): 267–270.
[93] Hu J.M., Wu Y.Y., Yang D.Z., et al. A study on the effects of the proton flux on the irradiated degradation of GaAs/Ge solar cells. *Nuclear Instruments and Methods in Physics Research B*, 2008, 266: 3577–3582.
[94] Hu J.M., Wu Y.Y., Qian Y., et al. Damage of electron irradiation to the GaInP/GaAs/Ge triple-junction solar cell. *Acta Physica Sinica*, 2009, 58(7): 683–688.
[95] Zhang X., Hu J.M., Wu Y.Y., et al. Direct observation of defects in triple-junction solar cell by optical deep-level transient spectroscopy. *Journal of Physics D Applied Physics*, 2009, 42(14): 145401–145405.
[96] Hu J.M., Wu Y.Y., Song H., et al. Evaluation of orbital behaviors of national GaAs/Ge solar cells. *Acta Energiae Solaris Sinica*, 2010, 31(12): 1568–1573.
[97] Li X.Y., Geng H.B., Liu C.M., et al. Radiation effects on bipolar junction transistors and integrated circuits produced by different energy Br ions. *Nuclear Instruments and Methods in Physics Research Section A: Accelerators, Spectrometers, Detectors and Associated Equipment*, 2009, 612(1): 171–175.
[98] Li X.Y., Geng H.B., Liu C.M., et al. Combined radiation effects of protons and electrons on NPN transistors. *IEEE Transactions on Nuclear Science*, 2010, 57(2): 831–836.
[99] Li X.Y., Geng H.B., Lan M.J., et al. Radiation effects on silicon bipolar transistors caused by 3–10 MeV protons and 20–60 MeV bromine ions. *Physica B Condensed Matter*, 2010, 405(6): 1489–1494.

[100] Li X.J., Geng H.B., Lan M.J., et al. Radiation effects on MOS and bipolar devices by 8 MeV protons, 60 MeV Br ions and 1 MeV electrons. *Chinese Physics B*, 2010, 19(5): 056103.

[101] Li X.J., Geng H.B., Lan M.J., et al. Degradation mechanisms of current gain in NPN transistors. *Chinese Physics B*, 2010, 19(6): 066103.

[102] Li X.J., Xiao J.D., Liu C., et al. Ionization damage in NPN transistors caused by lower energy electrons. *Nuclear Instruments and Methods in Physics Research Section A: Accelerators, Spectrometers, Detectors and Associated Equipment*, 2010, 621(1–3): 707–712.

[103] Wang X.Y., Geng H.B., He S.Y., et al. Effect of geometric dimension on the stress in solar panel under temperature field. *Materials Letters*, 2005, 59(16): 2064–2069.

[104] Xiaoyan W., Hongbin G., Shiyu H., et al. Thermal stress evolution of solar panel structure under thermal cycle. *Acta Solar Energy Sinica*, 2007, 28(4): 6.

[105] Wang X.Y., Geng H.B., He S.Y. Effect of thermal expansion coefficient on stress in solar panel under temperature field. *International Journal of Adhesion and Adhesives*, 2007, 4(27): 288–297.

[106] Wang X.Y., Geng H.B., He S.Y. Effect of elastic property of material on stress in solar panel under alternative temperature. *Materials Letters*, 2006, 60(28): 3354–3360.

[107] Wang X.Y., He S.Y., Zheng S. Stress–strain analysis of photovoltaic panel structure. *Acta Energiae Solaris Sinica*, 2005, 26(5): 631.

[108] Wang X.Y., Geng H.B., He S.Y., et al. Effect of thermal expansion coefficient on the stress distribution in solar panel. *International Journal of Adhesion and Adhesives*, 2007, 27(4): 288–297.

[109] Zhang L.X., Xu Z., Wei Q., et al. Effect of 200 keV proton irradiation on the properties of methyl silicone rubber. *Radiation Physics and Chemistry*, 2006, 75(2): 350–355.

[110] Zhang L.X., He S.Y., Xu Z., et al. Damage effects and mechanisms of proton irradiation on methyl silicone rubber. *Materials Chemistry and Physics*, 2004, 83(2–3): 255–259.

[111] Zhang L.X., He S.Y., Yang D.Z., et al. Synergistic effect of protons and electrons on radiation damage of methyl silicone rubber [C]. Proceedings of Protection of Materials and Structures from the Space Environment. Netherlands: Springer, 2006: 35–41.

[112] Zhang Y., Liao B. Atomic oxygen degradation of a fluorine-containing colorless polyimide. *Polymer Degradation and Stability*, 2023, 2015: 110443.

[113] Liu L., Miao C., Song D., Shi J., Liu W. A reactive molecular dynamics simulation of the atomic oxygen impact on poly(p-phenylene-terephthalamide). *Computational Materials Science*, 2023, 228: 112341.

[114] Zhao W., Qiang W., Huang C., Zhu Y., Hu N. Dependence of incidence angle and flux density in the damage effect of atomic oxygen on Kapton film. *Polymer*, 2022, 14: 5444.

[115] Jiang D., Wang D., Liu G., Qiang W., Atomic oxygen adaptability of flexible Kapton/Al_2O_3 composite thin films prepared by ion exchange method. *Coatings*, 2019, 9(624):1–16.

[116] Tong P., Qiang W., Hu N., Chen X. Asynchronous synergistic damage effect of atomic oxygen and space micro debris on Kapton film. *Coatings*, 2022, 12(179): 1–14.

[117] Qin Y.C., He S.Y., Yang D.Z. Effect of thermal–mechanical cycling on thermal expansion behavior of boron fiber-reinforced aluminum matrix composite. *Materials Chemistry and Physics*, 2004, 86(1): 204–209.

[118] Qin Y.C., He S.Y., Yang D.Z. Effect of thermal–mechanical cycling on damping capacity of B/Al composite. *Journal of Materials Science*, 2004, 39(11): 3845–3847.
[119] Qin Y.C., He S.Y., Yang D.Z. A study in improvement on strength of B/Al composite. *Transactions of Nonferrous Metals Society of China*, 2004, 10(5): 36–43.
[120] Qin Y.C., He S.Y., Yang D.Z. Effects of thermal–mechanical cycling on microstructure and tensile properties of B/Al composite. *Materials Science and Engineering: A*, 2008, 472(1–2): 130–135.
[121] Gao Y., Jiang S.L., Yang D.Z., et al. A study on radiation effect of< 200 keV protons on M40J/epoxy composites. *Nuclear Instruments and Methods in Physics Research Section B: Beam Interactions with Materials and Atoms*, 2005, 229(2): 261–268.
[122] Gao Y., Jiang S.L., Sun M.R., et al. Effect of irradiation with <200 keV electrons on AG-80 resin. *Radiation Physics and Chemistry*, 2005, 73(6): 348–354.
[123] Gao Y., Sun M.R., Yang D.Z., et al. Changes in mass loss and chemistry of AG-80 epoxy resin after 160 keV proton irradiations. *Nuclear Instruments and Methods in Physics Research Section B: Beam Interactions with Materials and Atoms*, 2005, 234(3): 275–284.
[124] Prameela S.E., Pollock T.M., Raabe D., Meyers M.A., Aitkaliyeva A., Chintersingh K.-L., Cordero Z.C., Graham-Brady L. Materials for extreme environments. *Nature Reviews Materials*, 2023, 8: 81–88.
[125] Wang J., Qiang W., Mao Z., Li Y., Zhang T., Liu H., Hu N., Yang Z. Digital twin application in ground simulating space debris system with laser-driven flyer technology. *IEEE Access*, 2022, 10: 105249–105262.

10 Structural Changes and Damage Behavior of Polymer in Aerospace Caused by Radiation

Hongxia Li, Bo Tian, and Jinfeng Li

10.1 INTRODUCTION

Studies of the radiation effects of polymer materials date back to 1896 when Becquerel observed X-rays on photographic film latex, caused by the radiation effect of the chemical changes after some scattered research. After the Second World War, the development of nuclear reactors and various kinds of accelerators for research work provides a variety of strong radiation sources, at the same time, the development of the reactor itself requires that the polymer materials used have stability under irradiation, so that the research field has made great progress. Since the 1960s, there has been a wide application of radiation technology, such as radiation processing of polymer (cable and heat shrinkable products and rubber vulcanization, etc.), agriculture (radiation breeding, etc.), national defense (nuclear weapons, etc.), new energy (nuclear power), environmental space technology and space (space radiation, etc.). Current polymer radiation chemical development has become an independent branch different from pure radiation chemistry and polymer chemistry. Radiation effects on polymer to produce free radicals, ionization or excitation further the following chemical reaction after primary reaction, a new chemical bond is formed between macromolecules, namely radiation cause crosslinking reaction. Polymer chain break, that is, radiative degradation reaction, unsaturated (double bond) changes, oxidation occurs in the presence of oxygen, which produces multiple reaction gas products and other reaction processes (e.g., isomerization, disproportionation, recombination reaction, etc.). Affecting polymer nature is the most basic reaction crosslinking, and degradation is often a result of existing reaction, depending on which is the dominant reaction. It is of great practical significance to study the aging effect and service life of polymer materials under the action of the radiation field.

Radiation may degrade the mechanical properties of materials so that they are no longer mechanically suitable. Polymer materials are increasingly used for technical applications under irradiation environmental conditions. In addition, radiation processing has the potential to play an expanded role in polymer manufacturing, since ionizing radiation is a powerful means of modifying polymers. For these reasons, it is

DOI: 10.1201/9781003386223-10

becoming increasingly important to describe, quantify and understand the effects of polymer exposure to ionizing radiation. For example, polymers are used as insulators, cable systems or packaging materials in nuclear power plants and high-strength proton accelerators. In addition, polymer-based medical devices are often sterilized by gamma rays and electron beam irradiation.

Polymers may exhibit a variety of radiation effects. The formation or breaking of chemical bonds often leads to irreversible effects that alter the chemical, thermal, or mechanical properties of the material. The reactions that occur in irradiated polymers can be broadly classified into two types, namely crosslinking and chain breaking. Crosslinking causes new chemical bonds to form between polymer molecules, and breaking a chain means the polymer molecules are broken. In general, crosslinking increases the strength of the polymer material, while the opposite happens in the case of chain breaking. However, the degree and direction of change are extremely sensitive to the material.

10.2 POLYMER MATERIALS IN AEROSPACE

10.2.1 Preparation Process of Polymers

Although there are many kinds of polymer materials with different sources of raw materials, such as petrochemical route, natural gas route, coal chemical route and agricultural and forestry by-product route, their production process and process equipment are very similar. First, the raw materials are processed and prepared, and then the monomer is prepared by chemical reaction. Then the monomer is made into polymer by various polymerization methods under the action of certain temperatures, pressures and catalysts, and then it is combined into various polymer materials. Plastics, synthetic rubber, synthetic fiber and other polymer materials are made by injection, molding, pouring, blowing, calendering and drawing. It can be summarized as four links: synthesis, polymerization, coordination and molding.

According to the mechanism of polymerization, polymers can be divided into polycondensation and polycondensation.

10.2.1.1 Preparation of Polyaddition Polymer

The monomers (raw materials) of these polymers are mostly ethylene and diene compounds and their derivatives. These monomers are polymerized by the addition reaction of double bonds, so these polymers are called additive polymers. The relative molecular mass of addition polymers is generally high (between tens of thousands to tens of millions), and most of them are thermoplastic polymers. The preparation methods of such polymers mainly include emulsion polymerization, bulk polymerization, suspension polymerization and solution polymerization.

10.2.1.2 Emulsion Polymerization

Emulsion polymerization is usually a reaction in which monomers are dispersed in water to form emulsions by means of emulsifiers and mechanical stirring. Emulsifiers play an important role in emulsion polymerization. The emulsifier is actually a surfactant, and the most commonly used are higher fatty acid metal salts such as sodium stearate (potassium). Adding a certain amount of emulsifier (0.2%–0.5% of the monomer mass) to water will reduce the surface tension of water, thereby increasing the dispersion capacity of water to the monomer (oil phase), and the emulsifier molecules surround the oil droplets formed by the monomer due to stirring, forming a stable emulsion. When the emulsifier dissolved in water exceeds a certain concentration, emulsifier monomer micelles are formed, and the micelles can be agglomerated or striped. Studies have shown that the initiation and polymerization of monomers are usually carried out in micelles. After solubilization and crosslinking, monomer polymer latex particles are formed, and finally polymer latex particles are formed. Then the emulsion was destroyed, the polymer micelles were separated, and the powder product was obtained after washing and drying.

Emulsion polymerization is one of the most commonly used methods for industrial preparation of thermoplastic polymer products. (1) The polymerization speed is fast, and the heat transfer and temperature control are easy. (2) The polymerization reaction temperature is low, and the relative molecular mass of the polymer is high. (3) The viscosity of the emulsion polymerization is still very low, especially suitable for the preparation of high viscosity polymers, such as synthetic rubber. Emulsion polymerization is suitable for the preparation of polymers in a variety of polymer materials, such as synthetic rubber, plastics, coatings, adhesives, and polymer treatment agents. When used for the preparation of the latter three, the emulsion can be used directly, thus saving the production process. The disadvantage of emulsion polymerization is that in addition to water, monomers, initiators and emulsifiers, it is sometimes necessary to add co-dispersants (such as gelatin) and relative molecular weight regulators. Therefore, the product has low purity, poor dielectric properties and transparency.

10.2.1.3 Ontology Aggregation

Bulk polymerization is carried out in the monomer. In addition to a small amount of initiator, no solvent or dispersion medium is added, and no other components are added. According to whether the polymer can be dissolved in the monomer phase (or whether a new phase is formed), bulk polymerization can be divided into homogeneous and heterogeneous. Homogeneous bulk polymerization is suitable for the preparation of transparent polymers such as polymethyl methacrylate (PMMA) and polystyrene (PS). Heterogeneous bulk polymerization systems include polyethylene (PE) and polyvinyl chloride (PVC).

The advantages of bulk polymerization are, (1) less components and high product purity, which is a common method for preparing polymer transparent materials, (2) the preparation process is simple and the operation is easy. Bulk polymerization also has the following disadvantages, because no solvent or medium is added. The viscosity of the polymerization system is large, the polymerization heat is not easy to disperse, the reaction temperature is difficult to control, easy to produce local overheating, the reaction is not uniform, the formation of bubbles, and even explosive polymerization. However, these shortcomings can be overcome by mitigating the reaction or dissolving a small amount of polymer before polymerization and strictly controlling the reaction temperature (gradually increasing temperature). In addition, since the density of the polymer is usually greater than the monomer density, bulk polymerization is usually carried out in a closed mold. Therefore, the problem of volume shrinkage in the polymerization process is more prominent. If not handled properly, wrinkles will occur due to uneven shrinkage, which will affect the optical and mechanical properties of the product.

10.2.1.4 Suspension Polymerization

Suspension polymerization is carried out by using water as medium and adding dispersant. Under strong stirring, the monomer is dispersed into numerous liquid beads, which are initiated by the initiator dissolved in the monomer. The size of the small liquid beads varies with the type and amount of dispersant and the stirring speed, and the diameter is generally 0.001–1cm. Because of the use of oily initiator, and the polymerization is carried out in the monomer droplets, so the suspension polymerization is actually many small bulk polymerization. The difference is that these small bulk polymerizations are carried out in water and are easy to dissipate heat, which fundamentally overcomes the shortcomings of bulk polymerization that are difficult to dissipate heat.

Suspension polymerization has the most characteristics. (1) The system has fewer components and higher product purity, so the product transparency and electrical properties are better. (2) Due to the larger specific surface area of small droplets and the larger specific heat capacity of water, heat dissipation and temperature control are much better than bulk polymerization. Usually suspension polymerization products are beaded, so it is also called beaded polymerization. (3) Separation and purification of suspension polymerization products are easy. However, when the polymerization conversion rate reaches about 25 percent, the viscosity of the system increases, and there is a strong viscosity and coalescence tendency between the polymer beads, which makes the process operation difficult and needs to be strictly controlled. Otherwise, it will stick to the pot and agglomerate, causing accidents. Therefore, suspension polymerization is not suitable for the preparation of synthetic rubber with high relative molecular mass, but is mostly used for the preparation of products such as polyethylene nitride, polystyrene and polyacrylate. Sometimes, suspension polymerization products can be used directly, such as ion exchange resins.

10.2.1.5 Solution Polymerization

The method of dissolving monomer in solvent for polymerization is called solution polymerization. According to whether the generated polymer can be dissolved in the solvent, the solution polymerization is divided into homogeneous and heterogeneous, which is also called precipitation polymerization. The polymerization of acrylonitrile in dimethylformamide is homogeneous, while the polymerization of acrylonitrile in water is heterogeneous.

The most prominent advantage of solution polymerization is that a large number of solvents exist as heat transfer media, so that the polymerization heat is easy to remove and the reaction temperature is easy to control. For homogeneous solution polymerization, the product is less sticky, and it is easy to realize large-scale and continuous production. However, due to the presence of solvents, the solvent chain transfer reaction of the active macromolecular chain is often caused, which reduces the relative molecular mass of the product and initiates branching. This is the main disadvantage of solution polymerization. The chain transfer rate constants of the same active macromolecular chain to different solvents vary greatly, so in practice, the requirements of relative molecular mass can be met by selecting appropriate solvents and controlling concentration.

10.2.2 Structure and Classification of Polymers

Polymer materials have become essential in aerospace applications due to their advantages, like light weight, ease of manufacturing, cheap cost and high strength-to-weight ratio. It has been used in high performance structural application due to their ability to sustain loads, fulfill stringent strength requirements and resistance against the environmental degradation over the metals.

There are many different types of polymers that can be differentiated from the arrangement of repeating units and the different arrangements of molecular segment. A polymer prepared from one kind of monomer is called homopolymer. A polymer prepared from more than one kind of monomer is called copolymer, including random copolymer, alternating copolymer, block copolymer, and graft copolymer. These homopolymers and copolymers also can be prepared into polymers with different arrangements of molecular segment, such as star polymer, comb polymer, ladder polymer, dendrimer, and so on.

The selection of materials for aerospace applications is done based on the purpose, design, load carrying capacity and eco-friendly nature of the ensuing structure. It is always required to understand the complete structure of the aircraft so that the most suitable material can be chosen. The main content of this chapter is to briefly describe the types and structures of polymers used in the aerospace field.

10.2.2.1 Classification of Polymers

Polymers can be classified from different perspectives, such as monomer source, synthesis method, final use, heating behavior, polymer structure, and so forth.

According to the element structure of the main chain, the polymer can be divided into carbon chain, heterochain and element organic. The main chain of carbon chain polymer macromolecules is completely composed of carbon atoms. The vast majority of vinyl and diene polymers belong to this category, such as polyethylene, polystyrene, polyvinyl chloride, and so forth. In addition to the carbon atoms, there are oxygen, nitrogen, sulfur and other heteroatoms in the main chain of the polymer. Such as polyether, polyester, polyamide, polyurethane, polysulfide rubber, and so forth. Engineering plastics, synthetic fibers and heat-resistant polymers are mostly heterocyclic polymers. There is no carbon atom in the main chain of the element organic polymer macromolecule, which is mainly composed of silicon, boron, aluminum and oxygen, nitrogen, sulfur, phosphorus and other atoms, but the side group is composed of organic groups, such as methyl, ethyl, vinyl and so on. Organic silicone rubber is a typical example. Elemental organics are also called semi-organic polymers with hetero-chains. If the main chain and the side group have no carbon atoms, they become inorganic polymers.

10.2.2.2 The Structure of Polymer

The structure of polymer can be divided into chain structure and aggregation structure. Molecular chain structure chain structure is divided into short-range structure and remote structure. The short-range structure includes structure and configuration, the type and arrangement of atoms, the type of substituents and end groups, the arrangement order of monomer units, the type and length of branched chains, and so forth. Configuration refers to the arrangement of substituents of an atom in space. The short-range structure belongs to the chemical structure, also known as the primary structure. The remote structure includes the size and shape of the molecule, the flexibility of the chain and the conformation of the molecule in various environments. Remote structure is also called secondary structure. Chain structure refers to the morphology of a single molecule.

Polymer size: The most commonly used measure of polymer size is molecular weight. Due to the complexity of the polymerization reaction, the molecular weight of the polymer is not uniform and can only be expressed by statistical averages, such as number-average molecular weight and weight-average molecular weight. The molecular weight has an important influence on the mechanical properties and processing properties of polymer materials. Only when the molecular weight or degree of polymerization of the polymer reaches a certain value can it show the applicable mechanical strength, which is called the critical degree of polymerization.

Polymer internal rotation: The main chain of the polymer is very long, usually not straight; it can be curled up, so that the molecules appear in various forms, from the whole molecule, it can be curled into an ellipsoid, and also be stretched into a rod. From the local point of view of the molecule, it can be jagged or helical, which is caused by the internal rotation of the single bond, and the molecules exhibit different spatial forms. These forms can vary with conditions and environments.

Compliance of polymer chains: The property that polymer chains can change their conformation is called flexibility, which is the main reason why many properties of polymers are different from those of low molecular substances. The main chain structure has a significant effect on the flexibility of the polymer. For example, due to the large Si-O-Si bond angle, the Si-O bond grows and the internal rotation is relatively easy. Therefore, polydimethylsiloxane is very flexible and is a good synthetic rubber. Because the aromatic heterocyclic ring cannot rotate internally, the polymer chain containing the aromatic heterocyclic structure in the main chain has poor flexibility and high temperature resistance. The polarity of the side group has a great influence on the flexibility of the polymer chain. The weaker the polarity of the side groups is, the greater the interaction between them is, and the internal rotation of the single bond is difficult, so the flexibility of the chain is poor. The length of the chain also affects the flexibility. If the chain is very short, the number of internally rotating single chains is small, and the number of molecular conformations is small, rigidity will inevitably occur.

10.2.2.3 Polymer Used for Aerospace Applications

Thermosets and thermoplastics are two important types of polymer for these applications. Thermosets can be with stand at elevated temperature, and it is impossible to change them into liquid or powder form again. Thermosets polymer has strength and stability at high temperature because of complex cross-linking. thermoplastics are soft and ductile and melt at higher temperatures. Thermoplastics have less applications as compared to thermosets because they have high viscosity. Mostly used thermosets are epoxies, vinyl esters, phenolic resins, polyesters, and urethanes. Different kinds of fibers are recently used to fabricate composites. Glass fibers, carbon fibers, basalt fibers, aramid fibers, and natural fibers are mostly used [1].

Glass fibers are lightweight, very strong, robust fibers. Glass fibers are used due to their low cost, and higher strength as compared to metals. Different forms of glass fibers like woven fabrics, yarns, and textiles are present in market. Glass fibers have versatility and easy processing with low-cost. In addition, glass fibers have high strength and good fire resistance Glass fibers remain unchanged when exposed to high or low temperatures, and they show low moisture absorption. Hence, they are a good for ballistic and aerospace applications. Glass fibers are non-flammable, inorganic material [2]. Although glass fibers are chemically non-reactive, they are highly sensitive to alkaline medium and phosphoric acids. Glass fibers have high thermal conductivity and a low coefficient of thermal expansion. These materials are really important due to their light weight and high strength properties in field of aerospace and ballistic components and glass fiber reinforced possess great mechanical behavior like high impact resistance & enhanced tensile strength as compared to steel & less brittle than carbon fiber, and good dimensional stability. In comparison to other fibers, glass fibers are heavier, and this limits their applications as a reinforced material [3], [4] as shown in Figure 10.1.

Basalt is a material which is available in the earth. Basalt material is made up of black volcanic magma. Volcanic magma is made after fast cooling of molten lava. The performance of basalt material is based on the cooling rate, source of lava and geographical location of formation of lava. Basalt fibers are a very thin fibrous form of basalt rock and having excellent mechanical behavior and physical attributes and good environmental stability. Basalt fibers are considered a green industrial material due to their non-toxic nature and easy to recycle. Basalt fibers are low cost and easy in processing, which makes them useful in bulk production of composites. Although carbon fiber shows better mechanical behavior than basalt fiber but it comes with cost constraints. Basalt fibers are natural fibers without any kind of addition of other materials. Basalt fiber exhibits high strength and modulus, great resistance to temperature and corrosion, and outstanding chemical stability. Although the composition of basalt fibers is similar to that of glass fibers, the properties of basalt fibers are far better than those of E-type glass fibers. Basalt fibers possess high breaking load and high Young modulus along with excellent environmental stability [6].

Aramid fibers are fibers which are having high strength and stiffness, and they are able to replace steel. Along with mechanical properties that are superior to steel or metal wires of equivalent weight, aramid fibers possess outstanding heat resistance. Aramid fibers are used to make composites with advanced mechanical and flame resistant properties such as protective clothing, aircraft, automobiles, rope for off-shore oil rigs, industrial fillers, and bullet proof jackets. Aramid fibers can be used

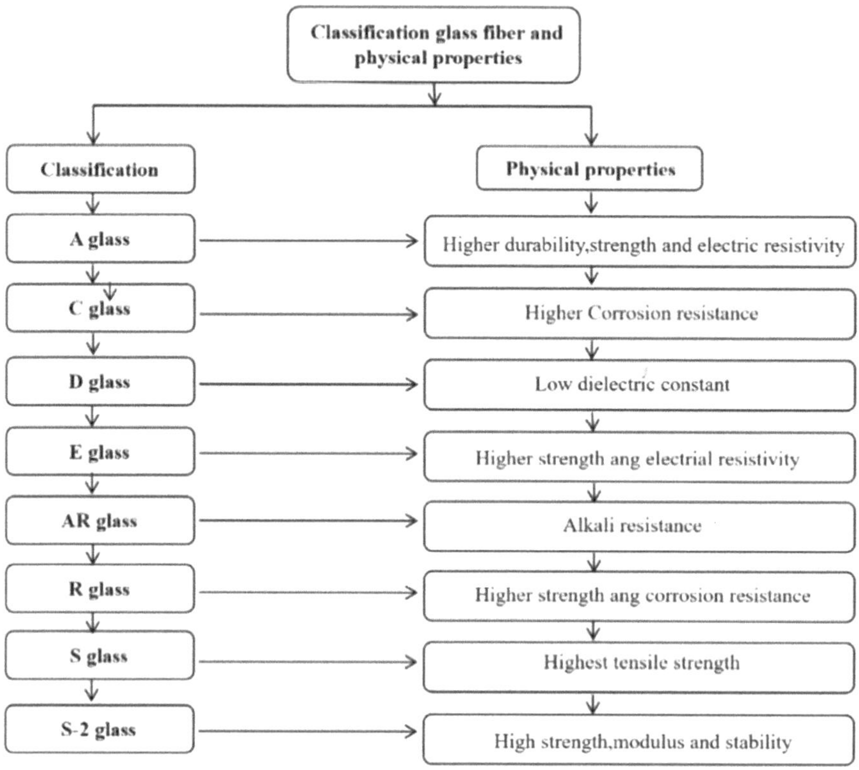

FIGURE 10.1 Classification and physical properties of various glass fibers [5].

in load bearing applications because they possess a higher tensile strength than steel and high stiffness. Aramid fibers are low in density and show creep rate. The energy absorption at failure and shock resistance are attractive features of aramid fibers. Additionally, flame retardation and non-conductivity make aramid fibers suitable for safety clothes and insulator materials, respectively. The impact resistance of aramid fibers is better than that of other fibers. However, there are some negative points of aramid fibers including their sensitivity to UV rays, poor compression strength in comparison to carbon fibers, and problems associated with cutting the fibers. Due to their properties, aramid fibers can be used in protective clothing and helmets, bullet-proof jackets, body armor, drumheads, speaker woofers, ballistic applications, and flame-resistant suits and gloves.

10.2.3 Physico-chemical Properties of the Polymer

The chemical and physical properties of polymers are dependent on their chemical and physical structures and molecular weight. Their relationship related to the chemical, mechanical, thermal, electrical, and optical properties [7–9] are discussed below.

10.2.3.1 Chemical Properties of Polymer

The chemical properties of polymer include chemical resistance, chemical permeation, and weather resistance. The polymer is hydrophobic generally. They are usually resistant to polar solvent but attacked by nonpolar solvent. The extent of chemical resistance can be classified into four categories: none, swelling or softening, dissolving, and reacting. Most polymers are corrosion resistance. Fluorine has proved to be an element that imparts both water and solvent resistance to a variety of polymers due to its low surface energy.

Crystalline polymers are more resistant than amorphous polymers toward solvent resistance, because crystalline polymers are more densely and closely packed that reduce the permeation of solvent. Crosslinked polymers are more resistant toward chemicals than linear polymers due to less free volume available in this type of polymer.

The trend of chemical permeates through polymers should be similar to the extent of chemical resistance of polymer. Thus, a polymer with many polar groups is sensitive to a polar chemical. Conversely, a nonpolar polymer would be a barrier to polar gases and liquids. For instance, polyethylene has very low water permeation but relatively high oxygen permeation. Amorphous polymer exhibits higher gas and liquid permeation than that of crystalline polymer due to the higher density of crystalline. Small molecules can work their way through the polymer much more easily than large molecules. Size effect can, therefore, outweigh all other permeation effects. Microorganisms degrade polymers by catalysing hydrolysis and oxidation. By combination of sunlight and microorganisms, the degradation of polymers will be more effective.

10.2.3.2 Mechanical Property of Polymer

The mechanical properties of polymer can be expressed by the stress-strain plot. Four important quantities characterize the stress-strain behavior of a polymer: modulus which is the resistance to deformation as measured by the initial stress divided by $\Delta l/l$ (ε), ultimate strength or tensile strength that is the stress required to rupture the sample, ultimate elongation that is the extent of elongation at the point where the sample ruptures, and elastic elongation that is the elasticity as measured by the extent of reversible elongation.

As the temperature is increased, the modulus will remain high until the glass transition temperature is reached, at which point the modulus drops sharply. The sample remains rubbery until the temperature is sufficiently high to cause flow. A thermoplastic amorphous polymer shows expected molecular weight dependence in the flow region. The higher the molecular weight, the higher the temperature necessary to overcome the increased molecular entanglements. A crosslinked polymer, on the other hand, does not flow. The higher the crosslinking density, the greater will be the modulus in the rubbery state. Semicrystalline thermoplastic polymers behave much like crosslinked polymers below the melting temperature, because of the very strong intermolecular forces arising from close chain packing. How sharp a break occurs at the glass transition temperature depends on the degree of crystallinity. Above melting temperature, the crystalline polymer will flow.

10.2.3.3 Thermal Properties of Polymer

Thermal stability of polymer is primarily determined by the bond energy of chemical bonds in the polymer chain. When the temperature increases to the point where vibration energy causes bond rupture, the polymer degrades. Polymer with cyclic repeating units can exhibit high thermal stability. The breaking of one bond in a ring does not lead to a decrease in molecular weight, and the probability of two bonds breaking within one ring is low. Thus, ladder or semi-ladder polymers are expected to have higher thermal stabilities than open-chain polymers. The bond strength data will provide information to determine how easy the bond will break at high temperature.

The stability of polymer in air is called thermo-oxidative stability. The presence of oxygen has little effect on the initial decomposition temperature; hence bond rupture is primarily a thermal process. Bond breaking is one step process. The oxidation is two steps process including bond breaking and bond formation.

10.2.3.4 Electrical Properties of Polymer

Most of polymers are insulators; because they are formed by covalent bonds without freely moving electrons or ions. However, they will form conductive paths under electrical stress. The tendency to form a conductive path under different conditions for every polymer is different. The electrical properties of polymers can be expressed by many properties including resistivity, dielectric strength, arc resistance, dielectric constant, and dissipation factor. The resistivity is the resistance that polymer presents to the flow of electric charge. The dielectric strength is the electric breakdown voltage divided by the thickness of the sample. The arc resistance is the property that measures the ease of formation of a conductive path along the surface of a sample. The dielectric constant or permittivity is a measure of how well the insulator will act as a dielectric in a capacitor. This constant is defined as the ratio of the capacitance of the sample in question to the capacitance of a vacuum. A high dielectric constant polymer indicates that the use of thin dielectric sample in capacitors is permitted. The dissipation factor measures the tendency of the polymer to dissipate internally generated thermal energy resulting from an applied alternating electric field. This heating is caused by the movement of the electrons and atoms within the polymer in response to the changing polarity of an alternating electrical field. It is indicated that the nonpolar polymers exhibit higher resistivity, higher dielectric strength, lower dielectric constant, and lower dissipation factor. The polar group has a higher tendency to move electrons.

Most of conjugated polymers are *P*-type semiconductors. After doping with oxidants, the electrons release. Although a complete understanding of the conduction mechanism remains elusive, molecule containing an extended conjugate backbone structure is necessary to have delocalized π electrons upon doping for conductivity. Kivelson proposed [10] the following conducting mechanism for charge transport involving mobile neutral solutions.

The dopant can be either electron acceptors or electron donors. The configurational structure and conformational structure of polymer also have great influence on the conductivity of polymer. Another mechanism involving the formation of mobile charge, polaron, was proposed. The removal of one electron from the polythiophene chain produces a mobile charge in the form of a radical cation. The positive charge tends to induce local atomic displacements, leading to the polaronic behavior. Further oxidation can either convert the polaron into a spinless bipolar or introduce another polaron. In either case, introduction of each positive charge also means introduction of a negatively charged counter ion.

When the common polymers are used as insulator or capacitor, the polymers function as passive material and does not transport charge carrier. On the other hand, when the conducting polymers are used for transistor, battery, photovoltaic device, etc., the polymers function as active material and are capable of transporting charge carrier.

10.2.3.5 Optical Properties of Polymer

The optical properties of polymers [11, 12] can also be passive or active depending on their molecular structures. When the polymers used for transmission, reflection, and so on, their optical properties do not change with any external force, they are functioned as passive material. On the contrary, the optical properties of polymers can be changed under external force such as light or electricity. They are called as active materials.

Insulating polymers are colourless because they do not contain delocalized πelectrons. Amorphous polymers are colourless and transparent in visible light, because their refractive indexes are close to air and does not contain easily delocalized π electrons. The refractive index of polymer depends on their chemical structure and conformation. If the polymer can be easily packed to dense structure through intermolecular force, a high refractive index will be resulted. If there is large difference in refractive index between polymer and air, then the light will reflect at the interface. Since crystalline polymers contain crystallite with dimension larger than the wavelength of visible light, the polymer will reflect or scatter the light.

Conducting polymers which contain conjugated π electrons can absorb light in visible range. The absorption wavelength depends on the ease of delocalization of π electrons in polymers and the morphology of the polymer. The λmax of conducting polymer in film is usually longer than that in solution. In addition to the intrinsic delocalization of π-electrons within the molecule, the moving of π-electrons in the solid film also occurs through the hopping between the molecules. Whole colour spectrum has been discovered from different conducting polymers.

Polymers which exhibit nonlinear dipole moment response or polarization under an applied electric field are called nonlinear optical polymers [13]. They are active function polymers. The nonlinear optical properties can be expressed either by the change in dipole moment.

The polarizable chromophores may be incorporated into the polymer as low-molecular weight compounds dissolved in the polymer matrix or, more commonly, they are designed into the polymer backbone or into a side chain. Crosslinking may also be employed to stabilize the system. A large number of nonlinear optical (NLO) polymers have been reported; examples having NLO chromophores in the backbone and side chain are the polyester 9 and the polyimide 10, respectively [7]. They usually have the structures of electron withdrawing and donating groups in the polymers to exhibit polar and non-central symmetrical characteristics.

10.2.4 Applications of Polymer Materials in Aerospace

The aviation and space industry has generally been a pacemaker for improvement and innovation of new materials frameworks and advances in their creation [14]. The development of aerospace materials can be traced back to the first day of flight, in the year 1903, when the airframe was a wooden structure. After 1927, Al-based alloys achieved the dominant position in aerospace materials with the development of cladding and anodizing technologies. Al-based alloys are dominant in aerospace materials over 80 years [15–17]. However, this situation has been changed in recent few years. Alloys, such as Mg-based alloys, Ti-based alloys, and Ni-based alloys, are developed for aerospace industry with outstanding advantages [18–20].

The key concern of these advanced developments are material properties, available manufacturing options for fabrication, and finally cost. Structural performance in the aerospace sector is predominantly influenced by mechanical properties such as strength, stiffness, and damage tolerance, as well as by physical and chemical properties such as density, and corrosion resistance at ambient and high temperatures. Life cycle costing has also been documented as an important tool to assess the economic feasibility of materials. The prime factors for materials development in the aerospace industries are weight reduction and increased temperature capability. The significance of weight reduction in aerospace systems is a key factor. Based on a rule of thumb, each pound of direct weight reduced over a primary structure brings about another pound saved indirectly in another part of the aircraft [21]. Although the alloy has excellent mechanical properties and heat and corrosion resistance, its weight is always one of the factors limiting the large size of spacecraft. In recent years, polymer materials and composite materials have become the focus of attention in the field of aerospace materials. Polymer materials have key advantages over other conventional metallic materials that are used in different aerospace components. One of the key properties is being light weight while providing specific strength, at the same time thereby reducing overall weight by around 20–40 percent. Other properties include potential for rapid process cycles, ability to meet stringent dimensional stability, lower thermal expansion properties, and excellent fatigue and fracture resistance. This chapter mainly introduces the application of polymer materials in the field of aerospace.

10.2.4.1 Advantages of Polymer Materials

The most common advanced composites are polymer matrix composites. These composites consist of a polymer thermoplastic or thermosetting reinforced by fiber (natural or synthetic). These materials can be fashioned into a variety of shapes and sizes [22, 23]. They provide great strength and stiffness, along with resistance to corrosion. Polymer materials have advantages over conventional materials that are used in different aerospace components [24].

One of the key properties is being light weight while providing specific strength, at the same time thereby reducing overall weight by up to 40 percent. The other properties include potential for rapid process cycles, ability to meet stringent dimensional stability, lower thermal expansion properties, and excellent fatigue and fracture resistance [22]. Modern military fighter aircraft have reduced 30 percent of their weight using application of polymer composite materials. The polymer composites constitute up to 80 percent of modern launch vehicles meant for satellites, comprised of several vital satellite components like the honey comb structures, equipment panels, cylinder support structures, solar array substrates, antennas, etc. The rocket motor cases of the space shuttle solid booster comprise 30 tonnes of graphite-reinforced epoxy composites [17, 25–27].

Other properties include potential for rapid process cycles, ability to meet stringent dimensional stability, lower thermal expansion properties, and excellent fatigue and fracture resistance.

Current aerospace applications have concentrated on replacing the secondary structure with fiber polymer composites, where the reinforcement media have either been carbon, glass, Kevlar or hybrids of these [28]. The need for high performance and capability to design material characteristics to meet specific requirements has made the polymeric materials a first choice for many aerospace applications. Such materials can be tailored to give high strength coupled with relatively low weight, corrosion resistance to most chemicals, and to offer long-term durability under most environmentally severe conditions.

10.2.4.2 Advanced Composite Materials Components

Polymeric composites have numerous applications in the aerospace industry. Often applications do come around where cost is only a secondary concern, as generally advanced polymer materials for specific applications are expensive.

Such applications include cockpit and crew gear, space optical instruments, heat-shrinkage tubing, solar array substrates, high temperature and pressure flare housing, shrouds, and nozzles, appliance moldings, space durable mirrors, high-precision detectors, space optical pipes, multifunctional satellite bus structures, aircraft interiors, and space structural equipment [29].

One focus in the development of advanced composite materials has been to address their vulnerability to impact damage. The orthotropic nature of composite materials results in a relatively low through-thickness strength which is often mitigated by creating damage tolerant designs for structural components. Williams et al. [28] in their study investigated the effect of the embedded Hollow Glass Fiber (HGF) on the host carbon fiber-reinforced epoxy polymer (CFRP) mechanical properties, and also the healing efficiency of the laminates after they were subjected to quasi static impact. Specimens were tested in the undamaged, damaged, and healed conditions using a commercial two-part epoxy healing agent. The results show that the material has good self-healing and mechanical properties.

10.2.4.3 Applications of Polymeric Matrix Composites in Aerospace

Polymeric matrix composites (PMCs) are conceived in such a way that the reinforcement provides stiffness and rigidity that is absent in conventional matrix material

[30]. The reinforcement is mainly required to carry the load, whereas the matrix keeps the constituent of the reinforcement together and distributes loads. This last class is the most widely known, or better, is the class one generally refers to when speaking about composites in a general sense. The class involves several different types of composites and, then, it can be subdivided again, owing to the type of matrix, which is in two main categories: thermosets and thermoplastics.

Thermosets include, unsaturated epoxies, polyesters, and phenolics. Before polymerization, thermosets behave like low-viscosity resin, that cures (polymerizes, cross-links) gradually at a relative low temperature (20 °C to 200 °C). The raw uncured resin molecules are cross-linked through a catalytic chemical reaction to create extremely strong bonds, which are permanently insoluble and infusible. The resin changes state from a liquid to a solid polymer, and cannot be reprocessed by reheating. This means that, once a thermoset composite is formed, it cannot be remolded or reshaped. The recycling of thermoset composites is extremely difficult, because the thermoset resin itself is not directly recyclable. Cured epoxy resins are reasonably stable to chemical attack and are excellent adhesives, having low shrinkage during curing and no emission of volatile gases. Since all these characteristics result in a material with high mechanical properties and high corrosion resistance (coupled with a quite simple curing process), they are the most popular amongst composite matrices. In contrast, epoxies are quite expensive, cannot be stored for a long time, and so are mainly used in high technology areas such as the aeronautical field.

Thermoplastics are also linear polymers, which are composed of chainlike molecules, they may be high-viscosity resins with varying degrees of crystallinity; a number of them can be dissolved in certain liquids, and soften, or melt, upon heating above a certain temperature (100 °C to 400 °C) for additional processing. Some types of thermoplastic resins include: polypropylene (PP), polyvinylchloride, polyether imide (PEI), polyether ether ketone (PEEK), polystyrene, and polyphenylene sulfide (PPS). There are two major advantages of thermoplastic composites: an increased impact resistance to comparable thermoset composites, and the ability to reform. At room temperature they are in a solid state and undergo a physical change, not a chemical reaction as with thermosets, under high temperature and pressure. This allows thermoplastic composites to be reformed and reshaped, allowing also for their recycling at the end of their life. However, some disadvantages exist. Thermoplastic resin occurs naturally in a solid state, it is therefore much more difficult to easily impregnate reinforcing fibers. The resin must be heated to melting point, pressure is required to impregnate fibers, and the composite must then be cooled under this pressure. This is complex and very different from traditional thermoset composite manufacturing, requiring special tooling, techniques, and equipment, many of which may be expensive. What is better still requires debate, also because advances in both thermoset and thermoplastic technologies are constantly happening, and their future may produce unexpected changes.

It is dramatic noting that actually the most used composites include ingredients (matrix, or fibers, or both) mostly coming from petrochemical resources. This, in addition to the safety-at-work concerns, poses serious problems of waste disposal, with risk of environmental pollution. And then, today, in the era in which words like: green, eco-, bio-, environmental protection, etc., are becoming increasingly popular, and the population awareness towards the environment is growing, the development of environmentally-friendly materials becomes mandatory. This is turning attention towards natural resources (e.g., the vegetable world). Polylactic acid (PLA) is an example of a biodegradable polymer which can be derived from renewable resources like corn starch, tapioca, or sugarcane, and which may replace the more conventional plastics (polypropylene, polyethylene) in many application fields. In particular, PLA may be used alone, or as a matrix for the production of composite materials [31]. Indeed, the possibility of reinforcing a natural matrix with natural fibers like flax [32], bamboo [33], or jute [34], opened the way to the production of natural composites. Undoubtedly, replacing conventional composites with biocomposites may allow many environmental issues to be solved [35] since, for example, PLA naturally degrades in a few years, as compared to the hundreds, or thousands, required by the most common plastics. However, a broad introduction of these materials requires thorough knowledge of their properties and performance in-service.

10.2.4.4 Sustainable Biocomposites for Aerospace Components

The aim to reduce CO_2 emissions, usage of fossil resources, and the need for cost control has driven efforts to identify renewable materials [36]. The material selection decision has to be taken during product design, and it should meet market demand, legal requirements, cost, and environmental sustainability [37]. Thou synthetic fiber-reinforced polymer composites have superior mechanical properties, thermal stability, corrosion resistance, and fatigue strength, they are not biodegradable. Biocomposites contain reinforcement which is made from natural resources. Most of the aircraft components have been manufactured from synthetic fiber-reinforced polymer composites [38].

Hybrid composites which are made from two different natural fibers are less common compared to natural/synthetic fiber, but these are also useful materials with respect to environmental issues [39]. Complete replacement of existing synthetic fiber with natural fiber is not possible, because of the superior properties of synthetic fibers, but partial replacement of synthetic fiber with natural fiber will reduce the harmfulness and increase the biodegradability of existing synthetic fiber-based composites.

Jacob John et al. [40] developed the commercial panels for cabin interiors from woven flax and phenolic resin. Haris et al. [41] reviewed the properties of five different natural fibers, namely bamboo, oil palm, banana, kenaf, and pineapple leaf fiber as a reinforcement for bio composites for aircraft radome applications. Boegler et al. [42] have used the ramie fiber composites for an aircraft wing box. While using ramie fiber composites, it was found that the decrease in weight was around 12~14 percent.

Also, fiber-matrix adhesion was promoted by surface modifications of fiber using an alkaline treatment and trimethoxy silane coupling agent. Sepehe et al. [43] studied the mechanical properties of woven hemp fiber reinforced with epoxy matrix composites, and the mechanical test results showed that silane treatment of hemp fibers improves the tensile and flexural properties. To meet the specific mechanical and other properties similar to synthetic fiber composites, most of the fibers are chemically modified in order to increase the compatibility between the fiber and matrix.

However, there is still a lack of research concerning the subject of aircraft structure design, especially regarding the selection of natural fiber, manufacturing processes for specific aircraft component with respect to natural fiber, and performance compared to existing synthetic fiber-based composites, as well as environmental concerns [44].

10.2.4.5 Sustainable Composites for Aerospace Applications

Despite the advanced technologies and techniques available for improving the material used in many industrial applications such as aircraft, civil, mechanical, and the piping industry, defects and damage still occur in components or in areas where repair application and services are needed. The probability of the whole structure being damaged is high, and if the repair work cannot be carried out this may lead to higher costs for a new replacement.

Repairs can be categorized into non-patching procedures for minor damage and reinforcement, or patch procedures to restore structural capability. The specialized technique for repairing cracked structures using high strength advanced composite materials is known as Crack Patching. For damaged or weakened structures, the composite reinforcement or patch procedures can be applied via mechanical fastener or adhesive bonding. There are numerous advantages of using adhesively bonded composite patches as a method of repair compared to mechanically fastened repair methods, including the reduction of installation cost, increase in strength and fatigue life, which leads to effective crack retardation.

The most vital characteristics of a patch repair are less repair time, corrosion resistance, light weight, high stiffness, and elimination of unnecessary fastener holes in an already weakened structure, and stress concentrations at fasteners. Generally, metal, polymers, ceramics, or composites are the materials selected for the patch, and within these classes are many different materials with various advantages and disadvantages associated with their use. Metals are strong materials, but are heavy and susceptible to corrosion; both ceramics and glasses are light but brittle in nature, whereas polymers are malleable and vulnerable to extreme weather [45]. Hence, composites are more preferred for patch repair application, due to their advantages compared to metal and others. A composite material is made by combining two or more materials to give a unique combination of properties [46]. They come in two typical configurations, which are monolithic and sandwich structures. These composite materials have high strength, high stiffness, resistance to corrosion, and durability.

Kenaf fiber (KF) is the most commonly used polymer material in aerospace. Depending on their origins, natural fibers can be classified into seed fibers (cotton, kapok), bast fibers (kenaf, flax, hemp, jute, and ramie), leaf or hard fibers (sisal, pineapple), and fruit fibers (coconut coir). The bast and leaf fibers are the most commonly used fibers in composite applications [47–49].

KF has been proposed to be mixed with other composites such as polypropylene (PP), ethylene-propylene-diene-monomer (EPDM), thermoplastic natural rubber (TPNR), and maleic anhydride polypropylene (MAPP). The attractive features of these fibers include low cost, light weight, high specific modulus, renewability, and biodegradability [50].

The fabrication of high-strength kenaf fiber-reinforced plastics can be obtained by taking the kenaf fibers acquired from the section of the plants that is closest to the

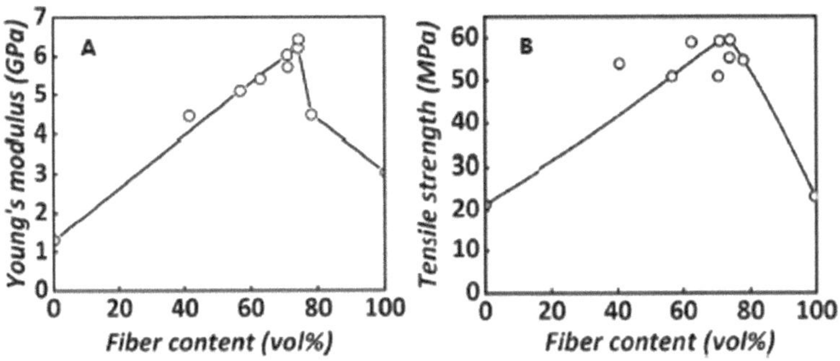

FIGURE 10.2 Relationship between Young modulus, the tensile strength, and the kenaf fiber content of kenaf/PLA composite [53].

ground. It was demonstrated that unidirectional biodegradable composites fabricated using an emulsion-type PLA resin and kenaf fibers at a fiber content of 70 percent have high tensile and flexural strengths of 223 and 254 MPa, respectively. Additionally, for 80 percent of volume fraction kenaf fiber, the tensile strength and tensile modulus are 225 MPa and 22 GPa, respectively, while the flexural strength and flexural modulus are 260 MPa and 22 GPa, respectively. It has been proven in this study that kenaf fiber exhibits higher strength values in terms of tensile and flexural properties as compared to other natural fibers when reinforcing PLA [51].

Generally, optimum tensile properties and Young modulus are dictated by the volume of reinforcing fiber used for the composites [52]. As can be seen in Figure 10.2, both properties increased with increasing fiber content.

The addition of long kenaf fiber as reinforcement with wood flour-pp composite has significantly increased tensile strength and Young modulus [53]. The alkalization will improve the tensile properties of the fiber. The 0.25 volume fraction of kenaf fiber in polypropylene has improved the thermal stability of the polypropylene matrix using DSC and DMA techniques. This shows that the addition of fibers has not only improved the thermal stability, but also the viscoelastic behavior of the composite. Kenaf fiber has also been used as reinforcement in bio-composite material. The objective was to compare the mechanical properties of short kenaf, bast, and core fiber-reinforced unsaturated polyester composites with varying fiber weight fractions at 0%, 5%, 10%, 20%, 30%, and 40%. The results showed that the optimum fiber content for achieving highest tensile strength for both bast and core fiber composites was 20wt% [54].

Natural fiber composites also provide positive environmental benefits, such as reduced dependence on nonrenewable energy and material sources, lower pollutant emissions, lower greenhouse gas emissions, enhanced energy recovery, and end-of-life biodegradability of components. Kenaf has been identified as an excellent source of fibers for the manufacturing of a large range of paper products. Kenaf accumulates carbon dioxide at a significantly higher rate compared with hemp, maize, and sorghum [55]. In terms of social-economic issues, the use of natural fibers as a source of raw materials is beneficial, because it can generate an economic development opportunity for nonfood farm products.

10.3 FREE RADICAL OF POLYMER CAUSED BY RADIATION

Radiation polymerization, also known as radiation-initiated polymerization, is a polymerization reaction that occurs by applying high-energy ionizing radiation (α-rays, β-rays, γ-rays, x-rays, electron beams) to the monomer to generate ions or free radicals and form active centers. Polymerization radiation is the monomer under the action of high-energy ionizing radiation, the polymerization system can produce radicals, anions and cations at the same time. Therefore, the radiation polymerization mechanism includes free radical mechanism, cation mechanism and anion mechanism. Since there is a cage effect in the cloud, free radicals can be rapidly compounded among themselves and between anions and cations, and only a small portion can escape from the cloud and form steady-state conditions [56]. Usually free radicals have a much higher chance of escaping from the cloud than anions and cations, so the vast majority of polymerization systems without special drying and with relatively low dose rates are based on free radical polymerization. Radiation polymerization does not require the addition of initiator and catalyst, the resulting polymer is more pure, the penetration ability of the radiation is very strong, uniform distribution in the irradiated system, radiation solid polymerization can be carried out, radiation polymerization at room temperature or low temperature, easy to control, the reaction rate and molecular weight [57]. Can be quantitatively controlled by adjusting the reaction start and termination of the measurement and dose rate and other factors.

Radiation polymerization is the application of ionizing radiation energy to initiate the polymerization reaction of organic monomers (mainly vinyl monomers) so that polymeric compounds can be obtained. The main difference between it and ordinary polymerization reactions lies in the chain initiation method. For chain polymerization reactions, radiation polymerization is initiated by ionizing radiation to reactive particles (radicals or ions) [58,59], rather than adding various types of initiators; the difference between the chain growth and chain termination stages and the ordinary polymerization method is less obvious. Compared with initiator initiation, thermal initiation and other polymerization, radiation polymerization has the following characteristics:

1. The resulting polymers are purer and free of initiator residues. This is especially important for the synthesis of biomedical polymers.
2. The polymerization reaction is easy to control, and with the penetrating γ-ray, the polymerization reaction can be carried out uniformly and continuously, preventing local overheating and uneven reactions.
3. It can be carried out at room temperature or low temperature. The chain initiation activation energy (Ei) in the radiation polymerization reaction is very low and independent of the reaction temperature.
4. The molecular weight and molecular weight distribution of the resulting polymer can be controlled by polymerization conditions such as dose rate.

Research on radiation polymerization started relatively early. a large number of experimental results have been accumulated in the basic and applied aspects. By the end of the 1960s, more than 350 organic monomers had been studied. Later, the scope of radiation polymerization research continues to expand, the reaction mechanism, the impact of various types of radiation polymerization reaction factors have a clearer understanding of the industrial scale application of radiation polymerization examples are still few, but a number of projects have been or are being intermediate tests, some of which are promising for industrial production.

The importance of radiation aggregation can be viewed from two perspectives.

(1) The ionizing radiation is regarded as a practical technique for initiating polymerization reactions to study the conditions of polymerization reactions under the action of ionizing radiation, the regularity and the characteristics of the resulting polymers (especially those polymerization reactions that are difficult to carry out by conventional methods), and most of the existing literature on radiation polymerization belongs to this category.
(2) In general, the study of radiation polymerization is based on the principles of radiation chemistry and theoretical and experimental methods of conventional polymerization reactions. It has been recognized that the research results of radiation polymerization of certain monomer systems can also deepen the understanding of the basic processes of radiation chemistry and enrich and deepen the understanding of conventional polymerization reactions.

There are many methods of radiation polymerization, including gas-phase polymerization, liquid-phase polymerization, and solid-phase polymerization. Polymerization reactions can take place in homogeneous or inhomogeneous phases. It can be a one-component system, or two-component, or even multi-component system. Different polymerization methods have their own characteristics, and the mechanism is not the same.

Radiation polymerization is mostly studied with vinyl monomer, and there are few examples of ring-opening polymerization and heterogeneous polymerization. such as ethylene at various pressures and temperatures [60–62], in solution or in supercritical CO_2 [63] and at pilot-scale [64,65]. It was confirmed that short chain branching is considerably reduced by γ-ray initiation at low temperature [66].

Liquid homogeneous radiation polymerization has been studied the most. Since most monomers are liquid at room temperature. Therefore, the mechanism and kinetic analysis of liquid-phase radiation polymerization reactions have been studied in depth and are somewhat representative.

Advanced studies on the γ-ray induced polymerization of styrene under pressure were conducted from 1970 to 1983 to clarify mechanistic aspects and to determine kinetic constants and thermodynamic parameters such as the activation volume, as a function of temperature and pressure in the bulk state and in emulsion. Rubbers based on butadiene were also studied quite intensively from the late 1950s to the early 1990s in solution and in emulsions [58,67].

Access to specialty polymers was explored with vinyl acetate and butyl acrylate polymerized in emulsion [68,69], or butyl acrylate copolymerized with butadiene, or polymerized at the surface of rubber particles to improve their compatibility with host matrices [70]. Water soluble poly(acrylamide) was obtained by γ-ray induced polymerization in various hydro-organic solutions to improve its efficiency as a flocculant [71]. The unique properties of fluorinated polymers stimulated studies on new synthetic approaches based on radiation-induced homo- and copolymerization of tetrafluoroethylene [72], hexafluoropropene [73], methyl trifluoroacrylate, in bulk conditions or in emulsion have been studied [74].

Radiation-induced polymerization also proceeds in bulk solid state in spite of the limited molecular diffusion for the propagation steps and of an expectable poor energetic efficiency. Topochemical effects and confinement in solid guest materials have stimulated various fundamental investigations. Contrasting behaviors were observed depending on monomer orientation and on the gap between the reactive functions within the crystal lattice. In some systems, polymerization seems to be initiated within crystalline defects, and continues as the formation of amorphous polymer domains affects the original crystalline lattice. In some favorable systems, the relative orientation of monomers prefigures the order of repeating units of the resultant polymer [75].

10.3.1 Excited State Processes

10.3.1.1 Excited Processes

The total energy of a molecule is a combination of electronic energy, vibrational energy, rotational energy and affine energy. However, as far as the radiative excited states are concerned, the changes of electronic and vibrational energies are mainly studied.

The vibronic intensity (f) of the electron leap is related to the total area under the absorption band. The relative intensity of each vibrational band is then related to the difference between the equilibrium geometrical configuration of the excited state and the equilibrium geometrical configuration of the ground state. The example of an atomic bimolecule is shown by a two-dimensional Morse curve. The horizontal coordinate indicates the interatomic distance (r_{AB}), which increases the potential energy of the system as the repulsive force increases, but quickly, while increasing the interatomic distance makes the bond between A and B elongate, slowly increasing the potential energy. Finally, when the bonds elongate repeatedly and the total molecular energy exceeds a certain limiting value, the bonds break and the molecules dissociate in the area of the potential energy curve The horizontal lines intersecting the potential energy curve represent the different vibrational energy levels of the molecules. The pre- and post-excitation states are discussed as an example. The radiative leap obeys the Franck-Condon principle. This principle states that when an electron leaps, the nuclear spacing in the molecule is constant. In other words, the leap is a vertical process. Because the redistribution of electrons (change of orbitals) is very fast, about 10~15 seconds, while the molecule vibrates in 10~13 seconds, the former is two orders of magnitude faster than the latter. Therefore, the geometric configuration of the molecule remains unchanged before and after the absorption of the radiation energy. The relative odds of various vibrational jumps are called Franck-Condon factors.

The above is a light approximation approach to the formation of excited states by radiation leaps. There is no major difference in the nature of the action of light and ionizing radiation. The difference lies in the fact that ionizing radiation acting on matter can realize light forbidden leap and can also appear as super ground state molecules.

10.3.1.2 Deactivation Process of the Excited State

A molecule is in the electronically excited state Sx or Tx and is at a higher vibrational energy level, and then the de-excitation process, this molecule will first undergo vibrational relaxation, i.e., loss of vibrational energy to surrounding energy, while in the range of 10~13 and 10~12 seconds to the vibrational energy level of Sx and Tx with v=0. What continues is the electronic relaxation, the drop to lower energy levels by radiative energy or non-radiative means, and analogously continues, eventually returning the molecule to the ground state. The relaxation can be a photophysical process or a chemical reaction.

10.3.2 Formation and Reaction of Free Radicals

When most systems are irradiated, free radicals are also one of the primary products. They come from excited molecules and ions or are generated through secondary reactions, so the determination of the free radical primary yield requires the addition of appropriate scavengers to eliminate the effects of secondary reactions.

The radicals are generally characterized as relatively reactive, such as small molecule radicals Some molecules have triplet ground states, such as O_2 with two spin-parallel uncoupled electrons, and show double radical behavior in chemical reactions. The reactive radicals have a short lifetime and can easily initiate chemical reactions in the system. Low-temperature condensed states may exist for a long time due to diffusion difficulties and fall into potential wells created by molecules; such radicals, called fallen radicals, can sometimes exist for several years and still have initiating activity after thawing. There is also a class of radicals that are inherently more stable, for example, trimethyl-tert-butyl radicals, allyl and benzyl radicals may acquire a very large stability due to resonance ionization of electrons. Free radicals are generally characterized as more reactive.

Regarding the charged radicals are more commonly found in radiation chemistry, such as the ionization of the formation of M^+ as a positive ion radical. In all cases ionic radicals can be seen as formed by the gain or loss of an electron from the parent organic molecule. Examples include, positive ion radicals, benzophenone negative ion radicals and ion radicals formed by the irradiation of alkene monomers.

10.3.2.1 Radiation-initiated Free Radical Polymerization Reaction

Simplified kinetic analysis of polymerization reaction

Vinyl monomer free radical polymerization is a typical chain reaction, and its reaction course consists of chain initiation, chain growth and chain termination stages [22].

(1) Chain initiation

Take initiator chemical initiation as an example, the first step of initiation process is the generation of primary radicals by initiator under the influence of external factors (temperature). The second step is the interaction of the initial radical with the monomer to open the double bond and form a new radical, that is, the monomer radical.

(2) Chain growth

The chain growth process starts when the first monomeric radical is generated. Since the activation energy required for the chain growth reaction is about 20~30 kJ/mol, and the radical polymerization reaction is an exothermic process, the chain growth process is very rapid.

(3) Chain termination

There are two main ways of chain termination. Double base composite and disproportionation reaction. The main difference between monomer ionization radiation polymerization and the above polymerization process is in the chain initiation stage. In the radiation polymerization, no initiator is added, and the ionizing radiation can initiate the polymerization reaction by ionization and excitation processes to generate reactive particles (radicals, ions). In the chain growth and termination stages, there is no significant difference.

It should be noted here that in addition to the above three radical reactions, the free radical polymerization reaction also includes chain transfer reactions. The chain transfer reaction is characterized by the transfer of the active center to other molecules, and the total concentration of free radicals in the system does not change. The chain transfer process has no effect on the rate of polymerization regardless of whether the active center is transferred to the monomer, initiator, solvent or to the resulting macromolecule, but it does reduce the molecular weight of the resulting polymer.

10.3.2.2 Spatial Distribution of Radiation-induced Radicals

At least in the condensed phase, after receiving ionizing radiation energy, the original generation of free radicals is distributed along the radiation trail, completely separated from each other in the cloud or sting, free radical distribution is not uniform for the whole system. Initially, the concentration of radicals can be very high locally (in clouds and stingers), and within a short time ($\approx 10^{-7s}$), the radical concentration decreases rapidly due to thermal diffusion or complexation, and due to diffusion, the original radical clouds expand and overlap and complex with adjacent clouds, which eventually leads to a uniform distribution of radicals in the system. The spatial distribution of radicals is very important from the perspective of polymerization reaction kinetics, which can be broadly classified into three cases as follows.

(1) During the whole growth chain lifetime, each radical cloud can be considered as an isolated system. All polymerization chains started in the cloud are terminated only by the interaction with the radicals generated in the same cloud. Thus, the number of polymers generated per cloud is a statistically fixed value, and the polymerization rate of the overall polymerization system is proportional to the number of clouds generated per unit time, i.e., proportional to the dose rate. At this point, the molecular weight of the polymer is determined only by the reaction kinetics in individual clouds, while the number-average polymerization degree ($\overline{X_n}$) is independent of the number of clouds per unit time, independent of the dose rate.
(2) During the entire growth chain lifetime (τ), all radical clouds overlap each other. At this time, the distribution of free radicals in the whole polymerization system can be considered as uniform. and normal reaction kinetics apply, with the polymerization rate proportional to the root of the dose rate $\dot{D}$ and $\overline{X_n}$ proportional to the reciprocal of the square root of the dose rate.
(3) The case in between is the transition state from the first case to the third case, where the radical clouds only partially overlap. In this case, the dose-rate index of the polymerization rate should be between 1 and 0.5, and the dose rate index of $\overline{X_n}$, on the other hand, ranges from 0 to -0.5.

The first case is perhaps suitable for very low dose rates, while the second case is suitable for high dose rates. However, it has been experimentally demonstrated that for most rays with low linear energy transfer (LET value) (X-rays, γ-rays or fast electron beams), the distribution of growth chains during radiation polymerization can generally be regarded as homogeneous. This is because the lifetime of the growth chains is relatively long, generally in the order of seconds, and even longer in some systems. The number of clouds is large in time, and neighboring clouds have sufficient time to overlap with each other. Even at very low dose rates, it is reasonable to treat the kinetics with a uniform distribution of radicals, because the lifetime of the radical chains increases with decreasing dose rates.

10.3.2.3 Polymerization Reaction Initiated by Polymer Radicals

When the polymerization conversion rate is high and the resulting polymer accumulates to a certain extent, the resulting polymer can also absorb radiation energy to generate polymer radicals. The latter can also trigger the polymerization of monomers to produce long branched polymers

$$\mathrm{P} \rightarrow 2\mathrm{R}\cdot(\text{or } \mathrm{R}\cdot + \mathrm{P}\cdot)$$

Their initiation rates are $kG_R^M\dot{D} + kG_R^P\dot{D}$. Where, G_R^P is the polymer radical yield, G_R^M is the monomer radical yield, and k is the numerical conversion factor [23].

In general, since monomers contain double bonds, even conjugated structures (e.g. styrene), while double bonds disappear in polymers. So the monomer is more stable than the polymer, generally $G_R^P > G_R^M$. If $G_R^M = 0.69$ for styrene, while G_R^P for polystyrene increases to 1-3. In this case, the initiation rate should increase continuously with the monomer conversion. Another consequence of this effect is that the polymers generated by polymer radical initiation have higher molecular weights, contain branched polymers, and even cross-linked gels, typical examples being the radiation polymerization of vinyl acetate.

10.3.3 Structural Changes

Due to radiation-induced activation of polymer molecules, which results in a wide variety of chemical or physical changes, the main types are, (1) formation of chemical bonds between molecular chains-radiation crosslinking, (2) breakage of the macromolecular backbone and decrease in the average molecular weight-radiation lecture, (3) changes in the concentration of unsaturated structures and special microscopic, (4) oxidation and other processes, (5) release of gas reactions, (6) isomerization and cyclization, (7) grafting and block copolymerization, etc. The chemical and structural changes lead to changes in the physical and mechanical properties of the polymer. Such as crystallinity, melting point, specific gravity, solubility properties, electrical properties, modulus of elasticity, tensile strength, hardness, permeability, and so forth. It is experimentally proven that the occurrence of chemical and physical changes is mainly determined by the chemical composition and structure of the polymer, and has little to do with the type of radiation. A number of laws have been generalized on the basis of a large number of experimental facts. More often than not, radiation crosslinking and degradation occur simultaneously, it just depends on the conditions and which one takes precedence under certain conditions. For hetero-chain polymers this rule does not necessarily apply.

10.4 RADIATION DAMAGE BEHAVIOR OF POLYMER

10.4.1 Electrical Behavior Caused by Radiation

10.4.1.1 The Influence Law of Gamma Ray Irradiation on the Dielectric Properties of Polyimide

The variation curves of the dielectric constant, dielectric loss coefficient and insulation resistance of polyimide before and after gamma ray irradiation show in Figure 10.3. As shown in Figure 10.3(a), under different irradiation dose conditions, the change trend of the dielectric constant of polyimide with frequency is the same. It decreases as the frequency increases in the entire frequency range. Under the same frequency conditions, with the increase of irradiation absorption dose, the dielectric constant of polyimide decreases significantly. This is mainly due to the fact that the molecular chain of polyimide breaks during the gamma ray irradiation process. As shown in Figure 10.3(b), under different irradiation dose conditions, the dielectric loss coefficient of polyimide slightly increases with the increase of irradiation dose, especially in the low frequency stage. As shown in Figure 10.3(c), under different irradiation dose conditions, the insulation resistance of polyimide decreases with the increase of irradiation dose. This indicates that gamma ray irradiation causes the insulation performance of polyimide to deteriorate.

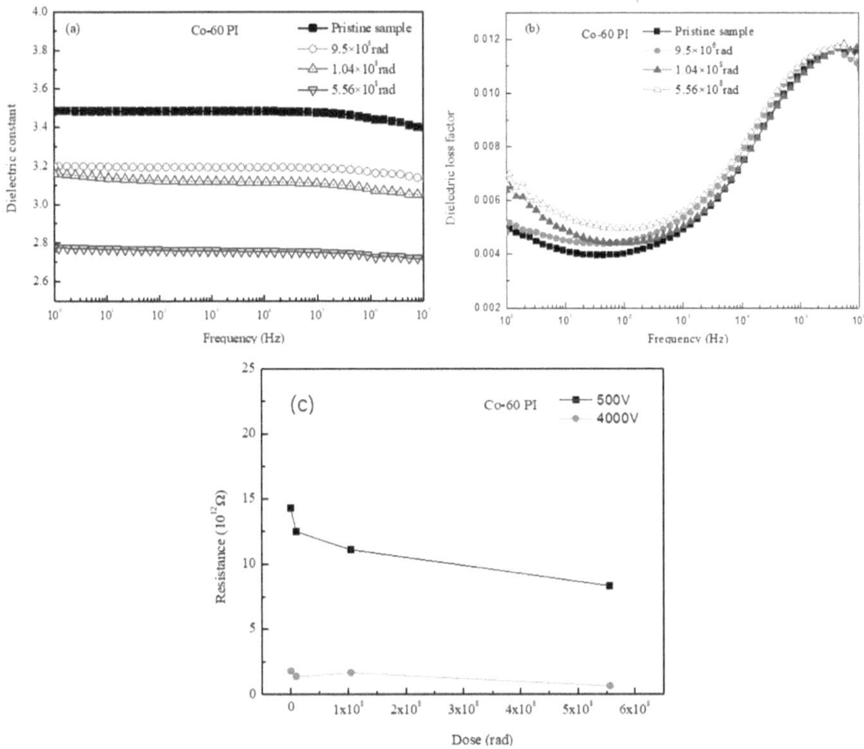

FIGURE 10.3 Evolution curves of dielectric constant, dielectric loss coefficient and insulation resistance of polyimide before and after gamma ray irradiation; (a) dielectric constant, (b) dielectric loss coefficient, (c) insulation resistance.

10.4.1.2 The Influence Law of Gamma Ray Irradiation on the Dielectric Properties of Cross-linked Ethylene Tetrafluoroethylene (XETFE)

Figure 10.4 shows the variation curves of the dielectric constant, dielectric loss coefficient, and insulation resistance of cross-linked ethylene tetrafluoroethylene (XETFE) after gamma ray irradiation. As shown in Figure 10.4(a), under different gamma ray irradiation doses, the trend of the dielectric constant of XETFE with frequency is similar, decreasing with the increase of frequency throughout the frequency range. Under the same frequency conditions, the dielectric constant of XETFE increases with the increase of irradiation dose. As shown in Figure 10.4(b), the variation of the dielectric loss coefficient of XETFE with frequency is the same under different irradiation doses.

Under the same frequency conditions, the dielectric loss of XETFE increases with the increase of irradiation dose. The insulation resistance test analysis of cross-linked ethylene tetrafluoroethylene after gamma ray irradiation shows that the insulation performance of XETFE is significantly degraded due to gamma ray irradiation.

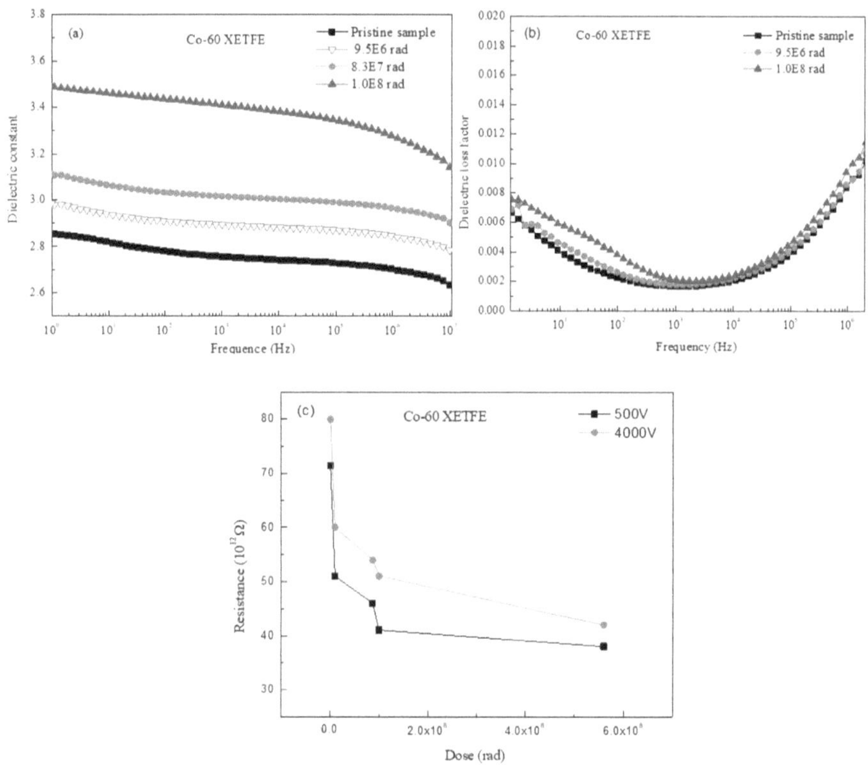

FIGURE 10.4 Variation curves of the dielectric constant, dielectric loss coefficient, and insulation resistance of cross-linked ethylene tetrafluoroethylene copolymer material after gamma ray irradiation. (a) Dielectric constant, (b) Dielectric loss coefficient, (c) Insulation resistance.

10.4.1.3 The Influence Law of Electron Irradiation on the Dielectric Properties of Polyimide

Figure 10.5 shows the evolution curves of the dielectric constant, dielectric loss coefficient, and insulation resistance of polyimide before and after 1 MeV electron irradiation as a function of frequency. Comparing Figure 10.3(a) and 4(a), it can be seen that the variation trend of the dielectric constant of polyimide with frequency under gamma ray and 1 MeV electron irradiation conditions is similar, decreasing with the increase of frequency throughout the frequency range. Under the same frequency conditions, the dielectric constant of polyimide decreases with the increase of irradiation dose. By comparing Figure 10.3(b) and 4(b), it can be seen that the dielectric loss coefficient of polyimide under gamma ray and 1 MeV electron irradiation conditions exhibits obvious different characteristics with the change of frequency. Under gamma ray irradiation conditions, the dielectric loss coefficient at low frequencies slightly increases with the increase of irradiation dose, while under 1 MeV electron irradiation conditions, the dielectric loss coefficient at high frequencies first

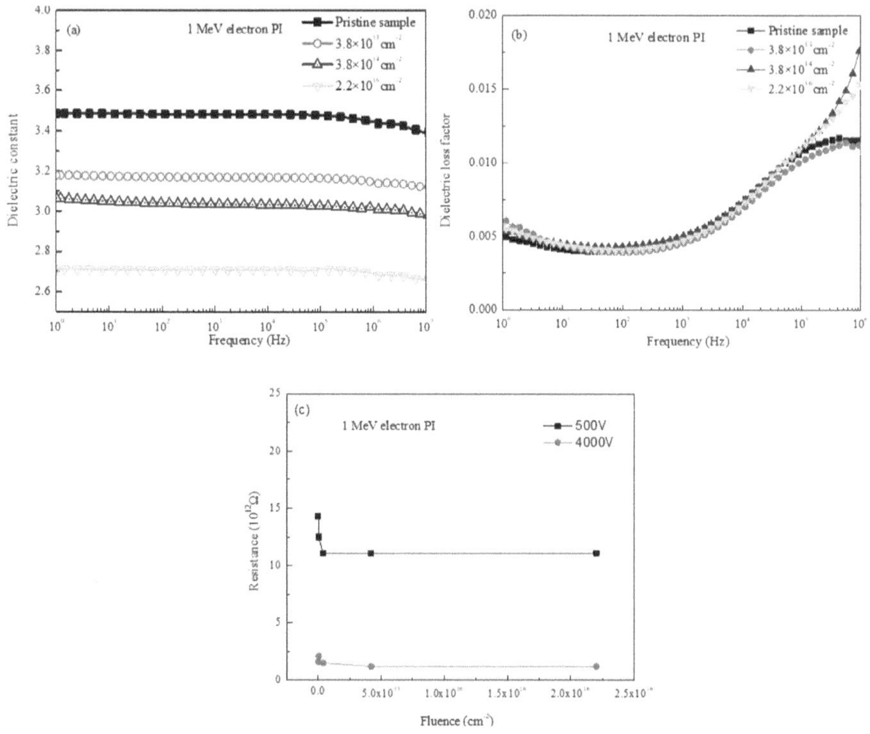

FIGURE 10.5 Variation curves of the dielectric constant, dielectric loss coefficient, and insulation resistance of polyimide before and after 1 MeV electron irradiation. (a) dielectric constant (b) dielectric loss coefficient (c) insulation resistance.

increases and then decreases with the increase of irradiation dose. This indicates that the polarization mode of polyimide changes when the irradiation dose reaches a certain level. Figure 10.5(c) shows that with the increase of the irradiation dose, the insulation resistance of polyimide decreases, and the resistance decrease tends to saturate when the dose reaches 3.8E14 e/cm^2. These results indicate that 1 MeV electron irradiation can deteriorate the insulation performance of polyimide.

Figure 10.6 shows the variation curves of the dielectric constant, dielectric loss coefficient, and insulation resistance of crosslinked ethylene tetrafluoroethylene before and after 1 MeV electron irradiation. By comparing Figure 10.4 and Figure 10.6, it can be seen that the characteristics of the variation in dielectric constant, dielectric loss coefficient, and insulation resistance of crosslinked ethylene tetrafluoroethylene are similar after gamma ray and 1 MeV electron irradiation. This may be due to the same damage mechanisms caused by the two irradiation sources to crosslinked ethylene tetrafluoroethylene. Figure 10.6(a) shows that the variation trend of the dielectric constant of crosslinked ethylene tetrafluoroethylene with frequency under different irradiation doses of 1 MeV electrons is similar, decreasing with the increase of frequency throughout the frequency range. Under the same frequency conditions, the dielectric constant and capacitance of crosslinked ethylene tetrafluoroethylene

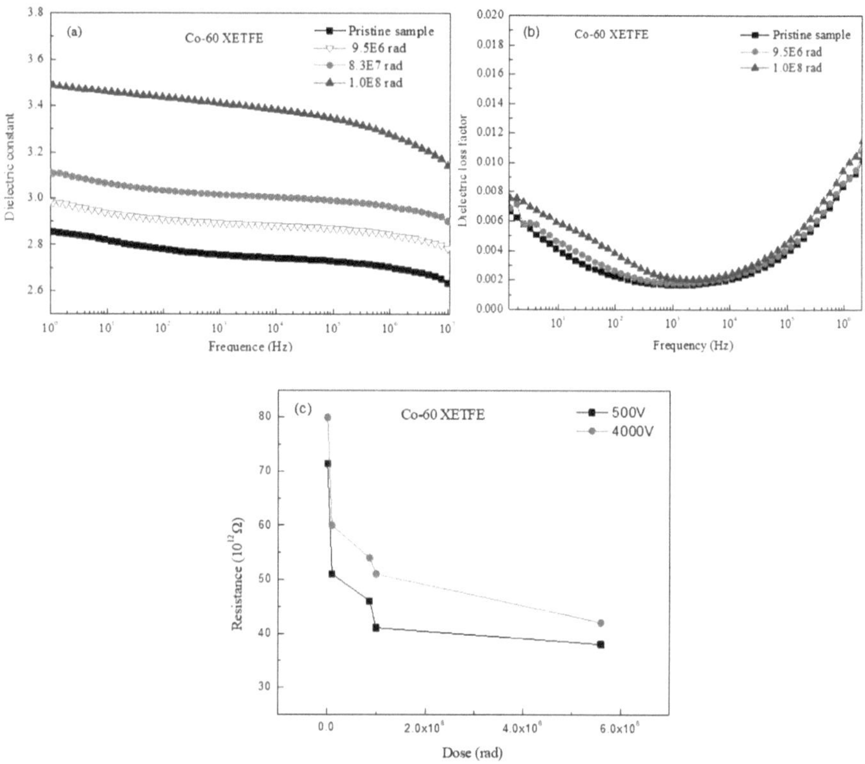

FIGURE 10.6 Evolution curves of the dielectric constant, capacitance, and conductivity of crosslinked ethylene tetrafluoroethylene copolymer materials as a function of frequency after 1 MeV electron irradiation, (a) dielectric constant, (b) dielectric loss coefficient, (c) insulation resistance.

increase with the increase of irradiation dose. Figure 10.6(b) shows that the dielectric loss coefficient of crosslinked ethylene tetrafluoroethylene slightly increases with the increase of irradiation dose. Figure 10.6(c) shows that the insulation resistance of crosslinked ethylene tetrafluoroethylene significantly decreases with the increase of irradiation dose, indicating that 1 MeV electron irradiation can deteriorate the insulation performance of crosslinked ethylene tetrafluoroethylene.

10.4.1.4 The Influence Law of Low-Energy Proton Irradiation on the Dielectric Properties of Polyimide

Figure 10.7 shows the evolution curves of the dielectric constant of polyimide as a function of frequency after proton irradiation at energies of 70 keV, 110 keV and 170 keV. As observed in the figure, after irradiation with the three different energies of protons, the dielectric constant of the polyimide increases with the increasing proton irradiation dose, exhibiting a similar evolution trend. When the proton dose is 1E15 cm^{-2}, the dielectric constant decreases overall. However, when the dose

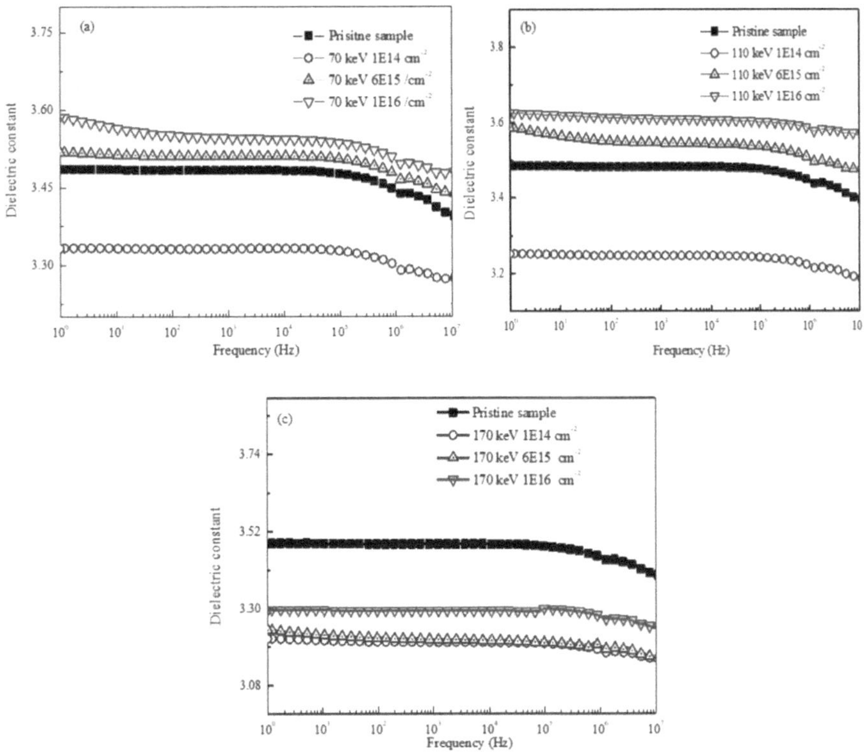

FIGURE 10.7 Evolution curves of the dielectric constant of polyimide as a function of frequency after low-energy proton irradiation at 300 K.(a) Proton irradiation at 70 keV, (b) Proton irradiation at 110 keV, (c) Proton irradiation at 170 keV.

increases to 6E15 cm^{-2}, the dielectric constant increases overall. Due to low-energy proton irradiation damage, the degradation of the bisphenol-A dianhydride groups in the polyimide molecules leads to a decrease in the content of dipole groups C=O and C-N in the material, which will cause a decrease in the polarization ability of the polyimide and thus a decrease in its dielectric constant. However, at high doses, the dielectric constant of polyimide after low-energy proton irradiation increases. It can be inferred that there are other factors at this temperature that can enhance the polarization ability of polyimide. Since low-energy protons cannot penetrate the sample, the maximum damage will occur at the end of their trajectory, resulting in a significant non-uniform damage in the interior of the material. Moreover, with the increasing irradiation dose, the degree of non-uniform damage will monotonically increase. The interfacial polarization caused by this non-uniform damage will affect the dielectric behavior of the material. At low doses, due to weak irradiation damage, this non-uniform damage is not the main factor dominating the medium polarization. The decrease in dipole groups is the main mechanism leading to the decrease in the dielectric constant of polyimide. As the proton irradiation dose increases, the degree of internal damage in the polyimide intensifies, and the polarization ability of the material caused by the interfacial polarization at the end of the low-energy proton range significantly improves, thus dominating the dielectric polarization behavior. As a result, the dielectric constant of polyimide increases at higher irradiation doses.

10.4.1.5 The Influence Law of Low-energy Proton Irradiation on the Dielectric Properties of Crosslinked Polyethylene Tetrafluoroethylene

Figure 10.8 shows the variation curves of the dielectric constant, dielectric loss factor, and insulation resistance of crosslinked ethylene tetrafluoroethylene after 70 keV proton irradiation. It can be observed from the figure that under the conditions of 70 keV proton irradiation, the variation trend of the dielectric constant of crosslinked ethylene tetrafluoroethylene is similar with the increase in frequency causing a decrease across the entire frequency range. Under the same frequency conditions, the dielectric constant of crosslinked ethylene tetrafluoroethylene slightly increases with the increase in irradiation dose, only changing within a small range. As shown in Figure 10.8(b), the characteristics of the variation in dielectric loss of crosslinked ethylene tetrafluoroethylene under different irradiation doses are different with the

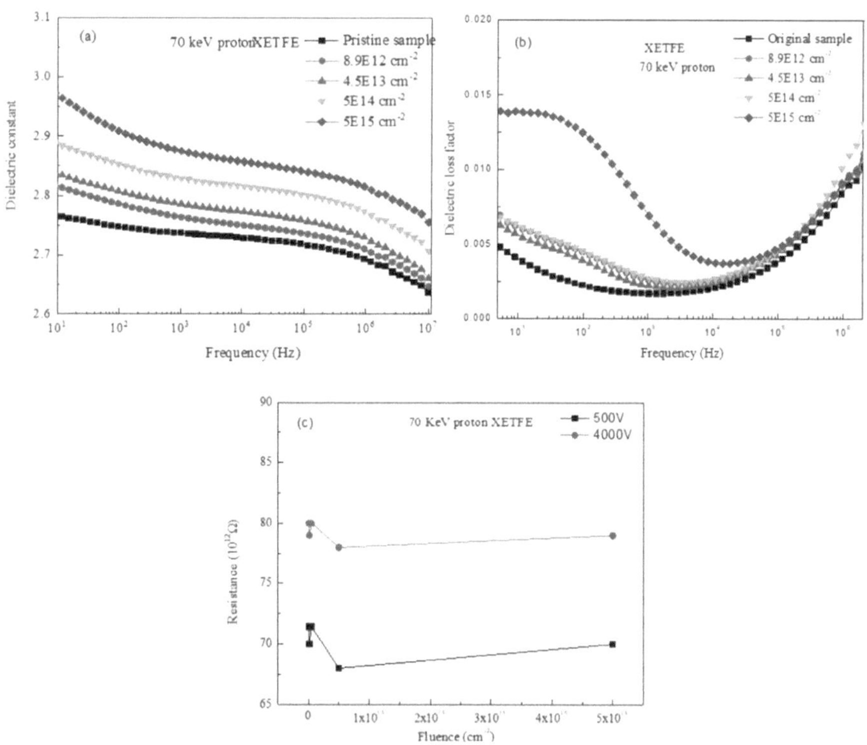

FIGURE 10.8 Curves of the dielectric constant, dielectric loss coefficient, and insulation resistance of crosslinked ethylene tetrafluoroethylene copolymer materials after 70 keV proton irradiation. (a) Dielectric constant, (b) Dielectric loss coefficient, (c) Insulation resistance.

irradiation dose greater than 5E14cm^{-2} causing a significant increase in the low-frequency range due to non-uniform interface polarization relaxation caused by radiation. Figure 10.8(c) shows that the insulation resistance of crosslinked ethylene tetrafluoroethylene is slightly decreased after 70 keV proton irradiation.

10.4.2 Mechanical Behavior Caused by Radiation

Figure 10.9 presents the stress-strain curves of polyimide under γ-ray irradiation. As shown in the figure, there is no significant change in the stress-strain curves of polyimide before and after irradiation, both exhibiting two distinct stages. In the first stage of deformation, the stress gradually increases and reaches a peak, showing strain strengthening characteristics. There is no significant strain softening feature, and it directly enters the second stage. In the second stage of deformation, the stress increases again until fracture occurs. It is worth noting that the irradiation dose significantly affects the tensile strength. When the irradiation dose is below 1.0E8rad, the tensile strength of the samples after irradiation is significantly reduced compared with that before irradiation, while the tensile strength slightly increases when the irradiation dose reaches 5.56E8rad, but the elongation at break significantly decreases.

Figure 10.10 illustrates the stress-strain curves of crosslinked ethylene tetrafluoroethylene copolymer under γ-ray irradiation. It can be observed from the Figure 10.9 that there is no significant change in the stress-strain curve characteristics of crosslinked ethylene tetrafluoroethylene copolymer before and after irradiation when the irradiation dose is less than 5.6E8rad, both exhibiting two distinct stages. The tensile yield strength and elongation at break of ethylene tetrafluoroethylene copolymer are significantly affected by γ-ray irradiation, with increasing irradiation dose leading to an increase in the yield strength and a decrease in the elongation at break. When the irradiation dose reaches 5.6E8, the stress-strain curve of the material undergoes significant changes, and fracture occurs after only the first stage.

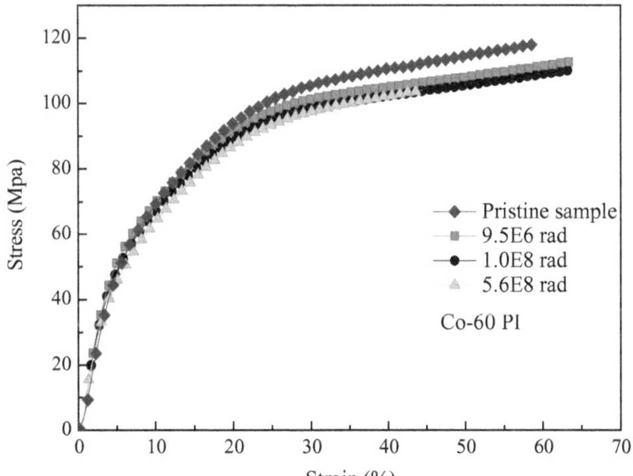

FIGURE 10.9 Stress-strain curves of polyimide under γ-ray irradiation.

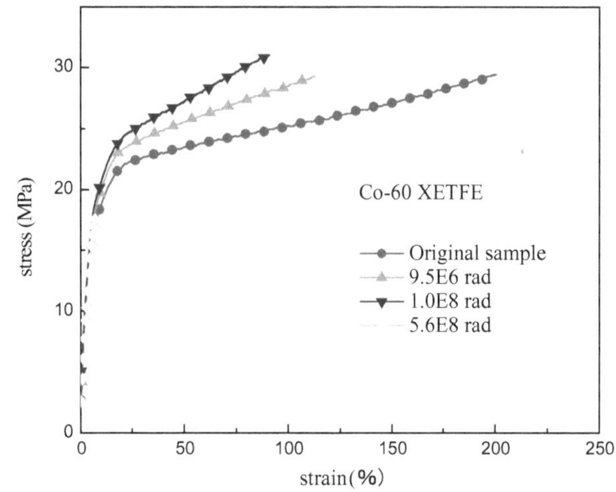

FIGURE 10.10 Stress-strain curve of crosslinked ethylene tetrafluoroethylene material after γ-ray irradiation.

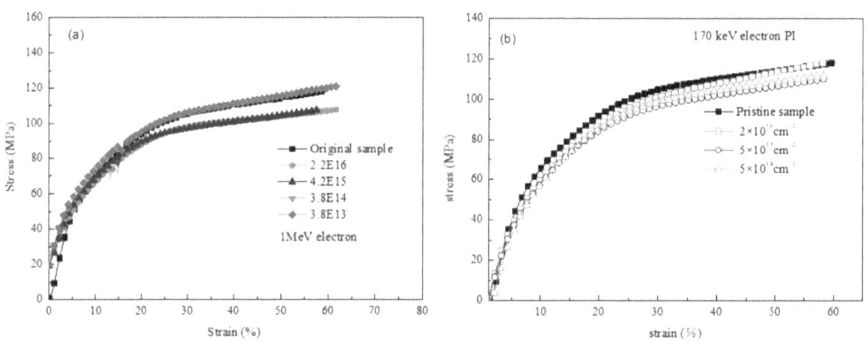

FIGURE 10.11 Stress-strain curves of polyimide under 1MeV and 170keV electron irradiation conditions.

Figure 10.11 presents the stress-strain curves of polyimide under 1MeV electron and 170keV electron irradiation conditions. It can be seen from the figure that there is no significant change in the stress-strain curve characteristics of polyimide before and after electron irradiation at different energies, both exhibiting two distinct stages. The irradiation dose has no significant effect on the tensile strength and elongation at break of the material. A comparative study of the effects of high and low energy electron irradiation and γ-ray irradiation on the tensile properties of polyimide shows that electron and γ-ray irradiation have no significant effect on its tensile properties. This indirectly indicates that the material has good resistance to ionizing radiation.

Figure 10.12 presents the stress–strain curve of polyimide under 170keV proton irradiation conditions. It can be seen from the figure that there is no significant change in the stress-strain curve characteristics of polyimide after irradiation at different

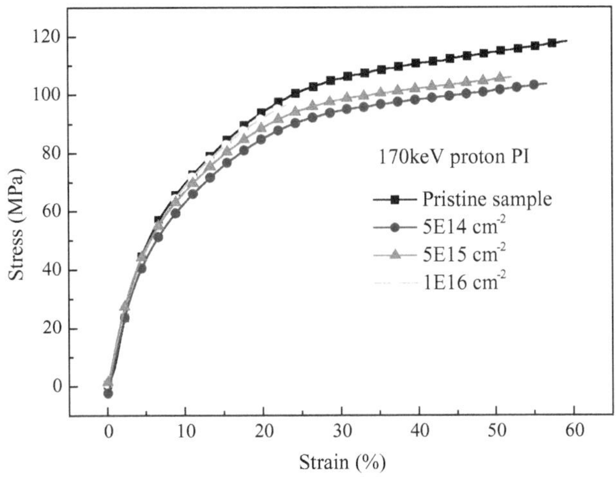

FIGURE 10.12 The stress-strain curve of polyimide under 170keV proton irradiation conditions.

doses when the dose is less than 1E6 p/cm^2, both exhibiting two distinct stages, and the effect of irradiation dose on the tensile strength and elongation at break of the material is not significant. When the dose reaches 1E6 p/cm^2, the elongation at break of polyimide is significantly reduced. This may be due to the fact that 170keV protons can only affect the surface layer of the material, forming an interface between the proton-affected region and the unaffected region. When the dose reaches a certain level, the presence of this interface will promote material fracture.

10.5 SUMMARY AND PERSPECTIVE

This chapter introduces the radiation damage behavior of polymer materials in space radiation environment. The preparation process, physical and chemical properties of thermoset and thermoplastic materials and their application in aerospace field are briefly described. The excitation of polymers induced by irradiation, the formation of free radicals and their structural changes were introduced. The radiation damage behavior of thermoplastic and thermosetting polymers was studied by radiation tests. It was found that the mechanical properties of polymers were damaged by different radiation doses of electrons and protons. However, the dielectric constants of polymer materials are different for different irradiation doses and sources.

In the future, with the continuous development and deepening of aerospace technology, the research on the effect of radiation on polymer structure and damage behavior will become more important. First, with the increase of space exploration missions, the working time of space devices in high-radiation environments may become longer. Therefore, it is necessary to further study the mechanism of long-term radiation exposure on the properties of polymer materials in order to better predict and evaluate the lifetime of aerospace devices. Secondly, with the gradual development of human exploration of deep space, the radiation environment will be more severe. Therefore, it is necessary to strengthen the research on the damage behavior of polymer structures caused by high-energy radiation, and explore the corresponding radiation protection and repair technology to ensure the reliability and safety of aerospace devices. In addition, with the development of materials science and engineering technology, new types of radiation-tolerant polymer materials may gradually emerge. Future research should focus on the design, synthesis and performance evaluation of these new materials to provide more durable and radiation-resistant material solutions for the aerospace field. In general, future research should be aimed at deepening the understanding of the mechanism of radiation effects on polymer structures in aerospace, strengthening the research and development of radiation protection and repair technologies, as well as the development of new radiation-tolerant polymer materials to improve the performance, reliability and safety of aerospace devices, and promote the continuous development and progress of the aerospace field.

REFERENCES

[1] Vidya, Mandal L, Verma B, et al. Review on polymer nanocomposite for ballistic & aerospace applications. *Materials Today: Proceedings*, 2020, 26: 3161–3166.
[2] Fu S, Sun Z, Huang P, et al. Some basic aspects of polymer nanocomposites: A critical review. *Nano Materials Science*, 2019, 1(1): 2–30.
[3] Mittal G, Rhee K Y, Mišković-Stanković V, et al. Reinforcements in multi-scale polymer composites: Processing, properties, and applications. *Composites Part B: Engineering*, 2018, 138: 122–139.
[4] Velmurugan R, Manikandan V. Mechanical properties of palmyra/glass fiber hybrid composites. *Composites Part A: Applied Science and Manufacturing*, 2007, 38(10): 2216–2226.
[5] Sathishkumar T P, Satheeshkumar S, Naveen J. Glass fiber-reinforced polymer composites -A review. *Journal of Reinforced Plastics and Composites*, 2014, 33(13): 1258–1275.
[6] Dhand V, Mittal G, Rhee K Y, et al. A short review on basalt fiber reinforced polymer composites. *Composites Part B: Engineering*, 2015, 73: 166–180.
[7] Stevens M P. Polymer chemistry. New York: Oxford university press, U.S., 1990.
[8] Portney L G, Watkins M P. *Foundations of Clinical Research: Applications to Practice.* Upper Saddle River, NJ: Prentice Hall, 2000.
[9] Odian G. Principles of polymerization. John Wiley & Sons, U.S., 2004.
[10] Kivelson S. Electron hopping in a soliton band: Conduction in lightly doped (CH)*x*. *Physical Review B*, 1982, 25(6): 3798.
[11] *Bowden M J. Polymers for electronic and photonic applications. Springer, Dordrecht 1988.*
[12] Wong C P. Encapsulation: Process techniques and materials. Polymers for Electronic & Photonic Application, 2013: 167.
[13] Su W F. *Principles of Polymer Design and Synthesis* . Berlin, Heidelberg, Germany: Springer Berlin Heidelberg, 2013.
[14] Ezugwu E O. Key improvements in the machining of difficult-to-cut aerospace superalloys . *International Journal of Machine Tools and Manufacture*, 2005, 45(12-13): 1353–1367.

[15] Starke E A, Staley J T. Application of modern aluminum alloys to aircraft . *Progress in Aerospace Sciences*, 1996, 32(2): 131–172.

[16] Dursun T, Soutis C. Recent developments in advanced aircraft aluminium alloys . *Materials & Design (1980-2015)*, 2014, 56: 862–871.

[17] Mouritz A P. *Introduction to Aerospace Materials*. Cambridge: Woodhead Publishing, 2012.

[18] Zhang X, Chen Y, Hu J. Recent advances in the development of aerospace materials. *Progress in Aerospace Sciences*, 2018, 97: 22–34.

[19] Czerwinski F. Controlling the ignition and flammability of magnesium for aerospace applications. *Corrosion Science*, 2014, 86: 1–16.

[20] Boyer R R, Briggs R D. The use of β titanium alloys in the aerospace industry. *Journal of Materials Engineering and Performance*, 2005, 14(6): 681–685.

[21] Peters M, Leyens C. Aerospace and space materials. *Materials Science and Engineering*, 2009, 3: 1–11.

[22] Bakis C E, Bank L C, Brown V L, et al. Fiber-reinforced polymer composites for construction - State-of-the-art review. *Journal of Composites for Construction*, 2002, 6(2): 73–87.

[23] Breuer O, Sundararaj U. Big returns from small fibers: A review of polymer/carbon nanotube composites. *Polymer Composites*, 2004, 25(6): 630–645.

[24] Ratna D, Karger-Kocsis J. Recent advances in shape memory polymers and composites: A review. *Journal of Materials Science*, 2008, 43(1): 254–269.

[25] Njuguna J (editor). *Lightweight Composite Structures in Transport*. Woodhead Publishing, U.K., 2016.

[26] Cantor B, Assender H, Grant P. Aerospace materials. CRC Press, U.S., 2015.

[27] Njuguna J, Pielichowski K. Polymer nanocomposites for aerospace applications: Properties. *Advanced Engineering Materials*, 2003, 5(11): 769–778.

[28] Williams G, Trask R, Bond I. A self-healing carbon fibre reinforced polymer for aerospace applications. *Composites Part A: Applied Science and Manufacturing*, 2007, 38(6): 1525–1532.

[29] Leng J, Lan X, Liu Y, et al. Shape-memory polymers and their composites: Stimulus methods and applications. *Progress in Materials Science*, 2011, 56(7): 1077–1135.

[30] Dutton S, Kelly D, Baker A. *Composite Materials for Aircraft Structures*, 2nd edn. Marietta: American Institute of Aeronautics and Astronautics, 2004.

[31] Plackett D, Andersen T L, Pedersen W B, et al. Biodegradable composites based on l-polylactide and jute fibres. *Composites Science and Technology*, 2003, 63: 1287–1296.

[32] Wong S, Shanks R A, Hodzic A. Effect of additives on the interfacial strength of poly(l-lactic acid) and poly(3-hydroxy butyric acid)-flax fibre composites. *Composites Science and Technology*, 2007, 67(11): 2478–2484.

[33] Lee S-H, Wang S. Biodegradable polymers/bamboo fiber biocomposite with bio-based coupling agent. *Composites Part A: Applied Science and Manufacturing*, 2006, 37(1): 80–91.

[34] Acha B, Marcovich N, Reboredo M. Physical and mechanical characterization of jute fabric composites. *Journal of Applied Polymer Science*, 2005, 98: 639–650.

[35] Acha B A, Marcovich N E, Reboredo M M. Physical and mechanical characterization of jute fabric composites. *Journal of Applied Polymer Science*, 2005, 98(2): 639–650.

[36] Kling U, Empl D, Boegler O, et al. Future aircraft wing structures using renewable materials. Deutsche Gesellschaft für Luft-und Raumfahrt-Lilienthal-Oberth eV, 2015.

[37] Farag M M. Materials and process selection for engineering design. CRC press, U.S., 2020.

[38] George M, Bressler D C. Comparative evaluation of the environmental impact of chemical methods used to enhance natural fibres for composite applications and glass fibre based composites. *Journal of Cleaner Production*, 2017, 149: 491–501.
[39] Jawaid M, Abdul Khalil H P S. Cellulosic/synthetic fibre reinforced polymer hybrid composites: A review. *Carbohydrate Polymers*, 2011, 86(1): 1–18.
[40] John M J, Anandjiwala R D, Wambua P, et al. Bio-based structural composite materials for aerospace applications. South African International Aerospace Symposium, 2008.
[41] Mohd Haris M Y, Laila D, Zainudin E S, et al. Preliminary review of biocomposites materials for aircraft radome application. *Key Engineering Materials*, 2011, 471 – 472: 563–567.
[42] BöGler O, Kling U, Empl D, et al. *Potential of Sustainable Materials in Wing Structural Design*. Deutscher Luft- und Raumfahrtkongress 2014, Augsburg 2015.
[43] Sepe R, Bollino F, Boccarusso L, et al. Influence of chemical treatments on mechanical properties of hemp fiber reinforced composites . *Composites Part B: Engineering*, 2018, 133: 210–217.
[44] Santos C V D, Leiva D R, Costa F R, et al. Materials selection for sustainable executive aircraft interiors . *Materials Research - Ibero-american Journal of Materials*, 2016, 19: 339–352.
[45] Ngo T. *Introduction to Composite Materials*. Springer, Dordrecht, 2020.
[46] Rowell R M. A new generation of composite materials from agro-based fiber. In Prasad P N, Mark J E, Fai T J (editors), *Polymers and Other Advanced Materials: Emerging Technologies and Business Opportunities*. Boston, MA: Springer US, 1995: 659–665.
[47] Djafari Petroudy S R. Physical and mechanical properties of natural fibers. In Fan M, Fu F (editors), *Advanced High Strength Natural Fibre Composites in Construction*. Elsevier Publishing, the Netherlands, 2017: 59–83.
[48] Katnam K B, Da Silva L F M, Young T M. Bonded repair of composite aircraft structures: A review of scientific challenges and opportunities. *Progress in Aerospace Sciences*, 2013, 61: 26–42.
[49] Williams G I, Wool R P. Composites from natural fibers and soy oil resins. *Applied Composite Materials*, 2000, 7(5): 421–432.
[50] Mohanty A, Misra M, Drzal L, et al. *Natural Fibers, Biopolymers, and Biocomposites: An Introduction*. CPC press, Taylor & Francis Group, U.S,, 2005: 1–36.
[51] Ochi S. Mechanical properties of kenaf fibers and kenaf/PLA composites. *Mechanics of Materials*, 2008, 40(4): 446–452.
[52] Nishino T, Hirao K, Kotera M, et al. Kenaf reinforced biodegradable composite. *Composites Science and Technology*, 2003, 63: 1281–1286.
[53] Ramnath V, Arvind R, Dinesh I, et al. Review on natural fiber composites . *IOP Conference Series: Materials Science and Engineering*, 2018, 390: 012062.
[54] Ishak M R, Leman Z B, Sapuan S M, et al. Mechanical properties of kenaf bast and core fibre reinforced unsaturated polyester composites. *IOP Conference Series: Materials Science and Engineering*, 2010, 11: 012006.
[55] Amaducci S, Amaducci M T, Benati R, et al. Crop yield and quality parameters of four annual fibre crops (hemp, kenaf, maize and sorghum) in the North of Italy. *Industrial Crops and Products*, 2000, 11(2): 179–186.
[56] Tabata Y, Sobue H, Oda E. Radiation-induced ionic polymerization of butadiene. *Journal of Physical Chemistry*, 1961, 65: 1645–1647.
[57] Ishigure K, Oneill T, Stahel E P, et al. Radiation-induced polymerization and copolymerization of butadiene in emulsion. *Journal of Macromolecular Science Part A: Chemistry*, 1974, 8: 353–372.
[58] Radzinskii S A, Sheinker A P, Abkin A D. Radiation-induced polymerization of butadiene . *Vysokomolekul Soedin Seriya A*, 1977, 19: 2066–2072.

[59] Verdurmen E M, Verstegen J M, German A L. Particle growth in butadiene emulsion polymerization. 4. The promoting effect of mercaptans. *Macromolecular Chemistry and Physics*, 1994, 195: 647–659.
[60] Henley E J, Ng C C. Gamma-radiation-induced solution polymerization of ethylene. *Journal of Polymer Science*, 1959, 36: 511–517.
[61] Colombo P, Chapman R N, Steinberg M, et al. High-pressure ethylene polymerization reactions with cobalt-60 gamma radiation. *Transactions of the American Nuclear Society*, 1964, 7: 313–3214.
[62] Wiley R H, Parrish C F, Lipscomb N T, et al. Kinetics of gamma-radiation-induced polymerization of ethylene in alkyl chlorides. *Journal of Polymer Science Part A: Polymer Chemistry*, 1964, 2: 2503–2511.
[63] Hagiwara M, Mitsui H, Machi S, et al. Liquid carbon dioxide as a solvent for radiation polymerization of ethylene. *Journal of Polymer Science Part A: Polymer Chemistry*, 1968, 6: 603–608.
[64] Takehisa M, Machi S, Watanabe H, et al. Radiation-induced polymerization of ethylene in a pilot-plant. 1. Bulk process. *Journal of Applied Polymer Science*, 1979, 24: 853–864.
[65] Takehisa M, Watanabe H, Kurihara H, et al. Radiation-induced polymerization of ethylene in a pilot-plant. 2. Development of wet-wall process. *Journal of Applied Polymer Science*, 1979, 24: 865–882.
[66] Machi S, Tamura T, Hagiwara M, et al. Short-chain branching in gamma-radiation-induced polymerization of ethylene. *Journal of Polymer Science Part A: Polymer Chemistry*, 1966, 4: 283–291.
[67] Guarise G B, Palma G, Siviero E, et al. Polymerization induced by gamma radiation of styrene under pressure. *Polymer* 1970, 11: 613–625.
[68] Challa R R, Drew J H, Stahel E P, et al. Radiation-induced emulsion polymerization of vinyl-acetate in a pilot-plant reactor. 2. Kinetics . *Journal of Applied Polymer Science*, 1986, 31: 27–38.
[69] Xu X L, Zhang Z C, Zhang M W. Microemulsion polymerization of butyl acrylate initiated by gamma rays. *Journal of Applied Polymer Science*, 1996, 62: 1179–1183.
[70] Yu H, Peng J, Zhai M, et al. Radiation-induced copolymerization of styrene/n-butyl acrylate in the presence of ultra-fine powdered styrene-butadiene rubber. *Radiation Physics and Chemistry*, 2007, 76: 1736–1740.
[71] Okada T, Ishigaki I, Suwa T, et al. Synthesis of cationic flocculant by radiation-induced co-polymerization of methyl-chloride salt of n,n-dimethylaminoethyl methacrylate with acrylamide in aqueous solution. *Journal of Applied Polymer Science*, 1979, 24: 1713–1721.
[72] Tabata Y, Ito W, Oshima K. Radiation-induced polymerization of tetrafluoroethylene. *Journal of Macromolecular Science Part A: Chemistry*, 1970, 4: 789–799.
[73] Matsuda O, Watanabe T, Tabata Y, et al. Radiation-induced co-polymerization of methyl trifluoroacrylate with ethylene. *Journal of Polymer Science Part A: Polymer Chemistry*, 1979, 17: 1789–1793.
[74] Matsuda O, Watanabe T, Tabata Y, et al. Radiation-induced co-polymerization of methyl trifluoroacrylate with propylene. *Journal of Polymer Science Part A: Polymer Chemistry* 1979, 17: 1795-1800.
[75] Charlesby A. Solid-state polymerization induced by radiation. *Reports on Progress in Physics* 1965, 28: 463–518.

11 Wire Arc Additive Manufacturing of Nickel-based Superalloy and Stainless Steel Dissimilar Material Component

Tianhao Zhou, Zhijun Qiu, Yuxing Li, Yan Ma, Tao Wei, Bosheng Dong, Zengxi Pan, and Huijun Li

11.1 SUPERALLOYS AND STAINLESS STEELS

11.1.1 SUPERALLOYS

The employment of metallic materials, tracing back to the Bronze Age, marks a significant chapter in human history. These materials are characterized by a plethora of advantages: their malleability facilitates widespread application in both professional settings and everyday life, while their superior mechanical properties have been instrumental in advancing human civilization. However, the contemporary era presents a set of challenges, particularly in the demand for high-performance metallic materials by various industries. Critical applications in extreme conditions, such as in aerospace, submarine operations, and nuclear reactors, necessitate materials that can withstand severe environments, often characterized by temperatures exceeding 500 °C. Consequently, these industries have prioritized the development of materials with exceptional thermal stability – maintaining performance close to their melting points, resistance to creep deformation – ensuring longevity under stress, and enhanced corrosion resistance – to counteract the detrimental effects of harsh conditions, including exposure to corrosive gases and seawater. Alloys that embody these remarkable attributes, capable of operating under such extreme conditions, are commonly referred to as superalloys.

Despite regular usage, industries have raised concerns regarding the requirements for high-performance metallic materials, especially in demanding conditions such as those found in aerospace, submarine, and nuclear systems. These challenging environments frequently involve high temperatures (above 500 °C), necessitating materials that demonstrate exceptional mechanical performance at elevated temperatures.

 DOI: 10.1201/9781003386223-11

Therefore, it's crucial for the metallic materials employed in these settings to exhibit a high degree of thermal stability – maintaining robust performance even near their melting points, resistance to creep deformation – withstanding degradation over extended periods, and corrosion resistance – essential in harsh operational conditions like exposure to corrosive hot gases and seawater. Alloys that possess these remarkable qualities and are capable of operating under such extreme conditions are commonly referred to as superalloys.

The market already benefits from a significant array of superalloys, with notable series like Inconel, Hastelloy, and Waspaloy playing a crucial role across aerospace, nuclear, and marine industries. These superalloys, diverse in composition, fall into three main categories based on their principal elements: nickel-based, cobalt-based, and iron-based [1].

Nickel-based superalloys stand out as the most critical and widely favoured materials, particularly for their application in turbine engines. This is attributed to their exceptional mechanical properties and corrosion resistance, qualities that remain superior even at working temperatures that can exceed 800 °C [2]. Their unparalleled performance makes nickel-based alloys the prime choice for environments demanding high temperature resilience.

However, these alloys are not without their challenges. While their excellent mechanical properties contribute to their high performance, they also complicate the manufacturing process. The intricate processing required for these superalloys can, if not executed with precision, lead to a reduction in mechanical integrity or cause alterations to the metallic structure [3, 4].

A specific variant of nickel-based superalloys, known as Inconel® epitomizes the strength and versatility of these materials. This alloy, which will be discussed in further detail in subsequent sections and used as the selected material in case study section, exemplifies the high standards set by nickel-based superalloys in terms of performance and durability in demanding conditions.

Cobalt-based superalloys, primarily composed of cobalt, are celebrated for their exceptional characteristics, including high strength, corrosion resistance, excellent wear, and heat resistance. These materials excel in demanding environments due to their robust performance, akin to that of nickel-based alloys. This superior performance can be attributed to the Gamma (γ) phase, enhanced by secondary element additions, and the Gamma Prime (γ') phase, deriving from the L12 structure of Co_3Ti or face-centred cubic Co_3Ta [5]. These phases contribute to solid solution strengthening, bolstering corrosion and thermal resistance. Furthermore, carbide phases within the alloy induce precipitation hardening, enhancing its overall resilience.

Despite these advantages, cobalt-based superalloys face challenges, including the presence of topologically close-packed (TCP) phases, which can detract from the alloy's properties. While cobalt-based superalloys offer numerous benefits and possess similarities with nickel-based alloys in terms of advantages, the latter have largely supplanted cobalt alloys in the market. This shift can be attributed to the higher cost and narrower application range of cobalt alloys compared to their nickel counterparts. Even though cobalt alloys demonstrate superior strength at elevated temperatures, the substantial cost difference often makes nickel-based alloys a more attractive choice for most industrial applications.

Today, despite of some specific areas, the primary value of cobalt-based superalloys lies within research, serving as a critical resource for theoretical support in the study of superalloys. Their unique properties continue to contribute valuable insights into material science, helping to advance the development of high-performance alloys for future applications [6–8].

Fe-based superalloys constitute one of the primary categories of superalloys, characterized not by an iron-nickel composition but rather by iron being the dominant element. While iron-nickel alloys exist and may contain iron and nickel in varying proportions, Fe-based superalloys are distinct in that they might include nickel, chromium, and other elements to enhance their properties. These alloys typically derive their structure from austenitic stainless steel, featuring a face-centred cubic (FCC) matrix. They achieve strength through solid-solution and precipitation strengthening mechanisms. Utilizing iron in superalloys offers economic advantages and simplifies the manufacturing process due to the relative abundance and cost-effectiveness of iron. These alloys exhibit comparable creep and oxidation resistance to some nickel-based superalloys under certain conditions. The performance of Fe-based superalloys also relies on the formation of gamma (γ) and gamma-prime (γ') phases, which contribute to solid-solution strengthening and precipitate strengthening, respectively. This enhances their corrosion resistance and thermal stability, making them suitable for use in harsh environments. However, it's important to note that not all iron-based alloys qualify as superalloys. Only those with an FCC matrix, like Nitronic®®® and Multmet®®, fit the criteria for superalloys due to their exceptional properties. Research indicates that, in some severe environments, although Fe-based alloys have a superior corrosion and oxidation performance compared with might underperform compared to nickel-based counterparts, particularly when exposed to high-temperature water vapor, potentially reducing their lifespan [9]. Furthermore, Fe-based superalloys may exhibit a higher propensity for forming detrimental phases such as $\eta(Ni_3Ti)$, $\sigma(Fe_xCr_y)$, G ($Fe_6Ni_{16}Si_7$), μ (Fe_7Mo_6), and Laves phases compared to nickel- and cobalt-based superalloys, which could affect their stability and performance [10].

11.1.1.1 Nickel-based Superalloys

Nickel-based superalloys have been prominently developed and utilized for over half a century, particularly in gas turbines where operational temperatures often exceed 500 °C [11]. These superalloys are especially crucial for the most critical engine components in aircraft, which can withstand temperatures up to 800 °C – more than 70 percent of their melting point – due to their exceptional creep and rupture resistance at high temperatures. In contrast, other materials like titanium and aluminium, which are used in cooler engine parts, may experience significant creep at temperatures just above 150 to 350 °C [12]. The superior mechanical properties and thermal stability of nickel-based alloys have made them highly sought after in additive manufacturing (AM) industries [13].

One of the primary reasons nickel-based alloys excel in high-temperature environments is their face-centred cubic (FCC) crystal structure. This structure contributes significantly to the alloy's toughness and ductility, thanks to the bonding of outer d electrons which provide cohesive energy. Additionally, the FCC configuration ensures stable thermal performance from room temperature up to the melting point, preventing dimensional instability due to phase transformations. This stability is crucial for components that operate under varying thermal conditions [14]. Other transition metals with similar FCC structures, such as those in the platinum group (Pt, Ru, Ir), are either prohibitively expensive or, in cases like osmium (Os) and technetium (Tc), toxic, although cobalt also finds application in superalloys but often at a higher cost and with similar performance characteristics compared to nickel-based options [15].

Furthermore, the strength and durability of nickel alloys are enhanced through mechanisms such as solid solution and precipitate strengthening. Elements like chromium (Cr) and molybdenum (Mo) integrated into the alloy contribute significantly to its thermal durability. Additionally, the presence of carbides and other secondary components within the precipitates increases both the strength and ductility of the material [16].

Nickel superalloys are broadly divided into three main types, each distinguished by its unique strengthening mechanisms: solid solution strengthened, precipitation strengthened, and specialty alloys. The first two types are extensively used across various industries, whereas specialty alloys are tailored for specific, advanced applications.

Solid solution strengthened alloys enhance the properties of nickel by integrating various alloying elements into its matrix. Copper (Cu), for example, dissolves entirely in nickel, forming a homogenous single-phase alloy that boosts the creep strength of the matrix. This category also includes elements like tungsten, molybdenum, and titanium, known for their high melting points and hardening capabilities, though they have a relatively low solubility in nickel. Even elements with lesser hardening efficiency such as tantalum, niobium, and vanadium are included in small quantities. These alloys are prized for their improved mechanical properties and corrosion resistance but require precise welding techniques to prevent defects such as porosity [17].

Precipitation strengthened alloys incorporate additional elements like titanium (Ti), aluminium (Al), and niobium (Nb), which form precipitates within the matrix. These precipitates block the movement of dislocations, significantly enhancing the alloy's strength post heat treatments. The relationship between the precipitates and the matrix is typically coherent, involving phases known as Gamma prime (γ') or Gamma double prime (γ''). A notable example is Alloy 718 (Inconel 718), extensively employed in aerospace for parts like gas turbine shafts, recognized for its excellent resistance to oxidation and corrosion at high temperatures. This alloy benefits from an advanced heat treatment process that optimizes the properties of the gamma double prime phase to enhance its weldability and reduce susceptibility to strain age cracking [17].

Specialty alloys, designed for high-temperature applications, include materials like Alloy MA6000, which achieves remarkable creep resistance and stability through a combination of precipitate strengthening and oxide dispersion strengthening. Nickel-Aluminides, another group within this category, are derived from compounds such as NiAl and $NiAl_3$, noted for their superior performance and corrosion resistance, though they present challenges in weldability. Enhancements with elements such as boron (B), chromium (Cr), and molybdenum (Mo) improve their applicability, elevating these alloys to the status of superalloys due to their exceptional properties [17, 18].

These categorizations highlight the versatility and specific applications of nickel superalloys, underscoring their critical role in modern engineering and technological advancements.

11.1.1.2 Producing Nickel Superalloys

The initial step in crafting nickel-based superalloys traditionally involves casting, much like other metals. However, casting can lead to grains with random orientations, adversely affecting the alloy's performance. This issue has sparked interest in grain refinement techniques among materials science researchers. To address this, directional solidification was introduced in the 1970s as a method to refine the grain structure. This technique involves solidifying molten metal from bottom to top using a controlled temperature gradient and solidification speed, effectively eliminating grains that are oriented perpendicular to the desired direction and thus enhancing the crystalline structure alignment [19].

Advancements in the 1980s led to the development of the single crystal casting process, which is still prevalent. This method builds upon directional solidification principles but adds a crucial step of removing misaligned grains before casting through the use of a grain selector. This selector restricts the number of crystal orientations allowed, ensuring the metal solidifies as a single crystal, greatly reducing the risk of deformation [20].

Nickel-based superalloys are initially produced from a metal ingot that is combined with appropriate components. The manufacturing involves a three-stage melting procedure designed to minimize elemental segregation, remove unwanted elements like Oxygen and Sulphur, and refine the ingot's microstructure. After casting, the large ingot is annealed at high temperatures to prevent segregation and improve the metal compounds' qualities, notably through precipitation strengthening. Subsequent processing, involving casting and deformation under hydraulic pressure, alters the ingot's diameter and microstructure. This ingot is then shaped into a metallic disk and undergoes forging [20].

As alloy compositions become more complex to enhance strength, casting and deformation processes pose increased challenges for these advanced alloys. An alternative production method, known as "powder processing," has been explored. This begins with melting the alloy in a crucible, followed by atomization using pure argon gas. The atomized metal is then sieved to ensure uniform particle size before being compacted at temperatures above 1000°C and pressures around 100 MPa, resulting in a shape close to the final desired form [21].

Nickel-based superalloys can be processed through various pathways, starting with their formation into large ingots, which are then either cast, wrought, or first remelted into powder before wrought processing. This section introduces different remelting techniques applicable to nickel-based and other superalloys, highlighting the diverse methods available for superalloy production.

11.1.2 Machinability of Superalloys

Superalloys commonly used in aerospace, such as nickel-based and titanium alloys, along with other high-performance materials like structural ceramics and tantalum, are integral in manufacturing components for industries including aerospace, electronics, defence, and medical sectors like dental and orthopaedic applications, as well as environments involving seawater. These materials are chosen for their superior properties, which include high strength at elevated temperatures, resistance to chemical degradation, and wear resistance. The inherent toughness and thermal resistance of these materials make them challenging to machine, categorizing them as difficult-to-cut alloys [22].

The machining of these tough materials often involves high energy consumption, with a significant portion of this energy converted into heat. This heat generation primarily occurs during the deformation process and through friction at the tool-chip and tool-workpiece interfaces, resulting in high temperatures that can complicate the machining process. Specifically, alloys such as titanium (with a thermal conductivity of about 15 W/m °C), nickel (about 11 W/m °C), and silicon nitride (about 13 W/m °C) have lower thermal conductivity compared to conventional steels or cast iron, exacerbating temperature increases at the cutting tool and workpiece [23, 24].

Furthermore, the austenitic matrix of these superalloys tends to cause rapid work hardening. They can react with tool materials under atmospheric conditions and are prone to issues such as the formation of a built-up edge, welding to cutting tools, and the presence of abrasive carbides in their microstructure. These characteristics contribute to rapid flank wear, cratering, and notching during the machining process, presenting additional challenges for manufacturers.

Aerospace superalloys are typically available in several forms, including cast, wrought and forged, and sintered (powder metallurgy), each suited to different manufacturing needs and performance criteria. Cast components are favoured for their exceptional creep strength and toughness, though this combination often complicates machining due to difficulties in chip segmentation. On the other hand, wrought and forged components offer superior strength, alongside enhanced fatigue and fracture resistance, but they tend to be more abrasive and may cause more significant wear and deformation to machining tools [22].

For producing more complex and near-net-shape components, the powder metallurgy technique is preferred. This method, while innovative, results in parts that are challenging to machine due to their extremely low machinability and high abrasiveness. As aerospace components are pushed to operate at higher temperatures to meet the demands of modern engine designs, the machinability of these alloys is expected to decline further. Innovations such as the use of single crystal alloys and directionally solidified eutectics reinforced with long carbide fibres are gaining traction, although these materials are particularly difficult to machine [23, 24].

Further compounding the challenge, recent advancements in manufacturing techniques such as isothermal forging and other thermal-mechanical processes are aimed at producing parts that are closer to final net shapes with improved thermal and structural stability and uniformity. While these developments enhance the performance and durability of aerospace components, they also tend to reduce machinability, setting a complex scenario for future manufacturing processes in the aerospace industry.

11.1.2.1 Machinability of Nickel-based Superalloys

Approximately 50 percent of the materials used in aero-engine manufacturing are nickel-based alloys, due to their superior strength-to-weight ratio compared to denser materials like steel [25]. These alloys are not only pivotal in aerospace engineering but also in other critical sectors such as marine equipment, nuclear reactors, petrochemical plants, food processing equipment, and pollution control systems. Their widespread application across such varied and demanding environments is largely due to their outstanding resistance to corrosion, mechanical and thermal fatigue, shocks, creep, and erosion at high temperatures.

However, the machining of various alloys reveals significant differences in ease and efficiency, particularly when comparing nickel-based alloys like Inconel 718 with other metals. Inconel 718, known for its robustness and high temperature resistance, is notably the most difficult to machine among common alloys used in industry. In contrast, stainless steels and mild steels are considerably easier to work with, possessing a machinability rate that is approximately 5 to 6 times higher than that of Inconel 718. This disparity highlights the challenges and considerations needed when machining high-performance materials, which require more specialized techniques and tools due to their superior mechanical and thermal properties [23].

11.1.3 Nickel Superalloys in Aerospace

Nickel-based superalloys are primarily utilized in the manufacturing of turbine blades for aircraft and helicopter gas-turbine engines, which operate under extreme conditions. These materials are chosen for their high strength and exceptional thermal creep resistance, essential for components that face very high temperatures, severe mechanical stresses, and corrosive oxidizing environments [26].

Since the 1960s, the development of single crystal (SC) blades using high-performance nickel-based superalloys has markedly increased the temperature capability of these components and prolonged their service life, significantly enhancing engine performance. This improvement is reflected in greater specific power, reduced fuel consumption, and lower cost of ownership. By the early 1980s, SC vanes and blades became increasingly common in advanced military and civilian aircraft engines.

These advanced components are made from complex nickel-based superalloys that contain substantial quantities of refractory elements such as molybdenum (Mo), tungsten (W), tantalum (Ta), and rhenium (Re). The structure of these superalloys is defined by a nickel-based face-centred cubic γ matrix, which is strengthened by a high-volume fraction of ordered L12 γ'-Ni_3Al-type precipitates, contributing up to 70% of the material's composition.

Historically, turbine blades were initially produced using conventional melting and casting processes, which resulted in a polycrystalline equiaxed grain structure. However, this structure was prone to high-temperature creep and centrifugal stress, leading to grain boundary cracking that significantly reduced the lifespan of the blades. To address this, an intermediary development involving directional solidification processes allowed for the production of columnar grain components. This process aligns the grain boundaries parallel to the main centrifugal stress direction, thereby reducing but not entirely preventing grain boundary damage [26].

The ultimate advancement in this field has been the creation of blades from a single crystal, eliminating grain boundary cracking entirely. This was made possible through a specialized directional solidification process that includes a grain selection system.

Alongside these developments in casting processes, the chemistry of nickel-based superalloys suited for SC blade applications has also evolved. There has been a gradual increase in the content of γ'-former elements such as aluminium (Al), titanium (Ti), and tantalum (Ta), which enhances the volume fraction of the γ'-strengthening phase. Additionally, the introduction of higher amounts of solid-solution refractory elements like Mo, W, Ta, Re, and possibly ruthenium (Ru) has occurred.

While equiaxed and columnar grain blades are still employed in the intermediate and low-pressure sections of gas turbines where temperatures are comparatively moderate (around 600–900 °C), single crystal blades and vanes made from these advanced nickel-based superalloys are essential in the high-pressure sections. In these critical areas, temperatures may reach up to 1150 °C, demanding the exceptional properties provided by these specialized superalloys [26].

11.1.4 Stainless Steels

Stainless steels, as well known as an indispensable class of alloys, play a crucial role in a broad spectrum of applications, from everyday household items to sophisticated aerospace components. This chapter explores the history, development, and diverse uses of stainless steel, highlighting its essential properties and the innovations that have shaped its current applications.

The term "stainless steel" is derived from its ability to resist staining, rusting, or corroding in environments where ordinary steel would fail. This resistance is primarily due to the presence of chromium, which must be added at a minimum of about 11 wt. percent to form a protective, self-healing chromium oxide layer on the surface of the steel. In more aggressive environments or in the presence of elements like carbon, higher chromium contents are necessary to prevent pitting and rusting [27, 28].

Beyond chromium, modern stainless steels also contain a variety of other alloying elements such as molybdenum (Mo), which enhances resistance to pitting, and nickel (Ni), which promotes an austenitic structure. The inclusion of these elements has given rise to several types of stainless steels, such as super-ferritic and super-austenitic, which are known for their generous alloying element contents and are often categorized as heat-resisting alloys.

Stainless steel microstructures, which include ferritic, austenitic, and martensitic types, can be engineered by adjusting the steel's chemical composition. This versatility allows for the development of duplex, precipitation-hardening, and Mn-N substituted austenitic stainless steels, each tailored for specific properties and applications [28].

The historical development of stainless steel began in the early twentieth century, when it was recognized that the addition of chromium to iron alloys could significantly enhance their resistance to corrosive environments. Although early experiments in the nineteenth century by scientists such as Robert A. Hadfield indicated that chromium could reduce an alloy's corrosion resistance when combined with carbon, it was not until between 1905 and 1912 that researchers in several countries accidentally developed what we now recognize as stainless steel. These alloys proved to be highly resistant to corrosion, a property that was initially applied in the manufacture of cutlery [29].

The use of stainless steel expanded rapidly, with significant developments during World War I when Harry Brearley developed a stainless-steel blade for cutlery that significantly resisted wear and corrosion. This innovation marked the beginning of widespread recognition and use of stainless steel, particularly in the aerospace industry where its properties were ideal for combating the rigorous demands of airplane manufacturing [29].

11.1.4.1 Austenitic Stainless Steel

Austenitic stainless steels represent the largest group within the stainless-steel family, comprising about two-thirds of all stainless-steel types. These steels are distinguished by their FCC microstructure, similar to that of nickel, which is achieved by adding elements such as nickel, manganese, and nitrogen. Unlike iron-based steels, austenitic stainless steels are non-magnetic and cannot be heat-hardened due to their inherent microstructure, though they can be strengthened through cold working [30].

There are two main subgroups within the austenitic stainless-steel category: the 200 series and the 300 series. The differentiation between these series primarily lies in their formulation for stabilizing the austenite phase. The 300 series relies mainly on nickel to stabilize the austenite structure, while the 200 series uses nitrogen and manganese as cost-effective substitutes for nickel [31, 32].

Austenitic stainless steels are highly valued for their exceptional corrosion resistance, excellent formability, and the increased strength they exhibit when cold worked. Among the most prevalent is Type 304 stainless steel, also known as 18-8 stainless steel, which contains 18 percent chromium and 8 percent nickel. This type is widely used for cookware, cutlery, and various kitchen appliances, making it one of the most recognized forms of stainless steel.

For environments that are particularly harsh and more prone to corrosion, Type 316 stainless steel is often chosen because it includes molybdenum, which enhances its corrosion resistance. Similarly, Alloy 20, or C-20, offers superior corrosion resistance compared to Types 304 and 316 due to its additional copper content, making it particularly suitable for use in environments that involve exposure to sulfuric acid [33].

In addition to these corrosion-resistant variants, there are also heat-resisting austenitic steels like Types 309, 310, and 314, which are used in environments with temperatures exceeding 600 °C. Their heat resistance is enhanced by higher chromium content along with aluminium and silicon, similar to superalloys. These elements help form a thin, passive oxide film on the surface of the steel, providing protection at high temperatures.

11.1.4.2 Other Stainless Steels

Despite of austenitic steels, several types of stainless steels can be categorized distinguished by their specific microstructures and alloying elements, tailored to meet diverse industrial requirements.

Ferritic stainless steels are known for their content of 11.5 percent to 18 percent chromium with a low carbon component, often under 0.2 percent, and occasionally include nickel. These steels are magnetic, not heat-hardenable, but can be strengthened by cold working. They offer substantial corrosion resistance but tend to become brittle in sub-zero temperatures, much like the carbon steels implicated in the Titanic's sinking. Common types within this group include 430, used broadly in domestic and industrial applications, and 409, typically employed in automotive exhaust systems [34, 35].

Martensitic stainless steels, on the other hand, feature 12–17 percent chromium and carbon levels ranging from 0.1 percent to 1.2. Their usage varies greatly with carbon content; lower carbon varieties are utilized for their high strength in applications such as pumps and valves, while higher carbon types are favoured in wear-resistant applications like surgical blades and cutlery. These steels are notable for their ability to be hardened by heat treatment, which enhances their mechanical properties [32, 36].

Duplex stainless steels are characterized by a dual-phase microstructure that combines austenitic and ferritic characteristics. This combination generally results in superior strength and exceptional resistance to various forms of corrosion, particularly in chloride-containing environments, making them ideal for marine and chemical processing applications. These alloys are often chosen for their balanced properties, offering twice the strength of typical austenitic grades like 316 or 304 and enhanced resistance to stress corrosion cracking [37].

Precipitation hardening (PH) stainless steels are engineered to achieve high strength through heat treatment, capable of reaching three to four times the yield strength of austenitic steels. These alloys are divided into low-carbon martensitic, semi-austenitic, and austenitic groups, each designed to provide specific properties. For instance, 17-4 PH, a low-carbon martensitic PH steel, is extensively used in chemical and aerospace sectors for its excellent oxidation resistance and high strength. In contrast, A286, an austenitic PH steel, is recognized for its formidable mechanical properties and creep resistance at high temperatures, often classed alongside superalloys for its performance [38, 39].

Through these developments, stainless steels continue to serve a wide range of applications, demonstrating a remarkable ability to meet the evolving demands of industries from household goods to high-stress environments in aerospace and energy sectors.

11.1.5 Stainless Steels in Aerospace

Stainless steel can be used in the aerospace industry, though its application is more limited compared to materials like titanium and aluminium alloys, or advanced composites, which are preferred for their lighter weight and high strength-to-weight ratios, but the higher strength, corrosion resistance, durability, and heat resistance comparing with aluminium and titanium alloys, making it suitable for specific uses including engine and exhaust components, landing gear and key superstructure joints where these properties are essential.

Steels have traditionally concerned as airframe materials, who played a pivotal role in the structural integrity of commercial metallic frameworks, particularly in aerospace applications such as landing gear, flap tracks, and actuation components. Among these, the 4340M (or 300M) alloy, classified as a high-strength low-alloy (HSLA) steel, has been particularly prominent. This chromium-molybdenum steel alloy, renowned for its impressive minimum tensile strength of 1930 MPa and toughness of approximately 60 MPa m1/2, was the material of choice for most landing-gear structures before the 1990s. However, the issue of corrosion necessitated frequent, costly maintenance cycles every 7–10 years, involving the complete removal and refurbishment of landing gear components to address rust and wear [40].

In more recent developments, β-titanium alloys have begun to replace steel in many airframe applications, driven by their superior resistance to corrosion. Concurrently, there has been a marked increase in the use of stainless steels in airframes, particularly since 2000, with the advent of higher-strength grades aimed at extending maintenance intervals. Stainless steels like the 15-5PH and PH13-8 have gained traction for their corrosion resistance and comparatively high strength, traditionally ranging from 1035 to 1520 MPa.

Pushing the envelope further, Carpenter Technology Corporation developed Custom 465 stainless steel to achieve even greater strength levels up to 1930 MPa, positioning it as a viable substitute for 4340M steel. This alloy not only addresses the perennial issue of corrosion but also eliminates the need for cadmium plating for corrosion protection. Custom 465 has been successfully heat-treated to achieve tensile strengths of approximately 1655–1795 MPa, proving its mettle as a "true" stainless steel and finding diverse applications across major aircraft manufacturers for parts such as torque tubes, pneumatic cylinders, and flap tracks. Progress continues with the development of another promising alloy, Custom 565, poised to meet or even exceed the 1930 MPa strength threshold, signifying a significant advance in the application of stainless steels in aerospace engineering [41].

In addition to its role in airframe construction, stainless steel has found significant applications in aeroengine designs, both historically and in contemporary settings. In the realm of modern large engines, stainless steel is particularly utilized in critical components of fan blades. While these blades are primarily composed of polymer matrix composites (PMC), typically carbon fibre reinforced with a toughened epoxy matrix, their leading edges are often fortified with stainless steel. This strategic reinforcement is vital as it provides the blades with enhanced durability, enabling them to withstand severe impacts, such as bird strikes, without succumbing to catastrophic failure. This use of stainless-steel underscores its importance in enhancing the resilience and safety of critical aerospace components.

Notably, in engineering applications, particularly when considering the risk of galvanic corrosion, material selection is of paramount importance. Stainless steels, titanium-, and nickel-based alloys are regarded as noble materials in the galvanic series, which means they exhibit superior corrosion resistance compared to more active metals like aluminium alloys and low-alloy steels.

When two dissimilar metals come into contact in a moist environment, galvanic corrosion can occur, with the more active metal (such as aluminium or low-alloy steel) corroding at a higher rate due to its lower position in the galvanic series. Stainless steels, however, when paired with each other or with titanium- and nickel-based alloys, generally do not initiate significant corrosion issues in the presence of moisture. This makes them highly favourable in diverse applications where moisture exposure is inevitable.

However, even these noble metals are not completely immune to corrosion. Under certain environmental conditions, such as the presence of chlorides or other aggressive ions, localized pitting can still occur. This underscores the importance of understanding not only the basic properties of materials but also how they interact under specific environmental conditions.

This shift in focus towards stainless steel as a leading material in preventing galvanic corrosion highlights its critical role in ensuring the durability and integrity of engineered structures and components, making it a primary choice for applications demanding high corrosion resistance [41].

11.2 JOINING DISSIMILAR MATERIALS

Joining technologies are recognized as crucial enablers in the field of innovative and sustainable manufacturing [42]. Despite advancements in material properties and the evolution of processes like additive manufacturing, which aim to minimize the necessity for joins, the complexity of products and their functional demands often make a completely joint-free production unfeasible. Products are typically assembled from multiple components, making joining processes indispensable for ensuring product functionality and enhancing manufacturing efficiency [43]. The quest for improved product features and performance, particularly in hybrid structures that incorporate diverse material classes, further underscores the necessity of joints. As a result, a deep understanding of joining technologies is essential in manufacturing, driving continuous developments and enhancements of both new and existing processes.

11.2.1 Methods of Joining Dissimilar Materials

The EU Manufacture Technology platform's Sub-Platform on Joining defines joining as the process of creating a bond between materials or components to achieve a specified physical performance. This bond, essential for constructing functional structures, can manifest in various forms, combining with other research, they can be categorized broadly into four main types:

1. Mechanical Joining: This involves forming a joint using mechanical means, such as fasteners or interlocking parts. Mechanical joints are characterized by the physical engagement of materials without altering their inherent properties.
2. Chemical Joining: This type of joining occurs through chemical reactions between materials, resulting in a bond. Adhesives are a common example where the joint is formed by the chemical cure of the bonding agent.
3. Thermal Joining: This method uses thermal energy to create a bond, typically through processes like welding, soldering, or brazing, where the heat is applied to melt or reshape materials to form a joint.
4. Hybrid Joining: It combines techniques such as laser welding with traditional brazing to capitalize on the precise energy control of lasers and the metallurgical benefits of brazing. This integration facilitates the manufacturing of components with superior structural integrity, especially in heterogeneous material combinations like composites bonded to metals. However, owing to its limited scope of use, it will not be detailed introduced in this chapter.

Understanding these joining processes is crucial as they significantly influence the design, functionality, and durability of assembled products.

11.2.1.1 Mechanical Joining

Mechanical joining encompasses a range of traditional methods that have long been fundamental in construction. Compared to other techniques, mechanical joining offers unparalleled flexibility in material selection and boasts a straightforward operational process. These attributes make it the most prevalent method encountered in everyday scenarios, firmly establishing its role as a cornerstone of daily construction and assembly tasks. In the text below, three representatives of them will be introduced.

The using of threaded fasteners (or screws) is a kind of highly accepted joining method that has been used for a long time, screws of various types are particularly useful for one-sided joining scenarios where access is restricted to one side of the assembly. Conversely, bolt and nut combinations are typically employed when it is possible to access both sides of the components being joined. For assemblies that may require disassembly, threaded fasteners are often favoured because they can be removed without damaging the components. However, when dealing with thin gauge sheet metals, screws alone may not provide adequate load-bearing length. In such cases, alternative methods or additional elements like flow drill screws, collar forming, or spring nuts are utilized to enhance the load-bearing capacity of the joint [44].

Flow Drill Screws serve as a pivotal single-sided mechanical fastening method for structural automotive joints, now widely accessible as a commercial technology suitable for both manual and automated assembly. Traditionally, automotive Body in White processes relied on gas metal arc welding, which often led to significant dimensional distortion and was less effective for joining dissimilar materials. Flow

Drill Screws, however, heats and rotates the screw to soften the joining material for easier penetration, which can be done with or without a pre-drilled hole depending on the screw design. As it penetrates, the screw extrudes the material into a funnel shape, enhancing the joint's strength through greater interlock depth. Despite its advantages, the process has limitations, such as challenges with thin gauge materials and the need for precise application to avoid screw skidding. Flow Drill Screws is frequently used in conjunction with adhesives to optimize the durability and robustness of the joint in automotive constructions [45].

Clinching is a mechanical joining technique that securely fastens sheet metal parts together through local plastic deformation, creating an interlock between the sheets without the need for additional fastening elements like rivets or external fasteners. This process requires the joined materials to be sufficiently ductile, meaning they can stretch significantly without breaking, which ensures effective deformation and bonding under pressure. During the clinching process, two overlapping metal sheets are pressed together using a specialized punch and die set within a press machine. The action of the punch and die deforms the sheets, forming a joint characterized by a distinct pit on the punch side and a corresponding rise on the die side. The clinching method is valued for its simplicity and cleanliness, producing no heat, splashes, or emissions, making it an environmentally friendly and efficient option for joining metal sheets [46, 47].

11.2.1.2 Chemical Joining Processes

Chemical joining, which is primarily aimed at creating adhesive bonds between dissimilar materials, involves initiating chemical reactions to achieve adhesion. This process is supported by various theoretical frameworks such as Adsorption theory, Chemical bonding theory, and Diffusion boundary theory etc. However, it's important to note that no single theory completely encapsulates all the nuances of adhesive interactions or joints [48].

During the bonding process, a variety of mechanisms and specific surface interactions may be involved. These interactions result in the formation of a contact zone, also known as the intermediary zone, between the adhesive and the adherent. For effective bonding, the adhesive must completely wet the bonding area. Yet, challenges such as irregular material surfaces and potential voids can complicate this requirement [49].

The characteristics of the bond are influenced by multiple factors including the composition of the adhesive, the nature of the contact zone, the viscosity of the adhesive, and the surface energy of the substrate. These elements collectively determine how well the adhesive will perform in terms of durability and effectiveness. Understanding the interplay of these factors can help in optimizing the bonding process for various industrial applications.

11.2.1.3 Thermal Joining

The thermal joining process is closely associated with the fusion technique, predominantly encompassing methods such as welding, brazing, and sintering. These methods involve a continuous heat source to melt the material, facilitating the bond.

This type of joining creates a "permanent" bond at the microstructural level, especially when it involves the joining of dissimilar materials. This permanence arises from the material fusion that occurs during the process, which, once cooled, typically cannot be reversed without damaging the original materials.

Electric arc welding is a prevalent fusion joining technique, notably including methods such as Gas Tungsten Arc Welding, Shielded Metal Arc Welding, and Gas Metal Arc Welding. These methods are particularly effective for joining dissimilar metals that are relatively easier to bond, such as carbon steel to stainless steel. However, one of the main challenges in electric arc welding is managing the relatively large Heat Affected Zone (HAZ) and molten pool. These larger zones can lead to the formation of brittle intermetallic compounds when welding dissimilar metals, which may compromise the structural integrity of the weld [45].

The selection of appropriate filler materials plays a crucial role in addressing this challenge and is critical for achieving high-quality dissimilar welds. The choice of filler material must be compatible with the mechanical properties and thermal expansion coefficients of the base metals to minimize stress and prevent the formation of IMCs. Proper selection ensures a stronger, more reliable weld and mitigates potential weaknesses in the joint area.

Plasma sintering, originally viewed more as a brick forming technique than a metallic joining process, has roots dating back to the early twentieth century, gaining scientific attention in the 1920s and 1930s. The first significant use of metallic powders in sintering occurred in 1933. The process took a significant leap forward in the 1950s when Lockheed Missile Company introduced 'spark sintering,' utilizing electrical discharge to enhance the technique. This innovation has since evolved into what is known today as 'Spark Plasma Sintering' (SPS), although the exact physics of the process are still not fully understood, with some debate regarding the presence of actual plasma during sintering. Due to this uncertainty, some in the field have opted to refer to it as Pulsed Electric Current Sintering. This renaming reflects a shift towards a more technically accurate description of the process as it is understood today [50–53].

SPS is recognized as a highly effective method for consolidating a diverse range of materials, including intermetallic compounds, metal and ceramic matrix composites, nanostructured materials, amorphous materials, and even highly refractory metals and ceramics. This versatility is highlighted in its ability to process materials such as tungsten carbide effectively. The SPS technique is particularly valued for its rapid sintering capabilities facilitated by self-heating, which not only speeds up the process but also reduces power consumption significantly [54]. Additionally, it is known for its safety, reliability, and overall performance, making it a preferred method in various advanced material manufacturing scenarios. These attributes collectively underscore the growing importance of SPS in modern material science and engineering fields [55, 56].

High energy beam welding encompasses two primary welding techniques: laser beam welding and electron beam welding. Both methods operate at exceptionally high energy densities, typically ranging from 10^{10} to 10^{13} W/m^2. Welding is generally performed in one of two modes: conduction mode or keyhole mode. In the conduction mode, heat is conducted through the material, melting it to create a bond. This mode is particularly effective for welding thin sheets, as it allows for precise control of heat application and minimizes the risk of material distortion. When it comes to keyhole mode, it is characterized by a higher energy density, keyhole mode vaporizes a cavity in the material while melting the surrounding area to create a strong joint. This mode is especially suitable for welding thicker materials, where the vaporized cavity allows for deeper penetration. The keyhole process may require the addition of filler material to ensure a strong, durable joint.

Brazing and soldering are techniques used for joining materials through the application of a filler material that melts at a temperature lower than the base materials being joined. This allows for the joining of the parts without melting them, making these methods particularly suitable for combining dissimilar materials, such as ceramics to metals or different types of metals that may not be compatible for welding due to their complex geometries or material properties. In general, brazing involves the use of a filler material that melts above 450 °C. It is often employed to join dissimilar metals and is renowned for producing joints that are exceptionally strong – comparable in strength to natural connections between materials [57]. This strength makes brazing a preferred method in many industrial applications where robustness and durability are required. However, soldering: In contrast, soldering uses fillers that melt below 450 °C. While it is also useful for joining dissimilar materials, the bonds formed through soldering are typically not as strong as those created through brazing. However, soldering is highly valued for its ability to make precise and delicate joins, especially in electronics and small-scale fabrications [58, 59].

Resistance Spot Welding (RSW) is a mainstay in the automotive industry, especially prevalent in the production of steel-dominated automotive bodies-in-white due to its ability to be easily automated and its flexibility in operation. The weld gun electrodes in RSW effectively clamp and level small gaps between sheets, enhancing manufacturing efficiency [60]. Extensively applied beyond steel, RSW has been researched for its capability in joining a diverse range of dissimilar material combinations including aluminium-steel, magnesium-steel, and even glass-steel, among others. This adaptation extends to even more complex combinations such as copper-nickel plated steel, stainless steel with titanium, and titanium with aluminium [61]. The key advantage of RSW, particularly for aluminium-steel joints, is its potential to utilize existing infrastructure, thereby allowing the continued use of established equipment and processes. This capability significantly reduces the need for costly retooling for new joining techniques, offering an economically advantageous solution for modern automotive manufacturing that frequently involves various metals and materials. Thus, RSW not only maintains production standards but also drives cost efficiency by facilitating versatile material integration [62, 63].

11.2.1.4 Solid State Thermal Joining

The solid-state joining method is an advanced technique that evolves from traditional thermal processes but uniquely operates without melting the materials. This method typically involves using physical processes such as friction or ultrasonic energy to heat the materials sufficiently to make them pliable. Once heated, the materials can be deformed and fused together under applied force, creating a durable joint while maintaining the solid state of the base materials. This approach is particularly advantageous for materials that might lose desirable properties through melting, offering a robust alternative to more conventional welding techniques.

Friction Stir Welding stands out as one of the representative solid state thermal joining techniques, particularly adept at bonding light metals with steels. This process utilizes a rotating tool that generates friction heat upon contact with the target materials. Unlike conventional welding, FSW does not melt the materials; instead, it softens them sufficiently to enable mechanical mixing and bonding under pressure [64]. This approach not only preserves the bulk properties of the materials but also significantly limits the formation of detrimental intermetallic phases often associated with other welding methods. By preventing these phases and addressing compatibility issues between different materials, as well as disparities in their thermal properties and melting points, FSW enhances the joint quality, maintaining the integrity and performance of the materials involved [65, 66].

Friction welding is a solid-state joining technique where two workpieces are heated to a plastic state and welded using an upsetting force that displaces the material at the joint interface [45]. This process does not involve melting the materials; instead, it generates heat through the relative motion of the workpieces under compressive load. There are two main types of friction welding: rotary friction or spin welding, where one workpiece is rotated against another under pressure; and linear friction welding, where the motion involved is lateral oscillation rather than rotation. Both methods effectively create a strong bond by heating the materials to a point where they can flow and fuse under pressure [45].

11.2.2 Dissimilar Material Joining and Aerospace

In aerospace manufacturing, the development and application of advanced welding techniques are crucial for maintaining structural integrity and material compatibility. Techniques such as Gas Tungsten Arc Welding (GTAW) and Gas Metal Arc Welding (GMAW) are renowned for their precision and are particularly effective for welding aluminium alloys used in aerospace constructions like fuel and oxidizer tanks. Plasma Arc Welding (PAW) is favoured for its ability to handle higher temperature applications suitable for thicker sections in components such as solid rocket boosters.

While FSW has revolutionized the welding of high strength aluminium alloys, such as the 2xxx and 7xxx series, it's important to compare its effectiveness with traditional methods such as TIG. FSW avoids common defects like porosity and cracking seen in TIG welding, making it superior for critical aerospace applications where structural integrity is paramount. Its solid-state nature prevents porosity and cracking, making it essential for critical aerospace components like the Space Shuttle's external tanks. Linear Friction Welding and Inertia Friction Welding also play significant roles, with linear friction welding being ideal for creating high integrity joins in precision components like bladed disks for jet engines, and inertia friction welding suitable for cylindrical components needing high strength such as turbine shafts and compressor discs.

Further enhancing the aerospace sector's capabilities, Laser Beam Welding (LBW) and Hybrid Laser Welding provide high precision and control, critical for minimizing distortion and ensuring dimensional accuracy in sensitive materials and thin-walled structures. These methods allow for the welding of thick sections and dissimilar materials, improving filler material integration and weld pool control [67, 68].

Diffusion bonding, coupled with superplastic forming for titanium alloys, facilitates the creation of complex shapes, crucial for parts like wing access panels in aircraft such as the Airbus series. Brazing is extensively utilized to join dissimilar materials and complex parts, using fillers with lower melting points to avoid altering the microstructure significantly [68].

Challenges persist with new aerospace materials like metal matrix composites and oxide dispersion-strengthened alloys, which can degrade under traditional welding. Techniques like Transient Liquid Phase bonding and the use of interlayers help mitigate these issues, ensuring that welds meet necessary performance standards [67].

Overall, the continuous evolution of welding and joining techniques is integral to the aerospace industry, underpinning the performance, reliability, and durability of aerospace vehicles. For example, the joining of titanium to nickel-based superalloys using diffusion bonding has been particularly beneficial in manufacturing complex components like engine bypass ducts and compressor cases, where the thermal stability of titanium and the high-temperature strength of nickel alloys are crucial. Each method is critically evaluated for its ability to meet the stringent demands of aerospace materials and applications, emphasizing the sector's ongoing need for innovation and adaptation in material processing.

11.3 AM AND BIMETALLIC COMPONENTS

11.3.1 AM and Metallic Materials

Additive Manufacturing (AM), commonly referred to as 3D printing in certain contexts, has emerged as a significant focus within the field of materials science. The origins of AM date back to the 1980s when the first method known as Rapid Prototyping was developed, utilizing CAD technologies [69]. Initially, these technologies were nascent and not sufficiently advanced for large-scale industrial application, proving both costly and time-intensive due to the lack of reliable supporting technologies.

In recent years, advancements such as multi-nozzle printing heads and high-speed laser cutters have significantly enhanced the capabilities of AM processes. These improvements have led to increased industrial interest and application of AM technologies. Current 3D printing technologies not only facilitate the production of complete parts rather than mere models but also allow the entire manufacturing process to be computer controlled. This automation results in substantial reductions in cost and time, as well as decreased reliance on human labour, thereby enhancing efficiency [69].

Furthermore, AM technology empowers consumers to manufacture customized products using specific software, thereby enabling the production of tailored items ranging from rapid repairs to unique objects. The technology's adaptability extends to the use of diverse materials including metals, ceramics, polymers, and other unconventional materials such as paper and carbon fibre. This versatility renders AM far more flexible than traditional manufacturing methods [70].

Given its advantages in flexibility and efficiency, AM is poised to play a predominant role in the production of precision goods in the future. Wohlers et al. have projected that the AM industry will continue to expand in the coming years, with sales of AM products potentially exceeding AU$22.9 billion (US$15.8 billion) globally [71].

Additive Manufacturing (AM) of metallic materials represents one of the most promising avenues for the production of metallic parts, as recognized by the industry [72]. Initially, the adoption of metallic AM was slow due to concerns over its efficiency and cost-effectiveness. However, advancements in technology and dedicated research have transformed AM into a profitable and valuable approach for fabricating metallic components [73].

The American Society for Testing and Materials (ASTM) categorizes metallic AM technologies into seven distinct groups. Among these, five are predominantly utilized in the industry: Material Extrusion (ME) accounting for 10 percent of the market, Powder Bed Fusion (PBF) at 54 percent, Directed Energy Deposition (DED) with 16 percent, and Material Jetting (MJ) and Binder Jetting (BJ), which together make up 16% of the market. Each category encompasses specific techniques that exemplify the capabilities of AM in metal fabrication [71].

For instance, the PBF category includes advanced techniques such as Direct Metal Laser Sintering, Electron Beam Melting, Selective Laser Sintering, and Selective Laser Melting. In the DED category, notable methods include Electron Beam Freeform Fabrication (EBF3), Wire and Arc Additive Manufacturing (WAAM), and Laser Consolidation. The Material Extrusion group features techniques like Composite Extrusion Modelling. Moreover, Powder Bed and Inkjet 3D printing and Ultrasonic Consolidation are representative technologies for the Binder Jetting and Lamination categories respectively [74, 75].

11.3.2 WAAM

In the realm of additive manufacturing, WAAM stands out as a compelling alternative to traditional methods that often involve significant material waste and struggle with handling high-performance alloys. Traditional approaches like friction welding are limited by the shapes and sizes they can manage, while methods such as diffusion bonding and electron beam welding are hindered by lengthy process times and high operational costs. WAAM, in contrast, offers greater flexibility in fabrication, enhanced material utilization rates, a wider selection of materials, faster construction speeds, and lower overall costs, making it an increasingly preferred choice for metal manufacturing.

Wire arc additive manufacturing, a subset of Direct Energy Deposition (DED), has garnered significant industrial attention in recent decades. It is prized for its high deposition rate, which can reduce processing times by 40–60 percent compared to conventional methods, coupled with its low equipment costs that permit operation in diverse environments. Additionally, WAAM achieves material utilization rates of 90 to 100 percent, substantially reducing waste and associated costs [76].

Contrasting with the predominant metal AM technique, powder-bed based systems, WAAM showcases numerous benefits. Its higher deposition rates lead to shorter processing times; for example, the deposition rate for the Ti6Al4V alloy in laser powder bed fusion is approximately 0.1–0.18 Kg/h, and in electron beam powder bed fusion, it is 0.26–0.36 kg/h. In contrast, WAAM can achieve rates of 0.5–4 kg/h. WAAM's use of wire feedstock, as opposed to powder, generally results in fewer and smaller defects like porosity, which can compromise the integrity of the built material. This characteristic makes WAAM a more suitable choice for producing high-quality materials, with a reported material utilization efficiency of 90–100 percent even accounting for any necessary surface finishing [74], whereas powder-bed systems typically achieve around 50 percent efficiency throughout the process [77–79].

The technical foundation of WAAM is grounded in established atomized welding techniques such as Gas Metal Arc Welding (GMAW), Gas Tungsten Arc Welding (GTAW), and Plasma Arc Welding (PAW). GMAW is particularly prevalent in WAAM, utilizing a voltage-generated arc between a consumable wire electrode and the workpiece, shielded by either inert (MIG) or active (MAG) gases to protect the weld pool and material. GTAW and PAW, while similar, differ primarily in electrode placement and the type of arc used. PAW utilizes a constricted plasma arc that enables higher temperatures and energy densities, advantageous for large and complex parts. However, its complexity, cost, and the high skill level required for operation render it less flexible than GTAW, which is noted for its stability, user-friendliness, and precision, producing nearly defect-free results [80, 81].

11.3.3 WAAM and Aerospace

WAAM has become an essential technology in aerospace manufacturing due to its unique benefits over conventional methods. The aerospace industry, characterized by its need for lightweight yet strong components, benefits significantly from WAAM's capabilities. This manufacturing process supports the creation of large-scale parts with complex geometries, such as airframes and engine components, which are typically made from high-performance materials like titanium, nickel alloys, and aluminium. The ability of WAAM to handle these materials effectively addresses a critical industry challenge – minimizing material waste while maintaining structural integrity and performance [82].

One of the standout advantages of WAAM in aerospace applications is its cost efficiency and material conservation. Traditional manufacturing processes often involve a high buy-to-fly ratio, indicating substantial material wastage. In contrast, WAAM enables a more direct and efficient material deposition method, significantly reducing waste and, consequently, production costs. This efficiency is crucial for aerospace applications where material costs constitute a significant portion of overall manufacturing expenses. Additionally, the rapid deposition rates of WAAM mean that parts can be produced much faster than with other additive manufacturing technologies, further enhancing productivity and reducing lead times [83].

WAAM's adaptability and flexibility also play a crucial role in its applicability to the aerospace sector. The process is capable of producing both structural and non-structural components, offering the flexibility to create customized solutions for specific aerospace needs [84]. For example, WAAM has been used to fabricate critical parts like brackets, mounts, and even more complex assemblies that require precise mechanical properties and reliability under extreme conditions. The technology's ability to integrate seamlessly with existing manufacturing systems and to produce parts that meet rigorous aerospace standards makes it a valuable addition to the aerospace manufacturing toolkit.

These attributes demonstrate why WAAM is not just an alternative but often a preferable choice for aerospace applications, providing significant advantages in terms of cost, speed, and material efficiency [83].

11.3.4 Dissimilar Aerospace Material Joining via WAAM

11.3.4.1 WAAM-based Joining

The joining of dissimilar materials, essential for manufacturing functionally graded materials (FGM), can be achieved through various techniques. Conventionally, this is done using standard fusion welding, but methods such as casting, rolling, diffusion bonding, and powder metallurgy are also utilized. These traditional techniques, however, often come with significant limitations such as high costs, restricted shapes, and lengthy processing times [77]. Additionally, conventional fusion welding frequently encounters issues like intermetallic cracking, porosity, and reduced mechanical properties in the HAZ [85].

WAAM presents a versatile alternative, capable of fabricating bimetallic materials in both sequential and simultaneous orders. Research by Ahsan et al. [86] on the GMAW-based fabrication of low carbon steel to stainless steel 316L showed successful joint formation without defects, though there was a notable reduction in strength properties on the low carbon steel side. Similarly, Liming Liu et al. [87] demonstrated the feasibility of joining iron ER70S-6 and bronze SG-$CuSi_3$ using GMAW, resulting in an interface with excellent mechanical properties and no defects, underscoring WAAM's capability in joining dissimilar materials.

Thus, further aerospace material joining studies by A. Rajesh Kannan et al. [88] and Yinbao Tian et al. [89] explored other material combinations like stainless steel 904L with Hastelloy C-276, and Ti6Al4V with Al-6.25Cu, respectively. These studies found variations in microhardness and tensile strength, depending on the sequence of material deposition, highlighting the influence of alloying elements in the bimetallic interface. Ahsan et al. [90] designed a part combining nickel aluminium bronze with stainless steel 316L, which was free from defects but formed Fe_3Al intermetallic compounds, although occasional liquation cracking was observed in the HAZ.

Furthermore, Tanvir et al. [91] investigated the joining of bimetallic austenitic stainless steel 316L and Inconel 625 using an overlapping building strategy via GMAW. The findings indicated a uniform hardness distribution across the component, except in the first and last layers, and highlighted a significant presence of detrimental Laves phases in the Inconel part, suggesting the need for subsequent heat treatment to eliminate these phases. These studies collectively validate WAAM's high potential and adaptability in creating high-quality joints between dissimilar materials, offering a promising path forward in advanced manufacturing processes.

11.3.4.2 New Approach: Interweaving Building Strategy

In contrast, Bintao Wu's innovative approach using Wire Arc Additive Manufacturing (WAAM) introduces an interweaving building strategy that enhances the joining of dissimilar metals, specifically steel and nickel. This technique involves alternating layers of steel and nickel, which not only avoids the weaknesses typical of conventional welding but also enhances the mechanical properties of the joint. The interlocking microstructure created by this method increases the tensile strength at the interface beyond that of the individual materials used. This process has been shown to produce defect-free components with superior mechanical properties due to the solid solution strengthening effects of intermixed Fe and Ni atoms. This interweaving strategy holds great promise but requires further practical experimentation to refine and validate its effectiveness, especially in producing components with precise mechanical properties [92].

This innovative approach, while promising, underscores the need for more practical studies to explore its full potential. The subsequent case study aims to further investigate the practical applications and validate the capabilities of the interweaving building strategy in real-world manufacturing scenarios.

11.4 CASE STUDY: INTERWEAVING BUILDING STRATEGY ON DISSIMILAR AEROSPACE MATERIAL COMPONENTS

11.4.1 Abstract

In this case study, wire arc additive manufacturing (WAAM) process is used to build the nickel-based Inconel 718 superalloy (IN718) and stainless-steel 316L (SS316L) bimetallic components with an interweaving building strategy in a cost-effective manner. The bimetallic wall structures are built with different welding parameters, aiming to investigate the feasibility of fabricating a bimetallic component using these two alloys. The results show that typical columnar grains with elongated dendrites characterized the matrix of the manufactured components. A number of cracks are found in the interface between the two materials and mostly in the SS316 site of those samples, nevertheless, the enhanced cooling rate appeared to decrease cracks. In addition, the as-fabricated samples have a similar hardness value to the WAAM process fabricated IN718 and SS316, respectively. This research provides valuable knowledge in process parameters to produce high-quality IN718-SS316L bimetallic complex geometric components via WAAM to meet the demand for joining highly alloyed materials for specific structural applications.

11.4.2 Introduction

Nickel-based superalloys are typically applied wherever they are in alignment with severe environmental conditions, such as submarines, aerospace, and nuclear, given their high strength, thermal stability, corrosion and oxidation resistance [93]. At present, nickel-based superalloys can be used as a structural material for turbine blades, power plants or jet engine parts. For instance, nickel-based alloys are used in gas turbines where temperatures are typically over 500 °C [11]. Due to their outstanding creep and rupture resistance at high temperatures, nickel-based superalloys are also the best choice for aircraft engine components, where service temperatures usually arrive at 800 °C [94].

Stainless steel is one of the essential materials in the ferrous group alloys that contains a chromium content higher than 11 percent, providing the material with high temperature resistance and oxidation resistance [30]. The categories of stainless steel are various, depending on the specific element contents in the steel alloys, which include austenitic, ferritic, martensitic, duplex, and precipitation hardening stainless steel. As a relatively affordable and reliable material, stainless steel has seen wide applications. Austenitic stainless steel, which nearly occupies about two-thirds of the whole stainless-steel family, has a higher corrosion resistance than other stainless steels. Such preferential property can be ascribed to the face-centred-cubic (FCC) microstructure achieved by adding nickel, magnesium, and nitrogen [30]. There are two subgroups of austenitic stainless steels: 200 series and 300 series. The difference between them is based on the formation of austenite with different elemental contents, that is, the 300 series is mainly based on the nickel content, and the 200 series is achieved by adding nitrogen and magnesium to partially substitute for the lower amount of nickel. One of the outstanding materials in the 300 series is SS316/316L, which exhibits comparative corrosion resistance, high-temperature resistance and excellent mechanical properties thereby SS316/316L materials can be used as a more affordable corrosion-resisting material than some alloys, such as Ni-based alloys in some cases [95–97].

Joining dissimilar metals has already become an advanced technique for achieving functional and structural demands, with its main merits being weight reduction, cost control, and increment of toughness [77]. Materials in which composition and structure vary gradually with volume are called functionally graded materials (FGM), having seen widespread application in aerospace industries [98], medicine area [99], defence applications [100], energy devices [101], and optoelectronics [102]. The joining of dissimilar materials or FGM manufacturing can be achieved through several techniques. Traditionally, it can be accomplished via standard fusion welding, while some of the conventional processes, however, have drawbacks, such as high cost, limited shapes and long processing periods [77]. In addition, joining them via conventional fusion welding is also followed by many technical challenges, like the formation of intermetallic cracks, porosities or a decreased mechanical property in the heat-affected zone (HAZ) [85]. Recently, friction welding, diffusion welding, and laser beam welding were selected as alternatives. However, shortcomings, including lack of flexibility for friction welding, the long producing period of diffusion bonding, and the high cost of laser beam welding, severely limit their overall widespread utilization [92, 103–105].

Wire arc additive manufacturing (WAAM), a recently developed wire-based directed energy deposition (DED) manufacturing process, has drawn much attention from industry and academia due to its versatility, flexibility, and uniqueness in producing metallic components with complex geometry. As a low-capital investment manufacturing process, WAAM is especially favourable for low-volume production owing to the material and time saving, as well as a typical process for repairing and refurbishing models and dies. [106–107]. Many studies have demonstrated the prominence of joining dissimilar materials using WAAM because of their high-quality final products [86, 108–110]. Ahsan et al. [86] investigated the fabrication process of low carbon steel (LCS) to stainless steel 316 via the Gas Metal Arc Welding (GMAW) technique. The joining of those two materials succeeds without defects, while the LCS site exhibits a lower strength property. A defect-free hybrid nickel aluminium bronze (NAB)-stainless steel 316L part was produced by Dharmendra et al. [111], in which a series of Fe_3Al intermetallic compounds were found near the bonding interface.

Another advantage of WAAM technique is the achievement of comparable mechanical properties of the final product, compared with feedstock materials. Tian et al. [89] have explored the formation of different Ti6Al4V-Al-6.25Cu bimetallic components by cold metal transfer (CMT) welding, and tensile strengthening was observed in the fusion joint. Kannan et al. [88] explored the joining of stainless steel 904L and Ni-based Hastelloy C-276. A significant increase in microhardness was observed in the bimetallic area due to the diluted alloying elements. Liu et al. [87] examined the feasibility of joining Fe (ER70S-6) and bronze (SG-$CuSi_3$) via GMAW. The interface layer of the final part showed an outstanding mechanical property without any visible defects. These results illustrate the possibility of manufacturing nickel-based alloy-stainless steel bimetallic components via WAAM.

Apart from the conventional layer-by-layer WAAM manufacturing process, a new building strategy by alternative disposition of overlapped beads was investigated by Wu et al. [92] to fabricate steel (ER70S-6) and nickel (ERNi-1) bimetallic material, which is called interweaving deposition strategy. The results show that the interface region had an increased average tensile strength when compared to feedstock materials, which is caused by an interlocking microstructure of nickel and steel in the bimetallic interface. Additionally, a strong solid solution strengthening effect of the material was achieved because of the Fe-Ni gradient formed by the intermixed Fe and Ni.

This study aims at making high-quality nickel-based alloy-stainless steel bimetallic components by WAAM with the interweaving deposition strategy based on previous studies. A group of IN718-SS316L bimetallic thin walls were deposited with comparable mechanical properties to the feedstock material using an interweaving building strategy. By adjusting the welding parameters, including torch angle, welding sequence, interpass temperature and heat input, different macrostructures, microstructures, and properties have been drawn at as-built samples. The related microstructure of the representative samples was examined by optical microscopy (OM), scanning electron microscopy (SEM) and energy dispersive spectrometry (EDS). Special attention was given to the characteristics of the interfacial region to examine the stability of bimetallic bonding through the study of microstructure and related mechanical properties. This work explored the feasibility of building IN718-SS316L bimetallic components via WAAM, providing a new avenue for joining nickel superalloy and stainless steel with comparable mechanical properties.

11.4.3 Experimental Procedure

The WAAM system applied in this study is based on metal inert gas (MIG) welding with pulse welding technique, a tandem torch was applied for achieving the combination of two materials. Figure 11.1 shows the whole twin wires WAAM system for the study, including a control PC (1), robotic controller (2), two Fronius TPS 4000 welders (3), shielding gas system (4), and industrial ABB robot (5).

The feedstock materials were Inconel 718 1.2 mm commercial wire and stainless steel 316L 1.2 mm commercial wire (Table 11.1). The thin metallic walls with one-half nickel alloy and one-half steel alloy were manufactured onto a mild steel substrate with planned height and length at about 50 mm and 120 mm, respectively. High-purity argon gas was applied to isolate the weld pool from severe oxidation during welding. The tandem welding work is illustrated in Figure 11.2.

The welding process used a layer-by-layer building strategy (Figure 11.2a), followed by a to-and-fro repeating welding path for balancing the height of the start and end points of the as-built component (Figure 11.2b). The twin torch supplied the steel and nickel wire separately with two welding principles using a welding torch perpendicular to the substrate (Figure 11.2c) and a 60° to the substrate (Figure 11.2d)

TABLE 11.1
Chemical Composition of Feedstock Materials

	C	Mg	Cr	Ni	Mo	N	Fe	Nb	Ti	Al	Co
IN718	<0.08	<0.35	17~21	50~55	2.80~3.30	-	Balance	4.75~5.5	0.65~1.15	0.2~0.8	<1.00
SS316L	<0.03	<2.00	16~18	10~14	2.00~3.00	<0.1	Balance	-	-	-	-

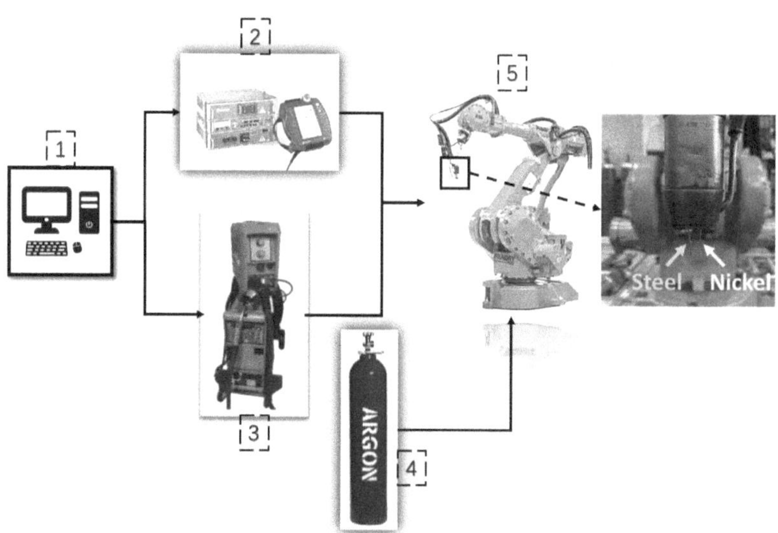

FIGURE 11.1 Illustration of GMAW welding system in this study.

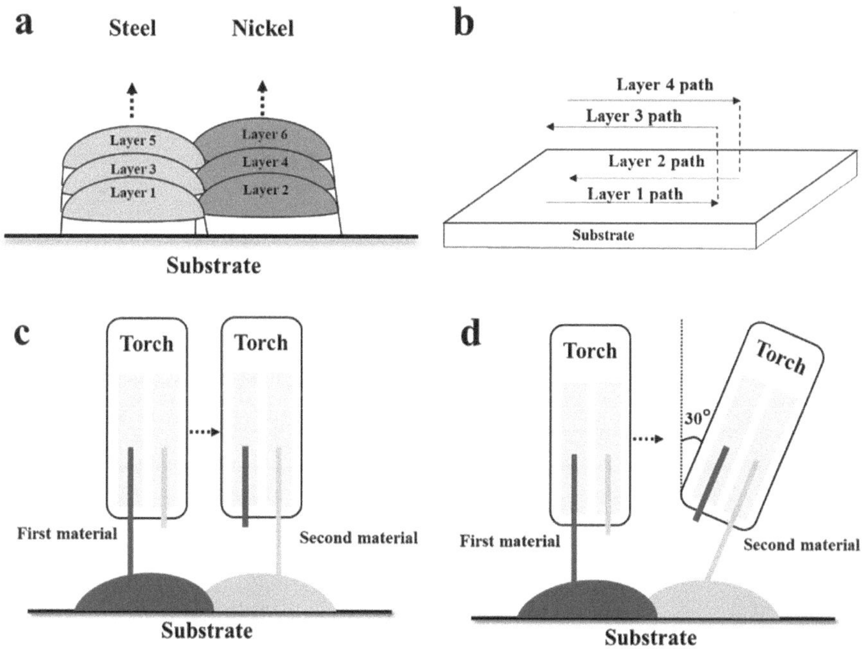

FIGURE 11.2 Experimental setup: (a) interweaving building strategy, (b) to-and-fro welding path, (c) welding method with perpendicular twin torch, (d) welding method with 30° angle applied twin torch.

for investigating the relationship between torch angle and bead geometry. Oxides and other contaminants were removed from each layer by a steel brush after every layer was deposited. Additional pressured air was used to increase the cooling rate for several selected walls. FLIR A655sc thermal imaging camera was applied for monitoring the temperature evolutions, and interpass temperatures between each layer were controlled at about 150 °C.

In this study, six bimetallic walls with different welding parameters, as shown in Table 11.2 and Table 11.3, were built-up with about 12 mm in length, 4 mm in width and 50 mm in height. The first five walls with the same heat input (calculated by the heat input formula for GMAW welding as follows [112]):

$$Q = Efficiency * \frac{Voltage * Amperage * 60}{Travel\ speed * 1000} \tag{11.1}$$

Additionally, the last wall with a lower wire feeding rate and travelling speed than the former five groups induced a lower heat input (Table 11.3). Figure 11.3 shows morphology of as-deposited walls and their cross-section with different welding parameters.

The cross-section specimens for metallographic observation are extracted from the middle of the metallic wall where the structures appear relatively homogeneous. These were polished by Struers automatic polisher Tegrapol 21 via a standard

TABLE 11.2
Comparative Experiments Plan with Modified Working Parameters for Welding in this Study

Wall No.	First Welded Material	Torch Angle Applied	Cooling	Interpass Temperature
1	IN718	Perpendicular	Natural Cooling	~150°C
2	SS316L	(Figure 11.3. c)	Natural Cooling	~150°C
3	SS316L	30° angle applied	Natural Cooling	~200°C
4	SS316L	(Figure 11.3. d)	Natural Cooling	~150°C
5	SS316L		Pressured Air cooling	~150°C
6	SS316L		Pressured Air cooling	~150°C
(Less heat input)				

TABLE 11.3
Processing Parameters of As-built Bimetallic Walls in this Study

	Current (A) (Ni/Fe)	Voltage (V) (Ni/Fe)	Travel speed (m/min)	Wire Feed speed (m/min)	Heat input for Ni	Heat input for Fe	Average heat input
Wall 1-5	191/158	21.6/22.8	0.3	6.5	660.10	576.38	618.24
Wall 6	152/120	20.2/22.3	0.25	5	589.52	513.79	551.65

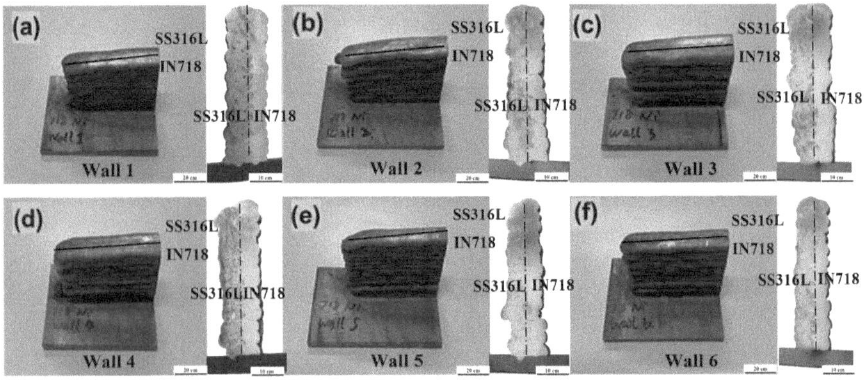

FIGURE 11.3 Macro-morphology of as-deposited walls and their cross-section with different welding parameters: (a~f) wall 1~6.

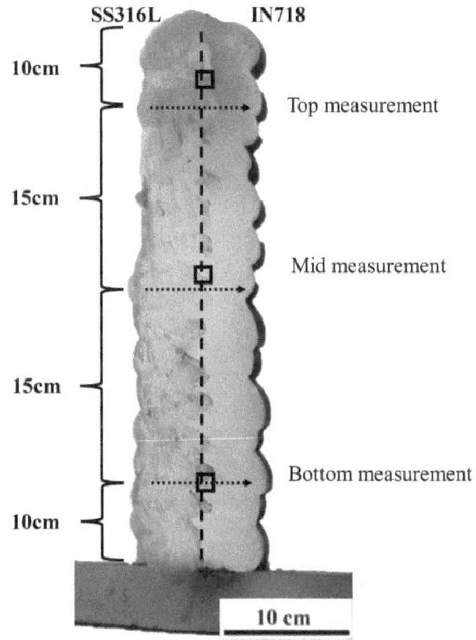

FIGURE 11.4 Hardness measuring map for a representative cross-section from wall 6.

metallographic polishing procedure and then electro-etched with a solution of 5 gm Oxalic acid mixed with 95 ml HCl (reagent grade) at ~ 3.5 VDC for 2-3 s for further detailed observation.

The microstructural observation was achieved by Leica DMR optical microscope to obtain the microstructural information, including dendrite growing direction, detailed structural information of interface, and number and morphology of microcracks. Further chemical distribution analysis was examined via JEOL 6000 operating SEM equipped with an EDS at 15 kV.

The hardness value of as-built specimens was tested via MATSUZAWA Vickers automatic hardness tester with a load of 1.96 N and a dwell time of 10s, and the results were measured from the top, middle, and bottom of the sample. Twenty points are tested in a row with an interval of 0.25mm with a direction going from the IN718 site to the SS316L site. Also, for studying the hardness distribution around the interface, a matrix of 5*5 with an interval of 0.25 mm as shown in Figure 11.4 was allocated across the interfacial region.

11.4.4 Results and Discussion

11.4.4.1 Microstructure

From the cross-section observed in Figure 11.3, it appears that the width of the wall structure is slightly decreased with enhanced cooling by the pressed air. This is because the overflow in the welding pool is highly controlled by the accelerated solidification and cooling rate [113]. Figure 11.5 (A)–(C) shows the microstructures

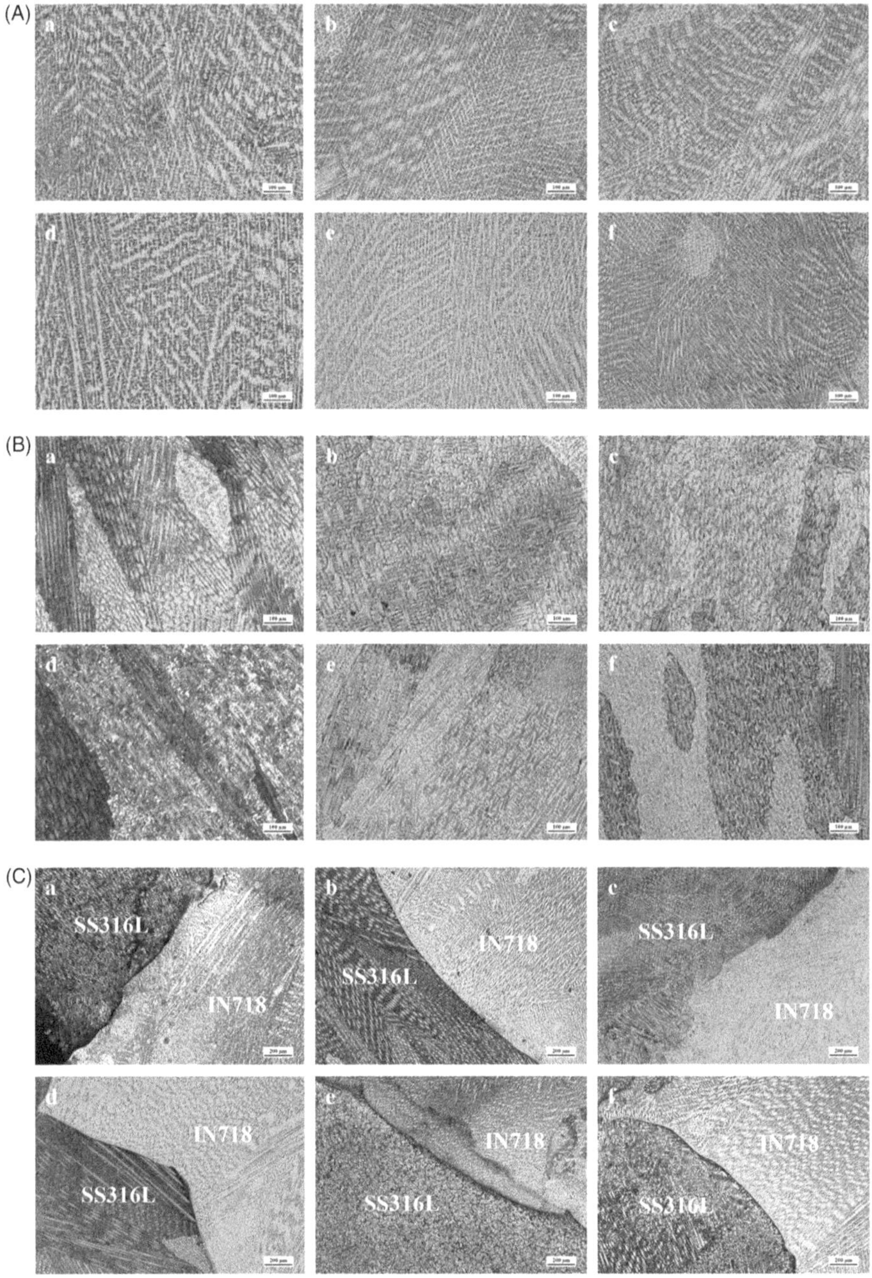

FIGURE 11.5 (A) Similar dendritic structures from the IN718 site observed from the top site of wall 1~6 (a~f). (B) Similar interfacial structures from the SS316L site observed from the middle region of wall 1~6 (a~f). (C) Similar interfacial structures from the IN718 site observed from the middle region of wall 1~6 (a~f).

of the cross-sections in six walls investigated via OM. In general, typical columnar grains with elongated isomorphic dendrites can be found throughout the matrix in the fabricated six samples at both alloy sides. The formation of dendrites happens during the solidification process, which begins with a spherical solid nucleus as a result of the elemental segregation. The solidified metal then grows in the preferred growth direction which is based on the nucleus, material, and the condition of depositions, following the highest temperature gradient [114]. As shown in Figure 11.5 (C), a clear interface induced by different growing directions of the dendrites in two alloys was observed from the middle between the deposition of IN718 and SS316L.

Owing to the similar microstructure, detailed microstructural images of a representative sample, wall 6, are shown in Figure 11.6, and some defects consisting mainly of microcracks were studied. From Figure 11.6a to 11.6c, a series of both cellular and columnar structures are observed on the SS316L side, which is consistent with literature [115–117]. The formation mechanism of different morphologies is generally consistent with compositional fluctuations and super cooling theory, being related to the ratio of liquid thermal gradient (G) and growth velocity (V) of

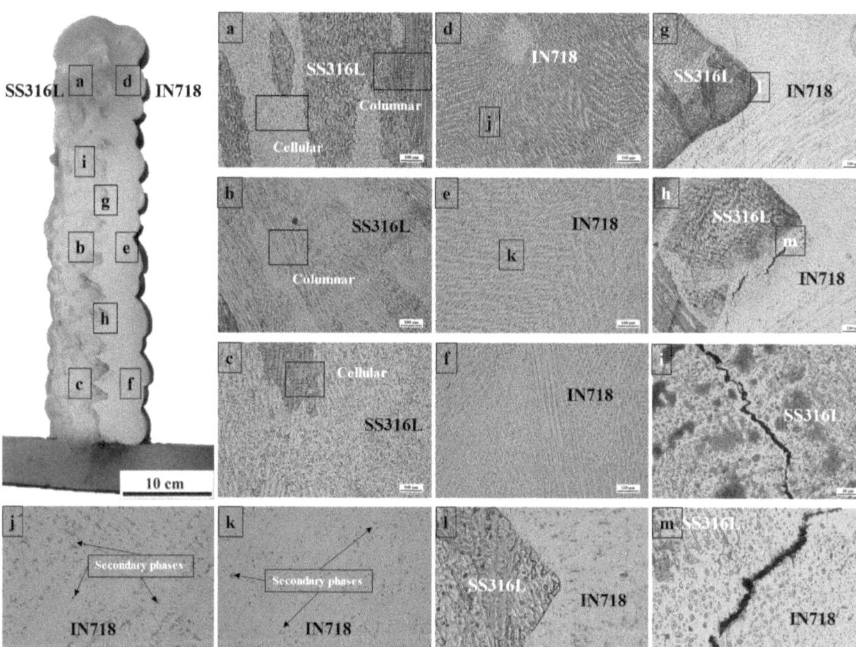

FIGURE 11.6 The cross-sectional macrograph and the representative microstructure in both SS316L and IN716 sides of wall 6.

the solid/liquid interface. When G increases and V decreases, it favours the growth of the equiaxed grains, while when G decreases and V rises, this results in favour of columnar dendrites with secondary or tertiary dendrite formations [118–119]. For the IN718 site, the main columnar grain matrix company with plenty of dendrites can be clearly observed in Figures 11.6d to 11.6f. A typical dendritic microstructure in all growing directions is displayed from the top to the bottom region of the specimen. A number of secondary phases grew along the dendrites in the IN718 site, which takes place due to the solute element segregation under a relatively quick cooling rate during WAAM [120]. The dendritic growth orientations in certain parts of the two alloy sides are different because of the change of the thermal gradient direction of each layer with varying sequences of welding during deposition, thus resulting in various dendrite growing directions [121].

Figures 11.6g and 11.6h reveal the optical microscope image of the bonding area between two alloys. An interface between the two materials is evident, showing an interlocking structure. Due to the interweaving building strategy, a relatively stable bonding has been created via WAAM, nevertheless, interface cracking still occurs. As shown in Figures 11.6h and 11.6m, this crack is located near the fusion line which is known as the weak area in solidification processes [122]. On the one hand, the weak area can be led by the solid-transmitted thermal stress that localized at the liquid films remained during solidification process, which results in a high possibility of high thermal stress concentration after the liquid becomes stable at low temperature. Additionally, the heat conduction from molten metal of the new deposition bead can dramatically impact the microstructure of HAZ of already deposited material and promotes the elemental segregation and formation of detrimental phases [123]. Also, Figure 11.6i shows a thermal crack at the SS316L site, which is determined to be a solidification crack based on its morphology and growth direction [124].

11.4.4.2 Compositional Variation Analysis of Interface Crack

In order to analyse elemental distribution, especially around the cracking area, SEM coupled with EDS analysis was conducted to analyse the compositional variation of two representative interface hot cracks from walls 5 and 6, respectively, as shown in Figure 11.7. The different concentration of Fe and Ni adjacent to interface areas is apparent. Moreover, high levels of segregated Nb elements and Mo occur in the interdendritic area and near the interfacial cracking zones of both wall 5 and wall 6, along the dendritic direction and parallel to the interface cracking. As with previous studies on welded IN718 [125, 126], this elemental segregation originates from the solidification into the eutectic phase, which is caused by the uneven distribution of Nb over dendritic microstructures. When the dendrite grows, the Nb is ejected from the solid phase to the liquid phase because of the different solute solubilities between the two phases. During solidification, a low concentration of Nb liquid leads to the formation of the austenite phase (γ). However, a high concentration of Nb in IN718 liquid can lead to the eutectic phase (γ + Laves phase), especially when Nb is isolated into spots by dendrite arms, where a high level of laves phase particles will be formed after solidification [125, 126].

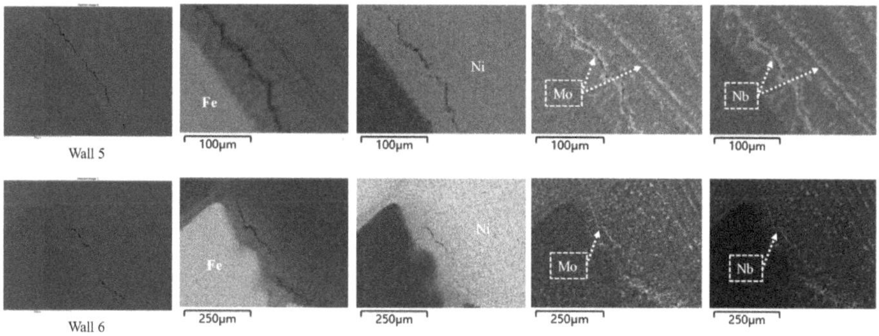

FIGURE 11.7 The EDS mapping of two interface cracks in wall 5 (first row) and wall 6 (last row).

11.4.4.3 Cracking Analysis

Typical cracks occurring in different walls are shown in Figure 11.8. The location of cracks that existed in as-built walls varies, and they are mainly concentrated around the SS316 L site, but very few cracks appear in the IN718 part. Throughout all the samples from this study, the cracking distribution from top to bottom is even, and the cracks are rarely clustered. Based on the recognition of cracks in steel side and interface areas, as well as morphologies of the cracking, the formation mechanism was analysed as follows.

The morphology of cracks depends on the location of formation, and those that formed in the steel part always grew along the homogeneous dendrites at grain boundary areas, as shown in Figures 11.8a to 11.8c. First of all, Figures 11.8a to 11.8b show two multi-segment cracks at the SS316L site between the two grains, behaving as a polyline (Figure 11.8a) or zigzag-shaped intergranular cracks (Figure 11.8b). These are considered to be ductility-dip cracks (DDCs) [127]. DDCs are generated along the mitigated grain boundaries, which is caused by a loss of intergranular ductility at high temperatures instead of the formation of the liquid film [128]. Ductility-dip cracking is a solid-state phenomenon and happens below solidus temperature under proper conditions when the thermal strain induced by manufacturing or welding intercepts the ductility dip temperature range [129–130]. As Figures 11.8a and 11.8b reveal, the morphological information of those cracks growing along the crystallographic grain boundaries is consistent with ductility-dip cracking characteristics.

Next, Figure 11.8c shows the typical straight cracking, growing alongside the dendrites, which appear to be solidification cracks [124], formed in the final step of the solidification stage. When the dendrites grow into grains and are separated only by a small amount of grain-boundary liquid films, the material is prone to generate hot cracks. Once the localised stress exceeds the resistance of cracking, cracking may happen in the grain-boundary fluid films or the grain-grain bridges [131]. This is the reason why solidification cracking often occurs along grain boundaries [131, 132].

Except for the cracks in the steel side, Figure 11.8d-f show that cracking near the bimetallic interface has grown along the borderline between two metals where the

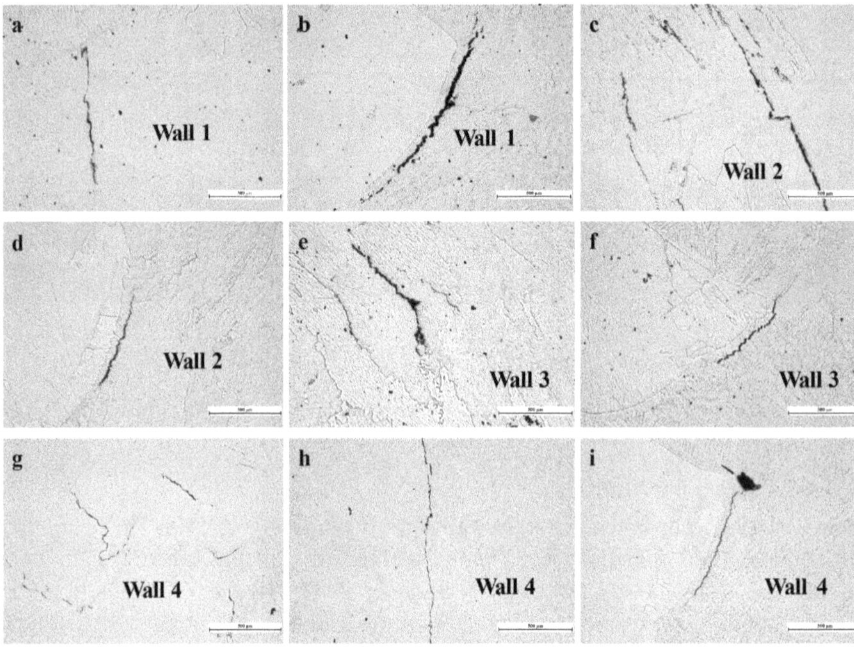

FIGURE 11.8 Representative hot cracks observed from cross-sections of as-built samples: (a) a crack grown across the grains in the stainless steel part from wall 1, (b) a deep and long hot crack happened on the interface region of wall 1, (c) a multi-segment crack grown along the grains in the stainless steel part from wall 2, (d) a small-scale crack grown along the grains in stainless steel part from wall 2, (e) two small cracks grown along the grain boundary from wall 3, (f) a interface cracking at the bottom of interlocking structure in wall 3, (g) a series of hot cracks grown along the grain boundaries in stainless steel part from wall 4, (h) a long crack grown in the stainless steel region from wall 4, (i) a crack with a minor porosity grown along the interface from wall 4.

HAZ locates. The geometry of the cracks is dependent on the shape of the interface contour, which are wavey and along the boundary in this study, being considered to be liquation cracks [123]. Those liquation cracking happen in the HAZ, always being parallel to the interface (Figure 11.8d to 11.8f). This area is where the solidified material can be affected by the welding technique when the material has been heated up above the eutectic temperature of its partially melted zone or solidus temperature for totally solidified material, as a result of the deposition of following layers. The grain-boundary liquation generated during this heating process, and the solidification shrinkage leads to a significant tensile strain induced by welding during the following cooling process, which makes the HAZ prone to crack [132–133]. For IN718 superalloy, the liquation cracks are mainly considered as the formation of constitutional liquation of Laves phase during eutectic in the solidification process, which is an Nb-based intermetallic phase [134]. Similar results about liquation cracks can also be found in the research by Chen et al. [123], where the current work differs in the interface crack observed in this study is parallel to the interface rather than perpendicular. This is less than ideal, as parallel interface cracking can easily deviate from the strongly bonded interface, thereby significantly compromising the stability of the material [134]. As Huang and Kou [133] demonstrated a similar result to this study, the cracks are parallel to the interface between two materials. Also, according to the results from Chen et al. and Vishwakarma et al. [123, 135], the Nb and Mo elemental enrichment can always be observed near those liquation cracks. Based on elemental distribution mapping with high Mo and Nb segregation (see Figure 11.7), it is highly likely that liquation cracks occur in those specific areas.

11.4.4.4 Microhardness Distribution

Figure 11.9 shows the microhardness distribution from the nickel site to the steel site sampled from the top, middle, and bottom regions of the as-deposited walls, all the walls are measured at approximately same positions with 40 points from IN718 to SS316, and the testing results are shown in Figure 11.9. In addition, for statistical analyse the relationship between hardness and welding parameters, the average hardness values of each site of each sample are listed in the Figure 11.10 and Table 11.4 below. Figure 11.9 (1) and Figure 11.11 highlighted a notable observation, whereby the top site of the sample shows no significant change, unlike other measurements. This is attributed to the distinct thermal coefficient between the two materials, which led to unbalanced bead modelling during welding. To address this, an additional layer of stainless steel was deposited to balance the height of each site.

The hardness values of walls 1, 2, 3, and 5 showed no significant differences, and there were no observable variations from top to bottom regions of each side of the sample, except for the sudden drops that indicated a testing point had passed through a cracking area. However, it is worth noting that the hardness value of each site in wall 6 is approximately 20HV higher than the aforementioned groups. Similar results can be found in Su et al. research [136], in which six groups of specimens built with different deposition parameters were tested, produced by the well-controlled wire feeding rate and travelling speed, and concluded a uniform hardness distribution from top to bottom upon the fabricated samples. In addition, Su et al. also found the relationship between traveling speed (TS), wire feed speed (WFS) and microhardness of the sample, when the WFS decreases, the microhardness of the as-tested sample has increased, which is because of a smaller grain size conducted by the slower WFS during cooling process, which is also a key role of decreasing actual cooling rate, mentioned by Mohammad [137], however the TS does not significantly affect the microhardness of the final product. This theory is in good agreement with the experimental data observed in this study, the experimental sample of wall 6 that has the slowest WFS has a higher microhardness than the others as per shown in the text.

Another obvious find is the hardness value of wall 4 is higher than any other groups, which is because a lower interpass temperature that controlled during welding process. According to Ma [138] and Wang [139], A higher interpass temperature can have a detrimental effect on the microhardness of the final product. Conversely, a lower interpass temperature has been observed to increase the grain size, as well as enlarge the reheated zone and heat-affected zone induced by deposition. Notably, both walls

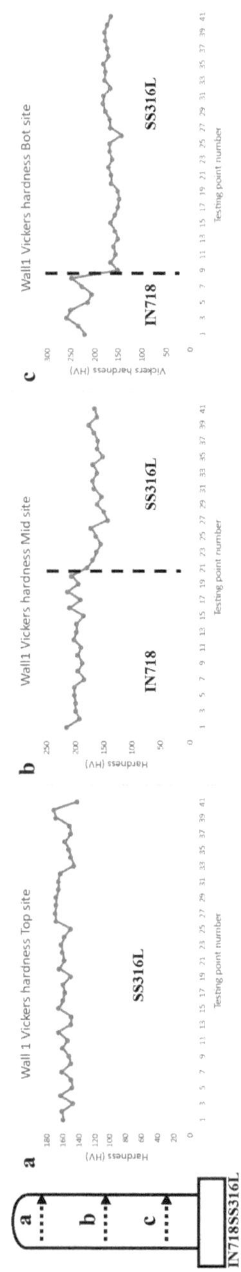

FIGURE 11.9(1) Hardness distribution of as-deposited bimetallic wall 1.

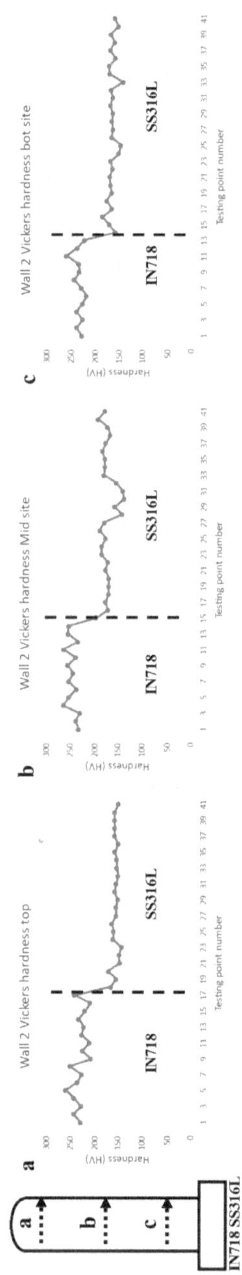

FIGURE 11.9(2) Hardness distribution of as-deposited bimetallic wall 2.

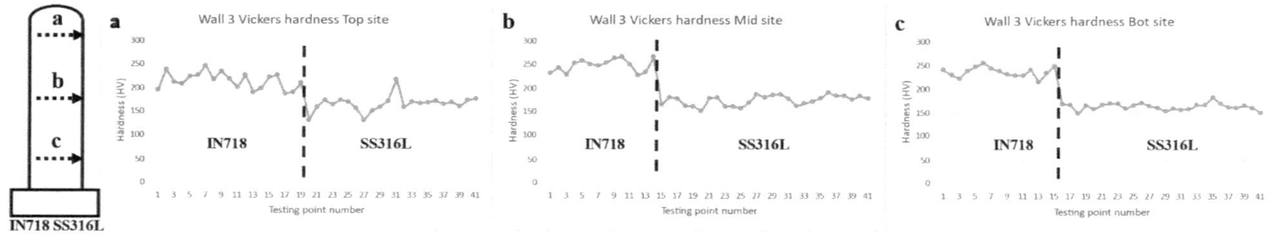

FIGURE 11.9(3) Hardness distribution of as-deposited bimetallic wall 3.

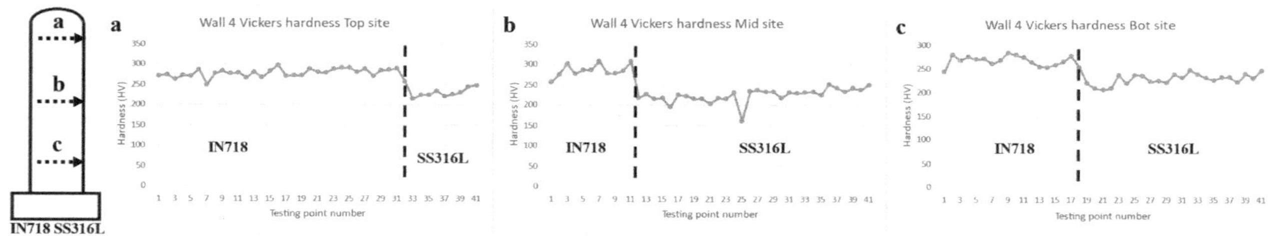

FIGURE 11.9(4) Hardness distribution of as-deposited bimetallic wall 4.

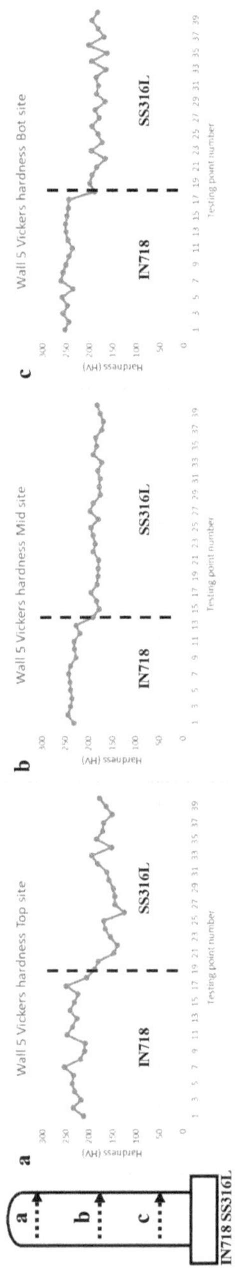

FIGURE 11.9(5) Hardness distribution of as-deposited bimetallic wall 5.

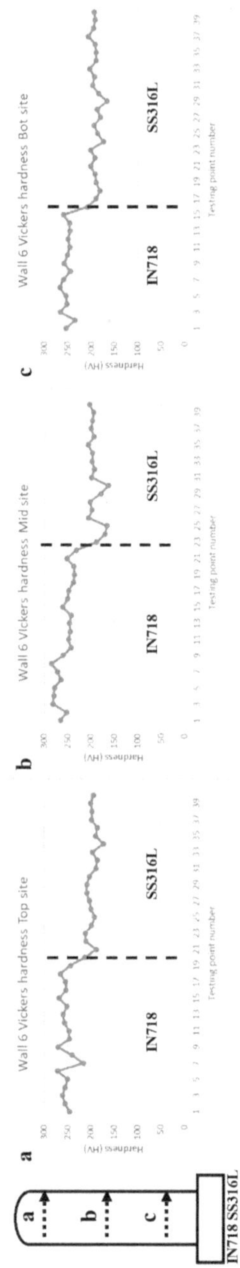

FIGURE 11.9(6) Hardness distribution of as-deposited bimetallic wall 6.

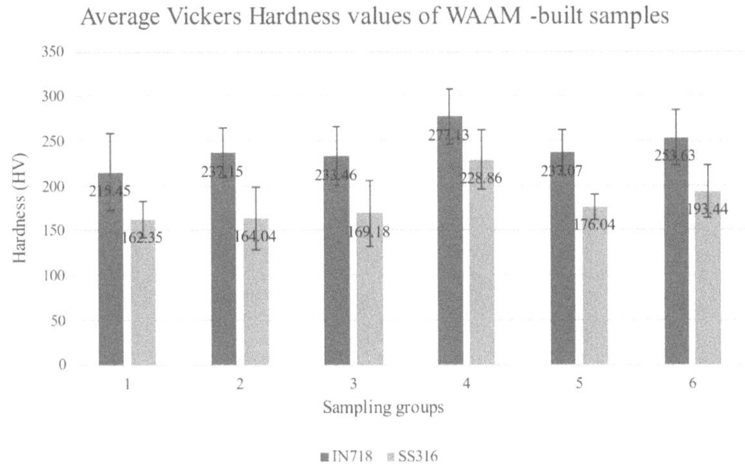

FIGURE 11.10 Average hardness of as-built samples, error bars represent the actual deviations of each single sample.

5 and 6 were subjected to a lower interpass temperature yet displayed a much lower hardness behaviour compared to wall 4. This underscores the importance of applying air cooling, which significantly increases the cooling rate of the material. However, it is worth noting that our results are inconsistent with those of other scholars, who concluded that a higher cooling rate can yield a finer grain size or precipitate size that increases the hardness value of both SS3316L [137] and nickel-based alloys [140, 141], which consequently results in a increased hardness value of the final product [137]. The reason for this discrepancy can be attributed to the distinctive air-cooling technique employed in this study, which differs from those employed in previous research. While prior studies utilized instantaneous cooling during welding to ensure comprehensive coverage of the solidification process, the approach adopted in this study involved applying pressured air cooling subsequent to the deposition of each layer. As a result, the solidification process was not entirely covered by this cooling method. Nonetheless, the material temperature was able to decrease at a faster rate, thereby mitigating the risk of cracking and reducing the overall welding duration.

While compared with the feedstock material or as-built samples by other WAAM processes, the fabricated component is identical to the separately built materials from previously reported research of ~190 HV for SS316L [142, 143] and 230~280 HV for IN718 [141, 144], which means the hardness behaviour of fabricated bimetallic components are comparable to the standard materials and acceptable for industrial usage.

Especially, for determining related harness distribution along the interfacial region, a hardness mapping of 5*5 pointer matrix with an interval of 0.25mm from IN718 site to SS316 site has been allocated to study the hardness variation at the interface.

As depicted in Figure 11.13 (a), the 3D hardness distribution map illustrates a relatively uniform hardness value in each site of the materials. Notably, the hardness value changes abruptly from the SS316L site to the IN718 site while crossing the interface,

TABLE 11.4
Average Hardness Value of As-built Sample

Hardness (HV)	Wall 1	Wall 2	Wall 3	Wall 4	Wall 5	Wall 6
IN718	215.45 ± 43.35	237.15 ± 27.25	233.46 ± 33.04	277.13 ± 30.87	237.07 ± 25.03	253.63 ± 31.07
SS316L	162.35 ± 20.35	164.04 ± 35.04	169.18 ± 36.98	228.86 ± 33.36	176.04 ± 14.04	193.44 ± 29.84

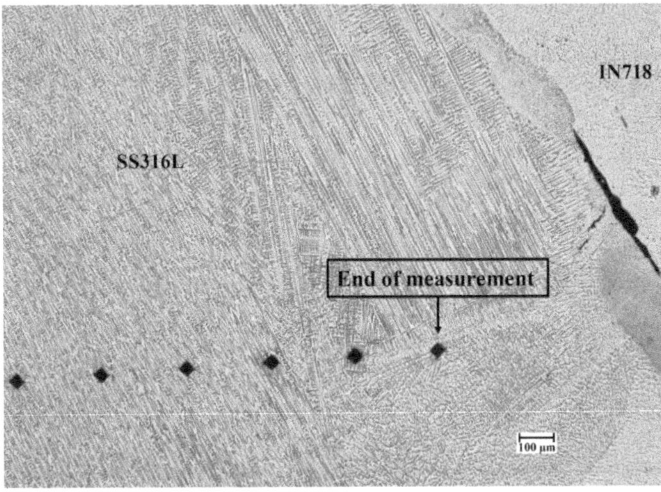

FIGURE 11.11 Demonstration of the end of hardness testing in wall 1 top site.

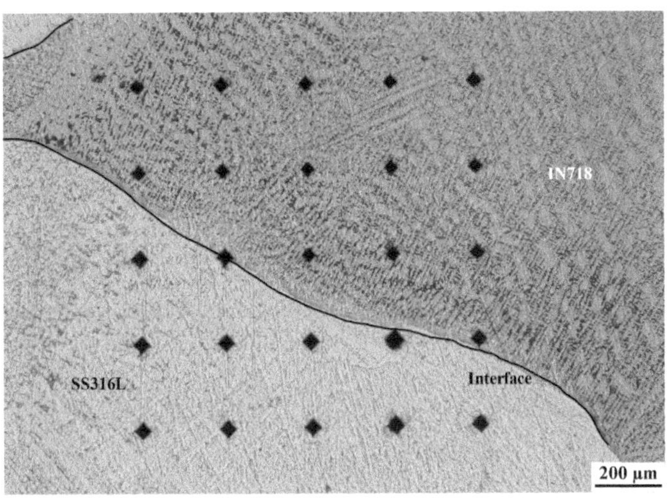

FIGURE 11.12 An example image demonstrates the hardness matrix measurement along the interfacial region.

without a discernible transition zone, suggesting that few large-scale penetrations occurred in the interface region of the as-deposited material. However, Figure 11.13 (b) that corresponds to wall 6 top map in Figure 11.13 (a) shows a typical penetration of approximately 500 µm in length through the interface, which resulted from spatter [145] or HAZ cracks [108], even though the crack may have been polished and hence not visible in this image. The corresponding image also displays a remarkable hardness value at the SS316L region, as highlighted by the square in each of

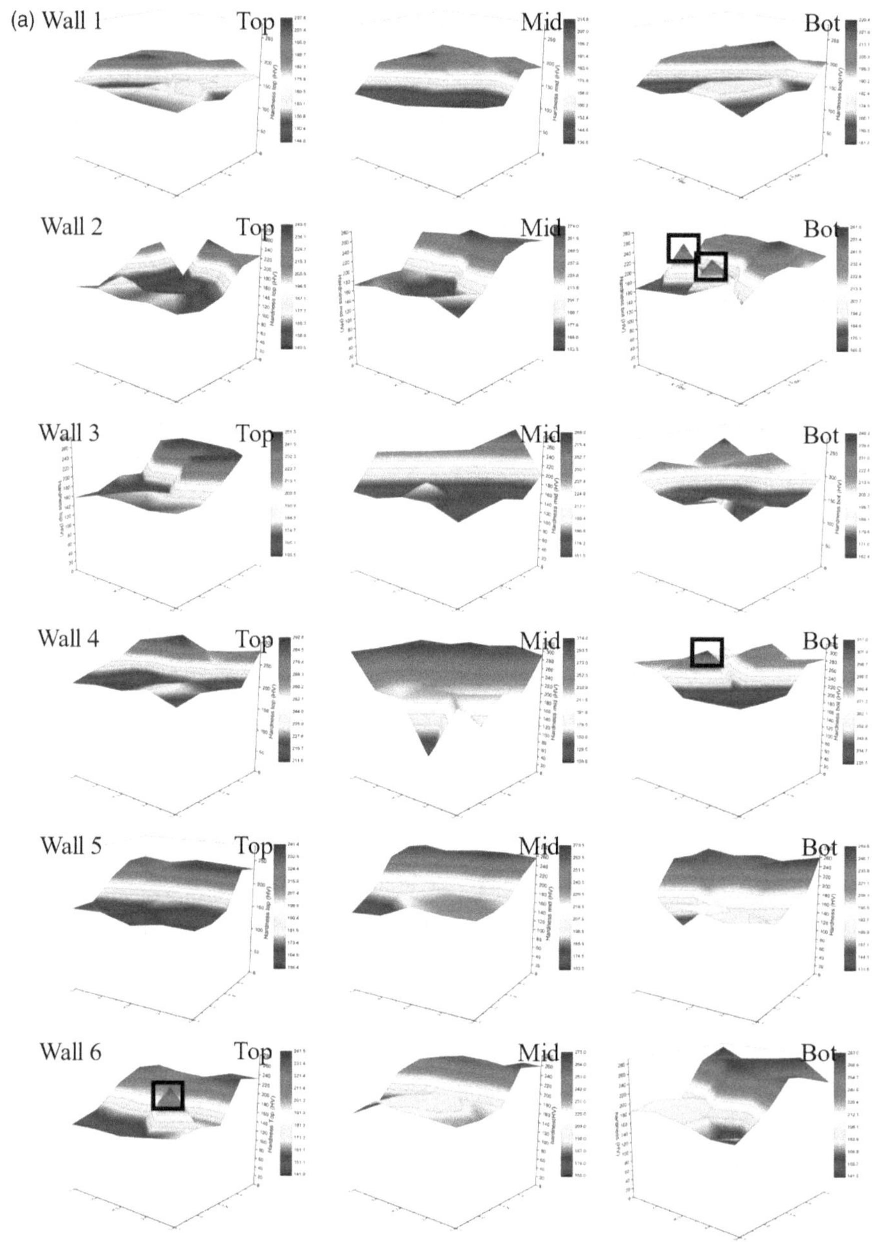

FIGURE 11.13 (a) 3D colour-coded map of hardness measured near the interface of as-fabricated bimetallic walls.

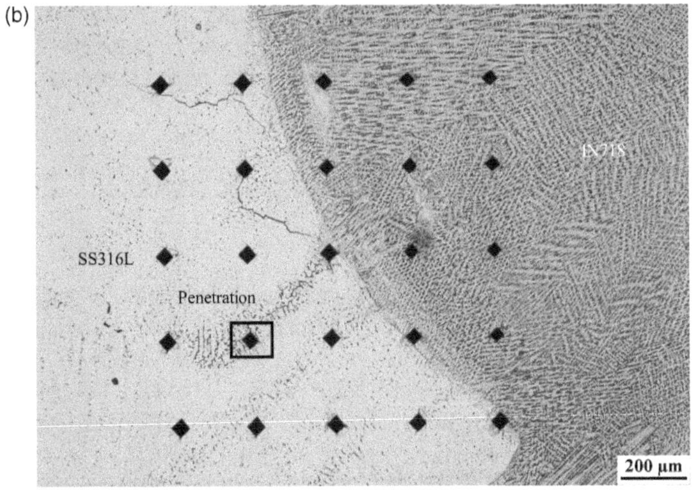

FIGURE 11.13 (Continued) (b) Image of actual hardness mapping of wall 6 top region that has a few IN718 penetrated into SS316L.

the figures. Comparable outcomes are also observed in the wall 2 bottom and wall 4 bottom regions, indicating that the hardness value of interfacial area of the WAAM built bimetallic components with an interweaving strategy are usually tidy and uniform without large-scaled fluctuates. However, in some cases, minor penetrations may occur as a result of spatter or HAZ cracks.

11.4.5 Case Study Conclusion

In this study, six IN718-SS316L bimetallic walls with different welding parameters were successfully built to investigate the feasibility of fabricating IN718-SS316L bimetallic materials. The microstructure and elemental distribution of the cross-sections of as-built walls were observed via optical microscopy and SEM-EDS analysis, respectively. The microhardness of selected specimens was tested. The following conclusions can be drawn based on the resulting analysis:

1. The feasibility of using WAAM to manufacture IN718-SS316L has been proven by six successfully built bimetallic thin walls using an interweaving building strategy. Microstructural inspection observed an excellent bond between two materials in the interfacial area.
2. Micro thermal cracks have been observed from all deposited components. Most of them appear in the SS316 part and some near the interface, determined as solidification cracking and liquation cracking, respectively. In addition, ductility-dip-cracking-like defects also occur in the SS316L site. However, none of them appear in the IN718 region. SEM-EDS analysis proved the cracks found near the interface are strongly related to the segregation of Nb and Mo in the HAZ.

3. The hardness of specimens is uniform on either the IN718 or SS316L side. These hardness values of each side are comparable to commercially produced monolithic alloys. The strength of the interfacial region is comparable to commercial stainless steel.

11.5 CONCLUSION AND FUTURE PERSPECTIVES

11.5.1 Overview of Findings

Throughout this chapter, we have explored the significant advancements and applications of superalloys, stainless steels, and a recent new joining method: WAAM, in joining dissimilar materials, particularly focusing on aerospace applications. WAAM has demonstrated substantial benefits over conventional manufacturing processes, including enhanced material utilization, reduced production costs, and the capability to produce complex geometries. Studies such as those by Ahsan et al., Liming Liu et al., and Bintao Wu et al. have showcased WAAM's versatility in achieving defect-free joins with improved mechanical properties across various bimetallic combinations. These successes highlight WAAM's potential to meet the rigorous standards of aerospace manufacturing, presenting a viable alternative to traditional methods that often struggle with efficiency and flexibility.

11.5.2 Contributions of Interweaving Building Strategy

Particularly noteworthy is the interweaving building strategy introduced by Bintao Wu et al., which has proven to enhance mechanical properties through the creation of an interlocking microstructure between different metals. This novel approach not only strengthens the joint interface but also optimizes material properties through solid solution strengthening effects. Despite its promising results, this technique is still in the nascent stages of development and requires more empirical data to refine its application and validate its effectiveness in industrial settings. However, issues such as hot cracks, residual stresses, and the enhancement of mechanical properties remain critical areas for ongoing research to optimize the process parameters and ensure the structural integrity of the manufactured components.

11.5.3 Future Perspectives

The joining of dissimilar aerospace materials, such as metals and composites, is crucial for enhancing the performance, reliability, and efficiency of aerospace structures. Advanced techniques like friction stir welding, laser beam welding, and hybrid methods ensure robust, high-quality joints while addressing the unique challenges posed by these diverse materials.

The future of dissimilar material joining will be mainly focusing on AM for those hard-to-normalized materials, in the realm of dissimilar material joining holds great promise. As we push the boundaries of what AM can achieve, the performance of bimetallic joins is expected to rise significantly. One of the critical advancements anticipated is the resolution of defects such as porosity and cracks that currently compromise the integrity of finished parts. This improvement is likely to come through refined pre-welding and post-welding treatments, as well as through the optimization of welding parameters that align more closely with the unique thermal and material characteristics of each component.

When it comes to AM in aerospace promises to revolutionize the industry by enabling the creation of complex components and dissimilar joints with unprecedented precision and efficiency. The integration of robotic technology with AM will bring innovative solutions to the skies, allowing for on-time recovery and replacement of intricate parts, thus enhancing the operational readiness and resilience of aerospace systems, which is the "on-the-fly" robot that with more limited size but more accuracy on the building, and those robotics will not only streamline production but also significantly reduce downtime and maintenance costs, pushing the boundaries of aerospace engineering.

Looking ahead, the development of more stable building strategies for bimetallic components promises to revolutionize their application across diverse fields. By creating new alloys or using similar compositions that bridge the properties of two distinct materials, we can facilitate a smoother transition within the joint area, enhancing the overall performance of the fabricated components. Additionally, the adoption of advanced techniques like ultrasonic welding could play a pivotal role in eliminating defects commonly associated with single-material fabrication. These methods, when adapted for bimetallic components, could significantly reduce issues such as porosity and cracking, thus improving the structural integrity and functionality of the final products.

Moreover, the integration of artificial intelligence into the AM processes presents a transformative potential for the field. AI's role extends beyond mere process optimization; it involves a comprehensive understanding of the welding dynamics through real-time monitoring and adaptive control systems. By harnessing AI to not only guide the welding process but also to simulate and predict the outcomes on the final components, manufacturers can achieve a higher degree of precision and repeatability. This proactive approach ensures that each layer of the build meets stringent quality standards, enhancing the structural integrity and functional performance of the components.

As these technological advancements converge, the role of WAAM in manufacturing dissimilar material components is set to become even more pivotal. The method's ability to efficiently produce complex parts with tailored mechanical properties – while minimizing waste and energy use – makes it an increasingly attractive option for industries ranging from aerospace to automotive. The ongoing development of AM technologies, coupled with deeper insights from AI and improved process controls, indicates a bright future for WAAM. This progression not only promises enhanced capabilities and broader applications but also heralds a new era of innovation in high-performance material manufacturing.

REFERENCES

[1] M. J. Donachie, S. Donachie, M. Donachie, and S. Donachie, *Superalloys: A Technical Guide*. A S M International, 2002.

[2] A.P. Mouritz ,"12 – Superalloys for gas turbine engines," *Introduction to Aerospace Materials*, Ed.: Woodhead Publishing, 2012, pp. 251–267.

[3] T. Nagayama, T. Yamamoto, and T. Nakamura, "Thermal expansions and mechanical properties of electrodeposited Fe–Ni alloys in the Invar composition range," *Electrochimica Acta*, vol. 205, pp. 178–187, 2016/07/01/ 2016, doi: https://doi.org/10.1016/j.electacta.2016.04.089.

[4] H. S. Mali and D. R. Unune, "Machinability of nickel-based superalloys: An overview," in *Reference Module in Materials Science and Materials Engineering*: vol. 1 pp. 1–5 Elsevier, 2017.

[5] D. Coutsouradis, A. Davin, and M. Lamberigts, "Cobalt-based superalloys for applications in gas turbines," *Materials Science and Engineering*, vol. 88, pp. 11–19, 1987/04/01, 1987, doi: https://doi.org/10.1016/0025-5416(87)90061-9

[6] Y. Lei, C. Li, and L. Wan, "High-temperature tensile properties of a cobalt-based Co-20Cr-15W-10Ni superalloy with a bimodal grain structure," *Crystals*, vol. 13, no. 2, p. 232, 2023 [Online]. Available: www.mdpi.com/2073-4352/13/2/232

[7] D. Davydov, N. Kazantseva, I. Ezhov, and E. Popova, "Effect of alloying on the γ-γ' microstructure of W-free Co-based superalloys," *Materials Today: Proceedings*, vol. 38, pp. 1971–1973, 2021/01/01, 2021, doi: https://doi.org/10.1016/j.matpr.2020.09.131

[8] D. Migas, G. Moskal, H. Myalska, and T. Mikuszewski, "The effect of alloying elements on oxide scale spallation of multicomponent Co-based superalloys," *Corrosion Science*, vol. 192, p. 109787, 2021/11/01, 2021, doi: https://doi.org/10.1016/j.corsci.2021.109787

[9] Y. Yamamoto *et al.*, "Development of creep-resistant, alumina-forming ferrous alloys for high-temperature structural use," in *ASME 2018 Symposium on Elevated Temperature Application of Materials for Fossil, Nuclear, and Petrochemical Industries, ETAM 2018*, 2018, doi: 10.1115/ETAM2018-6727 [Online]. Available: www.scopus.com/inward/record.uri?eid=2-s2.0-85068510470&doi=10.1115%2fETAM2018-6727&partnerID=40&md5=646f1c68bc8972c5289ad7e55efb132b

[10] T. J. Headley, M. M. Karnowsky, and W. R. Sorenson, "Effect of composition and high energy rate forging on the onset of precipitation in an iron-base superalloy," *Metallurgical Transactions A*, vol. 13, no. 3, pp. 345–353, 1982/03/01, 1982, doi: 10.1007/BF02643343

[11] R. M. German, "1 – Metal powder injection molding (MIM): key trends and markets," in *Handbook of Metal Injection Molding (Second Edition)*, D. F. Heaney Ed.: Woodhead Publishing, 2019, pp. 1–21.

[12] A. P. Mouritz "1 – Introduction to aerospace materials," in *Introduction to Aerospace Materials*, Ed.: Woodhead Publishing, 2012, pp. 1–14.

[13] K. Horke, A. Meyer, and R. F. Singer, "24 – Metal injection molding (MIM) of nickel-base superalloys," in *Handbook of Metal Injection Molding (Second Edition)*, D. F. Heaney Ed.: Woodhead Publishing, 2019, pp. 575–608.

[14] R. C. Reed, *The Superalloys: Fundamentals and Applications*. Cambridge University Press, 2006.

[15] A. P. Mouritz "2 – Aerospace materials: past, present and future," in *Introduction to Aerospace Materials*, Ed.: Woodhead Publishing, 2012, pp. 15–38.

[16] J. W. Martin, *Precipitation Hardening: Theory and Applications*. Butterworth-Heinemann, 2012.

[17] J. N. DuPont, S. D. Kiser, and J. C. Lippold, *Welding Metallurgy and Weldability of Nickel-Base Alloys*. John Wiley & Sons, 2009.

[18] V. Sikka, S. Deevi, S. Viswanathan, R. Swindeman, and M. Santella, "Advances in processing of Ni3Al-based intermetallics and applications," *Intermetallics*, vol. 8, no. 9–11, pp. 1329–1337, 2000.

[19] A. Wagner, B. Shollock, and M. McLean, "Grain structure development in directional solidification of nickel-base superalloys," *Materials Science and Engineering: A*, vol. 374, no. 1–2, pp. 270–279, 2004.

[20] M. T. Jovanović, B. Lukić, Z. Mišković, I. Bobić, I. Cvijović, and B. Dimčić, "Processing and some applications of nickel, cobalt and titanium-based alloys," Association of Metallurgical Engineers of Serbia, vol. 20, no. 9–11, pp. 92–106, 2007.

[21] G. R. Thellaputta, P. S. Chandra, and C. S. P. Rao, "Machinability of nickel-based superalloys: a review," *Materials Today: Proceedings,* vol. 4, no. 2, Part A, pp. 3712–3721, 2017/01/01, 2017, doi: https://doi.org/10.101/j.matpr.2017.02.266

[22] Z. Wang and K. Rajurkar, "Cryogenic machining of hard-to-cut materials," *Wear*, vol. 239, no. 2, pp. 168–175, 2000.

[23] E. O. Ezugwu, "Key improvements in the machining of difficult-to-cut aerospace superalloys," *International Journal of Machine Tools and Manufacture*, vol. 45, no. 12, pp. 1353–1367, 2005/10/01, 2005, doi: https://doi.org/10.1016/j.ijmachtools.2005.02.003

[24] D. Graham, "Turning difficult-to-machine alloys," *Modern Machine Shop (USA)*, vol. 75, no. 2, pp. 92–96, 2002.

[25] S. Miller, "Advanced materials mean advanced engines," *Interdisciplinary Science Reviews*, vol. 21, no. 2, pp. 117–129, 1996.

[26] M. Perrut, P. Caron, M. Thomas, and A. Couret, "High temperature materials for aerospace applications: Ni-based superalloys and γ-TiAl alloys," *Comptes Rendus Physique*, vol. 19, no. 8, pp. 657–671, 2018.

[27] Lula, R A. Stainless steel. United States: N. p., 1985. Web.

[28] J. Beddoes and J. G. Parr, "Introduction to stainless steels, 3," 1999. Beddoes, J, and Parr, J G. Introduction to stainless steels, 3. edition. United States: N. p., 1999. Web.

[29] H. M. Cobb, *The History of Stainless Steel.* ASM International, 2010.

[30] J. R. Davis and A. S. M. I. H. Committee, *Stainless Steels.* ASM International, 1994.

[31] J. Davis, *Stainless Steels.* ASM International, 2000.

[32] A. D. H. Series, "Design Guidelines for the Selection and Use of Stainless Steels.

[33] "Alloy 20 Datasheet." Rolled Alloys (accessed).

[34] "Stainless Steel" *Britannica.* (accessed 19 January 2022).

[35] T. Belesis (Winter 2018), "Ferritic stainless steels," *Australian Stainless Magazine*, 2018. Doi: www.assda.asn.au/blog/297-ferritic-stainless-steels

[36] "What is Martensitic Stainless Steel and What Can It Do for Your Business?," vol. 2022, ed. Ulbrich, 2020.

[37] "Duplex Stainless Steel, Lean Duplex and Super Duplex," vol. 2022, ed. Montanstahl, 2017.

[38] R. Brucker, "A Guide to Precipitation-Hardening Stainless Alloys," vol. 2022, ed. Ulbrich, 14/Jan.

[39] "17-7PH Stainless Steel: An Introduction to this Workhorse Stainless Alloy," vol. 2022, ed. Ulbrich, 16/Sep.

[40] P. M. Novotny and G. E. Maurer, "Ultra-high-strength steels vs titanium alloys: this study compares the strength, ductility, toughness, and fatigue properties of AerMet alloys, maraging steels, and titanium," *Advanced Materials & Processes*, vol. 165, no. 11, pp. 37–41, 2007.

[41] R. Boyer, J. Cotton, M. Mohaghegh, and R. Schafrik, "Materials considerations for aerospace applications," *Mrs Bulletin*, vol. 40, no. 12, pp. 1055–1066, 2015.

[42] A. Allison and R. Scudamore, "Strategic Research Agenda: Joining: Joining Sub-Platform." Available: www.slideserve.com/oralee/the-joining-sub-platform-and-the-strategic-research-agenda-powerpoint-ppt-presentation, 2014.

[43] S. J. Hu *et al.*, "Assembly system design and operations for product variety," *CIRP Annals*, vol. 60, no. 2, pp. 715–733, 2011.
[44] E. A. Association, "Joining–mechanical joining," *The Aluminum Automotive Manual*, 2002.
[45] K. Martinsen, S. J. Hu, and B. E. Carlson, "Joining of dissimilar materials," *CIRP Annals*, vol. 64, no. 2, pp. 679–699, 2015/01/01, 2015, doi: https://doi.org/10.1016/j.cirp.2015.05.006
[46] J. Varis, "Ensuring the integrity in clinching process," *Journal of Materials Processing Technology*, vol. 174, no. 1–3, pp. 277–285, 2006.
[47] K.-I. Mori, N. Bay, L. Fratini, F. Micari, and A. E. Tekkaya, "Joining by plastic deformation," *CIRP Annals*, vol. 62, no. 2, pp. 673–694, 2013.
[48] A. Baldan, "Adhesively-bonded joints and repairs in metallic alloys, polymers and composite materials: adhesives, adhesion theories and surface pretreatment," *Journal of Materials Science*, vol. 39, pp. 1–49, 2004.
[49] B. Yacobi, S. Martin, K. Davis, A. Hudson, and M. Hubert, "Adhesive bonding in microelectronics and photonics," *Journal of Applied Physics*, vol. 91, no. 10, pp. 6227–6262, 2002.
[50] Z. A. Munir, U. Anselmi-Tamburini, and M. Ohyanagi, "The effect of electric field and pressure on the synthesis and consolidation of materials: a review of the spark plasma sintering method," *Journal of Materials Science*, vol. 41, pp. 763–777, 2006.
[51] M. Suárez *et al.*, "Challenges and opportunities for spark plasma sintering: a key technology for a new generation of materials," *Sintering Applications*, vol. 13, pp. 319–342, 2013.
[52] Z.-H. Zhang, Z.-F. Liu, J.-F. Lu, X.-B. Shen, F.-C. Wang, and Y.-D. Wang, "The sintering mechanism in spark plasma sintering – proof of the occurrence of spark discharge," *Scripta Materialia*, vol. 81, pp. 56–59, 2014.
[53] M. Belmonte, J. González-Julián, P. Miranzo, and M. Osendi, "Spark plasma sintering: a powerful tool to develop new silicon nitride-based materials," *Journal of the European Ceramic Society*, vol. 30, no. 14, pp. 2937–2946, 2010.
[54] A. Miriyev, A. Stern, E. Tuval, S. Kalabukhov, Z. Hooper, and N. Frage, "Titanium to steel joining by spark plasma sintering (SPS) technology," *Journal of Materials Processing Technology*, vol. 213, no. 2, pp. 161–166, 2013.
[55] S. Diouf and A. Molinari, "Densification mechanisms in spark plasma sintering: effect of particle size and pressure," *Powder technology*, vol. 221, pp. 220–227, 2012.
[56] W. Liu and M. Naka, "In situ joining of dissimilar nanocrystalline materials by spark plasma sintering," *Scripta Materialia*, vol. 48, no. 9, pp. 1225–1230, 2003.
[57] M. Rohde, I. Südmeyer, A. Urbanek, and M. Torge, "Joining of alumina and steel by a laser supported brazing process," *Ceramics International*, vol. 35, no. 1, pp. 333–337, 2009.
[58] R. Thompson, *Manufacturing Processes for Design Professionals*. Thames & Hudson, 2007.
[59] I. Calliari, E. Ramous, K. Brunelli, M. Dabala, and P. Favaron, "Characterization of vacuum brazed joints for superconducting cavities," *Microchimica Acta*, vol. 147, pp. 141–146, 2004.
[60] H. Oikawa, S. Ohmiya, T. Yoshimura, and T. Saitoh, "Resistance spot welding of steel and aluminium sheet using insert metal sheet," *Science and Technology of Welding and Joining*, vol. 4, no. 2, pp. 80–88, 1999.
[61] W. Zhang, D. Sun, L. Han, and D. Liu, "Interfacial microstructure and mechanical property of resistance spot welded joint of high strength steel and aluminium alloy with 4047 AlSi12 interlayer," *Materials & Design*, vol. 57, pp. 186–194, 2014.

[62] G. Haynes and B. Jha, "Joining aluminum to steel with transition material," SAE Technical Paper, 0148-7191, 1999.
[63] K. Miyamoto, S. Nakagawa, C. Sugi, H. Sakurai, and A. Hirose, "Dissimilar joining of aluminum alloy and steel by resistance spot welding," *SAE International Journal of Materials and Manufacturing*, vol. 2, no. 1, pp. 58–67, 2009.
[64] T. Watanabe, H. Takayama, and A. Yanagisawa, "Joining of aluminum alloy to steel by friction stir welding," *Journal of Materials Processing Technology*, vol. 178, no. 1–3, pp. 342–349, 2006.
[65] T. Khaled, "An outsider looks at friction stir welding," *Federal Aviation Administration*, vol. 25, pp. 27–29, 2005.
[66] R. Coelho, A. Kostka, J. Dos Santos, and A. Kaysser-Pyzalla, "Friction-stir dissimilar welding of aluminium alloy to high strength steels: mechanical properties and their relation to microstructure," *Materials Science and Engineering: A*, vol. 556, pp. 175–183, 2012.
[67] T. Mohandas, "Welding technologies in aerospace applications," *Aerospace Materials and Material Technologies: Volume 2: Aerospace Material Technologies*, In: Prasad, N., Wanhill, R. (eds), Springer, Singapore. pp. 65–83, 2017.
[68] A. Zavadski, "Advanced welding technologies used in aerospace industry," Thesis Saimaa University of Applied Sciences, 2018.
[69] M. Attaran, "The rise of 3-D printing: The advantages of additive manufacturing over traditional manufacturing," *Business Horizons*, vol. 60, no. 5, pp. 677–688, 2017, doi: 10.1016/j.bushor.2017.05.011
[70] B. Blakey-Milner *et al.*, "Metal additive manufacturing in aerospace: a review," *Materials and Design*, vol. 209, 2021, Art no. 110008, doi: 10.1016/j.matdes.2021.110008
[71] A. Vafadar, F. Guzzomi, A. Rassau, and K. Hayward, "Advances in metal additive manufacturing: a review of common processes, industrial applications, and current challenges," *Applied Sciences (Switzerland)*, Review vol. 11, no. 3, pp. 1–33, 2021, Art no. 1213, doi: 10.3390/app11031213
[72] A. Chabot, M. Rauch, and J. Y. Hascoët, "Towards a multi-sensor monitoring methodology for AM metallic processes," *Welding in the World*, vol. 63, no. 3, pp. 759–769, 2019, doi: 10.1007/s40194-019-00705-4
[73] T. Iso-Junno, K. K. Mäkelä, K. Mäntyjärvi, and T. Jokelainen, "A short glance on metal 3D AM," *Key Engineering Materials*, vol. 786 KEM, pp. 348–355, 2018.
[74] D. Ding, Z. Pan, D. Cuiuri, and H. Li, "Wire-feed additive manufacturing of metal components: technologies, developments and future interests," *The International Journal of Advanced Manufacturing Technology*, vol. 81, no. 1, pp. 465–481, 2015.
[75] C. Lieberwirth, A. Harder, and H. Seitz, "Extrusion-based additive manufacturing of metal parts," *Journal of Mechanics Engineering and Automation*, vol. 7, no. 2, pp. 79–83, 2017.
[76] B. Wu *et al.*, "A review of the wire arc additive manufacturing of metals: properties, defects and quality improvement," *Journal of Manufacturing Processes,* vol. 35, pp. 127–139, 2018/10/01, 2018, doi: https://doi.org/10.1016/j.jmapro.2018.08.001
[77] L. P. Raut and R. V. Taiwade, "Wire arc additive manufacturing: a comprehensive review and research directions," *Journal of Materials Engineering and Performance*, vol. 30, no. 7, pp. 4768–4791, 2021/07/01, 2021, doi: 10.1007/s11665-021-05871-5
[78] T. A. Rodrigues, V. Duarte, R. M. Miranda, T. G. Santos, and J. P. Oliveira, "Current status and perspectives on wire and arc additive manufacturing (WAAM)," *Materials*, vol. 12, no. 7, p. 1121, 2019 [Online]. Available: www.mdpi.com/1996-1944/12/7/1121

[79] S. W. Williams, F. Martina, A. C. Addison, J. Ding, G. Pardal, and P. Colegrove, "Wire + arc additive manufacturing," *Materials Science and Technology,* vol. 32, no. 7, pp. 641–647, 2016/05/02, 2016, doi: 10.1179/1743284715Y.0000000073

[80] F. Martina, J. Ding, S. Williams, A. Caballero, G. Pardal, and L. Quintino, "Tandem metal inert gas process for high productivity wire arc additive manufacturing in stainless steel," *Additive Manufacturing*, vol. 25, pp. 545–550, 2019/01/01, 2019, doi: https://doi.org/10.1016/j.addma.2018.11.022

[81] C. Wu, L. Wang, W. Ren, and X. Zhang, "Plasma arc welding: process, sensing, control and modeling," *Journal of Manufacturing Processes*, vol. 16, no. 1, pp. 74–85, 2014.

[82] F. Xu *et al.*, "Realisation of a multi-sensor framework for process monitoring of the wire arc additive manufacturing in producing Ti-6Al-4V parts," *International Journal of Computer Integrated Manufacturing*, vol. 31, no. 8, pp. 785–798, 2018.

[83] H. Pant, A. Arora, G. S. Gopakumar, U. Chadha, A. Saeidi, and A. E. Patterson, "Applications of wire arc additive manufacturing (WAAM) for aerospace component manufacturing," *International Journal of Advanced Manufacturing Technology*, vol. 127, no. 11, pp. 4995–5011, 2023/08/01, 2023, doi: 10.1007/s00170-023-11623-7

[84] J. C. Najmon, S. Raeisi, and A. Tovar, "2 – Review of additive manufacturing technologies and applications in the aerospace industry," in *Additive Manufacturing for the Aerospace Industry*, F. Froes and R. Boyer Eds.: Elsevier, 2019, pp. 7–31.

[85] B. Onuike and A. Bandyopadhyay, "Additive manufacturing of Inconel 718–Ti6Al4V bimetallic structures," *Additive Manufacturing*, vol. 22, pp. 844–851, 2018.

[86] M. R. U. Ahsan, A. N. M. Tanvir, T. Ross, A. Elsawy, M.-S. Oh, and D. B. Kim, "Fabrication of bimetallic additively manufactured structure (BAMS) of low carbon steel and 316L austenitic stainless steel with wire + arc additive manufacturing," *Rapid Prototyping Journal*, vol. 26, no. 3, pp. 519–530, 2020. https://doi.org/10.1108/RPJ-09-2018-0235

[87] L. Liu, Z. Zhuang, F. Liu, and M. Zhu, "Additive manufacturing of steel–bronze bimetal by shaped metal deposition: interface characteristics and tensile properties," *International Journal of Advanced Manufacturing Technology*, vol. 69, no. 9–12, pp. 2131–2137, 2013.

[88] A. R. Kannan, S. M. Kumar, N. P. Kumar, N. S. Shanmugam, A. Vishnu, and Y. Palguna, "Process-microstructural features for tailoring fatigue strength of wire arc additive manufactured functionally graded material of SS904L and Hastelloy C-276," *Materials Letters*, vol. 274, p. 127968, 2020.

[89] Y. Tian, J. Shen, S. Hu, J. Gou, and E. Kannatey-Asibu, "Wire and arc additive manufactured Ti–6Al–4V/Al–6.25 Cu dissimilar alloys by CMT-welding: effect of deposition order on reaction layer," *Science and Technology of Welding and Joining*, vol. 25, no 1, pp. 73–80, 2020. doi:10.1080/13621718.2019.1629379

[90] M. R. Ahsan *et al.*, "Microstructures and mechanical behavior of the bimetallic additively-manufactured structure (BAMS) of austenitic stainless steel and Inconel 625," *Journal of Materials Science & Technology*, vol. 74, pp. 176–188, 2021.

[91] A. Tanvir *et al.*, "Heat treatment effects on Inconel 625 components fabricated by wire + arc additively manufacturing (WAAM) – part 2: mechanical properties," *The International Journal of Advanced Manufacturing Technology*, vol. 110, pp. 1709–1721, 2020.

[92] B. Wu *et al.*, "Enhanced interface strength in steel–nickel bimetallic component fabricated using wire arc additive manufacturing with interweaving deposition strategy," *Journal of Materials Science & Technology*, vol. 52, pp. 226–234, 2020.

[93] M. P. Brady, G. Muralidharan, D. N. Leonard, J. A. Haynes, R. G. Weldon, and R. D. England, "Long-term oxidation of candidate cast iron and stainless steel exhaust system alloys from 650 to 800 °C in air with water vapor," *Oxidation of Metals*, vol. 82, no. 5, pp. 359–381, 2014/12/01, 2014, doi: 10.1007/s11085-014-9496-1

[94] A. P. Mouritz, *Introduction to Aerospace Materials*. Elsevier, 2012.

[95] B. Weiss and R. Stickler, "Phase instabilities during high temperature exposure of 316 austenitic stainless steel," *Metallurgical and Materials Transactions B*, vol. 3, no. 4, pp. 851–866, 1972.

[96] I. Tolosa, F. Garciandía, F. Zubiri, F. Zapirain, and A. Esnaola, "Study of mechanical properties of AISI 316 stainless steel processed by 'selective laser melting', following different manufacturing strategies," *The International Journal of Advanced Manufacturing Technology*, vol. 51, no. 5, pp. 639–647, 2010.

[97] T. Olugbade and J. Lu, "Effects of materials modification on the mechanical and corrosion properties of AISI 316 stainless steel," in *Twelfth International Conference on Fatigue Damage of Structural Materials*, Hyannis, Massachusetts, United States. 2018.

[98] L. Marin, "Numerical solution of the Cauchy problem for steady-state heat transfer in two-dimensional functionally graded materials," *International Journal of Solids and Structures*, vol. 42, no. 15, pp. 4338–4351, 2005.

[99] W. Pompe *et al.*, "Functionally graded materials for biomedical applications," *Materials Science and Engineering: A*, vol. 362, no. 1–2, pp. 40–60, 2003.

[100] L. Lu, M. Chekroun, O. Abraham, V. Maupin, and G. Villain, "Mechanical properties estimation of functionally graded materials using surface waves recorded with a laser interferometer," *NDT & E International*, vol. 44, no. 2, pp. 169–177, 2011.

[101] E. Mueller, Č. Drašar, J. Schilz, and W. Kaysser, "Functionally graded materials for sensor and energy applications," *Materials Science and Engineering: A*, vol. 362, no. 1–2, pp. 17–39, 2003.

[102] R. M. Mahamood, E. T. Akinlabi, M. Shukla, and S. L. Pityana, "Functionally graded material: an overview," Proceedings of the World Congress on Engineering. vol. 3. 2012.

[103] V. Dhinakaran, J. Ajith, A. F. Y. Fahmidha, T. Jagadeesha, T. Sathish, and B. Stalin, "Wire arc additive manufacturing (WAAM) process of nickel-based superalloys – a review," *Materials Today: Proceedings*, vol. 21, pp. 920–925, 2020.

[104] X. You *et al.*, "Effect of solution heat treatment on the precipitation behavior and strengthening mechanisms of electron beam smelted Inconel 718 superalloy," *Materials Science and Engineering: A*, vol. 689, pp. 257–268, 2017.

[105] J. Brooks, "Forging of superalloys," *Materials & Design*, vol. 21, no. 4, pp. 297–303, 2000.

[106] C. E. Seow *et al.*, "Effect of crack-like defects on the fracture behaviour of wire + arc additively manufactured nickel-base Alloy 718," *Additive Manufacturing*, vol. 36, p. 101578, 2020/12/01/, 2020, doi: https://doi.org/10.1016/j.addma.2020.101578

[107] S. W. Williams, F. Martina, A. C. Addison, J. Ding, G. Pardal, and P. Colegrove, "Wire + arc additive manufacturing," *Materials Science and Technology*, vol. 32, no. 7, pp. 641–647, 2016.

[108] C. Dharmendra, S. Shakerin, G. J. Ram, and M. Mohammadi, "Wire–arc additive manufacturing of nickel aluminum bronze/stainless steel hybrid parts – interfacial characterization, prospects, and problems," *Materialia*, vol. 13, p. 100834, 2020.

[109] Y. Tian, J. Shen, S. Hu, J. Gou, and Y. Cui, "Effects of cold metal transfer mode on the reaction layer of wire and arc additive-manufactured Ti-6Al-4V/Al-6.25

Cu dissimilar alloys," *Journal of Materials Science & Technology*, vol. 74, pp. 35–45, 2021.

[110] T. Abe and H. Sasahara, "Dissimilar metal deposition with a stainless steel and nickel-based alloy using wire and arc-based additive manufacturing," *Precision Engineering*, vol. 45, pp. 387–395, 2016.

[111] C. Dharmendra, A. Hadadzadeh, B. Amirkhiz, G. J. Ram, and M. Mohammadi, "Microstructural evolution and mechanical behavior of nickel aluminum bronze Cu-9Al-4Fe-4Ni-1Mn fabricated through wire–arc additive manufacturing," *Additive Manufacturing*, vol. 30, p. 100872, 2019.

[112] J. Winczek, M. Gucwa, K. Makles, M. Mičian, and A. Yadav, "The amount of heat input to the weld per unit length and per unit volume," in *IOP Conference Series: Materials Science and Engineering*, 2021, vol. 1199, no. 1: IOP Publishing, p. 012067.

[113] Z. Qiu *et al.*, "Tailoring the surface finish, dendritic microstructure and mechanical properties of wire arc additively manufactured Hastelloy C276 alloy by magnetic arc oscillation," Additive Manufacturing, vol. 48, p. 102397, 2021/10/12, 2021, doi: https://doi.org/10.1016/j.addma.2021.102397

[114] T. Haxhimali, A. Karma, F. Gonzales, and M. Rappaz, "Orientation selection in dendritic evolution," *Nature Materials*, vol. 5, no. 8, pp. 660–664, 2006/08/01, 2006, doi: 10.1038/nmat1693

[115] J. Suryawanshi, K. Prashanth, and U. Ramamurty, "Mechanical behavior of selective laser melted 316L stainless steel," *Materials Science and Engineering: A*, vol. 696, pp. 113–121, 2017.

[116] D. Herzog, V. Seyda, E. Wycisk, and C. Emmelmann, "Additive manufacturing of metals," *Acta Materialia*, vol. 117, pp. 371–392, 2016/09/15, 2016, doi: https://doi.org/10.1016/j.actamat.2016.07.019

[117] Y. Zhong, L. Liu, S. Wikman, D. Cui, and Z. Shen, "Intragranular cellular segregation network structure strengthening 316L stainless steel prepared by selective laser melting," *Journal of Nuclear Materials*, vol. 470, pp. 170–178, 2016/03/01, 2016, doi: https://doi.org/10.1016/j.jnucmat.2015.12.034

[118] L. Nastac, J. Valencia, M. Tims, and F. Dax, "Advances in the solidification of IN718 and RS5 alloys," *Superalloys*, vol. 718, pp. 625–706, 2001.

[119] W. M. Tucho, V. H. Lysne, H. Austbø, A. Sjolyst-Kverneland, and V. Hansen, "Investigation of effects of process parameters on microstructure and hardness of SLM manufactured SS316L," *Journal of Alloys and Compounds*, vol. 740, pp. 910–925, 2018/04/05, 2018, doi: https://doi.org/10.1016/j.jallcom.2018.01.098

[120] C. E. Seow, H. E. Coules, G. Wu, R. H. Khan, X. Xu, and S. Williams, "Wire + arc additively manufactured Inconel 718: effect of post-deposition heat treatments on microstructure and tensile properties," *Materials & Design*, vol. 183, p. 108157, 2019.

[121] H. Yin and S. Felicelli, "Dendrite growth simulation during solidification in the LENS process," *Acta Materialia*, vol. 58, no. 4, pp. 1455–1465, 2010.

[122] M. C. Flemings, "*Proceedings of the Merton C. Flemings Symposium on Solidification and Materials Processing*", Tms (No Title) (2001) 2001.

[123] Y. Chen, K. Zhang, J. Huang, S. R. E. Hosseini, and Z. Li, "Characterization of heat affected zone liquation cracking in laser additive manufacturing of Inconel 718," *Materials & Design*, vol. 90, pp. 586–594, 2016/01/15, 2016, doi: https://doi.org/10.1016/j.matdes.2015.10.155

[124] J. Lippold and W. Savage, "Solidification of austenitic stainless steel weldments: Part III – the effect of solidification behavior on hot cracking susceptibility," *Welding Journal*, vol. 61, no. 12, p. 388, 1982.
[125] L. L. Parimi, G. Ravi, D. Clark, and M. M. Attallah, "Microstructural and texture development in direct laser fabricated IN718," *Materials Characterization*, vol. 89, pp. 102–111, 2014.
[126] P. Nie, O. A. Ojo, and Z. Li, "Numerical modeling of microstructure evolution during laser additive manufacturing of a nickel-based superalloy," *Acta Materialia*, vol. 77, pp. 85–95, 2014/09/15, 2014, doi: https://doi.org/10.1016/j.actamat.2014.05.039
[127] N. Chen *et al.*, "Microstructural characteristics and crack formation in additively manufactured bimetal material of 316L stainless steel and Inconel 625," *Additive Manufacturing*, vol. 32, p. 101037, 2020/03/01, 2020, doi: https://doi.org/10.1016/j.addma.2020.101037
[128] H. I. Ahn, S. H. Jeong, H. H. Cho, and H. W. Lee, "Ductility-dip cracking susceptibility of Inconel 690 using Nb content," *Journal of Alloys and Compounds*, vol. 783, pp. 263–271, 2019/04/30, 2019, doi: https://doi.org/10.1016/j.jallcom.2018.12.208
[129] N. E. Nissley, M. G. Collins, G. Guaytima, and J. C. Lippold, "Development of the strain-to-fracture test for evaluating ductility-dip cracking in austenitic stainless steels and Ni-base alloys," *Welding in the World*, vol. 46, no. 7–8, pp. 32–40, 2002, doi: 10.1007/bf03263388
[130] M. Collins and J. Lippold, "An investigation of ductility dip cracking in nickel-based filler materials – Part I," *Welding Journal*, vol. 82, no. 10, p. 288, 2003.
[131] J. Borland, "Fundamentals of solidification cracking in welds – II," *Welding and Metal Fabrication*, vol. 47, no. 2, pp. 99–107, 1979.
[132] S. Kou, "Solidification and liquation cracking issues in welding," *JOM*, vol. 55, no. 6, pp. 37–42, 2003.
[133] C. Huang and S. Kou, "Liquation cracking in full-penetration Al-Cu welds," *Welding Journal – New York*, vol. 83, no. 2, pp. 50-S–58-S, 2004.
[134] B. Radhakrishnan and R. Thompson, "Solidification of the nickel-base superalloy 718: a phase diagram approach," *Metallurgical Transactions A*, vol. 20, no. 12, pp. 2866–2868, 1989.
[135] K. R. Vishwakarma, N. L. Richards, and M. C. Chaturvedi, "Microstructural analysis of fusion and heat affected zones in electron beam welded ALLVAC® 718PLUS™ superalloy," *Materials Science and Engineering: A*, vol. 480, no. 1, pp. 517–528, 2008/05/15/ 2008, doi: https://doi.org/10.1016/j.msea.2007.08.002
[136] C. Su, X. Chen, C. Gao, and Y. Wang, "Effect of heat input on microstructure and mechanical properties of Al-Mg alloys fabricated by WAAM," *Applied Surface Science*, vol. 486, pp. 431–440, 2019/08/30/ 2019, doi: https://doi.org/10.1016/j.apsusc.2019.04.255
[137] M. H. Farshidianfar, A. Khajepour, and A. P. Gerlich, "Effect of real-time cooling rate on microstructure in laser additive manufacturing," *Journal of Materials Processing Technology*, vol. 231, pp. 468–478, 2016.
[138] Y. Ma, D. Cuiuri, C. Shen, H. Li, and Z. Pan, "Effect of interpass temperature on in-situ alloying and additive manufacturing of titanium aluminides using gas tungsten arc welding," *Additive Manufacturing*, vol. 8, pp. 71–77, 2015.
[139] X. L. Wang *et al.*, "Effect of interpass temperature on the microstructure and mechanical properties of multi-pass weld metal in a 550-MPa-grade offshore engineering steel," *Welding in the World*, vol. 61, no. 6, pp. 1155–1168, 2017/11/01, 2017, doi: 10.1007/s40194-017-0498-x

[140] H. Attar, M. Bermingham, S. Ehtemam-Haghighi, A. Dehghan-Manshadi, D. Kent, and M. Dargusch, "Evaluation of the mechanical and wear properties of titanium produced by three different additive manufacturing methods for biomedical application," *Materials Science and Engineering: A*, vol. 760, pp. 339–345, 2019.

[141] D. Van, G. Dinda, J. Park, J. Mazumder, and S. H. Lee, "Enhancing hardness of Inconel 718 deposits using the aging effects of cold metal transfer-based additive manufacturing," *Materials Science and Engineering: A*, vol. 776, p. 139005, 2020.

[142] C. Cunningham, J. Wang, V. Dhokia, A. Shrokani, and S. Newman, "Characterisation of austenitic 316 LSi stainless steel produced by wire arc additive manufacturing with interlayer cooling," in *2019 International Solid Freeform Fabrication Symposium*, 2019. University of Texas at Austin.

[143] L. Wang, J. Xue, and Q. Wang, "Correlation between arc mode, microstructure, and mechanical properties during wire arc additive manufacturing of 316L stainless steel," *Materials Science and Engineering: A*, vol. 751, pp. 183–190, 2019.

[144] G. Sujan, H. Li, Z. Pan, D. Liang, and N. Alam, "Application of wire arc additive manufacturing for Inconel 718 superalloy," in Kuşhan, M.C., Gürgen, S., Sofuoğlu, M.A. (eds) Materials, *Materials, Structures and Manufacturing for Aircraft*: Springer, 2022, pp. 367–410.

[145] Z. A. Young *et al.*, "Types of spatter and their features and formation mechanisms in laser powder bed fusion additive manufacturing process," *Additive Manufacturing*, vol. 36, p. 101438, 2020.

Index

Surface roughness, 51–52, 54, 89, 91, 97, 108, 110–114, 116, 128–131, 133–134, 136, 178, 187, 204–206, 213, 226, 228, 234, 242, 265, 273, 276

T

U

V

W

X

Y

Z

Kolibri 360 GmbH
Ferdinand-Jühlke-Straße 7
99095 Erfurt, Deutschland
produktsicherheit@kolibri360.de